Trends in Optics

Research, Development and Applications

Trends in Optics

Research, Developments and Applications

Edited by

Anna Consortini

ICO President
Università degli Studi,
Dipartimento di Fisica,
Florence, Italy

ACADEMIC PRESS
Harcourt Brace & Company, Publishers
San Diego   London   New York   Boston
Sydney   Tokyo   Toronto

Academic Press Inc
525 B Street, Suite 1900, San Diego, California 92101–4495, USA

Academic Press Limited
24–28 Oval Road, London NW1 7DX, UK

ISBN 0–12–186030–2

A catalogue record for this book is available from the British Library

Front cover image:
Single objective lens of Galileo Galilei, set in the ivory frame carved by
Vettorio Croster in 1677 (Courtesy of the *Instituto e Museo di Storia della Scienza,*
firenze; photo by Franca Principe).

Typeset by Mathematical Composition Setters, Salisbury Wilts.
Printed and bound in Great Britain at the University Press, Cambridge.

Contents

List of Contributors

Yoshihisa Aizu
Muroran Institute of Technology, Muroran, Japan

Yaakov Amitai
Head of Optics Group, Electro-Optics Industries Ltd, Rehovot, Israel

F. T. Arecchi
Istituto Nazionale di Ottica, Florence, Italy

Toshimitsu Asakura
Research Institute for Electronic Science, Hokkaido University, Sapporo, Japan

George Barbastathis
Graduate Student, California Institute of Technology, Pasadena, USA

S. Boccaletti
Istituto Nazionale di Ottica, Florence, Italy

Connie J. Chang-Hasnain
University of California, Berkeley, USA

Young Min Cho
Research Scientist, PhD, Korea Aero-Space Research Institute, Taejon, Korea

Daniele Contini
PhD Student, University of Florence, Italy

Christine De Mol
Maître de recherches, Belgian National Fund for Scientific Research, Université Libre de Bruxelles, Brussels, Belgium

Krzysztof Ernst
Professor, Institute of Experimental Physics, Warsaw University, Poland

Zhi-Liang Fang
NanKai University, Tianjin, China

Stanley M. Flatté
Professor, University of California at Santa Cruz, USA

Rod Frehlich
Research Scientist, Remote Sensing Program of CIRES, University of Colorado, Boulder, USA

A. A. Friesem
Professor of Optical Sciences, Weizmann Institute of Science, Rehovot, Israel

Carlos Gómez-Reino
Professor of Optics, Universidad de Santiago de Compostela, Spain

Joseph W. Goodman
William Ayer Professor of Electrical Engineering, Stanford University, USA

Christophe Gorecki
Laboratoire d'Ophique P. M. Duffieux (URA–CNRS 214), Université de Franche-Comté, Besançon, France
(Present address: LIMMS/CNRS, Institute of Industrial Science, University of Tokyo, Japan)

Vincenzo Greco
Istituto Nazionale di Ottica, Florence, Italy

Reginald J. Hill
Physicist, National Oceanic and Atmospheric Administration, Boulder, USA

Pierre Jaegle
Directeur de Recherche, Laboratoire de Spectroscopie Atomique et Ionique, CNRS, Université Paris-Sud and Directeur, X-Ray Laser Program, Laboratoire pour L'Utilisation des Lasers, Intenses, Palaiseau, France

Denis Joyeux
Chargé de Recherches, Institut d'Optique Théorique et Appliquée, Orsay, France

Dong Eon Kim
Vice Professor, Pohang University of Science and Technology, Kyungbuk, Korea

William J. Kozlovsky
Almaden Research Center, San Jose, USA

Kazuo Kyuma
Mitsubishi Electric Corporation, Hyogo, Japan

Jane C. Lam
Research Assistant, Stanford University, USA

Eberhard Lange
Senior Research Assistant, Mitsubishi Electric Corporation, Amagasaki, Japan

Sang Soo Lee
Professor Emeritus, Korea Advanced Institute of Science and Technology, Taejon, Korea

Emmett N. Leith
Professor of Electrical Engineering, The University of Michigan, Ann Arbor, USA

Michael Levene
Graduate Student, California Institute of Technology, Pasadena, USA

Anne L'Huillier
Lecturer, Lund Institute of Technology, Sweden and *CEN Saclay, Gif sur Yvette, France*

G. S. Li
University of California, Berkeley, USA

Jesús Liñares-Beiras
Professor of Optics, University of Santiago di Compostela, Spain

Fu-Lai Liu
NanKai University, Tianjin, China

Adolf W. Lohmann
Physikalisches Institut der Universität Erlangen-Nürnberg, Germany

José J. Lunazzi
Universidade Estudual de Campinas, Sao Paulo, Brazil

David A. B. Miller
Head of Advanced Photonics Research Department, AT&T Bell Laboratories, Holundel, USA

Giuseppe Molesini
Istituto Nazionale di Ottica, Florence, Italy

Guo-Guang Mu
NanKai University, Tianjin, China

Chang Hee Nam
Vice Professor, Korea Advanced Institute of Science and Technology, Taejon, Korea

Yoshikazu Nitta
Mitsubishi Electric Corporation, Hyogo, Japan

Kevin A. O'Donnell
Professor, Georgia Institute of Technology, Atlanta, USA

Jorge Ojeda-Castañeda
Universidad de las Americas, Puebla, Mexico

E. Pampaloni
Istituto Nazionale di Ottica, Florence, Italy

Demetri Psaltis
Thomas G. Meyers Professor of Electrical Engineering and *Executive Officer for Computation and Neural Systems, California Institute of Technology, Pasadena, USA*

P. L. Ramazza
Istituto Nazionale di Ottica, Florence, Italy

Alexander Rebane
Oberassistent, Swiss Federal Institute of Technology, Zürich, Switzerland

S. Residori
Istituto Nazionale di Ottica, Florence, Italy

Cheon Seong Rim
Research Scientist, PhD, Samsung Electro-Mechanics, Industry, Suwon, Korea

Patrick Sandoz
Laboratoire d'Optique P. M. Duffieux (URA–CNRS 214), Université de Franche-Comté, Besançon, France

Giuseppe Schirripa Spagnolo
Università delgi Studi di L'Aquila, Roio Poggio, Italy

Joseph Shamir
Israel Institute of Technology, Haifa, Israel

Jari Turunen
Professor, University of Joensuu, Finland

Emil Wolf
The University of Rochester, USA

Frank Wyrowski
Head of Department, Berlin Institute of Optics, Germany and *Professor, Friedrich–Schiller Universität Jena, Germany*

Shizhuo Yin
Pennsylvania State University, USA

Francis T. S. Yu
Evan Pugh Professor, Electrical Engineering and *Director of the Center for Electro-Optics Research, Pennsylvania State University, USA*

W. Yuen
University of California, Berkeley, USA

Giovanni Zaccanti
Researcher, University of Florence, Italy

Hong-Chen Zhai
NanKai University, Tianjin, China

Foreword

In the last three or four decades we have seen the return of optics to it's former position of scientific and engineering prominence. It was largely the development of the laser which proved to be the catalyst for this re-birth, and its consequent spawning of dependant technologies such as fibre optics, optical communications, optical computing, laser materials processing and holography. The impact of this 'new-wave' in optics has been so overwhelming, that it is now possible to perceive of it as a distinct academic discipline in its own right, the roots of which lie in classical optics, materials science, electromagnetism, electronics and computing. The practitioners of this discipline may be physicists, chemists, botanists, biologists, mechanical, civil and electrical engineers, and the applicability of their skills stretches across the entire spectrum of science and technology.

This series of books on Lasers and Optical Engineering is intended to reflect the systems and applied nature of modern optics and to affirm our belief that we are witnessing the evolution of a new breed of engineer: the optical engineer. Books will be featured which project the multidisciplinary and wide-ranging coverage of optical engineering and underpin this with a strong commitment to fundamental principles. The series is aimed predominantly at the advanced student and the practising optical engineer, whether in industry or academia.

This new book in the series continues with a broad overview of applied optics. Promoted by The International Commission for Optics (and the third book in the ICO series) this work, edited by Professor Anna Consortini, enhances our knowledge of recent developments in optics with reviews of topics such as optical interconnects, interferometry in space and compact blue-green lasers. Future planned volumes in the series 'Lasers and Optical Engineering' include, Laser Induced Breakdown Spectroscopy; Holography and Interferferometry; Diffractive Optics; Laser Safety; Non-linear Optics; Lasers in Medicine; Optical Fibre Sensing; Optical Storage Technologies; Optical Image and Information Processing; Imaging Science; Laser Design and Development; Semiconductor Laser Technology; Adaptive Optic Systems; Lens and Optical System Design and Robotics; and Machine Vision.

Dr John Watson
University of Aberdeen
Dr John Andrews
Xerox Corp.

Preface

This volume, ICO-Book 3, is the third book of the series 'Trends in Optics', of the International Commission for Optics (ICO), to be published every three years with the main purpose of promoting knowledge of recent developments in the field of optics. The distinguishing feature of the series is its wide range of short, readable and partly speculative articles. The first volume, entitled International Trends in Optics and edited by J. W. Goodman, was Published by Academic Press, USA, in 1991; and the second one, entitled Current Trends in Optics, edited by J. C. Dainty, was published by Academic Press, UK, in 1994. Publication of the volumes of the series follows the three year cycle of the organizational structure of the ICO. So far the task of editing the volumes has been taken up by the current ICO President. For those unfamiliar with the Commission, let us recall that ICO, which is part of the family of ICSU (The International Council of Scientific Unions), was founded in 1947, only two years after the end of World War II, to promote optics on an international basis.

Optics has undergone far reaching developments over the last few decades, not only in connection with the fundamental research based on the use of coherent light produced by the laser, but also for technology and practical applications. Many specialized books and papers are available on most of these subjects, while the aim of this series is to use a different style from that found in academic journals and conference proceedings. The book is a collection of 31 papers on different subjects, written by outstanding scientists and engineers, chosen worldwide in accordance with the international character of ICO. Although the presentation is still in the form of a technical paper the style is more informal, and therefore also accessible to readers who are not specialists on the specific topic. Open problems and authors' personal viewpoints are included. Thus also providing specialist readers with valuable state-of-the-art presentations.

In general, there is no connection between the different papers, apart from the fact that they all refer to up-to-date subjects of research or applications of modern optics and were not included in the previous two books. There are papers devoted to optics in biology, to different applications of diffraction, interferometry and holography, including tomography and old device inspection, to optical storage and signal processing and synthesis, to atmospheric optics, X-rays, new active and passive devices, including new lasers and applications of colour, to research on scattering, dynamics of patterns and molecular spectroscopy. I have organized them in an arbitrary sequence, starting with a paper describing, for the first time, the development of the 'secret research' at Willow Run Laboratories of the University of Michigan at the time of the so-called 'Cold War'. Some readers may be able to identify a thin interconnecting thread between some groups of papers.

I wish here to express my gratitude to all the authors for their contribution to the book, as well as for their patient willingness to comply with the editor's requirements. Thank are also due to the Museo ed Istituto della Scienza in Florence for permission to use the photo of the Galileo Galilei lens as the cover illustration and emblem of the book.

Anna Consortini
University of Florence

1 A short history of the Optics Group of the Willow Run Laboratories

Emmett N. Leith
The University of Michigan, Ann Arbor, Mich, 48109–2122

INTRODUCTION

The Radar Laboratory of the University of Michigan's Willow Run Laboratories has throughout its history performed work supported by the military, and for that reason its accomplishments have often been shrouded in secrecy. Its work has often found its way into the technical journals only long after the work had been done; this interval has been on occasion as long as ten years. This secrecy and publication delay have masked its accomplishments, which have been rather astonishing. Few organizations of its size have made a comparable impact on science and technology. Out of this laboratory came the most advanced forms of synthetic aperture radar, a range of practical applications for optical processing, the technique of carrier frequency, or off-axis, holography that solved the well-known twin image problem and led to high-quality holographic imagery, the invention of hologram interferometry, the construction of spatial filters with arbitrary amplitude and phase control, the phase conjugation method for imaging through inhomogeneities, and various other major advances in optical processing concepts. The impact of this work on optical technology has been enormous. Yet, although the results of this work are well known, the history of this development is little known, and there are occasional misconceptions. Our purpose here is to give a comprehensive overview of the history of this group, and to describe the context in which these accomplishments were made.

The accomplishments are due to a number of key individuals who were indeed good people, and who in addition had the good fortune to be at the right place at the right time. Since the accomplishments described here are principally optical, the major portion of the discussion relates to that subset of the Radar Laboratory called the Optics Group.

THE BEGINNINGS

The Radar Laboratory had its beginnings in 1953, when two Willow Run staff members, L. J. Cutrona and W. E. Vivian, began investigating a new kind of radar that had been invented a few years earlier. Synthetic aperture radar (SAR) was the solution to a basic problem that had generally been considered unsolvable. The desire was for an imaging radar that would yield resolution much finer than had heretofore been achieved – resolution perhaps comparable to that achievable from optical aerial reconnaissance. But since the radar wavelength might be 10 000 times longer than optical wavelengths, the antenna necessary to achieve such resolution would have to be about 10 000 times larger than the aperture of an aerial reconnaissance lens – or about 3000 metres. Clearly such an antenna could not be carried by an airplane.

The answer, suggested by C. Wiley of the Goodyear Corporation, was the synthetic antenna. While indeed the long antenna was needed, it was not necessary that all elements of the antenna exist simultaneously. A small antenna mounted on an airplane could be carried along the aircraft flight path, and would sequentially occupy the positions that would have been occupied by each element of a dipole-element antenna array. At each position, the radar would transmit and receive a pulse, and the return pulse would be stored both in phase and amplitude. Such storage was not too difficult; as opposed to optical signals, electronic signals could readily be recorded and stored in both phase and amplitude. The stored signals could then be processed in a manner that imitates the way the dipole array antenna would have processed the signals, namely coherently summing the signals received on each element.

The long antenna thus is eliminated. But in its place is a problem of storage and signal processing. One method, utilized by a group from the University of Illinois, considered that as the aircraft travels along its flight path, the signals reflected from the terrain would be Doppler shifted and that the magnitude of the Doppler shift from an object point is proportional to its angular position in the beam, which is directed to the side, normal to the flight path [1]. Thus at a given range there is a one-to-one correspondence between a frequency shift and a position in the beam. Passing the received signal through a filter bank achieves this separation of signals in different angular parts of the beam, i.e., the object points will be resolved. This process works as long as the signals are filtered for a time sufficiently short that the object points move only slightly. If the angular position of an object point changes significantly during this time, the Doppler shift will change, whereas the above system assumes a constant Doppler shift from each point. Nonetheless, the method demonstrated that objects could be resolved within the beam, thus verifying the concept of the synthetic antenna.

Figure 1(a), showing an aircraft flying past two object points, A and B, describes the data acquisition process. The object points are about to enter the beam. When they first enter, they have a radial velocity component, relative to the aircraft, directed toward the aircraft, and thus the radar signal reflected from them is Doppler-shifted upward in frequency. As the object points progress across the beam, the Doppler shift decreases, becoming zero when an object point is directly abeam of the aircraft. When the objects pass this point, they acquire a radial velocity component directed away from the aircraft, and the Doppler shift is downward. Finally, the object points leave the beam. This Doppler frequency shift as a function of time, called a signal history, is shown in Fig. 1(b). For a beam width of a few degrees, the change in frequency is linear with time. The two object points undergo the same shifts, except that the shift for B occurs at a later time, with the zero shift occurring at a time t_b instead of t_a. Thus, every object entering the beam at the range y_1 undergoes the same linear Doppler shift, but with a time displacement corresponding to the object coordinate. Figure 1(b) shows that, for an observation time sufficiently short, the Doppler shift is approximately constant, with each point in the beam having in this time interval a different Doppler shift, thus providing the basis for the Illinois Doppler filtering system.

However if a large synthetic antenna is to be generated, as when the entire signal reflected from an object point is to be utilized, this simple system no longer works. What is required is a focused antenna.

The Cutrona–Vivian program for synthetic aperture radar (SAR) carried the process much farther than any of the previous systems. They envisioned synthetic apertures so large that the object would no longer be in the Fraunhofer, or far-field regime as previous systems had been, but in the Fresnel or near field. Stated equivalently, the data collection time would be sufficiently long that the object points would move a significant distance, perhaps several hundred metres; consequently, the angular position of a point would change

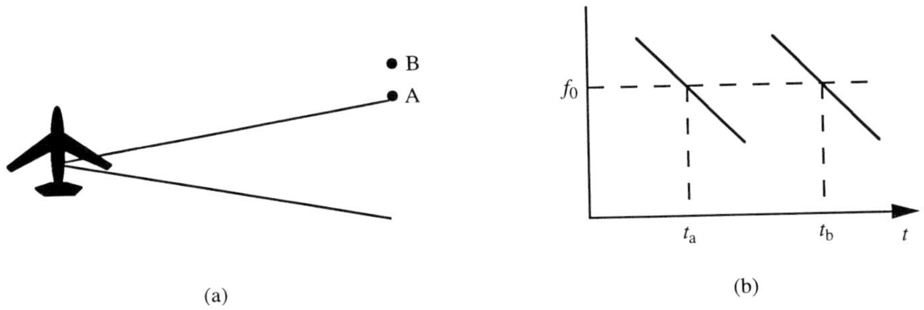

Fig. 1. (a) Aircraft with side-directed radar beam, flying past two point objects. (b) Doppler frequencies as a function of time for the two objects.

enough that its Doppler-shifted frequency would change significantly during the observation time. Thus, the data processing operation to produce the synthetic antenna was no longer a simple frequency filtering, but a focusing operation. The mathematical operation was formulated as the cross-correlation of the incoming signal with a reference function r that was a replica of the expected return from a point object at that range. The signal return from a point object at a range y from the aircraft flight path is readily shown to be

$$s = a \exp\{i[2\pi f_c t - 2\pi v^2(t - t_1)^2/\lambda y]\}, \tag{1}$$

where t_1 is the time at which the object is abeam of the aircraft, f_c is the radiated frequency, λ is the wavelength, and v is the aircraft velocity. The cross-correlation process is then

$$\chi(\tau) = \int s(t)\, r(t + \tau)\, dt. \tag{2}$$

The reference function is range dependent, so a different reference function is required for each range resolution element. Thus, if the resolution is three metres and the range interval to be imaged is 10 km then over 3000 separate reference functions must be utilized and 3000 separate cross-correlations must be performed. Such a processing operation was quite challenging, well beyond the capabilities of the computers of the time. Ways in which this monumental task might be accomplished were considered. Then one day Cutrona and Vivian returned from a trip, revealing that, in a discussion with R. Varian, one of the inventors of the klystron, the problem of carrying out the processing was possibly solved. The data could be recorded on photographic film with range being the dimension across the film and the so-called along-track dimension (the dimension along which the airplane flies) being along the length of the film. The film would then go into an optical processing system, where each range element would be correlated with a reference function proper for that range. The result would be a synthetic aperture that in effect was simultaneously focused at all ranges, and an incredibly sharp image should result.

The initial optical processing system was based on the use of incoherent light. Also, since the radar data was bipolar, both the signal and the reference function had to be written on a bias term. The equation that had to be evaluated at each range interval therefore became

$$\chi(\tau) = \int [s_b + s(t)][r_b + r(t + \tau)]\, dt. \tag{3}$$

Thus, the requirement to include a bias term led to three extraneous terms that must be calculated separately and subtracted out. In addition, theory indicated that the reference function had to be written as two separate masks, a sine and a cosine mask. Such a system is evidently not simple to implement, but a better alternative was not available.

The advanced SAR system concept was described in a document known as

5-T. It was classified secret and remained so for many years, finally being declassified in 1968. This document became the guidebook for the development of the proposed system. It described the basic theory, the implementation, the stability requirements, and various processing schemes. The authors were Cutrona, Vivian and myself. As a very junior person, my contributions were comparatively small.

Upon completion of this document in the summer of 1954, plans were laid for constructing this very challenging radar system. The group was enlarged, as engineers were brought in to carry out the many tasks envisioned in the 5-T report; the radar transmitter and receiver had to be designed and built, a stabilization system had to be developed to compensate for turbulence-produced irregularities in the aircraft flight path, a recording system had to be built, and finally, the optical processing system had to be developed.

OPTICAL PROCESSING

Given a choice of area to work in, I chose the optical processor, my choice being based on the principle of comparative advantage. As a physics major among mostly electrical engineers, I had considerably less electrical background than they, but I had four optics courses, which was considerable even for a physics major at that time: physical optics, two courses in spectroscopy, and a course in X-rays and crystal structure. These, it turned out, were all ideal for the optical processing project as it subsequently developed.

A simple optical processor was set up, similar to but somewhat simpler than the one shown in Fig. 2, which represented one of the two channels of the complete system. The recorded signal of Eq. (1) becomes

$$s(x, y) = a \cos[2\pi f_0 x - 2\pi p^2 (x - x_1/p)^2/\lambda q y], \tag{4}$$

where f_0 is a spatial carrier and p and q are scaling factors for the x and y dimensions, respectively. Reference masks of the form $r_b + r$, with $r = \cos(2\pi f_0 x - 2\pi p^2 x^2/\lambda q y)$, were constructed. In the absence of actual radar

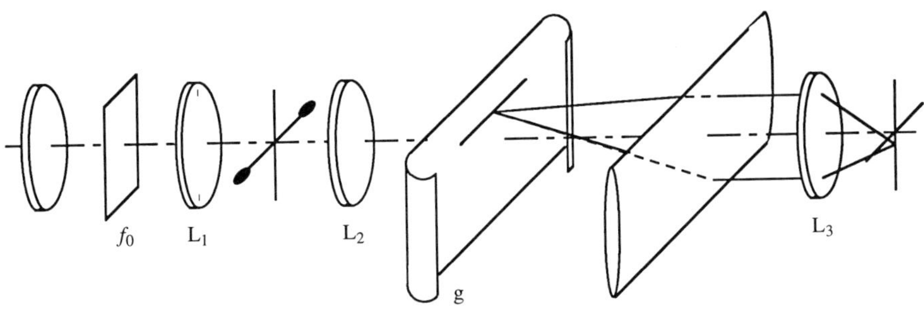

Fig. 2. A coherent optical system for processing of SAR data.

data, a reference mask was used as a simulated signal, and was moved through the aperture. The output light distribution was

$$\chi(x') = \int \ [r_b + r(x, y)][r_b + r(x + x', y)] \ dx \tag{5}$$

where the variable x' describes the movement of the simulated signal through the aperture.

We had expected many problems, but soon one arose that we had not anticipated. The system, based on geometrical optics, behaved as expected until the masks were reduced to sizes that would be required for a working system. Then geometrical optics failed and diffraction effects appeared. Of course, this should not have been surprising, since the masks were in fact diffraction gratings, although not of uniform spacing.

The straightforward solution was to enlarge the source, so that sufficient spatial incoherence was achieved, and then place a sampling slit at the output. The modification was successful. However, another possibility occurred to me: suppose that the diffraction pattern were sampled with a slit. Might it be that the desired information was not uniformly distributed in the pattern, so that one part of it could have most or all of the desired information? Expansion of the integrand in Eq. (5) leads to five terms, where the final terms result from the expansion $\cos A \cos B = \frac{1}{2} \cos (A + B) + \frac{1}{2} \cos (A - B)$. Four of the terms are extraneous, and the one of interest is $\cos(2\pi f_0 x' - 4\pi p^2 xx'/\lambda qy - 2\pi p^2 x'^2/\lambda qy)$. This term is simply a uniform periodic structure, and this part of the mask could thus be considered to be a simple diffraction grating of spatial frequency $2p^2 x'/\lambda y$, forming a $(+1)$ and a (-1) order. Each order by itself gives exactly the desired signal. The spatial frequency, and therefore the position of the diffracted order, is proportional to the displacement term, x', between the signal and reference beam. The picture thus presented is that each object point gives rise to two image points that march across the field in opposite directions, coming into coincidence on the axis for $x' = 0$.

This simple, heuristic viewpoint did not accord with the observed pattern, so a mathematical analysis was undertaken. At this point L. Porcello joined the group, and we spent the next two months analysing the optical cross-correlator under coherent (i.e., point source monochromatic illumination). The result was an extensive memo, about 30 pages, describing in detail the mathematics of the cross-correlator under coherent instead of incoherent illumination. The analysis confirmed the heuristic viewpoint, and the observed results upon careful inspection were found to be exactly in accordance with the theory.

Thus was born the coherent cross-correlator, which in fact solved most of the problems with the incoherent system. During this investigation Cutrona became interested in this work and related it to a paper that E. O'Neill had presented at a recent symposium in Ann Arbor. O'Neill's paper talked about optics and communication theory. We then incorporated the communication theory concept into our optical correlator process. Although the optical system could be completely described and understood in terms of classical physical optics, as indeed we had done in our memo, the communication theory added elegance to the process and also fitted in nicely with the radar theory.

Since the coherent cross-correlator seemed to solve all of the problems, the original incoherent version was abandoned. In retrospect, it seems likely that the problems with the incoherent version might have been solved, and it might have become just as successful as the coherent system was to become. In later years, especially in the 1960s, interest grew in incoherent optical processing. A. Lohmann was a prime mover in this surge of incoherent optical processing, as was G. L. Rogers, who wrote a book on incoherent optical processing [2]. Nonetheless, for us and for the entire field of optical processing of radar data, the die was cast; coherent optics would dominate.

THE HOLOGRAPHIC VIEWPOINT

Work on the optical correlator continued, and indeed, the development had yet a long way to run. While analysing the mathematics of the cross-correlator and of SAR data, I was intrigued by the similarity of the SAR process and the processing carried out by the coherent cross-correlator. I was quite suddenly struck by what I thought was a rather astonishing idea. The field that emerges from the coherently illuminated SAR photographic record is in fact a recreation of the field recorded by the radar system as it moves along the flight path. This recreation is a downscaling process; radar waves at 3 cm are regenerated as visible light waves at 0.000 05 cm, a 6000 : 1 shift. Geometrical factors are similarly scaled. Data collected along 1 km of flight path gets recorded on a few cm of photographic film. These miniaturized waves continue their propagation just as if they had never been interrupted.

I set about to develop a new theory of SAR based on this observation. It would be a radically different theory, based not at all on cross-correlation, Doppler filtering, or other viewpoints familiar to the radar world, but would instead be a strictly physical optics viewpoint. The recorded return from a point object at coordinate x_1, y_1 (Eq. 1) is illuminated with a coherent light beam $\exp(i2\pi f_1 t)$ (with f_1 the light frequency), producing a field $e^{i2\pi f_1 t}(\frac{1}{2} e^{i\phi} + \frac{1}{2} e^{-i\phi})$, where ϕ is the argument of the cos term in Eq. 1. The heterodyning process of recording and then reintroducing the wavefield, now an optical wavefield, has generated sum and difference phase terms, leading to two reconstructed wavefronts. The one wave is a miniaturized regeneration of the original wave, and continues on its original path, whereas the other describes a convergent wave. Each wave forms a focal point, one a real image, the other virtual. Each is an image representing exactly the fine resolution image that we sought, and that was to have been produced by a cross-correlation process. Now we found that the image forms without the need for a correlation process.

However, when considering the entirety of the object distribution and the resulting image, some problems arise. First, the image points form at a distance proportional to the distance from the flight path of the corresponding object points that produced them. Object points at close ranges produce signals that focus close to the film record, whereas object points at more distant ranges produce signals that focus farther from the signal record (Fig. 3). Thus, the image of the terrain forms on a highly tilted plane, as shown in Fig. 4.

Either the real or the virtual image could be recorded. However, the extreme tilt of the image makes recording unfeasible; the tilt has to be removed. Nor is this the only problem. The recorded signals, which are in fact Fresnel zone plates, have focal properties only in the x direction; in the range or y direction (the direction across the film record) the signals are without focal power, and in fact are sharply focused on the signal record itself. The similarity to the

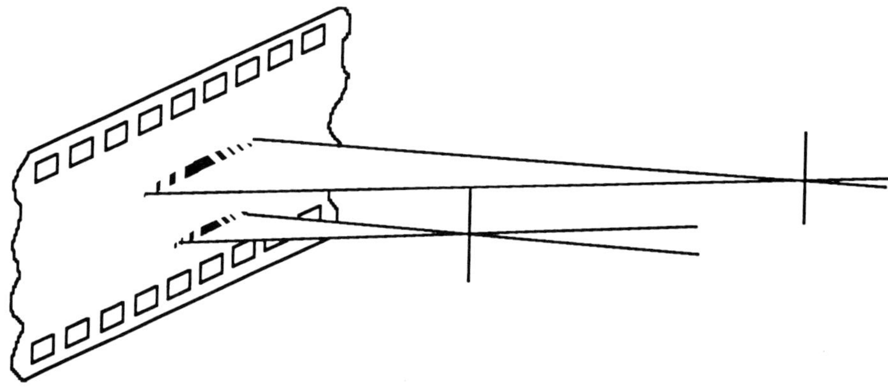

Fig. 3. Focal properties of the recorded signals.

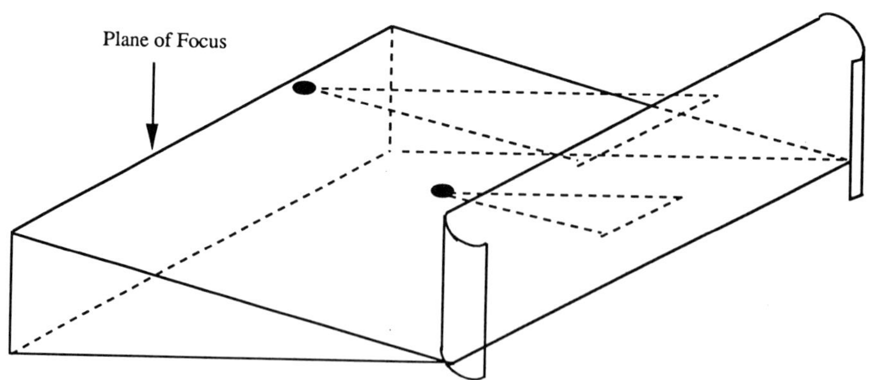

Fig. 4. Tilted plane image formed from signal record. Only virtual image plane is shown. A mirror-symmetric real image is also formed.

modern rainbow hologram is evident. Thus, while indeed the image forms without the need for image processing procedures, there is nonetheless a need for some corrective optics to remove the tilt and extreme astigmatism from the image; the azimuth focal plane must be untilted and then brought into coincidence with the range focal plane. Also, any non-unity aspect ratio (the ratio of the x and y scaling factors in recording the data) must be eliminated. All this can be done with corrective optics. First, one must overlay on the signal record a lens that, at any range position, has a focal length equal but opposite in sign to the focal power of the signal at that range. Thus such a lens would have focal power in one dimension only, and the focal power would be proportional to the vertical coordinate. Such a lens could be described as a conical lens. The conical lens images the tilted plane to infinity and in the process untilts it. Next, the plane of range focus must be moved to infinity to bring it into coincidence with the azimuth focal plane. This is implemented by placing a cylindrical lens, oriented so that its focal power is in the vertical direction, at a plane one focal length in front of the signal record. The focal length of the cylindrical lens is chosen to give a correct aspect ratio to the image. The image is now erect, free from astigmatism, and with unity aspect ratio.

Thus, although the desired image already existed and therefore did not have to be correlated into existence, it nonetheless required correction to compensate for its irregular properties. The correcting system that accomplishes this is indeed exactly the correlator that by the previous view correlated the image into existence, except that now the optical system function is described in an entirely different way. In particular, the reference function is a conical lens of the Fresnel zone plate or diffraction lens type. The development of this theory was completed in early 1956 and was summarized in a 36 page report that was then published as a progress report to the sponsor.

This new way of describing SAR in combination with optical processing is what today would be called a holographic viewpoint, in that the radar records wavefields along the flight path and the optical system carries out an optical reconstruction, and was a radical departure from the conventional theories of SAR, such as cross-correlation and Doppler filtering. I found the new viewpoint quite exciting, but no one else did; the response ranged from polite attentiveness to indifference. At that time electrical engineering and optics were just beginning to overlap, and the two areas were quite diverse. The optical viewpoint was not accepted.

This viewpoint was of course a recreation of Gabor's principle of wavefront reconstruction that had been published eight years earlier. During this development time I was not aware of Gabor's work, but I saw in October 1956 a paper on wavefront reconstruction by Kirkpatrick and El-Sum [3], which I read with considerable interest and then went on to study the referenced papers by Gabor and others. My feelings were mixed; there was some disappointment that the principle of wavefront reconstruction had already been invented, although in a wholly different context. This feeling was balanced with the knowledge that the concept of wavefront reconstruction was significant enough to have been published in the scientific literature.

In addition, this work had its points of originality, such as making the hologram with microwaves and reconstructing with visible light. This process was the opposite of Gabor's case, which had been to make the hologram at very short electron-beam wavelengths, with reconstruction at visible wavelengths. Also, this viewpoint introduced, for the first time in holography, the concept of the carrier frequency (or off-axis) method of separating the well-known twin images. The carrier was inherent in the SAR process. The Doppler signal arrives back on a high-frequency carrier – the microwave frequency that was radiated. In order to handle the signal, the high-frequency or r.f. is removed, and the signal downshifted to video frequencies. But as in all radio frequency situations, the frequency must not be reduced to the degree that aliasing, or folding-over of the frequencies occurs; if this happens, two different r.f. frequencies are mapped into the same frequency, and the signal is distorted. This is of course a basic requirement that pervades all communication systems, whether it be a radar system, an r.f. communication system, or whatever. And when the SAR process was reinterpreted in terms of wavefront reconstruction (holography) the carrier came along with the reinterpretation. Thus, the SAR holographic system included means for separation of the primary and conjugate images.

With this new interpretation of the reference function, the question immediately arose, why not use a refractive conical lens? This would be much more efficient in its light utilization, and in addition would shorten the system, since without the need for spatial filtering to remove extraneous diffracted orders, the conical lens could be placed directly against the signal recorded instead of being imaged on to it. The immediate conclusion was that such a lens did not exist and would be impractical to build, and the matter lay dormant for a few years.

However, in about 1957, library research revealed that a conical lens did exist; McLeod had described such a lens just about the time that we were

Fig. 5. The conical lens and the axicon lens from which it is derived.

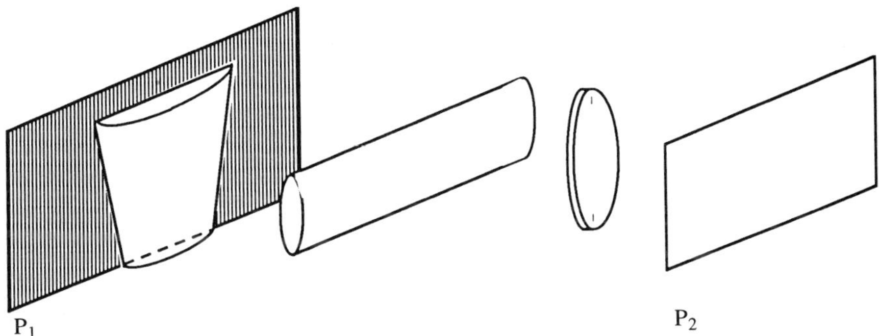

Fig. 6. Conical lens optical processor; collimating lens not shown.

inventing a need for it [4]. The basic axicon is a lens in the form of a right circular cone, with an apex angle rather large, in the range of 85° to 88° (Fig. 5). A pie-shaped slice from the axicon is, to an excellent approximation, just the conical lens required. It was then merely a matter of a visit to Eastman-Kodak to select from their stock an axicon of the proper specifications. This was done, and the optical processor now assumed the compact form shown in Fig. 6, consisting of only four lenses: collimator, conical, cylindrical, and conventional spherical. This lens system, one of the strangest ever developed, performed a task that no electronic data processing system of that day could have even come close to performing.

THE COMPLETED SYSTEM

Although the coherent optical correlator was well developed by 1956, there was yet no SAR data to process. The radar system was then still under construction, and would not be ready for operation until 1957. During this time the various groups of the Radar Laboratory were busily building their portions of the system. By 1957 the experimental system was ready for testing and was mounted in a C-46 aircraft and flown.

The results were disappointing. After the labour of about two-dozen people over a two-year period, the experiment seemed to end in failure. Flight after flight was made, and the recorded data yielded no image. The critics appeared to have been vindicated. There had been ever since the beginning of the project those who held the scheme to be unworkable, even though the more modest SAR efforts, using Doppler filter banks, had been successful. It had been said that the aircraft could never fly the stable path required to collect the radar signal without seriously degrading it with phase errors. Efforts were underway in Washington to stop the project funding.

Then one fine day, out of the blue sky (really, quite literally) came radar data that, when processed in our optical device, produced some startling results. The terrain was beautifully mapped, with detail such as roads, buildings, fields, etc., being clearly indicated. This flight, the ninth after eight failures, became known as Flight 9, and was for many years thereafter reverently remembered. A messenger was dispatched to the army sponsoring agency, just in time to avert the project termination. Now, elation was everywhere. The radar project personnel were excited, the sponsors were pleased, and the radar community took notice. The Michigan SAR system became famous, within of course the bounds of military secrecy.

The optical processing system was of central importance; without it the vast amount of complicated data generated by the system could not have been processed, and this success could not have been achieved. As a result, optical processing of radar data became fashionable; there arose a broad consensus that optical processing was the best, if not the only, way to process data from a high-resolution, high data rate SAR system, and optical processing became widely known and practised by the radar community.

In addition, experience with the optical processing system showed the utility of the physical optics viewpoint, which had for several years met with indifference. The SAR data verified all of the predicted optical properties. Now, signal film focal properties, conical lenses, and zone plates became bywords. By 1960, the wavefront reconstruction viewpoint had become dominant over the conventional theories of SAR. For example, the textbook *Introduction to Fourier Optics* by J.W. Goodman [5], describes SAR from this holographic viewpoint, exactly as it had been formulated years earlier.

The success with radar imaging continued, although not all flights were successful. The early error compensation system, consisting of accelerators to measure the various accelerations of the aircraft and the generation of compensating signals that were mixed with the recorded radar signals, was inadequate. It was soon learned that satisfactory results could be achieved with reasonable reliability only when the aircraft flew on cloudless nights at about 4 in the morning, before the atmospheric thermals from solar radiation had been generated. A fine start indeed for what was intended to be an all-weather surveillance system! Certainly the most difficult of all the problems in the development of the advanced SAR system was the stabilization of the system, and it was several years more until this problem was under control.

EXTENSIONS OF OPTICAL PROCESSING

Buoyed by the success of coherent optical processing for the SAR application, we turned our attention to other possible applications of coherent optical processing. New application ideas developed in profusion; most were not viable, but a significant number were. In 1957, just before the completion of our SAR system, I proposed the use of coherent optics to perform the compression of chirp pulses. In particular, the SAR system would radiate a chirp pulse instead of the conventional short pulse and the uncompressed pulse would be recorded on the SAR data record; thus, the signal from a point object became now a two-dimensional Fresnel zone plate (FZP) instead of the previous one-dimensional FZP. The data record now had focal properties in both dimensions, although in general the FZPs had different focal lengths in the azimuth and range dimensions; i.e., the recorded FZPs instead of consisting of circular contours, consisted now of either elliptical or hyperbolic contours. The two normally separate one-dimensional operations of SAR data processing and pulse compression were now combined into a single two-dimensional operation.

To change the optical system from an SAR processor into a combined SAR, pulse compression processor required no additional components, only the axial displacement of the cylindrical lens by an amount equal to the range dimension focal length of the combined SAR-chirp signal. Of course, the illumination had to be changed from a line source to a point source. Since lasers had not yet been invented, all the correlators were illuminated with Hg arc sources. As long as line sources were used the Hg arcs provided adequate levels of illumination, but when a point source was required and the slit on to which the source was focused became a pinhole, the light levels became significantly lower, but were still adequate, although data processing times were longer. By 1960, this method of chirp pulse compression had been incorporated into the Michigan SAR system, and the consensus grew that optical pulse compression was the preferred way to incorporate pulse compression into SAR systems.

Another successful application was the optical processing of underwater acoustic signals. Other applications under investigation included character readers and systems for analysis of seismic data. An atmosphere of optimism surrounded this exploration of new vistas for coherent optical processing. The optics group prospered. New and advanced optical systems were under development. Additional personnel were hired. By 1962 the optics group had grown to about 15 persons including engineers, scientists, technicians and other support staff. This was a thoroughly unique laboratory and probably the largest laboratory in the world devoted exclusively to coherent optical processing. The group possessed six or seven well-equipped laboratories with lenses, optical rails, benches, and tables. Central to each laboratory was a coherent light source, in the form of an Hg arc lamp, just as we had been using for the past five years. The laser had just been invented, and the commercial availability of suitable lasers such as the HeNe was over a year away. Yet, much was accomplished with the Hg source. Not only was radar data processed with this source but also early off-axis holography and early holographic spatial matched filtering, as well as other forms of coherent optical processing, were all performed with the Hg arc source. This fact may be surprising to many, since it is often assumed that optical processing requires laser illumination to be successful. This is a false belief, and our optical group was a stark testimony to its falsity. The truth, in fact, has two sides. On the one hand, it is true that the introduction of the laser greatly expanded the area of optical information processing, permitting many accomplishments that could be done only with considerable difficulty or not at all with a Hg source. On the other hand, skilled experimenters could get quite impressive results using the Hg source.

Interesting too was the makeup of the research staff. Only two members had had any coursework in optics, myself and an optical engineer who had been hired to package the optical system we had developed. We required for our new recruits a background in communication theory, and at that time optics schools had not reached the stage of turning out that kind of optical engineer. Our new optics recruits were taught optics through hands-on experience in the laboratory.

In the post Flight-9 years, our activities, though varied, fell into four main categories: holography, pattern recognition, optical processing of radar data as a support operation, and new concepts in SAR optical processing. We examine these.

OPTICAL HOLOGRAPHY

Following the discovery in 1956 of Gabor's papers and after studying them, I wondered how the carrier frequency method might be applied to Gabor's optical holography method to separate the twin images. The problem was clearly a classical aliasing situation. Two spatial frequency components of the Fresnel diffraction pattern, $a_1\exp(i2\pi f_1 x)$ and $a_1'\exp(-i2\pi f_1 x)$, were mapped into interference patterns of the same spatial frequency, $\cos(2\pi f_1 x)$, and thus could not be separated. The result was the twin image problem. The problem was in principle more than just cosmetic; the process failed, in fact, to record the entire wavefield. Let the object wavefield to be recorded be $a(x, y)\exp[i\phi(x, y)]$. Conventional photography records the amplitude a and discards the phase ϕ. Alternatively, writing the complex wavefield in rectangular rather than polar form leads to the equivalent expression $a(\cos \phi + i \sin \phi)$. Gabor's holographic process records only $a \cos \phi$, the $a \sin \phi$ portion being discarded. Again, as in conventional photography, only half of the field was recorded, only now the mixture was different.

This failure to record the entire wavefield was noted by Gabor, who had an ingenious solution. Instead of using the unscattered light of the object as the coherent background, or reference beam, he would use a separate reference beam, which gave more control. He could, by introducing appropriate phase shifts between the two beams, thus produce both the $a \cos \phi$ and $a \sin \phi$ components, recorded as separate holograms. The two components could then be combined in quadrature in an interferometer, thereby recreating the entire wavefield [6].

The method had problems and did not prove a viable method of holography.

If phase errors, due for example to lack of optical flatness of the hologram recording film, are of the order of $\pi/2$, the method fails. In addition, there are other extraneous terms that cause problems. The term $|u|^2$, the magnitude squared of the object beam, sometimes called the intermodulation product or self-interference term, adds to the background unless the reference wave is much stronger than the object beam. Also, a non-linear transmittance exposure characteristic of the recording film generates higher-order terms, which add to the background. In the off-axis case the principal effect of the non-linearity is the generation of higher diffracted orders that in the reconstruction process simply propagate away from the desired images.

Many of the early papers on holography that followed Gabor's seminal publications dealt with means for eliminating the twin image. Of particular note is the single sideband method of Lohmann (1956) [7]. This was a communication theory approach that closely paralleled our own thinking in optical processing.

It was clear that the ideal solution lay in placing the recorded holographic signal on a spatial carrier, thereby eliminating the aliasing. But how to accomplish this was a more difficult problem. What was natural and easy to do with electronic signals did not readily carry over to optics. The electronic signal processing methods operate on the instantaneous values of the signal, whereas interferometric processes in optics operate only with time-averaged intensity values. Sophisticated electronic techniques such as synchronous demodulation have as their optical correspondences only such crude methods as square-law detection. My initial conclusion was that it was not possible to transfer the technology of electronic systems to the optical problem.

For the next four years, 1956–60, I was ambivalent about the possible application of the electronic methods to optical holography. At an early stage I considered the introduction of a reference beam at an oblique angle to form a fringe pattern on to which to modulate the object beam signal. The first conclusion was that the process would simply produce, in each diffracted order, the usual terms, with inseparable twin images. Other possibilities were considered. The desired off-axis reference beam hologram, when illuminated with a plane wave, would produce a field consisting of a zero-order plane wave, as well as two images, one on either side, forming at different object positions. Therefore, one might reverse the process, placing two object distributions in appropriate positions, such that their Fresnel diffraction patterns, in combination with a coherent background, would sum to a positive real field distribution at the intended recording plane. Such a process seemed hopelessly complicated.

It was only in late 1960 that I decided seriously to investigate optical holography. In collaboration with Juris Upatnieks, a system for making holograms was set up. The off-axis reference beam technique was tried. The reference beam impinged on the recording plate at an angle sufficient to form a moderately fine fringe pattern (about 20 cycles mm^{-1}), well within the resolving power of conventional photographic emulsions. A diffraction grating was used to split the beam, two different diffracted orders were selected, and a simple object, just a wire, was placed in one of the two orders. At a plane where the beams recombined, a photographic plate recorded the hologram. In the reconstruction process, the beams separated only in the Fraunhofer regime, so a spatial filtering system was required.

In the reconstruction process, the zeroth order contained, as expected, the usual twin images, along with various noise terms. However, in one of the first diffracted orders, an image was formed, completely free from the competing twin image, which now appeared by itself in the other diffracted order. The process really worked.

Although questions about the theory were now put to rest, there remained practical questions. Possibly the price for the twin image separation was too high to make the technique usable. In the basic Gabor process, a coherent light beam was passed through a specularly transmitting transparency, and the Fresnel diffraction pattern was then recorded. In this new process, a second beam passed around the transparency and the two beams were combined interferometrically, producing a relatively fine fringe pattern superimposed on the original Fresnel diffraction pattern. The photographic record was thereby cluttered with many terms, all competing for the limited dynamic range of the film. The greater part of the dynamic range was utilized by the terms of the zeroth order, which were much stronger. And the diffracted orders could be separated only by a spatial filtering process, requiring two lenses and a spatial filter. The original Gabor process had been elegantly simple, and for this new process the original simplicity had been sacrificed. Experience showed, however, that the dynamic range problem was not severe, and the use of higher-resolution film and a steeper angle for the reference beam eliminated the need for the spatial filtering system. There remained other disadvantages of the off-axis method, but they were tolerable. There was of course the need for higher-resolution film. Also, the stability requirements are somewhat greater, since the off-axis method is a two-beam interferometric process, whereas the in-line method is not interferometry at all, except in the rather special sense that conventional imaging is interferometry. Aberrations are also more severe.

A widespread misconception relates to the coherence requirements. It is sometimes erroneously stated that the off-axis method requires greater coherence, and thus could be done only with the laser. Off-axis holography was first demonstrated with the Hg arc source, and for the system used, it can be shown that the coherence requirements need be no greater than for the in-line case.

The erroneous argument is that when two beams combine to form a fringe pattern, the path delay difference between the two beams increases by λ between successive fringes, and thus to get N fringes requires a source of coherence length $L = N\lambda$. This statement is true for some interferometers, but not for others. The interferometer used for this initial work used a diffraction grating as a beam splitter [8]. It is readily shown that no path differences are generated in the fringe formation process when the beams are split this way. Thus, the fringes were limited by the number of rulings on the grating, not by the coherence length of the light. Of course, for the initial demonstration of off-axis holography, the object was simple and the number of fringes generated was not large, perhaps several hundred, so that the experiment could have been carried out with any interferometer, since a low-pressure Hg arc source such as was used has a coherence length of about 1000 wavelengths. Thus, the fringe independence of spatial coherence was not needed, and the grating interferometer was not used out of necessity, but simply because the gratings were readily available.

Later, we reported the perfection of the process by making holograms from which high-quality images could be formed, not only from binary objects such as lettering, but also from continuous tone (or grey-scale) objects. During these experiments, we alternated between the newly available laser and the conventional Hg arc source, uncertain as to which to use. We ended up using the laser, although the decision could have gone either way. The laser was easier to use, since its enormous coherence length eliminated the need for careful path matching between the reference and object beams, but the penalty was that the great coherence of the laser, much more than was needed, tended to make the images noisier than when the Hg source was used. Technically, the decision was reasonable, but tactically it was a bad decision, since it fostered the view that these results required the use of a laser, which was not at all the case. In fact, the optimum course would have been to return to the grating beam splitter of the original off-axis work, as indeed we did a few years later, in 1966 [9]. We then repeated the experiments on high-quality holographic imagery, and for the one grey-scale object transparency that was common to the two sets of experiments, the imagery obtained with the grating beam splitter and the Hg source was of decisively better quality than the results obtained with the laser.

We have often been credited in the literature as being the first to apply the laser to holography. We have never claimed this to be the case, and it is in fact not true. Others, elsewhere, who had earlier access to lasers, used them to make holograms about a year before we did. The results had never been reported in the literature, but had been presented at technical meetings.

Finally in 1963 we reported the holographic imagery of arbitrary, three-dimensional reflecting objects. The results were dramatic, and more than any of our previous work created a world-wide interest in holography. Within a short time holography became one of the most active areas of optics. There came an onslaught of published papers that has continued unabated ever since. This work, of course, did indeed require the great coherence of the laser, primarily because the source coherence length has to be at least as great as the depth of the object scene, in order that all parts of the object could interfere with the reference beam. Thus, the source coherence requirement immediately jumped from less than a mm to several cm.

COMPLEX SPATIAL FILTERING

The development of the holographic complex spatial filter is one of the major accomplishments of our group. Holography permitted the easy construction of spatial filters with complete control over both the amplitude and phase. Arbitrary transfer functions could now readily be produced; in particular, spatial matched filters for a signal could be produced by making a Fourier transform hologram of the signal. This complex spatial filtering work involved the efforts of a number of persons in our optics group. It started in early 1961 with discussions about how we might practically construct some of the matched filters that we desired. The basic idea, due to C. Palermo, is to use as a signal an object transparency $s(x, y)$ placed side-by-side with its reflected signal $s(-x, -y)$, with a bright point of light in the middle, forming a transparency

$$t = s(x - a, y - b) + s(-x + a, -y + b) + \delta(x,y). \tag{6}$$

This object transparency is then Fourier transformed using a lens, producing a function that is both positive and real and can therefore be recorded photographically without loss of information. When this transparency is used as a spatial filter, it forms an image of the object in the centre of the field, the cross-correlation of the object with the filter impulse response s off to one side, and the convolution on the other side. This was just one of many significant ideas that this remarkably brilliant and creative individual contributed to the advancement of our optical processing program.

Upatnieks and I then recognized the similarity to our own holography work, and noted that the reflected signal was not needed. This made the filter construction process easier: just form the Fourier transform of s and bathe it in a reference beam, i.e., simply make an off-axis Fourier transform hologram.

An ambitious project was undertaken later in 1961 by A. Kozma, assisted by D. Kelley, to produce such a matched filter. The matched filter was to be used for compression of a coded radar pulse, of the specific type called a shift register sequence. Producing the filter was not an easy task. Path matching of the beams was more difficult than with the Fresnel holography that we were successfully pursuing. An alternative option was to generate the filter by computer. This method had the problem that the field of computer graphics was essentially non-existent at that time. Either method would be perfectly feasible but difficult. Since the shift register code is generated in a computer, the computer method was chosen, although the decision could have gone either way.

Making the filter was a laborious task, requiring a few months, but the result was spectacular. The filter worked perfectly, achieving compression ratios of the order 100. This achievement embodied several firsts; it was the first holographic spatial matched filter and it was the most sophisticated spatial filter ever produced up until that time. Also, Kozma had concluded that a phase-only filter, where the amplitude was discarded, would give only slight degradation, and this simplification made the filter construction easier. Thus it was also the earliest phase-only holographic spatial matched filter. The achievement was a major contribution to modern optics, but unfortunately, one that never received much recognition, since the work was classified and was released for publication only in late 1964 [*10*], at which time the impact of the work was greatly diminished.

During the period 1960–62 I had been working on a problem that arises in SAR when the resolution becomes very fine. As the aircraft sweeps past an object, the range to the object changes, so that the signal return from the object is recorded along an arc instead of a straight line. This effect is unimportant as long as the change in range is small compared to the range resolution, but becomes serious when the change in range is significantly greater. The signal must then be processed by integrating along the arc. The problem can be corrected by means of a matched filter, and the considerable success of the Kozma–Kelley work suggested that a holographic matched filter could be made for this application. In the spring of 1962, I succeeded in constructing the matched filter by making a Fourier transform hologram of a simulated signal. This time the filter was produced interferometrically, using a Hg arc for the light source. The task was not overly difficult because the filter was a rather simple one, and the required Fourier transformation was in one dimension only, so that the light source could be a line instead of a point. Again the result was quite good.

The matched filter work was then taken up by A. VanderLugt, who brought the method to its culmination [*11*]. Armed with the newly acquired HeNe laser, he produced spatial matched filters of much higher space-bandwidth product than had been made previously. His filters were two-dimensional, whereas the previous ones had been either one-dimensional (Kozma and Kelley), or two-dimensional but Fourier transformed in only one dimension (transfer function $H(x, f_y)$). He applied the technique to the broadly significant problem of pattern recognition, thus freeing this promising technique from its early ties with radar data processing. During the next several years VanderLugt and his colleague F. Rotz developed this complex spatial filtering technique to a very advanced level.

ADVANCES IN SAR OPTICAL PROCESSING

Although the conical lens optical processor was a remarkable and powerful instrument, it had an inherent defect that remained to be overcome. The image formed from the radar data had a magnification that was non-uniform across the field. The magnification in x is just the ratio of the focal length of the conventional lens to that of the conical lens, and the latter has a focal length proportional to the y dimension of the system. This distortion is corrected by placement of a slit at the recording plane, so that as the signal record passes through the processor aperture, the image scans across a slit. The process was wasteful of light. It was recognized that the conical lens system was in fact a spatial domain processor, in which the signal is correlated with the impulse response function, $h(x, y)$, of the conical lens. An alternative way would be to implement the system as a frequency domain processor, in which a filter $H(f_x, y)$, the Fourier transform of h, would be placed in the Fourier transform plane of a spatial filtering system. Analysis showed that such a system would form an image free from this distortion, and the scanning slit would be unnecessary. The Fourier transform of the conical lens transmittance function is another conical lens transmittance function, but one whose implementation was not in the form of a right circular cone. Its fabrication was considered unfeasible. It could, of course, have been constructed as a holographic spatial filter, but this was unacceptable, since most of the light would have been lost in other diffracted orders.

The matter was at an impasse until Martin and von Bieren of the Goodyear Corporation discovered that two closely spaced cylindrical lenses, one erect and the other tilted, would approximately generate the required phase function. They implemented this system, which came to be known as the tilted cylinder lens processor, or more simply, the tilted lens processor.

The culmination of this line of thought was the tilted plane processor, conceived by A. Kozma [12]. This system, the most elegant, sophisticated, and practical of all the optical processors, quickly replaced the conical lens processor. Its theory of operation, too complex to describe in detail here, is based on the manner in which afocal (or telescopic) optical systems image tilted planes into other planes of different tilt and without distortion. The tilted plane processor is a combination of two conventional lenses and two cylindrical lenses, all arranged in the afocal configuration, where the lens separation is the sum of the focal lengths. The result is that when the signal record is introduced on a tilt, the output image forms in another plane, of lesser tilt, the range and azimuth focal planes are brought into coincidence, any non-unity aspect ratio of the recorded data is corrected, and the image has a constant x dimension magnification over the entire field. This optical processor became the standard for SAR signal processing, and was eventually used world-wide.

The tilted plane and conical lens optical systems are certainly among the most unique optical systems ever devised. They also proved to be exceedingly practical systems within their specialized realm of application.

THE ARRIVAL OF THE LASER

CW lasers (HeNe) became commercially available about mid-1962, although some fortunate researchers had access to them before then. For many months we had been anxiously anticipating the arrival of the lasers and pondered on the impact they might have on our research. Indeed, the optics group was well-poised to exploit lasers, since it was already a powerful group, highly capable in optical processing and highly skilled with Hg arc source usage.

The radar data processing was by then reaching the end of the Hg source capability. The introduction of pulse compression into the radar data required the light source to be a point instead of a line, and the ever increasing space-bandwidth product of the signals meant greater coherence requirements, which translated into slower processing rates. The laser thus arrived at a convenient time.

However, when the laser was tried in place of the Hg source, the results were utterly distressing. The light at the recording plane was enormously brighter than before, the images a bit crisper, but the noise was abominable. The images, engulfed in this massive noise, were most unappealing. The disappointment was considerable, and there was concern about the laser ever being used for SAR optical data processing. Examination revealed that the noise had many causes, some of which could be eliminated. A principal source was spurious reflections, which for Hg light produced only a low-level uniform background, but the extreme coherence of the laser light converted the uniform background light into highly structured noise. Other noise sources resulted from dust on the lens surfaces, etc. One by one, these noise sources were discovered and either eliminated or reduced. Eventually, the noise was reduced to tolerable levels. The recently conceived tilted plane optical processor was thereupon designed to give minimal scatter; even the glass from which the lenses were to be made had severe specifications on the bubble content. Also, in the tilted plane processor, considerable noise smoothing occurred as the signal record moved across the aperture, thereby moving the image across the recording plane during the recording process, and smearing the noise. The tilted plane processor imagery was therefore relatively free from laser-induced noise.

On the other hand, the spatial matched filtering work was not troubled by the laser noise. There the problem was to find correlation peaks in a noisy background, and the laser noise was insignificant compared to the other noise. Furthermore, the high coherence of the laser allowed matched filters of previously unheard of space-bandwidth product to be made with ease.

For the holography project, the laser was of mixed value, as previously noted. The overwhelming advantage of the laser for holography came only later when three-dimensional reflecting objects were used. But it was primarily this type of object that gave holography its principal attraction, and made holography a major field.

The laser produced a vast world-wide expansion of optical processing activities, including holography. The optics group continued as a major player in this expanding activity, eventually growing to about 25 staff members by 1965.

PHASE CONJUGATION

The conjugate image of the holographic process had historically been one of its major problems. Early in our holography project we proposed a use for the conjugate image: to image through distorting media. It is ironic that the Leith–Upatnieks paper of 1962, which had presented an effective method for avoiding the conjugate image problem, also proposed, for the first time, a use for this image.

As shown in Fig. 7, a distorting medium lies between the object and hologram recording plane. The medium could be a turbulent atmosphere, a severely aberrated lens, or even a plate of frosted glass. The wavefront from an object point is distorted after passage through the medium, so that the resulting image will be degraded, or even obliterated. Let the medium be described by a transmittance $\exp(i\phi)$. The hologram forms a conjugate image $\exp(-i\phi)$. Placement of the original medium in the position where the conjugate image is formed results in the multiplication $\exp(i\phi)\exp(-i\phi)$. The distorting medium and the conjugate image superimpose and the phase of one is cancelled by the complementary phase of the other. The distorting medium seemingly disappears and a clear image of the obscured object is revealed.

It was not until 1965 that we found time to carry out an experimental demonstration of this idea. We chose an extremely severe distortion function, a piece of frosted glass, a choice that made the experiment quite difficult. After development, the hologram had to be replaced in its original position to within about a micrometre in all three coordinates, with similar constraints for the hologram orientation. Also, the readout beam had to very accurately retrace the path of the original reference beam. If there were even a slight misalignment in the hologram repositioning or in the alignment of the readout beam, the image was not merely degraded, it was obliterated. The experiment was carried out successfully, although in retrospect it seems pointless to have chosen such a

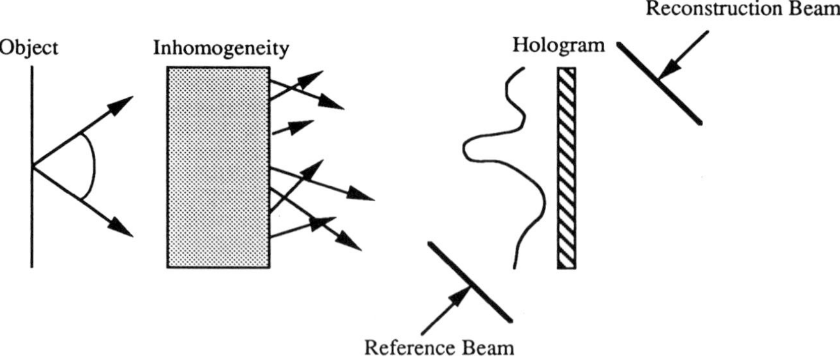

Fig. 7. Phase conjugation imaging with holography.

challenging distortion function for a proof of principle demonstration. The experimental results were described in a paper presented at an SPIE meeting in August 1965 and the written paper, distributed at the meeting, as was the SPIE practice then, was later published in the October–November 1965 issue of the *SPIE Journal* (the predecessor of *Optical Engineering*) [*13*].

Although the results were interesting, little use was found for this technique for many years. The major reason was that the most useful applications, such as imaging through a turbulent atmosphere, require a real time process, i.e., a time shorter than the correlation time of the medium, which is often several milliseconds. By 1970, non-photographic means for producing the conjugate wave, such as four-wave mixing in crystals, were found. Thereupon, a new area of optics, known as phase conjugation, rapidly developed. The major application of phase conjugation remains image formation through distorting media.

HOLOGRAM INTERFEROMETRY

Hologram interferometry, one of the major applications of holography and a major new advance in interferometry, was first described by Stetson and Powell in 1965 [*14*]. Their form of holography was called time average hologram interferometry, since it was produced by holographing a moving object, and the resulting hologram averages the interference pattern produced by the object at its various positions. The work turned one requirement of holography on its head. As holography became more sophisticated, the requirements on stability increased, until holography moved from its original place of performance on ordinary optical rails on to massive granite tables or specially designed vibration-isolated tables. If the object vibrated or otherwise moved during the exposure, the resulting image had regions of darkness corresponding to the regions of greatest motion. The hologram was thus ruined and was discarded. The viewpoint of hologram interferometry was that the hologram was not at all ruined, but instead offered important information about how the object had moved during the exposure. Later, the other two basic forms of hologram interferometry, the double exposure and real time methods, were invented. The double exposure method was invented almost simultaneously by several groups: Powell and Stetson of our group, Haines and Hildebrand also of our group, Wuerker *et al.* of TRW, and Burch of the National Physical Laboratory of Great Britain. In the ensuing patent competition, Burch emerged the winner in the US patent office as having the earliest date of conception.

Hologram interferometry became an important sub-area of holography and an area that stood on its own as a separate major area of optics. Conferences were organized on this topic and books were written. C. Vest, who had been a member of the optics group, as well as professor of mechanical engineering, for several years beginning about 1965, became an important figure in hologram interferometry, advancing the technical field in a significant way, and also writing the first book on hologram interferometry [*15*]. He eventually left the field to do administration, ultimately becoming President of the Massachusetts Institute of Technology. Even though removed from research, he still made a most momentous contribution, although a non-technical one, to holography by acquiring the assets of the bankrupt New York Museum of Holography and making them part of the MIT museum.

CONTINUATION AND CONCLUSION

The team of Haines and Hildebrand became well known in holography for a variety of important contributions, in hologram interferometry, in holographic contouring, in bandwidth reduction techniques, and other areas. A. Friesem, now professor at the Weizmann Institute of Israel, became another major figure in holography as a member of our group during the latter part of the 1960s, making important contributions in a variety of areas, including diffraction from thick media, holography in photochromics, and color holography. N. Massey, less well known than the others we have mentioned because he was not a paper publisher, was another one of the major figures of the laboratory because of his consummate skill in optical processing and for his development of a doctrine for carrying out optical processing. He also played a major role in the development of the tilted plane processor. The various names we have cited throughout this chapter will be recognized by holographers as among the major holography pioneers of the 1960s. It seems incredible that at one time they were all members of the same group.

W. Brown succeeded Cutrona as Head of the Radar Laboratory in 1962; he at once committed the laboratory to the goal of improving the resolution of SAR systems by a factor of 100 over the following decade. A brilliant analyst and radar specialist, he was a strong advocate of optics and was in large measure responsible for the development of a large and strong optics group. Although his research work was primarily in radar, he always kept one foot in the optics group, making significant contributions in optical processing. Under his leadership, the entire radar laboratory, but especially the optics group, prospered. He developed new concepts in coherent radar, which in turn led to new concepts in optical processing. Possibly his most significant contribution is the coherent radar imaging of rotating objects, which led to extremely high resolution through Doppler analysis of the return from various parts of the object. The data processing was done optically. J. Walker of the optics group led the development of this radar technique to a sophisticated state.

It is now many years later. The Willow Run Laboratories separated from the University of Michigan in 1972 and became the not-for-profit corporation called ERIM (Environmental Research Institute of Michigan). The personnel and organization of the Optics Group have undergone many changes and only a few of the early members still remain; however, the group is still identifiable as the descendant of that highly successful group. Its current membership includes well-known figures in holography and optical processing, who currently publish many papers. J. Fienup has brought the group fame for his work on phase retrieval. Others, including I. Cindrich, C. Alexsoff, L. Peterson, J. Marron, and A. Tai continue with significant innovations in optical processing. Clearly, the optics group still maintains a high standard of excellence. However, with holography and optical processing having become large, mature fields, it is no longer possible for one group to have the dominance that was possible then.

In the 1960s, while the Optics Group was busy making its innovations, there was a parallel activity conducted by A. Lohmann, who functioned as a one-man counterpart of our coherent optics group, busily applying communication concepts to optics and thereby developing holography and optical processing. Our conditions were different; we were a large group, extremely well funded, and with some very specific missions. The price paid for this enviable position was, at least until 1963, difficulty and long delays in publishing our work in the open literature. Despite the differences of operating modes, his and our work had strong parallels. It is interesting to speculate on what might have been accomplished had Lohmann been a member of this group.

REFERENCES

1. Sherwin, C.W., Ruina, J.P. and Rawcliffe, R.D. (1962) 'Some early developments in synthetic aperture radar systems,' *IRE Trans. Mil. Electron.*, **MIL-6**, 111–115.
2. Rogers, G.L. (1977) *Non-Coherent Optical Processing*. Wiley & Sons, New York.
3. Kirkpatrick, P. and El-Sum, H.M.A. (1956) 'Image formation by reconstructed wave front I. Physical principles and methods of refinement,' *J. Opt. Soc. Amer.*, **46**, 825–831.
4. McLeod, J., (1954) 'The axicon: a new type of optical element,' *J. Opt. Soc. Amer.*, **44**, 592–597.
5. Goodman, J.W. (1968) *Introduction to Fourier Optics*. New York: McGraw-Hill.
6. Gabor, D. and Goss, W.P. (1966) 'Interference microscope with total wavefront reconstruction,' *J. Opt. Soc. Amer.*, **56**, 849–858.
7. Lohmann, A. (1956). 'Optische Einseitenbandübertragung angewandt auf das Gabor-mikroskop,' *Optica Acta*, **3**, 97–100.
8. Leith, E. and Upatnieks, J. (1962) 'Reconstructed wavefronts and communication theory,' *J. Opt. Soc. Amer.*, **52**, 1123–1130.
9. Leith, E.N. and Upatnieks, J. (1967) 'Holography with achromatic-fringe systems,' *J. Opt. Soc. Amer.*, **57**, 975–980.
10. Kozma, A. and D. Kelly, D. (1964) 'Spatial filtering for detection of signals with additive noise,' *J. Opt. Soc. Amer.*, **54**, 1395–1402.
11. VanderLugt, A., (1964) 'Signal detection by complex spatial filtering,' *IEEE Trans. Information Theory*, **IT-10**, 139–145.
12. Kozma, A., Leith, E.N. and Massey, N.G. (1972) 'Tilted-plane optical processor,' *Appl. Opt.*, **11**, 1766–1777.

13. Leith, E.N. and J. Upatnieks, J. (1965) 'Hologram: their properties and uses,' *Soc. Photo-Optical Instrum. Engrs J.*, **4**, 3–6.
14. Stetson, K.A. and Powell, R.L. (1966) 'Hologram interferometry,' *J. Opt. Soc. Amer.*, **56**, 1161–1166.
15. Vest, C. (1979) *Holographic Interferometry*. Wiley & Sons, New York.

2 Bio-speckles

Yoshihisa Aizu
*Department of Mechanical System Engineering,
Muroran Institute of Technology, Muroran, Japan*

Toshimitsu Asakura
*Research Institute for Electronic Science, Hokkaido University,
Sapporo, Japan*

INTRODUCTION

Laser light scattered from diffuse objects produces granular interference patterns which are today well known as 'speckle phenomena'. If the diffuse object moves, the speckle grains also move and change their shape. The speckle pattern thus becomes time dependent. This property can be applied to measurements of velocity, vibration, displacement and so on. The dynamic statistical properties of speckles have been extensively studied by theory and experiment especially for velocity measurements [1]. However, most of the studies are related to simple scattering at the surfaces of inanimate objects. Dynamic speckle phenomena can also be observed with living objects. Such speckles have quite different characteristics from those of conventional speckles with inanimate objects, for example:

- speckles are usually produced by complex multiple scattering;
- speckles are often formed by a mixture of some moving speckles with different dynamics, including static speckles;
- space-time speckle dynamics strongly depend on the structure and activity of living objects and thus are quite complicated and inconsistent; and therefore
- theoretical treatments are generally difficult.

Due to these characteristic natures, speckles from living objects may exclusively be called 'bio-speckles' [2, 3].

With the recent increase of applications of laser light in the medical, physiological and biological fields, bio-speckles are receiving much attention from scientists [4, 5]. The fundamental and application-oriented studies on bio-speckles may be categorized under the heading of 'bio-speckle techniques'. Bio-speckles carry useful information about the biological or physiological activity of living objects, such as blood flow and motility. The statistical analysis of bio-speckle phenomena can, therefore, provide non-contact measuring methods for living objects. The physiological activity of living objects is often sensitive to changes in the state of the environment. From this point of view, bio-speckles may also play an important role as an indicator to give useful information about the environment, an application of which may be called 'bio-indication' [5]. With this potential, the development of studies on bio-speckles is now strongly expected to continue.

In this chapter, the fundamental statistics of bio-speckles are first reviewed briefly. They can be used for analysing bio-speckle dynamics and extracting useful information from living objects. Examples of bio-speckles obtained from various living objects are next introduced with close regard to their applications in the medical, physiological and biological fields. Finally some other properties peculiar to bio-speckles are discussed.

BIO-SPECKLE STATISTICS

Figures 1(a) and 1(b) [6] show photographs of typical bio-speckle patterns obtained from a human fingertip, under illumination of a direct beam from a He-Ne laser, with exposures of 0.5 sec and 1 sec, respectively. The speckle pattern (b) is seen to be blurred with increasing exposure time when compared with the pattern (a). This observation indicates that bio-speckles generally fluctuate in a space-time random fashion. This is due to the complicated structure and inconsistent activity of living objects, for example, skin blood flow in the case of Fig. 1. Thus the dynamic behaviour of bio-speckles should be analysed statistically. Some kinds of statistical approaches have been taken so far for the intensity fluctuation of bio-speckles while the phase statistics has

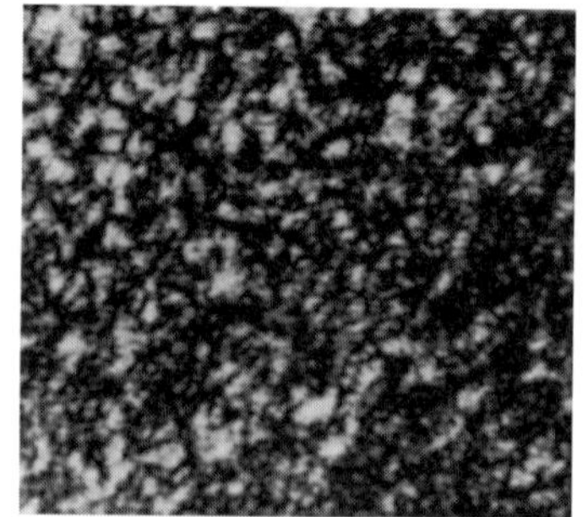
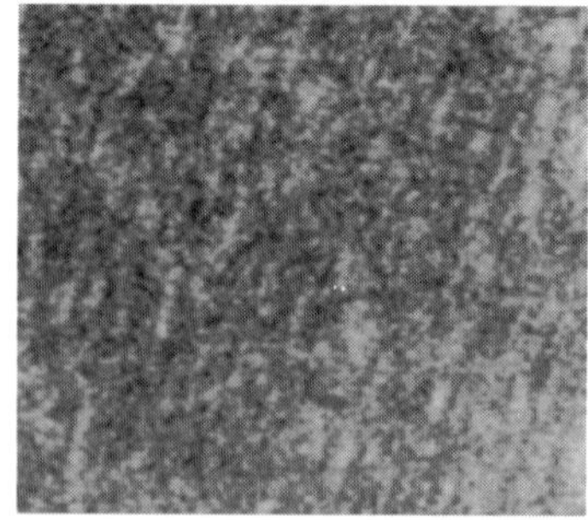

Fig. 1. Bio-speckle patterns recorded in (a) 0.5 sec and (b) 1 sec exposure times from the skin surface of a fingertip (with kind permission from [6]).

not been studied yet. The following seven categories of statistics can be used for studies on bio-speckles, but excluding the statistics of space-integrated or space-differentiated speckles which seem to be useless here.

Time-varying Bio-speckles

First-order temporal statistics

First-order temporal statistics provide us with the ratio of the standard deviation σ of intensity fluctuations to the mean intensity $\langle I \rangle$. This concept corresponds to the contrast of speckle patterns in the spatial statistics. The ratio $\sigma/\langle I \rangle$ is calculated from temporal signals recorded at a single detecting point on the bio-speckle pattern. The velocity of moving speckles or the frequency components of intensity fluctuations are not reflected in the ratio $\sigma/\langle I \rangle$, but it can estimate the relative magnitude of speckle fluctuations. A fully developed speckle pattern yields $\sigma/\langle I \rangle = 1$, but it can hardly be realized with living objects. This property originates from a mixture of static and moving speckle patterns, and will be discussed later. The ratio $\sigma/\langle I \rangle$ is used for bio-speckles from botanical specimens.

Second-order temporal statistics

Second-order temporal statistics are the most popular treatment which can be used for measuring the velocity or mobility of scatterers in various living objects. Usually, the autocorrelation function or power spectrum density function of temporal intensity fluctuations detected in the observation plane is used in this approach. These functions directly reflect the frequency components of speckle signals and, thus, the dynamics of objects. To relate these functions to the object velocity, a variety of evaluations is possible: the mean frequency $\langle f \rangle$, the cut-off frequency f_c, the ratio HLR of the high-to-low frequency components, and the blood flow parameter for spectrum analysis, and the correlation time τ_c and the speckle lifetime for autocorrelation analysis. They may be chosen according to the shapes of the functions and their variation. The main objective may be to determine how to estimate these functions, in order to extract necessary information effectively and sensitively from living objects.

The measuring techniques based on second-order temporal statistics are known also as dynamic light scattering spectroscopy, photon correlation spectroscopy, light beating spectroscopy or intensity fluctuation spectroscopy, and, in some cases, are closely related to laser Doppler velocimetry.

Spatial statistics

First-order spatial statistics of frozen speckle patterns yield the contrast of speckle patterns which is meaningless here for analysing bio-speckle dynamics. Another way to use first-order spatial statistics is to take a histogram of intensities at all the points of the bio-speckle pattern. The histogram shows approximately the probability density function which reflects a feature of the intensity distributions. Thus, the form of such a histogram may be dependent on individual cultures. Second-order spatial statistics give the average speckle size which is also unlikely to be useful. In bio-speckle techniques, however, the spatial autocorrelation function may contain some information about the state of living objects, such as the concentration of scatterers, because the number of scatterers has an influence on the coherent sum of their diffraction patterns.

Time-integrated Bio-speckles

First-order temporal statistics

With a certain integration time, the high-frequency speckle fluctuations are averaged out while the low-frequency fluctuations still remain. If necessary, the integration can be repeated many times to take the statistics during one measurement. Thus, the ratio of the standard deviation σ_i of time-integrated intensity fluctuations to the mean intensity $\langle I \rangle$ carries information about the object velocity. The integration time should be appropriately determined so that the necessary velocity information could be extracted effectively. The integration can be made electronically by digital processing with a computer.

First-order spatial statistics

The ratio $\sigma_i / \langle I \rangle$ in the temporal statistics above is analogous to the contrast of bio-speckle patterns in the spatial statistics. If the spatial standard deviation σ_{is} of the time-integrated bio-speckle pattern is taken, the contrast $\sigma_{is} / \langle I \rangle$ becomes velocity dependent. The rapidly moving and changing speckle grains, which are caused by the high velocity of scatterers, result in the low spatial contrast, while the slow movement yields the high contrast. Thus, the velocity distribution in the illuminated area can be observed as the contrast distribution. The integration can be performed optically on a photographic film and electronically on a CCD image sensor with digital image processing. No useful information can be extracted here from the second-order statistics of time-integrated speckles.

Time-differentiated Bio-speckles

First-order temporal statistics

Differentiation of high-frequency fluctuation signals produces a large amplitude while that of low-frequency fluctuations has a small amplitude. The ratio of the standard deviation σ_d of time-differentiated intensity fluctuations to the mean intensity $\langle I \rangle$ yields a velocity-dependent value. If the ratio is taken at all points of the observed area using many frames of the speckle pattern, such a result shows a velocity map. The statistics can be equivalently taken for the intensity difference $I(t) - I(t + \tau)$ between two successive time points, instead of the differentiated intensity. This is much easier than differentiation in signal processing. In this case, the ratio is considered as a so-called structure function $s(\tau)$ [7] that is given by

$$s(\tau) = \langle [I(t) - I(t - \tau)]^2 \rangle. \tag{1}$$

First-order spatial statistics

When the spatial standard deviation σ_{ds} of time-differentiated bio-speckle patterns is taken, the contrast $\sigma_{ds}/\langle I \rangle$ is also a function of the object velocity. The distribution of speckle moving (or changing) velocities is expressed by the contrast distribution. This technique is quite similar to the case of time-integrated bio-speckles (see above) although the velocity–contrast relation is opposite. No use of this method has been reported yet for bio-speckles. The second-order statistics of time-differentiated speckles is unlikely to be useful here.

EXAMPLES OF BIO-SPECKLES AND APPLICATIONS

Bio-speckles from Botanical Specimens

Briers [8] reported that laser speckle patterns obtained from tomatoes are observed to fluctuate. The fluctuation rate was estimated from the contrast of speckle patterns recorded on photographic films with exposure of 1.5 sec. The estimation is based on the first-order spatial statistics of time-integrated bio-speckles (above). The temporal fluctuations are caused by plastids or mineral particles in motion in the cells of tomatoes. The technique would be available for monitoring the cell activity *in vivo*.

The bio-speckle fluctuations tend to disturb holographic recording fringe patterns in holographic interferometry if they are made with living objects. Briers [4, 9, 10] proposed a screening test in holography in terms of the ratio $\sigma/\langle I \rangle$ in the first-order temporal statistics of bio-speckle fluctuations. A large value of $\sigma/\langle I \rangle$ indicates difficulty in obtaining holograms and interference fringes because of large decorrelation effects. Experiments were performed for tobaccos, wheat sprouts, and leaves of other plants [10].

Oulamara *et al.* [11] studied the temporal decorrelation effect of bio-speckles obtained from tomatoes, oranges and apples, by using digital image processing. The fluctuation rate was observed by a temporal pseudo-image which was constructed by encoding a line-scanning image on a videoscreen by writing successive scans one below the other. Various sets of two successive digital speckle images with different delay times were used also to calculate the correlation function, according to the second-order temporal statistics of speckle fluctuations. This technique can be a means for investigating quantitatively the living state or biological activity of botanical specimens, such as being fresh or old.

To evaluate all the data of the autocorrelation function of bio-speckle fluctuations, the use of logarithmic weighting or the 'equivalent rectangle' by Rabal *et al.* [12] was presented. Its usefulness was shown for monitoring the biological activity of fresh and drying apples, grapefruits and jasmines.

Bio-speckles from Skin Tissues and Internal Organs

From the standpoint of time-varying speckles, Fujii *et al.* [6] initially studied the dynamic intensity fluctuations of light scattered from human skin tissues

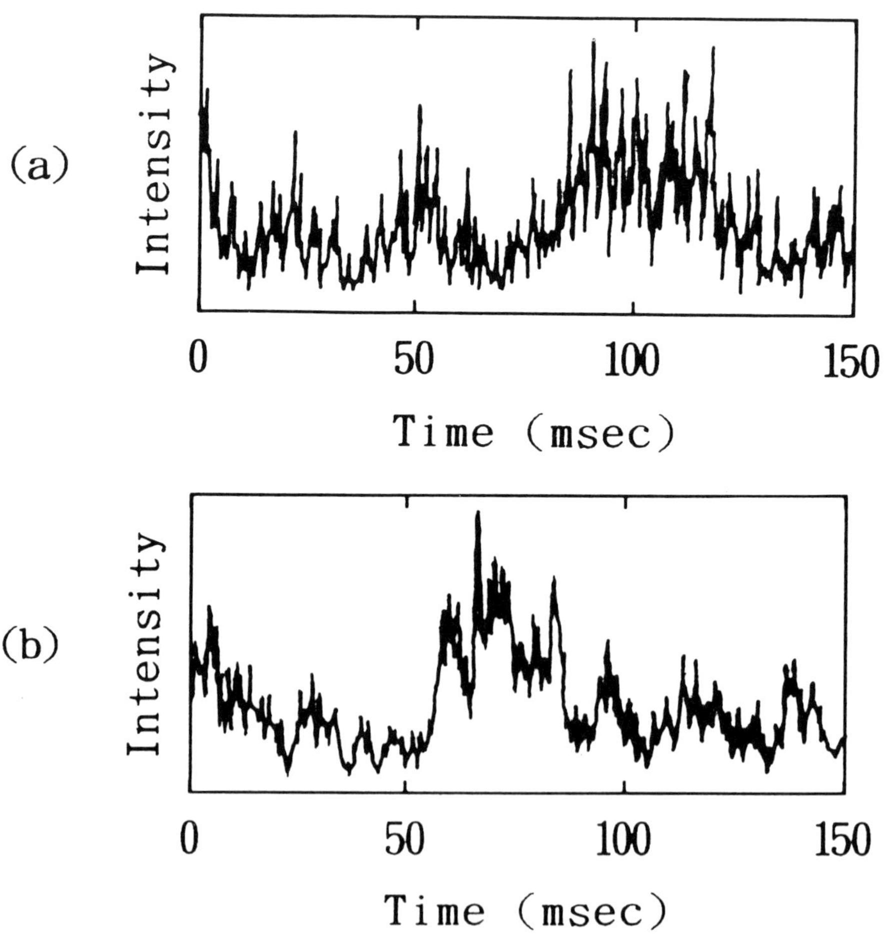

Fig. 2. Bio-speckle fluctuations (a) and (b) obtained from normal and reduced skin blood flows, respectively (with kind permission from [*3, 6*]).

while so far other studies were based on the Doppler effect for the same purpose. Figure 2 shows typical bio-speckle signals obtained from a fingertip, through a small pinhole placed in the diffraction plane. The signals (a) and (b) were recorded for the normal flow and the reduced flow with an inflated cuff, respectively. High-frequency components are clearly suppressed in Fig. 2(b) with the reduction of blood flow. The study was developed to construct a skin blood-flow monitoring system using an optical fibre probe, which is shown in Fig. 3. A multimode fibre F_1 and single-mode fibre F_2 are used for illumination and detection of laser light, respectively. Bio-speckle fluctuations are processed to obtain the power spectral distribution on the basis of second-order temporal statistics.

Figure 4 shows typical power spectra obtained from (a) a fingertip and (b) a leg. The difference between the two spectral curves implies that the fluctuation of bio-speckles varies with the type of skin tissues. A useful parameter HLR

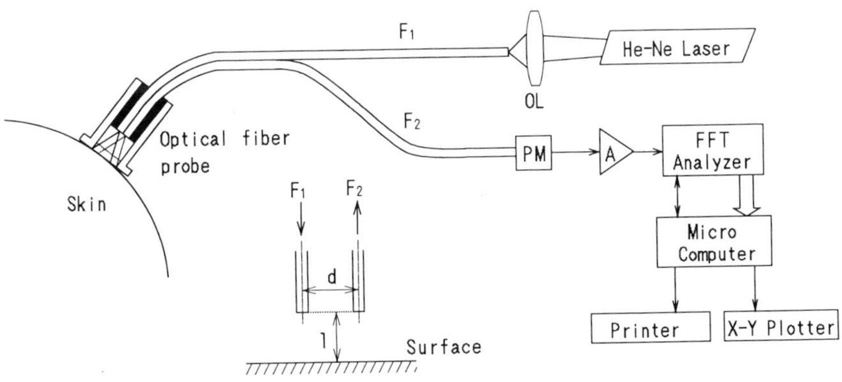

Fig. 3.. Schematic diagram of a blood-flow monitoring system [*3*].

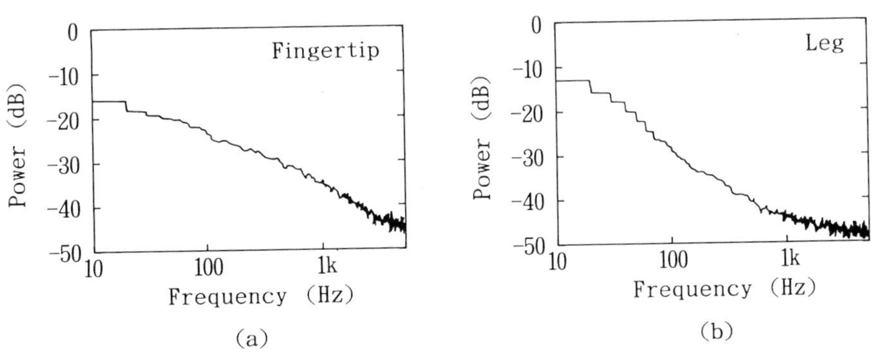

Fig. 4. Power spectra obtained from two different skin surfaces (with kind permission from [*3, 6*]).

[6] was introduced to estimate the skin blood flow, and is given by

$$\mathrm{HLR} = \frac{|V_{\mathrm{H}}|^2}{|V_{\mathrm{L}}|^2}, \tag{2}$$

where $|V_{\mathrm{H}}|^2$ and $|V_{\mathrm{L}}|^2$ indicate the powers of certain high and low frequency components of speckle signals obtained through corresponding band-pass filters. The parameter HLR was able to reflect quite sensitively the change of blood flow in the physiological test called 'reactive hyperaemia' (response of the blood flow change from reduced flow to re-opened flow).

The same technique can also be applied to internal organs [2, 3] using second-order temporal statistics. Figure 5(a) shows a typical power spectrum obtained from the gastric mucous membrane of an anesthetized albino rabbit. The parameter HLR is not suitable for analysing the power spectra in this case because of its different form of spectral curves. To evaluate suitably the spectra, the mean frequency $\langle f \rangle$ [3] is used here, and is defined by

$$\langle f \rangle = \frac{\sum_i f_i P(f_i)}{\sum_i P(f_i)}, \tag{3}$$

where $P(f_i)$ is the signal power at frequency f_i. The experiments show that the blood flow variation with vasoconstrictor and vasodilator can be successfully measured by the mean frequency $\langle f \rangle$. The intestinal mucous membrane was also found to be a suitable subject of this study.

Ruth [13–16] also studied the bio-speckles from skin tissues and their application to skin blood-flow measurements. The blood flow was estimated by the modified mean frequency M (called the blood flow parameter by Ruth *et al.* [17])

$$M = \left[\int T^2(f) P(f) \, \mathrm{d}f \right]^{1/2}, \tag{4}$$

where $P(f)$ is the signal power spectrum and $T(f)$ is the transfer function or weighting function. This is again based on second-order temporal statistics. It should be noticed that, strictly speaking, the blood flow parameter M is not proportional to the average blood flow but can be optimized by choosing an appropriate transfer function so that it could reflect most sensitively the change of blood flow. The choice of the best transfer function is made empirically [13] since it depends on the type of skin tissues and involuntary body movements.

Fujii *et al.* [18] developed a two-dimensional visualising system for skin microcirculation by means of bio-speckle statistics. The bio-speckle pattern in

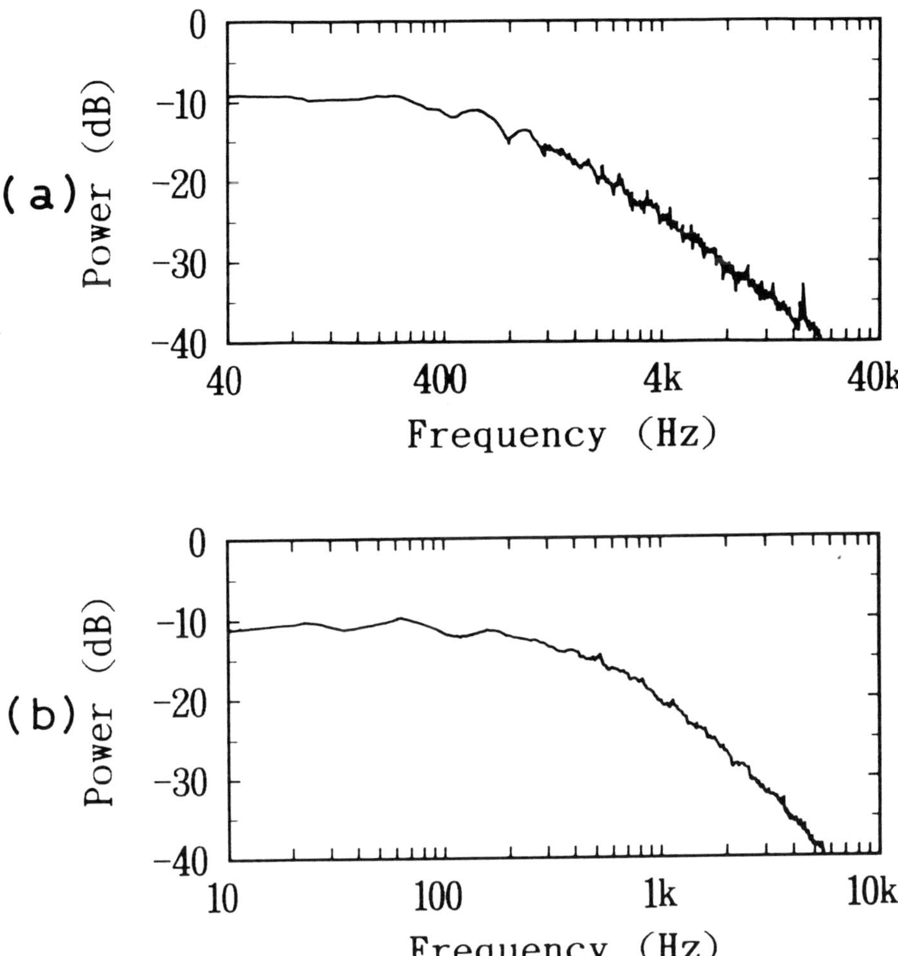

Fig. 5. Power spectra obtained from (a) gastric mucous membrane and (b) ocular fundus, of an albino rabbit [*3*].

the line image obtained from the skin surface is imaged on to a CCD linear image sensor. Figure 6(a) shows two successive scanning outputs of the one-dimensional line image. The profile of these signals shows the fluctuations due to the skin blood flow. A considerable difference is found between the two signals in the right-hand side where the skin blood flow is faster. To estimate the magnitude of the difference, the following parameter $D_{(n)}$ [*18*] is calculated on the basis of the first-order temporal statistics of time-

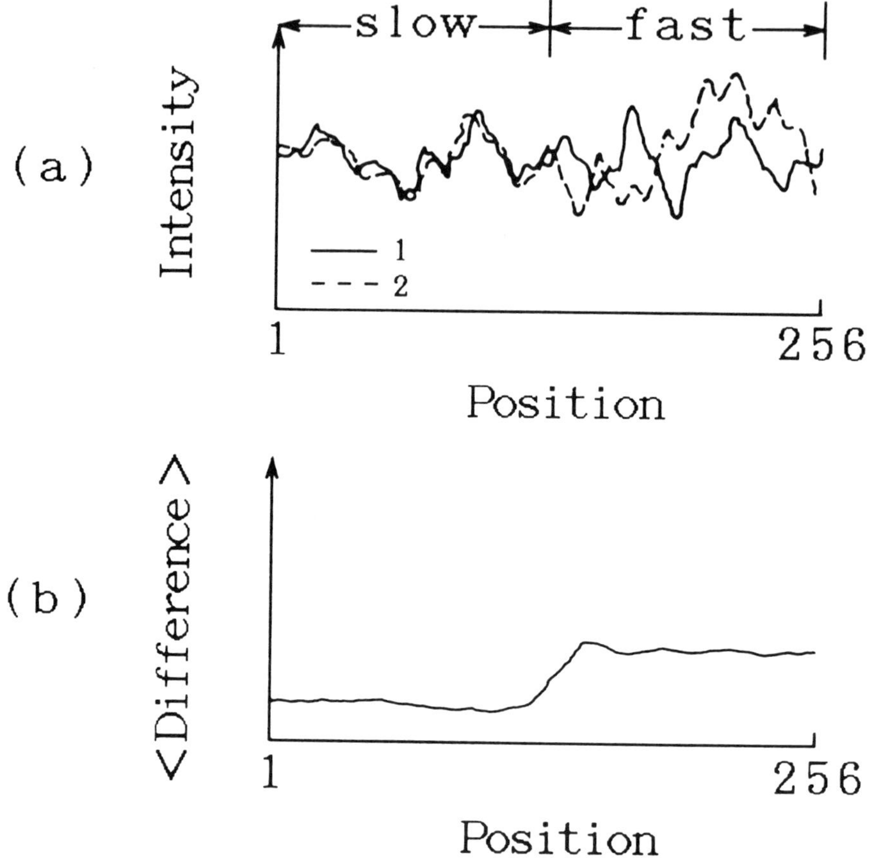

Fig. 6. (a) Two successive scanning outputs of a CCD image sensor and (b) one-dimensional blood flow map (with kind permission from [3]).

differentiated speckles:

$$D_{(n)} = \sum_{k=1}^{N} \frac{\left| I_k(n) - I_{k+1}(n) \right|}{\left| I_k(n) + I_{k+1}(n) \right|}, \tag{5}$$

where $I_k(n)$ denotes the amplitude of output signals at the nth pixel point of the kth scan in the CCD sensor. This process was repeated for more than 100 scans to average the data. Figure 6(b) shows a one-dimensional distribution of the cumulated difference value which corresponds to the blood flow distribution. By scanning orthogonally the one-dimensional line image, the technique can be carried out in a two-dimensional manner where the value of $D_{(n)}$ is displayed graphically in colours to visualize the blood flow map. Instead of scanning the line image, a two-dimensional area sensor may be more suitable for practical instruments. However, the scanning rate of commonly available sensors is not sufficiently high. Thus, some improvement is needed for treating high-frequency fluctuations. Studies of the application of this technique are now in progress to investigate retinal blood flow.

Bio-speckles from muscular tissues are also analysed [*19*] to estimate muscular activity which carries information about muscle fatigue. In this case also, the power spectral analysis (second-order temporal statistics) was a useful means for investigating the functional state of the brachial biceps of a patient. Further investigation is expected to be made for skeletal muscles.

Bio-speckles from Ocular Fundus

Bio-speckles can be observed from the ocular fundus illuminated with laser light. The intensity fluctuation of bio-speckles is due to blood flow in the retinal layer and, in some cases, the choroidal layer. Thus naturally the bio-speckles are utilized to evaluate those blood flows. A typical example using first-order spatial statistics of time-integrated speckles was presented by Fercher and Briers [*20–22*] as a suitable means for visualizing the retinal blood flow map. If a photograph of the fluctuating bio-speckle pattern is taken with a certain exposure time, a time-integrated speckle pattern is recorded. The contrast $\sigma_{\mathrm{is}}/\langle I \rangle$ of the recorded pattern becomes a measure of the retinal blood flow. Thus, the blood-flow distribution in the retinal plane is mapped as the variation in the speckle contrast. In this case, the exposure time is a key parameter. Some post-processing is necessary for converting the speckle contrast variation into the intensity distribution or colour graphic images [*23, 24*]. The maximum contrast obtained experimentally was about only 0.2, which probably means that bio-speckles from the ocular fundus are not fully developed.

On the basis of the second-order temporal statistics of bio-speckle fluctuations, Aizu *et al.* [*2, 3, 25–29*] studied another method to evaluate the fundus blood flow or the retinal blood-flow velocity. Figure 7 shows a schematic diagram of the measuring system which is constructed with a standard fundus camera. The He-Ne laser light illuminates a somewhat expanded area of the retina. The bio-speckle pattern can be observed in both the diffraction and image planes. If the bio-speckle fluctuation is detected in the diffraction plane by using a pinhole-type detecting aperture DA, the power spectrum of speckle signals reflects the overall activity of various blood flows in the illuminated spot area [*25–27*], because the detecting aperture DA receives the superposition of whole light fields scattered from every point in the spot. Figure 5(b) [*25*] shows a typical power spectrum obtained from the ocular fundus of an anesthetized albino rabbit. The form of the spectrum distribution is very similar to that for the gastric mucous membrane shown in Fig. 5(a). This probably implies that they have a similar type of tissue structure

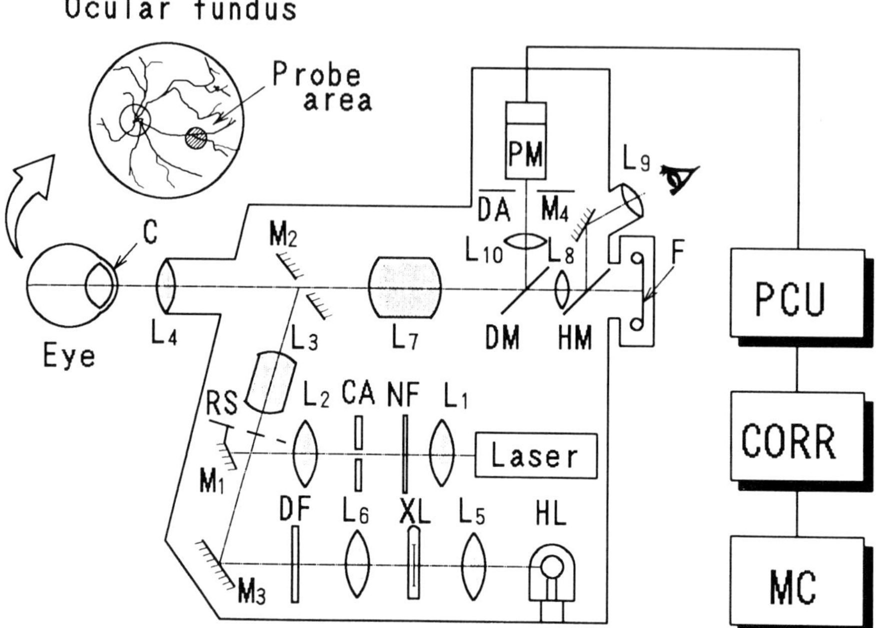

Fig. 7. Schematic diagram of the apparatus for measuring the blood flow at the ocular fundus [*3, 26, 27*].

and blood flow. For the result of Fig. 5(b), the mean frequency $\langle f \rangle$ can again be applied for spectrum analysis. Experiments showed that the mean frequency $\langle f \rangle$ was useful for monitoring the change of the ocular-fundus blood flow with the application of a CO_2–O_2 gas mixture.

Because of the weak-intensity speckles from the human ocular fundus, a photon correlation technique is necessary in practice [26, 27]. Figure 8(I) shows typical correlation functions obtained from (a) an area containing two major retinal vessels and (b) another area having no visible vessel, of the human ocular fundus. The correlation function attenuates more rapidly in (a) than in (b). This is due to more active blood flow in the area (a). The degree of the attenuation can be evaluated by calculating the correlation time τ_c which is defined by the delay time giving half the maximum value of the correlation function. Before calculating τ_c, a smoothing process may be recommended. The correlation time τ_c was measured at various points of the ocular fundus of volunteers to obtain the map of blood circulation states.

The image-plane detection of bio-speckle fluctuations results in measurements of the retinal blood-flow velocity, though data are given in relative values. A measuring point on a retinal vessel of interest is defined by positioning the detecting aperture DA in the magnified image plane. The blood-flow velocity is proportional to the reciprocal of the correlation time, $1/\tau_c$, which is

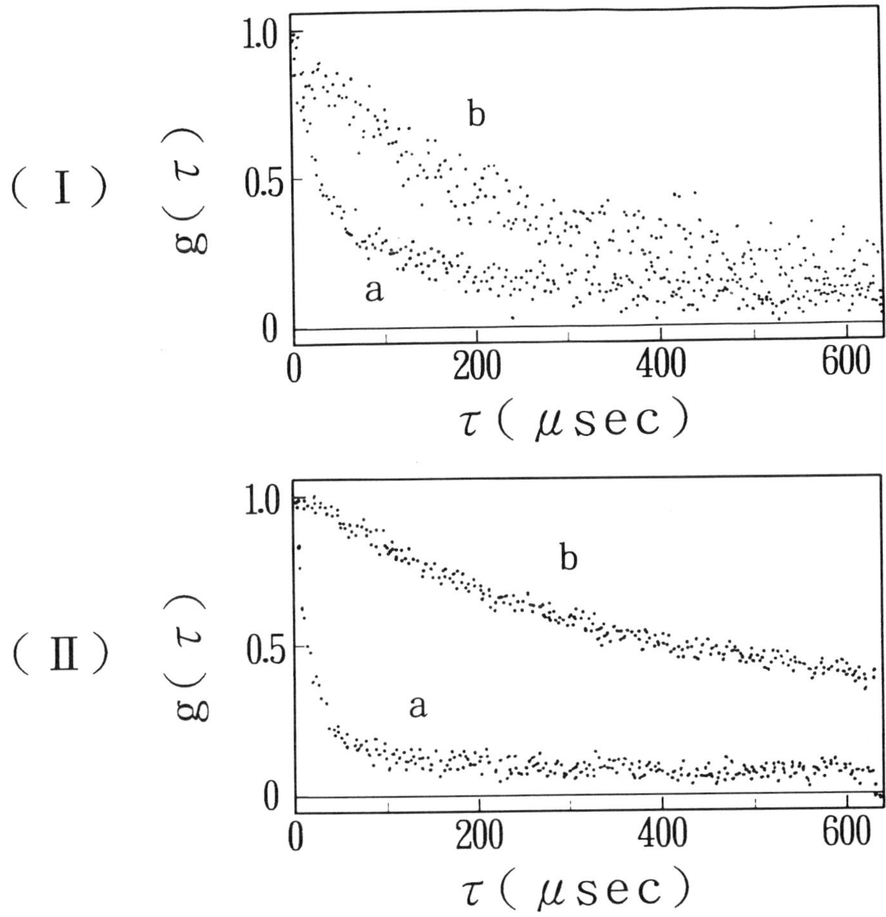

Fig. 8. Typical autocorrelation functions of (I) diffraction-plane and (II) image-plane bio-speckles, obtained from the ocular fundus (with kind permission from [*3, 26, 27*]).

calculated from the resultant autocorrelation function of detected bio-speckle fluctuations. The proportionality was verified by *in vitro* experiments [*3, 30*] using acryl latex and human-blood flows in glass capillaries that have various diameters. Figure 8 (II) shows typical correlation functions measured from (a) a point on a major retinal vessel and (b) a point on the surrounding tissue area, of the human ocular fundus. The correlation function (a) demonstrates an extremely early decay owing to the high blood-flow velocity expected in the vessel, but the function (b) shows a slow decay due to the low degree of capillary blood flow and perhaps the indirect choroidal blood flow. Various medical and physiological experiments are being carried out to investigate the applicability of the technique, such as measurements of pulsating blood flows or monitoring of blood flow states before and after photocoagulation treatments at the retina [3, 28, 31].

The state of bio-speckle fluctuations from the ocular fundus often changes during one measurement because of eye movements. This effect results in erroneous measurements of the correlation function. To solve this problem, a certain preprocessing [32] can be employed before calculating the correlation function, according to the first-order temporal statistics of time-integrated bio-speckles. Figure 9 gives a schematic explanation of the processing method. As shown in (a), speckle signals from a retinal vessel (case [I]) have a high-frequency fluctuation while those from tissues (case [II]) contain the low-frequency fluctuation. By integrating the sequential counting data over a gate time T_g shown in (b), the signal becomes rather smoothed in the case [I] but some fluctuation remains in the case [II], as shown in (c). The variance σ_i^2 of the integrated count fluctuation in a unit time T_u becomes small and large for the cases [I] and [II], respectively, as shown in (d). Thus, the ratio $\sigma_i^2/\langle n \rangle$ in which $\langle n \rangle$ is the mean number of counts in a gate time T_g can be used for discriminating whether a measurement in each unit time was made on a vessel or in tissues. After discrimination, the autocorrelation function is calculated safely from only the data for the vessel, which has been stored in memory.

Bio-speckles from Other Living Objects

Time-varying bio-speckles are also obtained from avian eggs [33]. The intensity fluctuation is used to monitor minute ballistic movements of the egg

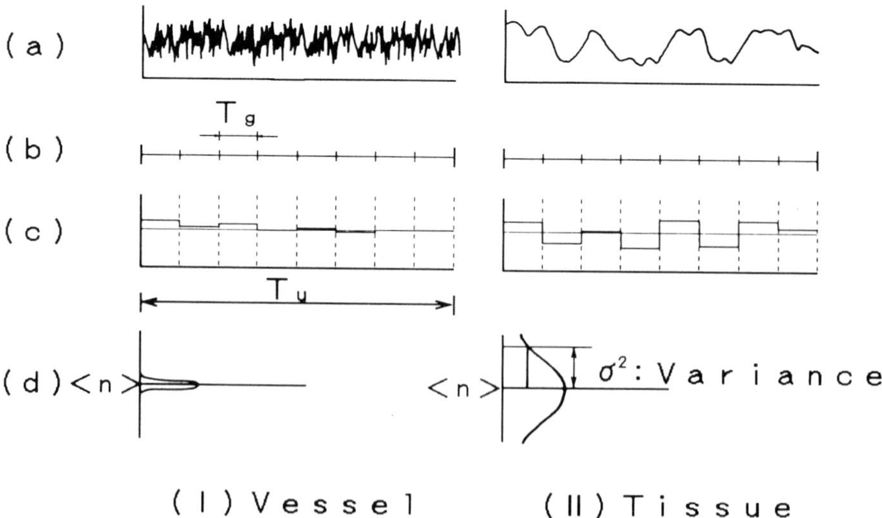

Fig. 9. Principle of signal processing based on the first-order temporal statistics of time-integrated bio-speckles [29].

which are closely related to the heart rate of avian embryos in it. Experiments verified that the bio-speckle fluctuation and the electrocardiogram yield the same periodicity, thus showing the potential applicability of the bio-speckle technique based on its non-contact method.

Laser light scattered by a population of tiny brine shrimps produces a bio-speckle pattern [*34*]. The time-sequential frames of a movie are used to perform the optical cross-correlation [*35*] which is based on the second-order space-time statistics of time-varying speckles. The speckle lifetime (close to the correlation time) is a suitable measure for monitoring the motility of organisms as a function of environmental parameters, such as temperature.

Bio-speckles obtained from cultures of motile microorganisms can be utilized to extract their feature information [*36*]. Temporal autocorrelation analysis is a possible way to monitor the culture vitality. The larger motility of microorganisms produces faster intensity fluctuations which result in a smaller correlation time τ_c. Thus it gives the ensemble average velocity or the average motility in the culture. By using digital image processing, different types of analysis are also performed for the bio-speckles, on the basis of the first-order or second-order spatial statistics of time-varying speckles. In the former case, after freezing the bio-speckle patterns, a histogram of grey levels is obtained for the speckle intensity difference between two frames of the same culture or of different cultures. A comparison of the histograms gives useful information about culture identifications. The latter case is the spatial autocorrelation function which will be discussed below.

SOME PECULIARITIES OF BIO-SPECKLES

Mixture of Different Speckle Dynamics

Bio-speckles are often formed by a superposition of differently fluctuating speckle patterns. The ratio $\sigma/\langle I \rangle$ of bio-speckles from botanical specimens is usually less than unity [*4*]. This means that the fluctuations are not fully developed. They are probably due to the superposition by a static speckle pattern that comes from objects not in motion. Thus, the ratio $\sigma/\langle I \rangle$ is utilized to estimate the relative contributions from moving and stationary scatterers [*37*].

Bio-speckles from some fruits have high- and low-frequency fluctuation components that result from both a fast random movement of mineral particles and a slow movement of plastids [*11*]. The same example is also found in the bio-speckles from skin tissues where the fast fluctuation due to the skin blood flow and the slow movement due to involuntary skin tissue displacement are present [*6, 14*]. The latter slow fluctuation results in the relative increase of low-frequency components in the power spectrum. A typical example [*3*] can be recognized when low-frequency components are compared between the spectra of Fig. 4 for skin tissues and of Fig. 5 for gastric mucous membrane or ocular fundus. The low-frequency components of the latter spectra show a nearly flat form. Metrologically speaking, some appropriate frequency filtering or frequency-dependent amplification may be necessary [16]. Some simulation studies [38] using ground-glass plates have been performed to approach this subject.

Conditions of Scattering Objects

In general, laser light is multiply scattered from retinal vessels. Thus, the resultant bio-speckle fluctuations may strongly depend on conditions such as vessel diameter, concentration of red blood cells (RBCs), background reflectance, and so on [31]. Some *in vitro* model experiments have been carried out to approach these subjects. In correlation analysis, the reciprocal of correlation time was found to be proportional to the square root of the vessel diameter [39]. This is used for measurements of retinal blood-flow volume [29]. The correlation time can be dependent on the concentration of RBCs, but usually in the case of less than a few per cent in volume. In the physiological range of *in vivo* hematocrit, the concentration is extremely high and, thus, the multiplicity of scattering by RBCs may be saturated sufficiently. Actually, the experiments showed that the variation in hematocrit did not cause a large difference in the value of the correlation time. A background having a reflectance of less than 10 per cent was found to affect the correlation time significantly. These results may be helpful to understand the measurement data on human subjects.

Another study [36] shows that the spatial autocorrelation function of bio-speckles from microorganisms is directly related to the culture concentration. The correlation length or the width of the correlation function becomes small with increasing concentration. The phenomenon is perhaps interpreted by optical diffraction theory. On the basis of the observation, an auxiliary method was proposed for detecting the growth ratio of the culture of microorganisms. Thus second-order spatial statistics could be a useful means in this case.

Wavelength and Intensity Dependence

Briers [8] reported that speckle fluctuations were more pronounced when the colour of laser light was the same as the colour of the botanical object, and were less pronounced when the colours were complementary. Complementary colours enhance the absorption of light while the same colours enhance reflection. The choice of the wavelength of light may enable us to measure selectively the movement of different kinds of scatterers within a tested living object. Selective measurement can be made, for example, with skins to explore the blood circulation in different skin tissue layers [3].

The energy of illuminating laser light may affect the biological activity of botanical specimens [*11*], because the absorbed energy is converted into heat. Thus the speckle fluctuation varies with the illumination energy. In this case, bio-speckle techniques do not promise non-destructive or non-disturbing measurements.

Detection Properties of an Optical Fibre Probe

The optical fibre probe is a most practical tool for skin blood-flow monitoring using bio-speckle techniques. Thus, the effects of the condition of the fibre probe should be investigated on the properties of detected signals, such as second-order temporal statistics, because these properties may be influenced by multiple scattering with living objects. Some experimental results are presented here. By using the optical system of Fig. 3, the power spectra were measured for three different inanimate objects in motion, (a) aluminium-coated ground

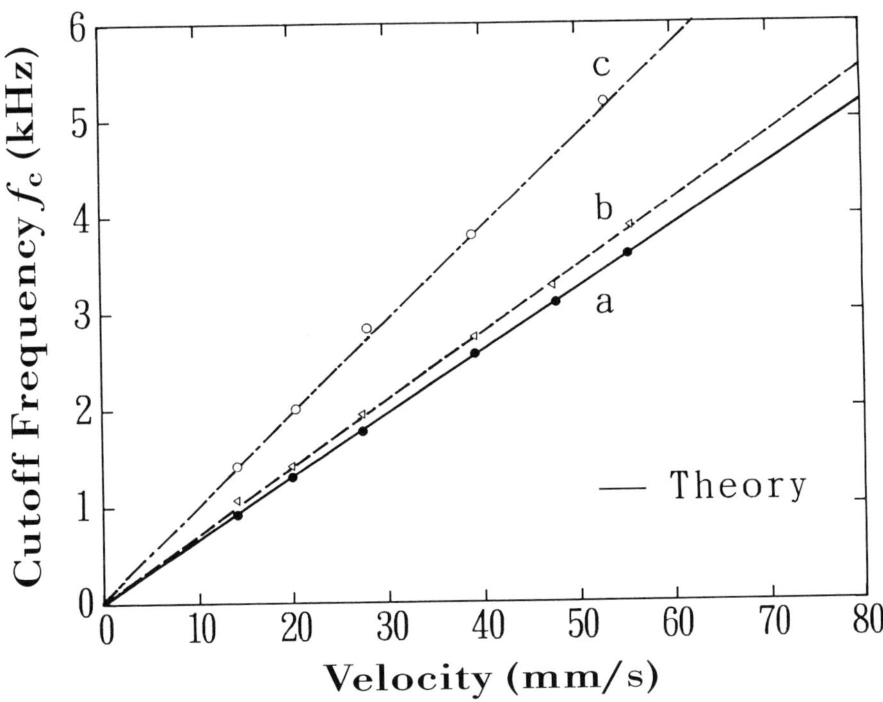

Fig. 10. Experimental relation between the cut-off frequency and the moving velocity for (a) aluminium-coated ground glass, and (b) and (c) white acrylic plates having normal and strong diffuseness, respectively.

glass, and (b) and (c) two white acrylic plates (4.5 mm thick) having normal and strong diffuseness, respectively. The velocity was evaluated by a cut-off frequency which is defined by the frequency giving half the value of the power spectrum at zero frequency.

Figures 10 and 11 show the experimental relation between the cut-off frequency and the moving velocity for three objects (a), (b), and (c), and the recorded intensity distribution of the illuminating spots on these objects. Good linearity is obtained for all the objects, but the cut-off frequency is larger in (c) than in (a). This result agrees with the theory [*40, 41*]. This difference may be due to the effective spot size shown in Fig. 11. The acrylic plate (c) scatters the light not only on its surface but also in its inside multiply. This makes equivalently the larger spot area of Fig. 11(c) which produces smaller speckle grains. The effect finally causes the higher-frequency fluctuations of detected speckle signals. This experiment concludes that speckle signals are dependent on the scattering structure of objects in which multiple scattering occurs.

Figure 12 shows the power spectra obtained from the acrylic plate, with different separation distances d between the illuminating and detecting fibre ends, and with two different probing distances l between the fibre ends and the plate. These distances are schematically shown in Fig. 3. In the result (a) for the larger distance l, the high-frequency components increase with the larger separation distance d, but these components increase almost independently of the distance d in the result (b) for the smaller distance l.

Figure 13 shows the variation of the cut-off frequency as a function of the distance l for three values of the distance d. With a larger value of l, the area illuminated by the fibre F_1 and the other effective area detected by the fibre F_2, both of which are determined by the numerical aperture NA of the fibres, become large. In this case, the overlapped portion of the two areas is a function of the distance d. For the smaller distance d, the detecting fibre F_2 receives

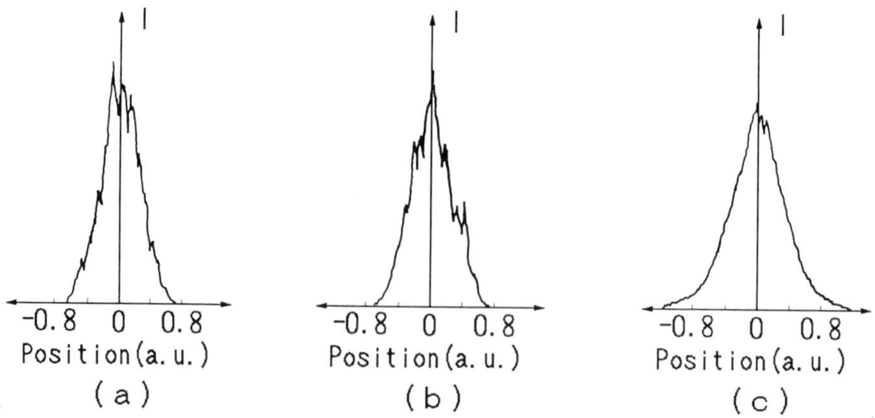

Fig. 11. Recorded intensity distributions of the illuminated spot on the three objects.

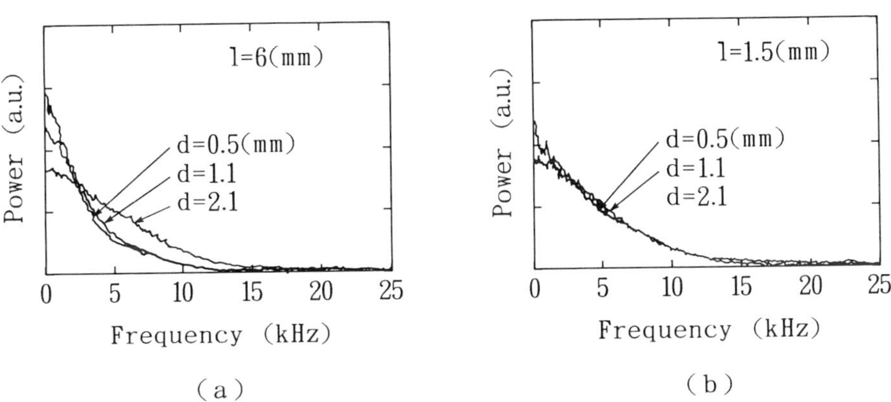

Fig. 12. Power spectra obtained from the acrylic plate for different separation distances *d* and probing distances *l*.

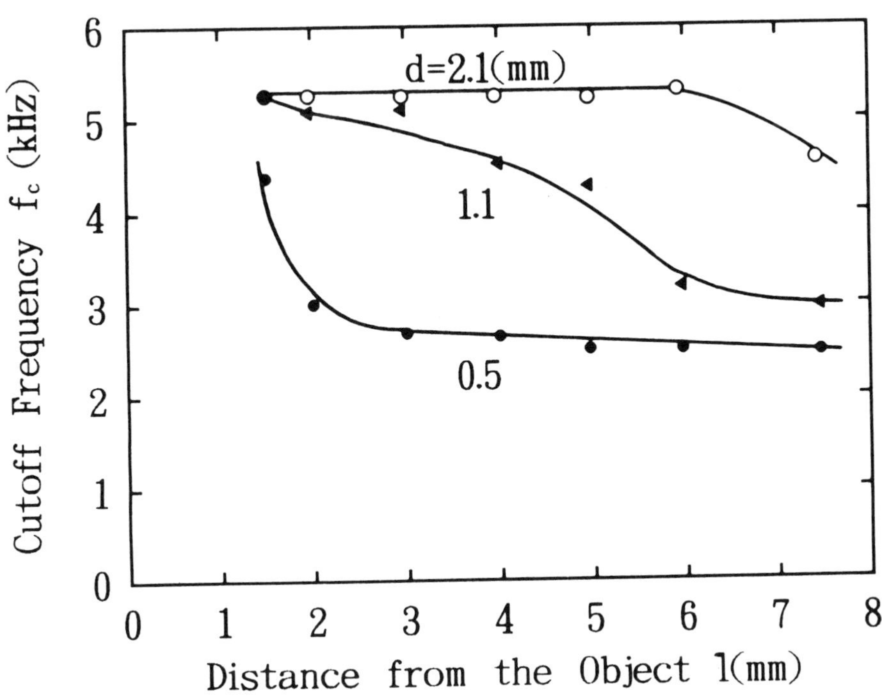

Fig. 13. Variation of the cut-off frequency f_c for different separation distances *d* and probing distances *l*.

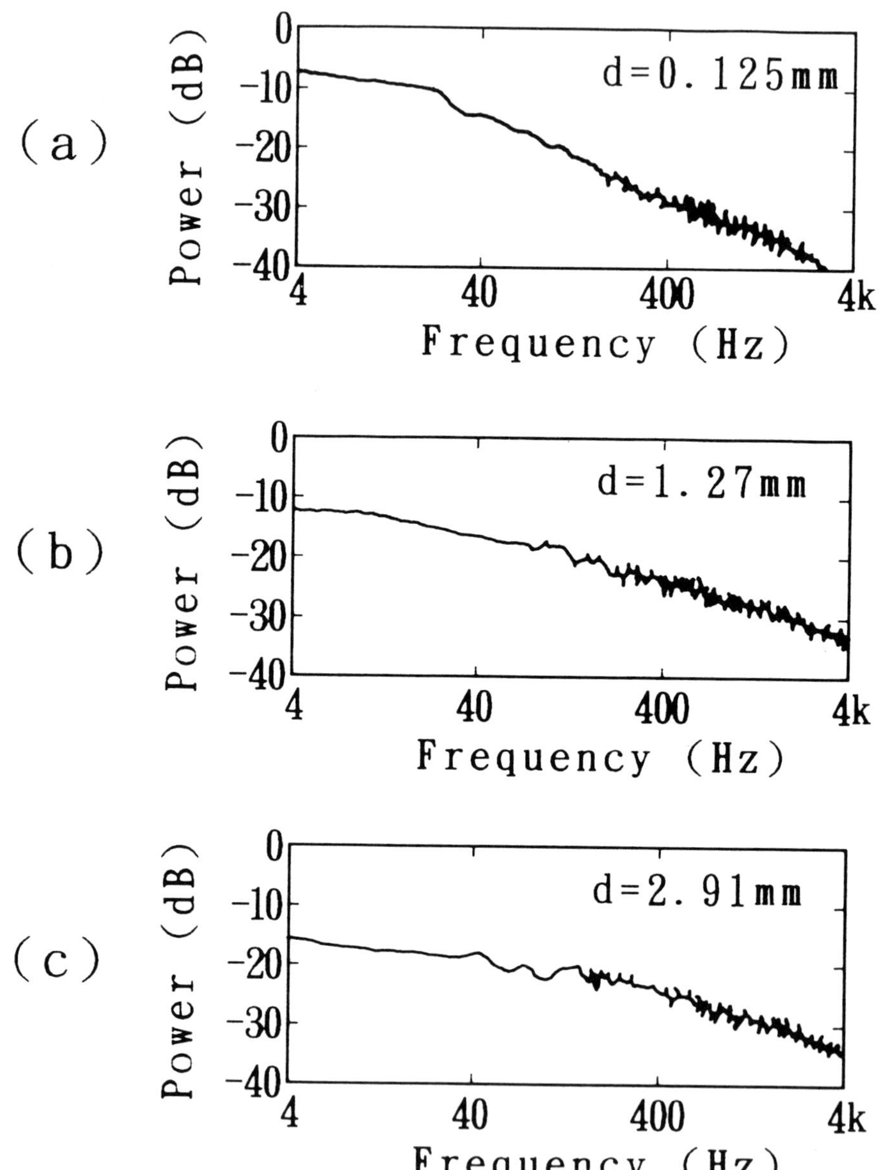

Fig. 14. Power spectra obtained from the palm of a hand, by using a fibre probe with different separation distances *d* (with kind permission from [3]).

mainly the translating speckles having a normal velocity from the plate surface. On the other hand, for the larger distance d, it detects dominantly the boiling speckles having an increased velocity which is produced by multiple scattering through a long light path inside the plate. In other words, the probe is able to detect the light coming from the deeper region inside the plate. If the distance l is small, both the illuminated area and the detected area also become small and they are not sufficiently overlapped even in the case of the smaller distance d. The detecting fibre almost always receives the high-frequency boiling speckles independently of the distance d. This tendency is well demonstrated in Fig. 13. Therefore, the distance parameters d and l are quite important and should be carefully optimized according to the scattering structure of living objects under inspection. Other investigations have been made theoretically and experimentally on certain parameters of the fibre probe [*14, 40, 41*].

The above simulation studies were applied to skin tissues. Figure 14 shows the power spectra obtained from the palm of a hand by using the fibre probe with different separation distances d while the probing distance l was fixed to be constant. With increasing distance d, the low-frequency components decrease and the high-frequency components increase. In the case of a small distance d, the detecting fibre F_2 mainly receives the scattered light propagated in the short path in and near the region of the epidermis that is close to the surface, or in some cases the light scattered by the surface with involuntary tissue movements. This light is modulated by the low degree of blood flow in papillary loops and, thus, shows the low-frequency fluctuations. For a large distance d, the light coming from the deeper dermal region dominantly contributes to the detected bio-speckles, and the more active blood flow in the dermis results in the increase of high-frequency fluctuations. This result agrees quite well with that obtained by the above simulation studies using inanimate objects.

CONCLUSION

Bio-speckles show various aspects depending on the living objects to be measured, and methods of their analysis and applications can be changed accordingly. Thus, general theoretical approaches to analysis of bio-speckles seem to be considerably difficult. A promising way is to observe the bio-speckle fluctuations in detail and to establish experimentally and empirically the relation between the fluctuations and the necessary information to be extracted. The field of bio-speckle techniques covers all the subjects of speckles obtained from various living objects: short- and long-range temporal fluctuations, spatial fluctuations, intensity and phase fluctuations, their statistical properties and fundamental modelling, and applications, including the design of measuring apparatus. The field can also be combined with conventional speckle metrology such as speckle photography and speckle interferometry [*5*].

REFERENCES

1. T. Asakura and N. Takai, *Appl. Phys.* **25**, 179 (1981).
2. Y. Aizu and T. Asakura, *SPIE Proceedings, Time-Resolved Spectroscopy and Imaging of Tissues* **1431**, 239 (1991).
3. Y. Aizu and T. Asakura, *Opt. Laser Technol.* **23**, 205 (1991).
4. J.D. Briers, *Opt. Eng.* **32**, 277 (1993).
5. K.D. Hinsch, *Optics for Protection of Man and Environment against Natural and Technological Disasters*, Elsevier Science Publishers B.V., Amsterdam, 267 (1993).
6. H. Fujii, T. Asakura, K. Nohira, Y. Shintomi and T. Ohura, *Opt. Lett.* **10**, 104 (1985).
7. Z. Jiang and W. Staude, *J. Exper. Botany* **40**, 1169 (1989).
8. J.D. Briers, *Opt. Commun.* **13**, 324 (1975).
9. J.D. Briers, *J. Exper. Botany* **28**, 493 (1977).
10. J.D. Briers, *J. Exper. Botany* **29**, 395 (1978).
11. A. Oulamara, G. Tribillon, and J. Duvernoy, *J. Mod. Opt.* **36**, 165 (1989).
12. H. Rabal, M. Trivi, R. Arizaga, G. Romero, and E. Alanis, *Optical Methods in Biomedical and Environmental Sciences*, Elsevier Science Publishers B.V., Amsterdam, 19 (1994).
13. B. Ruth, *J. Mod. Opt.* **34**, 257 (1987).
14. B. Ruth, *Opt. Laser Technol.* **20**, 309 (1988).
15. B. Ruth, *Int. J. Microcirc.: Clin. Exp.* **9**, 21 (1990).
16. B. Ruth, *Optics in Medicine, Biology and Environmental Research*, Elsevier Science Publishers B.V., Amsterdam, 131 (1993).
17. B. Ruth, D. Haina, and W. Waidelich, *Opt. Acta* **31**, 759 (1984).
18. H. Fujii, K. Nohira, Y. Yamamoto, H. Ikawa and T. Ohura, *Appl. Opt.* **26**, 5321 (1987).
19. L.T. Tanin, A.S. Rubanov, I.V. Markhvida, S.C. Dick, and L.I. Rachkovsky, *Optics in Medicine, Biology and Environmental Research*, Elsevier Science Publishers B.V., Amsterdam, 149 (1993).
20. A.F. Fercher and J.D. Briers, *Opt. Commun.* **37**, 326 (1981).
21. J.D. Briers and A.F. Fercher, *Invest. Ophthalmol. Vis. Sci.* **22**, 255 (1982).
22. J.D. Briers and A.F. Fercher, *Optics in Biomedical Sciences*, Springer-Verlag, Berlin, 158 (1982).
23. A.F. Fercher and M. Peukert, *SPIE Proceedings, International Conference on Speckle* **556**, 110 (1985).
24. A.F. Fercher, M. Peukert and E. Roth, *Opt. Eng.* **25**, 731 (1986).
25. Y. Aizu, K. Ogino, T. Koyama, N. Takai, and T. Asakura, *Laser Anemometry in Fluid Mechanics-III*, Ladoan, Lisbon, 55 (1988).
26. Y. Aizu, K. Ogino, T. Sugita, T. Yamamoto, and T. Asakura, *J. Clinical Laser Medicine & Surgery* **8**, 35 (1990).
27. Y. Aizu, K. Ogino, T. Sugita, T. Yamamoto, N. Takai, and T. Asakura, *Appl. Opt.* **31**, 3020 (1992).
28. Y. Aizu, T. Asakura, K. Ogino, T. Sugita, Y. Suzuki and K. Masuda, *Optics in Medicine, Biology and Environmental Research*, Elsevier Science Publishers B.V., Amsterdam, 146 (1993).
29. Y. Aizu, K. Ogino, T. Sugita, M. Suematsu, and T. Asakura, *Optics for Protection*

of Man and Environment against Natural and Technological Disasters, Elsevier Science Publishers B.V., Amsterdam, 233 (1993).

30. Y. Aizu, H. Ambar, T. Yamamoto, and T. Asakura, *Opt. Commun.* **72**, 269 (1989).

31. Y. Suzuki, K. Masuda, K. Ogino, T. Sugita, Y. Aizu, and T. Asakura, *Jpn. J. Ophthalmol.* **35**, 4 (1991).

32. Y. Aizu, A. Kojima, M. Suematsu, and T. Asakura, *Optical Methods in Biomedical and Environmental Sciences*, Elsevier Science Publishers B.V., Amsterdam, 159 (1994).

33. H. Tazawa, T. Hiraguchi, T. Asakura, H. Fujii, and G.C. Whittow, *Med. & Biol. Eng. & Compt.* **27**, 580 (1989).

34. J, Ebersberger and G. Weigelt, *Opt. Commun.* **58**, 89 (1986).

35. H. Böhm, A.W. Lohmann, and G. Weigelt, *Opt. Lett.* **6**, 162 (1981).

36. B. Zheng, C.M. Pleass, and C.S. Ih, *Appl. Opt.* **33**, 231 (1994).

37. J.D. Briers, *Opt. Quantum. Electron.* **10**, 364 (1978).

38. T. Iwai and T. Asakura, *Opt. Laser Technol.* **21**, 31 (1989).

39. Y. Aizu, T. Asakura, K. Ogino, and T. Sugita, *Opt. Commun.* **80**, 1 (1990).

40. H. Fujii, T. Asakura, and T. Okamoto, *Opt. Commun.* **55**, 393 (1985).

41. H. Fujii, T. Okamoto, and T. Asakura, *J. Opt. Soc. Amer.* **A4**, 1366 (1987).

3 Photon migration and imaging of biological tissues

Giovanni Zaccanti and Daniele Contini
*Dipartimento di Fisica dell'Università di Firenze Via S. Marta 3,
50139 Florence, Italy*

INTRODUCTION

Although a lot of useful information can be obtained by using currently available techniques for imaging and monitoring living tissues, there is still a strong request for new techniques, possibly non-invasive. Light offers a unique probe for studying living tissues. At visible and near-infrared (NIR) wavelengths light propagation is strongly influenced by many cromophores, present inside the biological tissue, that are related to the health status of the tissue. As an example, venous and arterial blood have a different colour due to the different degree of blood oxygenation. Therefore very useful information on the status of the tissue, like the blood oxygen saturation, can be obtained if the optical properties of the tissue can be properly measured. On the other hand, if the intensity of the radiation is not too large, light does not provoke any damage to the tissue: optical techniques are thus non-invasive.

There is hence a large potential for obtaining basic information on tissues by using non-ionizing radiation in the visible and near-infrared region of the electromagnetic spectrum, but there is relatively little currently available instrumentation based on light. The reason is that biological tissues are strongly scattering media and thus measurements of optical properties are difficult. Recent progress in optoelectronics aroused an interest in developing new instrumentation and provoked an impressive response from researchers and industries in developing tissue optics. Particularly important has been the use of laser systems emitting ultrashort pulses (picosecond lasers) and ultrafast detectors (streak camera, microchannel plate) [1, 2].

In the following sections we begin by introducing the optical parameters needed to characterize diffusing media and briefly describing theories and methods used to study photon migration through highly scattering media. Different techniques currently studied for imaging and monitoring biological tissues will be discussed.

OPTICAL PROPERTIES OF BIOLOGICAL TISSUES

Biological tissues are dense packed scattering media. The scattering effect is due to the small differences in the refractive index of different components of cells. The absorption effect is related to the absorption bands of different molecules.

The scattering properties are described by the scattering coefficient μ_s ($1/\mu_s$ represents the mean free path between two subsequent scattering events) and by the scattering function $p(\hat{s}, \hat{s}')$ which describes how the scattered light is redistributed in the various directions. The scattering function is usually assumed to be dependent only on the angle θ between the direction of motion of the photon before the scattering event ($\hat{s}'$) and the direction of motion after the scattering event ($\hat{s}$). The scattering coefficient of biological tissues at visible and NIR wavelengths is extremely large. As an example, measurements carried out on samples of brain showed $\mu_s \approx 200$ mm^{-1} (corresponding to a mean free path as small as 5 μm, i.e., comparable to the wavelength!) for white matter and $\mu_s \approx 20$ mm^{-1} (corresponding to a mean free path as small as 50 μm) for grey matter [3]. This means that photons propagating through the tissue undergo very frequent scattering events.

The scattering function exhibits a highly forward peak since the typical size of cells is usually large with respect to λ. Typically, more than 50 per cent of the scattered radiation is scattered at a $\theta < 3°$. The value of the asymmetry factor of the scattering function g ($g = \int_{4\pi} \cos(\theta)p(\theta)\,d\Omega$) is usually close to 1 ($g > 0.9$). Since the typical size of cells is usually larger than wavelengths in the visible and NIR range, the scattering properties of tissue slightly depend on λ.

When light propagates through highly turbid media the scattering properties are practically fully described by the reduced scattering or transport coefficient $\mu_s' = \mu_s(1 - g)$ instead of μ_s and $p(\theta)$ separately. $1/\mu_s'$ is the transport mean free path and represents the mean distance followed by photons along the original direction of motion before they 'forget' the original direction of motion. Typical values measured for μ_s' range from 0.5 to 1.2 mm^{-1} [3]. Therefore, photons migrating through the tissue 'forget' the original direction of motion at a penetration depth of the order of 1 mm.

The absorption properties are described by the absorption coefficient μ_a ($1/\mu_a$ represents the mean free path before absorption). For biological tissues the absorption properties are often related to the absorption properties of particular groups (called cromophores) within the molecules. The major absorbing components of many tissues are water, hemoglobin, and melanin. Water ($\approx 70\%$ per cent of the biological tissue) is more or less homogeneously distributed into the tissues, whereas melanin is found mainly in the skin (within a few μm) and hemoglobin in the blood. The absorption coefficient varies over many orders of magnitude when λ varies between 200 and 10 000 nm [4]. Absorption dominates light propagation (the mean free path before absorption is so small that photon migration is simply described by the Lambert–Beer law ($\mu_a > 10\ \mu_s'$)) when $\lambda < 300$ nm or $\lambda > 2500$ nm. Radiation at these wavelengths penetrates only a few μm into the tissue before absorption by melanin or by water. At NIR wavelengths ($650 < \lambda < 1200$ nm) the absorption coefficient is very small (typical value $\mu_a \approx 0.01$ mm^{-1}) and photon migration is dominated by scattering: photons propagating through the tissue follow chaotic zigzag trajectories due to frequent scattering events (the mean free path between two subsequent scattering events is of the order of 10 μm) and propagation occurs with modalities typical of a diffusion process. Although $\mu_a \ll \mu_s'$, the small absorption effect plays an important role on the propagation since photons, before emerging from the tissue, may follow trajectories having large lengths in comparison to the absorption mean free path.

At NIR wavelengths the absorption properties are strongly influenced by the blood content and by the degree of oxygenation of hemoglobin. At these wavelengths the absorption spectrum of oxygenated (HbO_2) and deoxygenated (Hb) hemoglobin shows significant differences. These differences are responsible for the different colour observed for arterial and venous blood. In particular Hb exhibits an absorption peak at 760 nm whereas 800 nm is an isosbestic wavelength (at this wavelength Hb and HbO_2 have the same absorption coefficient). Therefore, the knowledge of μ_a at two wavelengths enables us to determine the concentration of Hb and HbO_2: a very important parameter, like the tissue oxygenation, can thus be obtained from non-invasive measurements. Usually measurements are carried out at 760 and 800 nm. Many researchers are currently working to develop instrumentation to monitor blood oxygenation.

NIR radiation can follow long trajectories inside the tissue before being absorbed or before exiting from the tissue. At these wavelengths radiation can thus penetrate many millimetres before exiting from the tissue (radiation can be detected even after crossing a slab ≈ 10 cm thick). Therefore, from this radiation, information can be obtained on the properties of deep tissues.

THEORIES USED TO DESCRIBE PHOTON MIGRATION THROUGH BIOLOGICAL TISSUES

Photon migration through diffusing media is described by the radiative transfer equation:

$$\left\{ \hat{s} \cdot \nabla + \mu_a + \mu_s + \frac{1}{v}\frac{\partial}{\partial t} \right\} \Phi(\vec{r}, \hat{s}, t) = q(\vec{r}, \hat{s}, t) + \mu_s \int_{4\pi} p(\hat{s}, \hat{s}') \Phi(\vec{r}, \hat{s}', t)\, d^2 \hat{s}'$$

where: $\Phi(\vec{r}, \hat{s}, t)$ [W m^{-2} sr^{-1}] is the radiance, i.e., the power flux density within a unit solid angle at $\vec{r}$ in the direction $\hat{s}$ at time t; $q(\vec{r}, \hat{s}, t)$ is the source term; and $p(\hat{s}, \hat{s}')$ is the scattering function. There are no general solutions of the radiative transfer equation, but only approximate analytical theories or numerical methods.

Analytical Methods

Diffusion approximation

This is obtained by considering the lowest-order approximation of the expansion in spherical harmonics of the radiative transfer equation. In the diffusion approximation the scattering properties of the medium are described by the reduced scattering coefficient only. The diffusion equation has been solved for homogeneous diffusing media for different boundary conditions (infinitely extended medium, semi-infinite slab, finite slab, cylinder, sphere) [5–7]. The solutions obtained well describe photon migration when $\mu_a \ll \mu_s'$ and $\rho\mu_s' > 10$, i.e., when the distance ρ of the detector from the source is larger than about 10 transport mean free paths.

Random walk theory

The motion of a photon in the scattering medium is approximated as a random walk on a simple cubic lattice (the separation between two neighbouring lattice points is assumed to be equal to $\sqrt{2}/\mu_s'$); the photon proceeds through the lattice as steps to one of six nearest neighbouring lattice points. By using this simple scheme it is possible to obtain analytical relationships similar to the ones obtained using the diffusion approximation [8].

Numerical Methods

Finite Element Method (FEM)

This is a general method of solving the differential equations and it is very useful to study problems with complex boundary conditions and non-homogeneous optical properties [9]. The same simplifying assumptions made for the diffusion approximation and for the random walk are usually assumed. To avoid the use of big computers to solve full three-dimensional problems, two-dimensional models are often used.

Monte Carlo (MC) method

The Monte Carlo method provides a physical simulation of photon migration. The model assumes a non-deterministic, stochastic nature for light scattering and absorption of individual photons [10]. By using a random number generating routine the trajectories of emitted photons are chosen according to the statistical rules relevant for photon migration through the diffusing medium.

The trajectory of any emitted photon is followed until it exits from the scattering medium or it arrives at the detector. The probability of receiving a photon emitted by the source (i.e., the transmittance) is evaluated as the ratio between the number of 'useful' trajectories and the number of trajectories considered. Monte Carlo methods allow a full three-dimensional description of photon migration and the actual scattering properties of scatterers can be taken into account: the required parameters are μ_s, μ_a and $p(\theta)$. The main drawback with respect to the FEM method is the computation time, which may be very long.

OPTICAL IMAGING OF BIOLOGICAL TISSUES

The attenuation due both to the absorption and to the spread of radiation caused by scattering make optical imaging feasible only on small organs. Tissue optics especially investigates the possibility of obtaining imaging of the breast (to detect breast cancer) and of the newborn head (functional imaging). The examples reported in this section refer to a slab geometry. The simple geometrical scheme and the value of μ_s' are representative of a breast compressed between two plane parallel plates as occurs during mammography.

Photons migrating through a highly scattering medium undergo frequent scattering events; the resulting trajectories may be very different from a straight line. Light propagating through a diffusing medium travels on trajectories that are considerably longer than the geometrical distance between the source and the receiver. This causes a temporal spread of the received light. When photons are injected into the medium by means of an ultrashort laser pulse (a few picoseconds long) the received pulse after a few centimetres of biological tissue is a few nanoseconds long. To give an idea of this effect in Fig. 1 an example of the pulse received after 40 mm through a turbid medium having $\mu_s' = 0.5$ mm^{-1} is shown. The source was assumed to be a thin collimated laser beam with a temporal distribution described by a Dirac delta centred at the time $t = 0$. The received pulse with this kind of source is called Temporal Point Spread Function (TPSF). The medium was chosen to be not absorbing and to have a refractive index equal to 1. The geometrical scheme is also shown in the picture.

Figure 1 shows that there is a great temporal broadening of the pulse travelling inside the medium. Energy is received even 4 nanoseconds after light is delivered into the medium. This delay corresponds to trajectories 1200 mm long. Despite the geometrical distance between the source and the receiver being only 40 mm, the mean path length of the photons is 313.4 mm. This means that the radiation travels inside the medium, on average, along trajectories 7.83 times longer than the geometrical source–detector distance.

This temporal spreading is also followed by a large spatial spreading of light. When a laser beam is delivered on to the surface of a turbid medium the light,

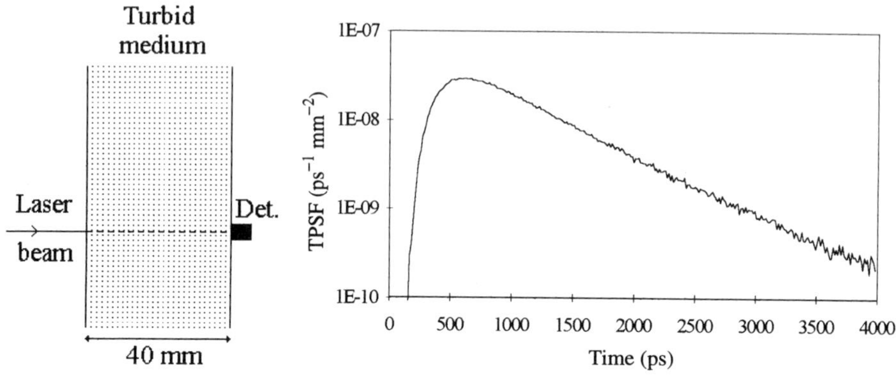

Fig. 1. Example of TPSF evaluated in transmittance through a slab having $\mu_s' = 0.5$ mm^{-1}, $\mu_a = 0$, $g = 0$ and $n = 1$. The geometry is shown in the picture.

recorded with a detector placed coaxially with the laser beam, passes inside the medium also in regions very far from the line joining the source and the detector. To give an idea of this effect, Fig. 2 reports the measured intensity versus the distance from the laser beam axis (the Point Spread Function, PSF).

Figure 2 (a) shows that the intensity received with a detector placed 30 mm away from the axis is still 1/10 of its maximum. In Fig. 2(b) the PSF obtained looking at the data presented in part (a) along one of the diameters is shown.

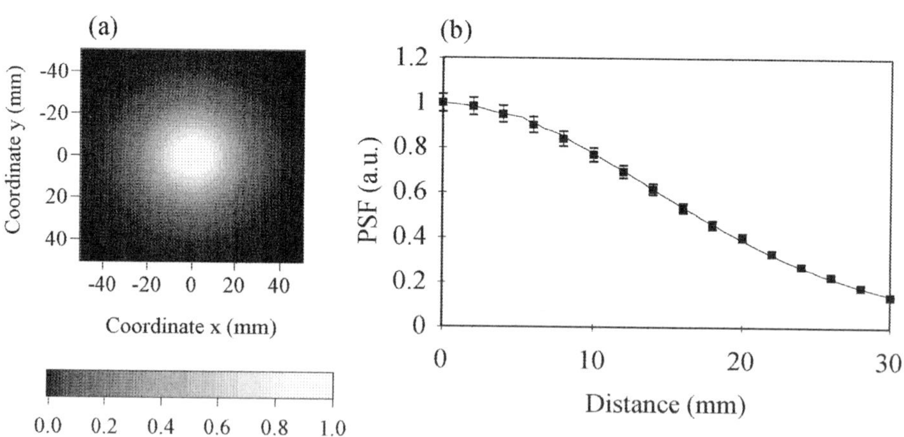

Fig. 2. Normalized PSF evaluated in transmittance through a slab having $\mu_s' = 0.5$ mm^{-1}, $\mu_a = 0.0088$ mm^{-1}, $g = 0$ and $n = 1$. (a) shows the distribution of light on the surface of the slab in a 256 grey levels scale; (b) represents the PSF obtained from the data shown in part (a) (evaluated along a diameter). The continuous line represents MC results and marks are experimental results.

The figure shows both Monte Carlo results (continuous line) and experimental results (marks).

The image of an object imbedded within the medium is blurred: scattering effects cause a decrease in contrast and in spatial resolution. To explain the effect of scattering on imaging of turbid media we refer to the results obtained by using a Monte Carlo code developed to study the region of the medium through which the received photons pass during migration from the source to the receiver. The Monte Carlo code stores, for any received photon, the coordinates of all the scattering points. By using the stored trajectories, maps were obtained representing the density of scattering points inside the turbid medium. The density of scattering points is proportional to the probability that received photons have to go through the different parts of the medium. The maps were obtained projecting the scattering points on the xz plane (the thin collimated light beam was assumed to lie along the z direction). The grey level was assumed to be proportional to the number of scattering points. Figure 3 shows a map representing the density of scattering points for photons migrating from the continuous wave (CW) source to the coaxial receiver for a non-absorbing medium. The figure shows that the volume of the medium occupied by the migration of photons from the source to the receiver is very broad: in the centre of the slab the broadening is about equal to the slab thickness. Very similar results were obtained for different values of μ_s'. Therefore, an inhomogeneity may affect the received signal even when it is placed at a significant distance from the straight line connecting the source to the receiver. Thus, the shadowgram of the inhomogeneity, obtained synchronously moving the source and the receiver, is broadened with respect to the one obtained in a non-scattering medium. Figure 3 also shows the shadowgram of a black cylinder (radius 3.75 mm) placed in the centre of the slab (broadened curve). The shadowgram expected for the non-scattering medium is also reported. The

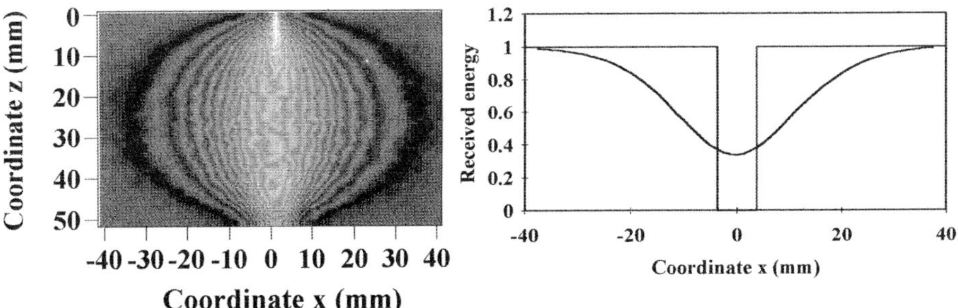

Fig. 3. Map representing the density of scattering points for photons migrating from the source to the coaxial receiver (radius = 4 mm). The figure refers to a homogeneous slab 50 mm thick having $\mu_s' = 0.4$ mm^{-1}, $\mu_a = 0$ and $g = 0$. The shadowgram referring to a black cylinder (radius 3.75 mm) placed in the centre of the slab for the non-scattering and for the scattering medium is also reported (broadened curve).

spatial resolution obtainable in these conditions by using a CW source is not better than 20–30 millimetres and is too poor for many medical applications. As an example, a spatial resolution of the order of 5–10 mm is desirable for optical mammography. It is thus necessary to use techniques capable of improving image quality.

Techniques used to Improve Optical Imaging

Different techniques have been investigated to improve image quality. The techniques used include measurements in the following:

Time domain

Measurements in the time domain involve a source emitting an ultrashort pulse (picosecond, femtosecond) and an ultrafast detector (streak camera, micro-channel plate) and make possible a direct measurement of the temporal point spread function $f(t)$ of the received pulse.

Frequency domain

Measurements in the frequency domain involve a sinusoidally intensity-modulated source and the determination of phase, and AC and DC components of the received signal. Analysis in the time domain and in the frequency domain are related via the Fourier transform. At least in principle the same information may thus be obtained by using these approaches. The main advantage of the frequency domain approach is the simpler and less expensive experimental setup.

Continuous wave (CW) domain

Measurements in the CW domain involve a simple continuous wave source. In general the information content obtainable with this approach is less than the one obtained with previous techniques. However, the same information content may be obtained if the dependence of the received signal on the absorption coefficient $(P(\mu_a))$ of the medium is available. In fact $P(\mu_a)$ is related to $f(t)$ via the Laplace transform [11].

The image quality can be improved by using two different approaches. The first approach involves discriminating between few marginally deviated photons and all other scattered light. These photons migrate through the medium remaining within a small volume around the source–receiver line. The narrowing of the spatial spread improves spatial resolution and contrast. The second approach involves tomographic measurements. In the following we will discuss in particular the time gating technique and time resolved optical tomography.

Time gating technique

When a short light pulse is used to illuminate the turbid medium the discrimination of few marginally deviated photons can be obtained on the basis of the time of flight. Earlier received photons, which are the least deviated from the laser beam axis, can be selected by using the time gating technique. With this technique only earlier photons received within a short temporal window Δt_g (gating time) are used for image reconstruction. To illustrate the benefits of the time gating technique we refer to Fig. 4. This figure shows examples of maps representing the density of scattering points referring to photons received within different gating times. When the gating time is decreased the volume of the medium occupied by photon migration becomes narrower and thus the spatial resolution is improved.

The spatial resolution obtainable with this method is ultimately limited by the strong attenuation of energy received during short gating times. When unscattered (ballistic) photons can be detected a high-quality image can be obtained. However, for strong scattering media like biological tissues, it becomes almost impossible to measure the received energy when the gating time becomes very short. Recent studies [12] have shown that the spatial resolution obtainable in typical physical and geometrical conditions for breast imaging is about one centimetre. A method to further improve the spatial resolution has been recently proposed [13]. This method involves the use of all information carried by TPSF: experimental results were fitted with an analytical model for the TPSF. A spatial resolution of about 5 mm was obtained by using the signal corresponding to very short gating times inferred from the analytical model.

Another possibility of selecting photons following short trajectories is given by absorption. Absorption reduces preferentially photons following long trajectories. The effect of strong absorption on imaging is similar to that of a

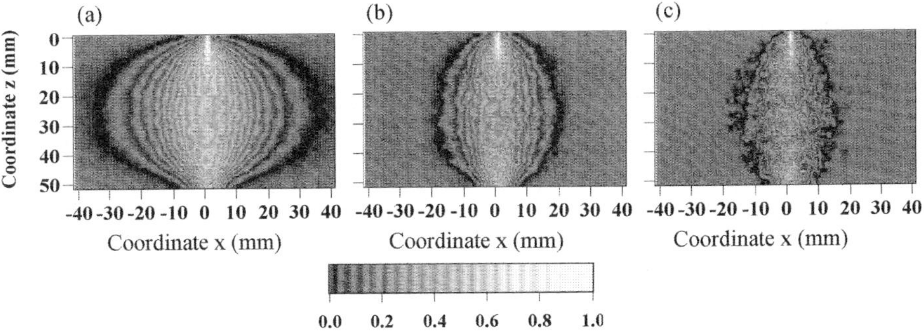

Fig. 4. Maps representing the density of scattering points for photons migrating from the source to the coaxial receiver (radius = 5 mm). The figure refers to a homogeneous slab 50 mm thick having $\mu_s' = 0.4$ mm^{-1}, $\mu_a = 0$, $g = 0$ and $n = 1$. Panels (a), (b), (c) refer to photons received during different gating times: continuous beam, $\Delta t_g = 625$ and $\Delta t_g = 250$ ps. When the gating time is shortened the volume of the medium involved in photon migration becomes narrower.

short gating time. An image quality similar to the one obtained by using the time gating technique can be achieved if the absorption coefficient of the medium is sufficiently high.

Time resolved optical tomography

Another possibility for improving the spatial resolution and the localization of the inhomogeneity hidden within the turbid medium is given by tomographic measurements. Time resolved measurements give us the possibility of discriminating among photons passing through different regions of the medium. To illustrate this effect we refer to Fig. 5. This figure shows examples of maps representing the density of scattering points referring to photons received within different temporal windows. The figure refers to photons received by a detector measuring the reflectance at 26 mm from the light beam. The volume of the medium occupied by migration of photons received at different times is significantly different. Photons remain within a hemisphere whose radius strongly depends on the transit time: the radius is about 15 mm for photons arriving within 600 ps; it becomes about 25 mm for photons arriving within 600–1400 ps, and about 35 mm for photons arriving within 1400–5800 ps. Photons arriving at different times are therefore differently affected by inhomogeneities inside the turbid medium. As an example, from Fig. 5 we can expect that an inhomogeneity placed between the source and the detector at a depth of 10 mm affects photons arriving with both short and long transit times. At a depth of 25 mm the inhomogeneity affects only photons arriving with long transit times. Photons arriving at different times carry information on different regions of the medium. An inhomogeneity within the medium is thus expected

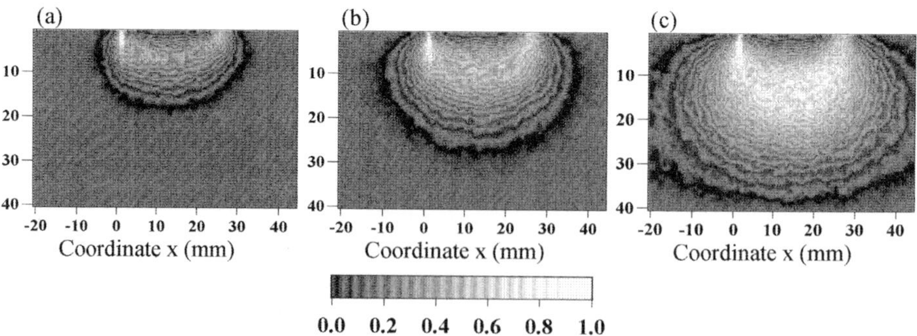

Fig. 5. Maps representing the density of scattering points for photons migrating from the source to the receiver (radius = 4 mm) measuring the reflectance at 26 mm from the source. The figure refers to a homogeneous slab 40 mm thick having $\mu_s' = 0.5$ mm^{-1}, $\mu_a = 0$, $g = 0$ and $n = 1$. Panels (a), (b), (c) refer to photons received during different temporal windows: 0–600 ps, 600–1400 ps, 1400–5800 ps. The volume of the medium occupied by photon migration becomes larger for photons travelling along long trajectories.

to modify both the intensity of the received signal and the shape of the TPSF. A lot of information is thus available from time resolved measurements carried out for different positions of the source and of the receiver with respect to the inhomogeneity. The big problem for optical tomography is the inverse problem, i.e., image reconstruction starting from the set of measured quantities. There is currently much effort in developing efficient reconstruction algorithms.

We have discussed in particular the time gating technique and time resolved optical tomography. Both these techniques are based on measurements in the time domain. Interesting results have been obtained also by using instrumentation based on measurements in the frequency domain.

REFERENCES

1. D.T. Delpy, M. Cope, P. van der Zee, S. Arridge, S. Wray, and J. Wyatt, *Phys. Med. Biol.* **33**, 1433–1442 (1988).
2. B. Chance, J.S. Leigh, H. Miyake, D.S. Smii, S. Nioka, R. Greenfeld, M. Finander, K. Kaufman, V. Levy, M. Young, P. Cohen, H. Yoshioka, and R. Boretsky, *Proc. Natl. Acad. Sci. USA* **85**, 4971–4975 (1988).
3. G. Zaccanti, A. Taddeucci, M. Barilli, P. Bruscaglioni, and F. Martelli, *Proc. SPIE* **2389**, 513–521 (1995).
4. B.C. Wilson, M.S. Patterson, S.T. Flock, and D.R. Wiman in: *Photon Migration in Tissues*, ed. B. Chance (Plenum), pp. 25–42 (1988).
5. M. Patterson, B. Chance, and B. Wilson, *Appl. Opt.* **28**, 2331–2336 (1989).
6. R. Arridge, M. Cope, and D.T. Delpy, *Phys. Med. Biol.* **37**, 1531–1560 (1992).
7. K. Furutzu and Y. Yamada, *Phys. Rev.* E **50**, 3634–3640 (1994).
8. A.H. Gandjbakhche, G.H. Weiss, R.F. Bonner, and R. Nossal, *Phys. Rev.* E **48**, 810–818 (1993).
9. S.R. Arridge, M. Schweiger, M. Hiraoka, and D.T. Delpy, *Med. Phys.* **20**, 299–309 (1993).
10. B. Bruscaglioni and G. Zaccanti in: *Scattering in Volumes and Surfaces*, ed. M. Nieto Vesperinas and J. C. Dainty (New York: Elsevier) (1990).
11. E.P. Zege, A.P. Ivanov and I.L. Katsev: *Image Transfer Through a Scattering Medium*, Springer-Verlag, Berlin (1991).
12. G. Zaccanti and P. Donelli, *Appl. Opt.* **33**, 7023–7030 (1994).
13. J.C. Hebden, D.J. Hall, and D.T. Delpy, *Med. Phys.* **22**, 201–208 (1995).

4 Direct image processing using artificial retina chips

Eberhard Lange, Yoshikazu Nitta and Kazuo Kyuma
Department of Neural & Parallel Processing Technology
Advanced Technology R&D Center, Mitsubishi Electric Corporation
1-1-18 Tsukaguchi-Honmachi, Amagasaki, Hyogo, 661, Japan

INTRODUCTION

Working towards the ambitious goal of implementing powerful vision chips, significant progress has been achieved in recent years, yielding both devices that excel at specific tasks, and devices that perform versatile processing functions [1–9]. In this context, we set out to explore to what extent artificial retinas based on arrays of variable sensitivity photodetectors (VSPDs) can be employed to provide fast and flexible image processing functions. We address the questions of how to utilize the device characteristics of VSPDs for image processing, and of how to take advantage of fast hardware neural networks for post-processing.

We first introduce approaches to direct image processing. Then we outline direct image processing in VSPD arrays. Both preprocessing operations such as filtering and transformation, and data reduction for fast, simplified postprocessing are described. Next we give some experimental results followed by discussion of a memory effect found in VSPDs, and its application to temporal processing. We conclude with an outlook on promising directions for future work, pointing out the potential that lies in the fusion of the optical and electronic approaches to vision processing.

DIRECT IMAGE PROCESSING APPROACHES

A traditional approach to image processing is to sense an image using a charge coupled device (CCD) based camera, digitize the sensed image, serially transfer it to a frame buffer, and finally process the image using specialized hardware, a general purpose computer, or a combination of both. Separating sensing and processing allows flexibility in processing functions, high computational accuracy, and good image quality, but often at the cost of a slow response time, large hardware volume and high power consumption.

In contrast, *focal plane processors, smart sensors* and *artificial retinas* which sense and process images both concurrently and in parallel show good potential for use in fast and compact low-power systems, at the cost of a lower flexibility and, in the case of analogue implementation, reduced computational accuracy. Such devices for direct image processing often draw their inspiration from vertebrate retinas, which extract essential features from images, form foci of attention, and reduce the processing load in the subsequent parts of the brain. Vertebrate retinas also serve as an excellent example of an analogue, rather low-precision approach that nevertheless supports outstanding system performance.

Image preprocessing and data reduction are two important motivations for the direct image processing approach. Image processing algorithms may involve applying tens or even hundreds of operations per pixel. When applied to large images at high frame rates, the resulting computational load is heavy, and any kind of on-chip image preprocessing that reduces the load is welcome. Data reduction goes one step further by trying to minimize the postprocessing load. The final result of processing often consists of only a few bytes of information, encoding, for example, the position of an object. Yet, it may be necessary to process a voluminous video data stream in order to extract these essential bytes. On-chip data reduction circumvents such processing bottle-necks. In application fields such as robotics or control, response time is a major issue, and methods for reducing the width of the data stream while preserving essential information are highly desirable.

Systems which employ optical components for processing circumvent computational bottle-necks by taking advantage of the inherent parallelism and speed of optics. If they are not run as stand-alone systems, however, but depend on an image processing system for postprocessing, then they can also derive benefit from the speed and processing capabilities of suitable smart sensors.

A great variety of devices with direct image processing capability have been implemented in recent years. We highlight here only a few approaches. For a comprehensive summary, see [9] and the references given therein.

On the one hand, there are devices with largely fixed functionality, designed to excel at specific processing tasks. The groundbreaking silicon retina design of Mahowald and Mead, for example, can operate over a wide range of intensity levels, enhance edges and detect motion by computing spatial and temporal derivatives [1]. The underlying basic function is spatio-temporal smoothing, or low-pass filtering, using a two-dimensional resistive grid. Computing the difference between original and the smoothed image yields an enhanced, high-pass filtered image. Another device, also based on resistive grids, extracts moments of the brightness distribution from an image [2]. The device uses these moments for the estimation of position and orientation of a bright object on a dark background and operates at the impressively high frame rate of 5000 frames per second. Linear and non-linear resistive networks are in general a useful tool for implementing vision algorithms that can be formulated as minimization problems [9]. Still other devices use the detector geometry for processing, taking advantage of the precision of the semiconductor manufacturing process. Applications include an alignment sensor for surveyors' marks that achieves 0.1% position accuracy [3], and a 3D-camera that acquires depth images with 128×500 pixels at video rate [4]. A device for spot intensity measurement protects the human retina against laser damage by quickly shutting a liquid crystal optical filter in case of too high exposure [5]. The device senses the maximum of the intensities on its 26×25 photodetectors within 20 μs. Fossum's group aims at replacing CCDs by less expensive active pixel image sensors in CMOS technology. They demonstrated an active pixel sensor with 256×256 pixels, featuring random access and the highly useful additional function of reading out either images or the differences between consecutive images [6].

On the other hand, there are some approaches which offer quite flexible image processing functions. An artificial retina with half-toning and subsequent neighbourhood processing in the digital domain has already been demonstrated [7]. Applying functions such as erosion and edge detection on binary images with 65×76 pixels takes only about 5 μs. Fossum discusses further CCD-based approaches to flexible focal plane image processing in [8].

All of these devices are subject to a tradeoff between the size of photosites and the area devoted to processing. Increasing the chip area devoted to processing decreases the size of the photosites and the sensitivity. The tradeoff affects in particular devices with flexible, area-consuming processing functions. Microlens arrays and three-dimensional integration techniques can alleviate the problem, but only at the expense of an increase in manufacturing complexity.

The next sections describe devices that have the advantage of combining large photosites with flexible processing functions. The devices utilize the properties of variable sensitivity photodetectors for processing. In conjunction with neural postprocessing, the approach allows a variety of fast image processing operations to be performed. Applications include convolution with separable kernels, image compression and character recognition.

DIRECT IMAGE PROCESSING USING ARRAYS OF VARIABLE SENSITIVITY PHOTODETECTORS

Figure 1 shows the basic configuration of an artificial retina system which uses an array of bipolar VSPDs as its core part. The term bipolar VSPD refers to a photodetector whose sensitivity s can be varied from negative to positive values by applying a voltage u across it,

$$s = f(u), \tag{1}$$

where the sensitivity s is defined as the current j that flows through the VSPD, divided by the radiant flux W on the VSPD:

$$s = \frac{j}{W}. \tag{2}$$

We set the sensitivities by applying a common voltage u_i^r to the first contact of the VSPDs in the ith row of the VSPD array, and applying a common voltage u_k^c to the second contact of the VSPDs in the kth column of the VSPD array.

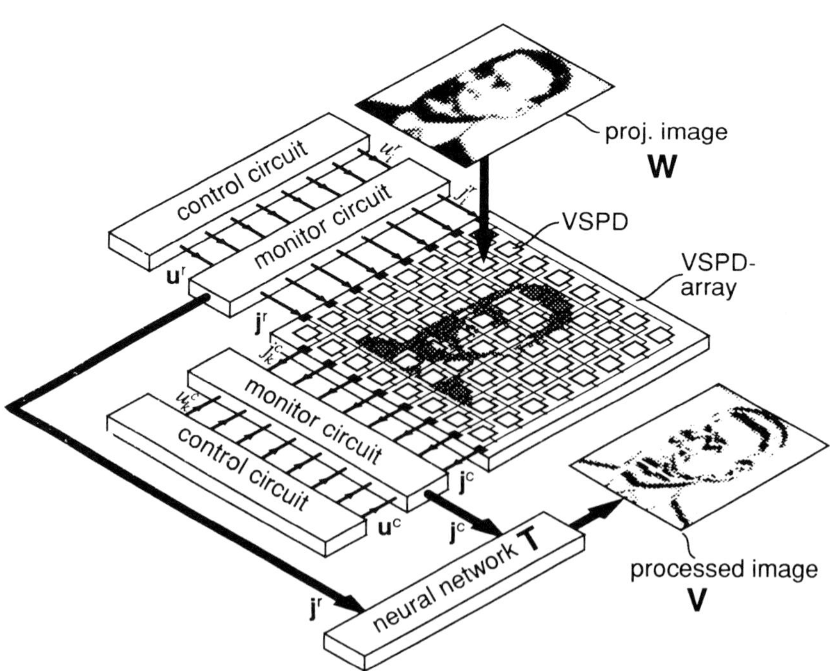

Fig. 1. System architecture of artificial retina system. Postprocessing in a neural network supplements processing in an array of bipolar VSPDs.

The current j_{ik} through a VSPD in the ith row and k^{th} column is proportional to the sensitivity of the VSPD and to the radiant flux W_{ik} on the surface of the VSPD,

$$j_{ik} = W_{ik} f(u_i^r - u_k^c). \tag{3}$$

The total current j_i^r flowing into the ith row of the VSPD array is the sum of the currents through the VSPDs in this row,

$$j_i^r = \sum_k j_{ik} = \sum_k W_{ik} f(u_i^r - u_k^c), \tag{4}$$

and the total current j_k^c flowing out of the kth column of the VSPD array is the sum of the currents through VSPDs in this column:

$$j_k^c = \sum_i j_{ik} = \sum_i W_{ik} f(u_i^r - u_k^c). \tag{5}$$

Control circuits apply vectors of voltages $\boldsymbol{u}^r$ and $\boldsymbol{u}^c$ to the VSPD array. Current monitors measure the two vectors of output currents $\boldsymbol{j}^r$ and $\boldsymbol{j}^c$ and feed proportional voltages into a neural network for postprocessing.

Direct image processing in arrays of VSPDs is aimed at reducing the postprocessing load significantly by means of image preprocessing and data reduction. In the case of image preprocessing, direct processing of an image with $N \times N$ pixels results in a processed image with $O(N^2)$ pixels. The benefit lies in a significant amount of on-chip processing, such as filtering. In the case of data reduction, on-chip processing extracts $O(N)$ or $O(1)$ features from images while preserving essential information. Data reduction eases the postprocessing load by curtailing the width of the data stream that leaves the device. Covering these two cases, we will now outline several operation modes of the artificial retina system shown in Fig. 1.

Filtering and Transforming Images

A very simple filtering function, the enhancement of horizontal edge components, may be achieved by applying a positive voltage to the first row of detectors and an equivalent negative voltage to the second row of detectors. All other rows and columns connect to ground. The voltage pattern results in a positive sensitivity in the first row of detectors and a negative sensitivity in the second row. The current vector $\boldsymbol{j}^c$ represents then the difference between the first and the second line of the image, a gradient operator which enhances horizontal edge components in this part of the image. Shifting the applied voltages cyclically through the detector rows yields the complete processed image in a time sequential, semi-parallel operation mode.

Other voltage patterns result in more general processing functions. Starting with the assumption that the columns of VSPDs are connected to ground,

$u_k^c = 0$, let us introduce a sensitivity vector s with the components

$$s_i = f(u_i^r). \tag{6}$$

Then, we can write the vector of vertical output currents j^c as a vector-matrix product,

$$j^c = sW, \tag{7}$$

corresponding to a weighted sum of image lines. Setting the sensitivities according to the rows of a sensitivity matrix S yields a full matrix-matrix product $J^c = SW$, again in a time-sequential, semi-parallel processing mode. Here, on-chip processing couples the pixels of an image only in the vertical direction. For additional coupling in the horizontal direction, a neural network may be employed, as shown in Fig. 1. A linear one-layer neural network with the connection matrix T, for example, assists in computing one line v of the processed image,

$$v = sWT, \tag{8}$$

resulting in the complete processed image $V = SWT$. Such pre- and postmultiplication of the image matrix W with two other matrices might seem to be quite a simple operation, but it covers a variety of common image processing functions. Demonstrated functions include feature extraction, filtering, and discrete cosine transformation with applications in image compression [10–16]. Transformations such as FOURIER, wavelet or discrete cosine transformation (DCT) may be applied to either the entire image, or to image blocks [15].

The speed of all filtering and transformation functions does not depend on the size of the convolution kernel or the complexity of the transformation. A convolution with a large Gaussian kernel, for example, is executed as quickly as simple edge enhancement.

The semi-parallel operation mode of the device represents a good trade-off between completely parallel readout, which would be much more difficult to implement, and strictly serial readout, which would be much slower. In comparison to continuous mode, parallel operation, semi-parallel scanning does not necessarily represent a disadvantage. In the absence of parallel readout, scanning is necessary for image retrieval, even in systems that otherwise operate in continuous mode. The interesting point is here to utilize the scanning process for performing a significant amount of processing.

So far we have discussed operation modes that implement coupling in the vertical direction between the image pixels on-chip, and relied on a neural network for coupling in the horizontal direction. To implement coupling in the horizontal direction on-chip, we need only to substitute u_k^r for u_i^r in Eq. (6), and connect the rows instead of the columns of VSPDs to ground, $u_i^c = 0$. Thus, for example both horizontal and vertical edge components may be enhanced without the help of neural postprocessing. Applying non-zero voltages to both rows and columns yields a further class of operation modes.

Pre- and postmultiplying image matrices with other matrices, as discussed here, is quite a flexible operation, but it does not allow convolution with non-separable kernels. Postprocessing using a multilayer neural network instead of a one-layer neural network offers one way to solve this problem. A difference of Gaussians (DOG) filter, for example, may be implemented in a two-step process using a two-layer neural network. A difference of Gaussians filter approximates the Laplacian of a Gaussian filter, which has proven to be useful for all-directional edge enhancement and edge detection. In the first step, on-chip processing in conjunction with postprocessing in the first layer implements Gaussian neighbourhood coupling using a wide kernel. The first layer of the neural network has the additional function of storing one line of the processed image. In the second step, on-chip processing in conjunction with postprocessing in the second layer implements Gaussian neighbourhood coupling using a narrow kernel. The second layer also receives input from the first layer and subtracts it to arrive at the final processing result, one line of a DOG-filtered image. As usual, scanning yields the complete processed image.

The operation modes discussed up to here result in complete processed images. The benefit of the approach lies in fast processing performed in the analogue domain. However, the bottleneck of a voluminous data stream that has to be processed is not removed. An array with $N \times N$ VSPDs still outputs $O(N^2)$ values per complete scan cycle. One way to circumvent this bottleneck is to find a focus of attention, and then restrict all or most of the processing and scanning to that area of interest. Other approaches try to extract meaningful features from the entire image so as to reduce the width of the data stream, and these approaches will be discussed next.

Circumventing the Processing Bottleneck

In applications that require fast response times, the processing of large images at high frame rates poses a serious computational bottleneck. The crucial problem lies in finding useful methods for reducing the wide data stream while preserving essential information. Even if response time is not the primary concern, sensing only the necessary amount of information for a given task is likely to result in compact, efficient systems. Accordingly, we will focus now on operation modes of VSPD arrays which extract $O(N)$ and $O(1)$ features from images with $N \times N$ pixels.

Projecting images constitutes one way to reduce the dimensionality of an image. Connecting the columns of a VSPD array to ground and the rows to the voltage u,

$$u_k^c = 0, \qquad u_i^r = u, \tag{9}$$

implements the two-directional projection of an image. Then the currents j_k^r and j_k^c flowing into the rows and out of the columns of the VSPD array represent the projections of the image in the horizontal and vertical directions, respectively:

$$j_i^r = f(u) \sum_k W_{ik}, \qquad j_k^c = f(u) \sum_i W_{ik}. \tag{10}$$

The two-directional projections are obtained concurrently at a speed corresponding to the response time of the photodetectors, allowing continuous mode operation without any scanning.

Replacing the uniform voltage pattern in Eq. (9) by non-uniform patterns allows the setting of foci of attention in conjunction with the extraction of projections. It allows also the implementation of various sensitivity profiles. The setting

$$u_i^r = (i/N - 1/2)^2, \qquad u_k^c = -u_i^r, \tag{11}$$

for example, yields an approximately parabolic sensitivity profile.

So far, we have explored processing modes which extract at least $O(N)$ features from images. For tasks such as position control, where an even smaller number of features suffices to solve the task, extraction of $O(1)$ features – a fixed number – is appealing. Computing the moments of a spatial intensity distribution has proven to be a useful means for finding the approximate position and orientation of a bright object on a black background, for example. Using VSPD arrays, the moments $M(l, m)$,

$$M(l,m) = \sum_{ij} (x_i)^l W_{ij} (y_j)^m, \tag{12}$$

may be computed using the settings

$$s_i^r = (x_i)^l = (i/N)^l, \qquad T_{j1} = (y_j)^m = (j/N)^m, \tag{13}$$

in Eq. (8). For $l = 0$ or $m = 0$, neural postprocessing is not required. Standley has demonstrated a CMOS VLSI chip for moment computation [2]. It uses resistive chains to integrate the boundary currents from a two-dimensional resistive grid. The approach allows the concurrent extraction of the moments corresponding to $1, x_j, y_i, (y_i)^2 - (x_j)^2$ and $x_j y_i$. In comparison, the approach based on VSPD arrays has the disadvantages that not all moments can be concurrently extracted, and that it does not feature a threshold mechanism that suppresses dark image areas. Its advantage lies in a greater flexibility, allowing, for example, the computation of the moments of the brightness distributions of all image lines and columns, or the restriction of the computation to a rectangular area within an image.

Significant potential lies also in the combination of the different operation modes discussed in this section. Using one of the fast data reduction modes, for example, makes it easy to seek certain events at high frame rates, and then trigger an action when the event occurs, using the same device and the same optical system for full or partial, direct or enhanced image readout.

EXPERIMENTS AND RESULTS

Many experimental results have been obtained using VSPD arrays for direct image processing. In this section, we discuss implementation details, show two applications in the fields of image compression and pattern recognition, and demonstrate a device with a novel hexagonal configuration.

Hardware Implementation

A simple proof-of-concept device integrated at first only 20×20 MSM-VSPDs on a GaAs substrate [10]. The advantage of MSM-VSPDs lies in their simple, planar structure, manufactured by depositing interdigital titanium/gold electrodes on a semi-insulating GaAs wafer. They also exhibit low leakage currents and fast response times. But most important for our applications is their variable, bipolar sensitivity [17]. If we apply a bias voltage and illuminate the detector, a photocurrent starts to flow. Photocurrent and sensitivity are a smoothly increasing function of the bias voltage, and reversing the bias voltage yields a reversed current flow and sensitivity. Positive and negative sensitivities mean that we can both add and subtract pixel intensities. Figure 2 shows a typical smooth response curve of a MSM-VSPD. The slight asymmetry of the curve is partly due to a slightly asymmetric device design.

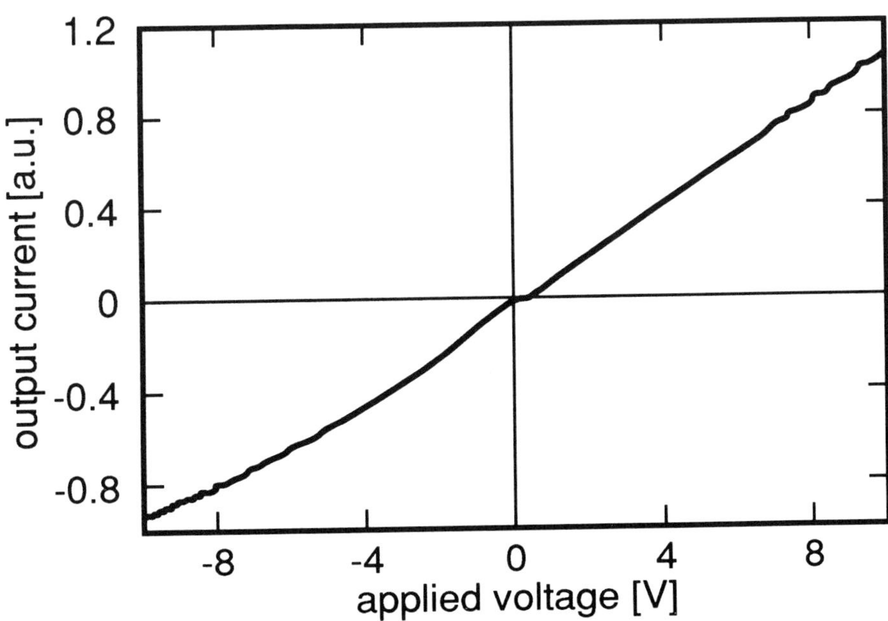

Fig. 2. Current/voltage characteristics of a MSM-VSPD [17].

Larger devices with 64×64 and 128×128 MSM-VSPDs allowed us to demonstrate a variety of high-speed processing functions, including feature extraction, discrete cosine transform, one-directional projection for character recognition and depth perception [11, 12]. The total size of these MSM-VSPD arrays is 10.24 mm $\times 10.24$ mm fabricated on a chip measuring 14.3 mm $\times 14.3$ mm.

Memory currents, discussed in detail below, can adversely affect the performance of the devices in some image processing applications. We therefore considered using alternative VSPD designs. VSPDs made from amorphous silicon have already been shown to be suitable for matrix-vector multiplication based image processing [18]. The sublinear response of the detectors to incident light followed a power law, implying that the sensitivity depends not only on the bias voltage, as in Eq. (1), but also on the illumination intensity. VSPDs in a pn-np structure exhibit good linearity over several orders of magnitude of illumination intensity and are not affected by memory currents [16, 19].

Processing modes which utilize only a small fraction of the detectors at any one time result in a relatively low total sensitivity – a problem in situations where light is scarce. Two types of enhanced VSPDs which address this issue will be briefly discussed in the last section.

The 128×128 pixel chips fit, together with control circuits and current monitors, into a compact casing equipped with a lens, allowing us to perform outdoor experiments. A TV monitor displays the processed image at video rate. The internal processing speed of the devices limits the maximum feasible frame rate to about 1 kHz. The neural network for postprocessing is currently implemented only in software but, depending on the demands of the application, it can either be implemented on-chip, or by using specialized external hardware. Fast analogue neural networks best match the processing speed of the VSPD arrays. The manufactured MSM-VSPDs respond to changes in bias voltage within about 3 µs, and smaller MSM-VSPDs can operate much more quickly. Intel's ETANN chip, for example, offers similar response times [20]. Small networks are also easily implemented using resistive grids, and implementing simple kinds of coupling, such as addition and subtraction, requires still less effort. Going into the digital domain by integrating D/A converters on-chip would allow the chip to serve as a fast, parallel front end for digital neurochips [21].

Image Compression

We have already mentioned the possibility of applying various transformations to image blocks. Figure 3 illustrates a DCT of image blocks. Such a DCT of image blocks is the computationally most expensive part of the Joint Photographics Expert Group (JPEG) compression algorithm for continuous tone still images. Figure 3(a) shows the image of a fingerprint as sensed in the

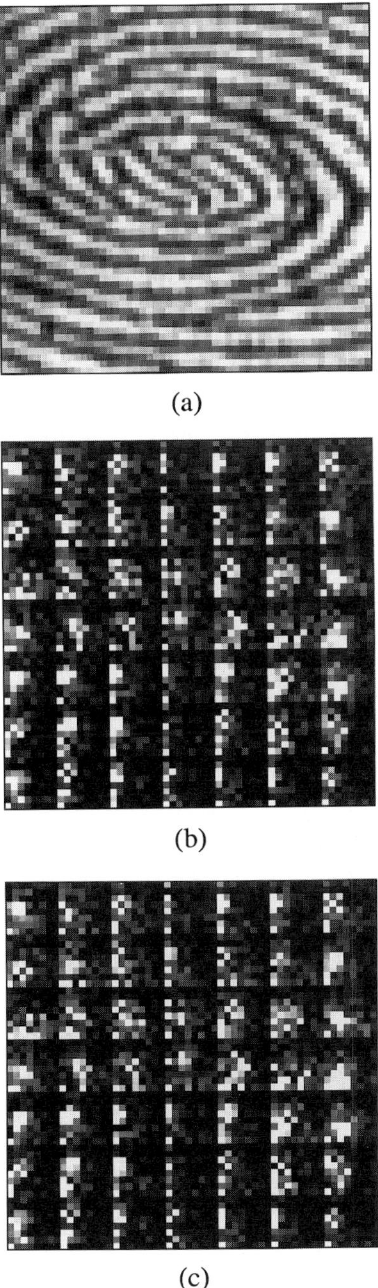

(a)

(b)

(c)

Fig. 3. Sensed image of a fingerprint (a) and its blockwise discrete cosine transforms. Simulation (b) and experiment using the artificial retina chip (c) are in good agreement [15].

TV-camera-like operation mode of the device. This operation mode simply scans the image by setting only one row of detectors to a positive sensitivity at any one time. The results of simulated DCT, Fig. 3(b), and DCT using on-chip transforms, Fig. 3(c), are quite similar. Computation and analysis of the effective on-chip DCT transform coefficients show a root mean square (RMS) deviation from the ideal of about 25%. Better adjustment of the sensitivity vectors halves the deviation, and the spatio-temporal fluctuation of the deviations is small. Thus, image compression using on-chip DCT has acceptable reconstruction quality, introducing an additional error which is a small fraction of the theoretically expected compression error [*16*].

Character Recognition

Figure 4 shows a VSPD array operating in projection mode applied to a character recognition task. Here, the horizontal and vertical projections serve as a compact representation of the sensed character and are fed into a shift and brightness tolerant neural network for recognition [*13, 16*]. Using a set of 1945 Sino-Japanese ideographic characters (Kanji) and two projections of 64 components each, the system achieves shift tolerant recognition rates of up to 99.99%. Adding Gaussian noise to the sensor output prior to processing allows us to evaluate the robustness of the approach, as illustrated in Fig. 5. The noise level is specified in terms of its RMS value, as a percentage of the maximum

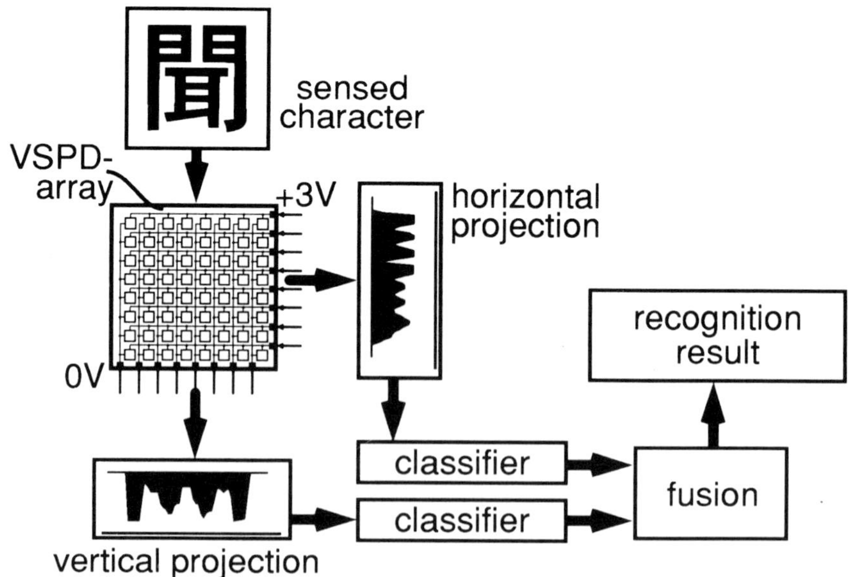

Fig. 4. Application of two-directional projection mode to character recognition. Horizontal and vertical projections represent the sensed character [*16*].

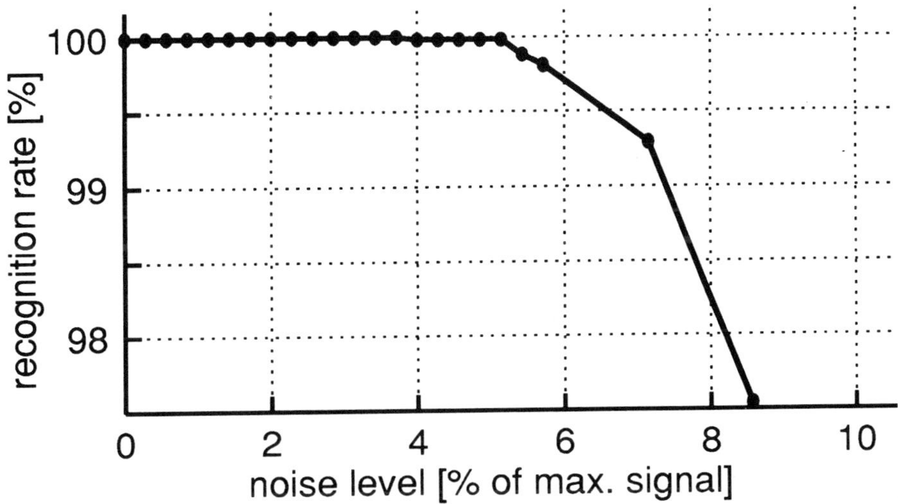

Fig. 5. Character recognition performance. Using a set of 1945 Sino-Japanese ideographic characters, the system achieves recognition rates of up to 99.99% [*16*].

signal level. Noise levels of up to about 7% still permit recognition rates in excess of 99%. The system noise caused by detector non-uniformity, subpixel shifts of the characters, noise in electric circuits, etc., corresponds to a noise level of approximately 2%.

The approach is not limited to character recognition, but may be applied to industrial object recognition tasks which require fast response times. Other applications of the two-directional projections include position sensing, object tracking and motion detection [*15*]. Simple subtraction suffices to remove fixed background patterns. In contrast with position sensitive devices, keeping track of multiple objects is possible: correspondence problems may be resolved by temporarily switching from continuous mode operation to scanning mode.

Hexagonal Configuration

We will conclude this section with a brief description of a device that allows not horizontal and vertical, but rather three-directional projections of an image to be concurrently sensed. We have already mentioned the correspondence problem encountered when tracking multiple objects using only horizontal and vertical projections. Using three-directional projections resolves many ambiguous situations both in tracking and recognition tasks. Figure 6(a) illustrates the overall design of a device for three-directional processing. A total of 1261 hexagonal cells, each measuring 240 μm × 277 μm, are arranged in a hexagonal configuration. In a rectangular VSPD array, two groups of scanning lines run through the array at angles of 0 and 90 degrees, and each VSPD

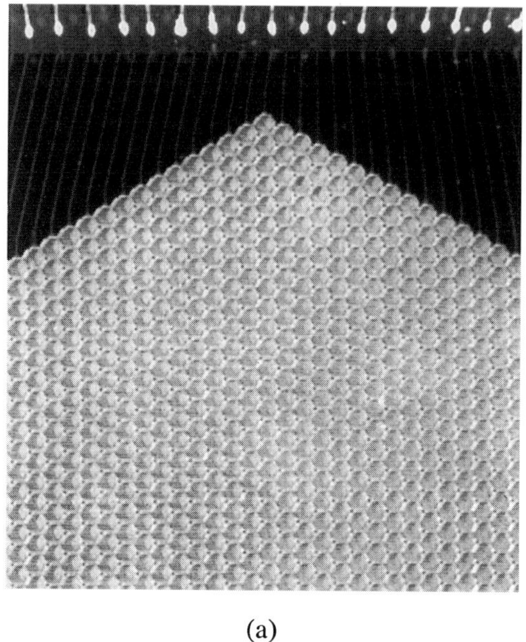

(a)

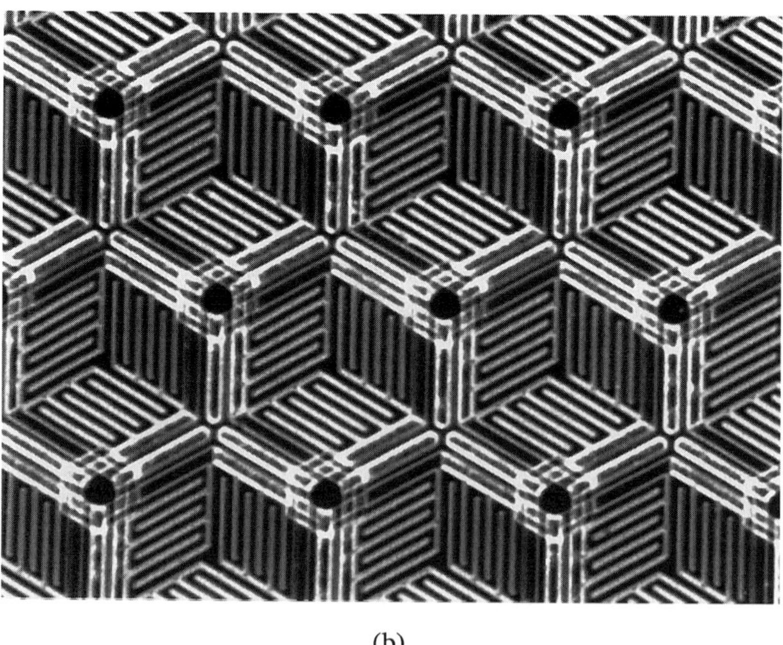

(b)

Fig. 6. Hexagonal MSM-VSPD device with three-directional scanning. (a) Device micrograph, (b) design of hexagonal cells, and (c) circuit schematics of the device.

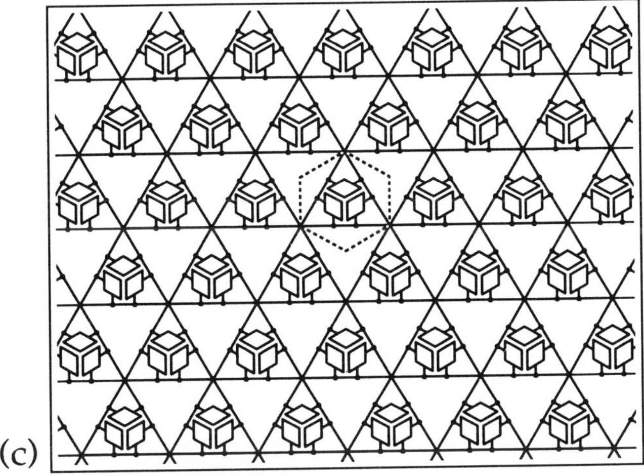

(c)

Fig. 6. (*continued*).

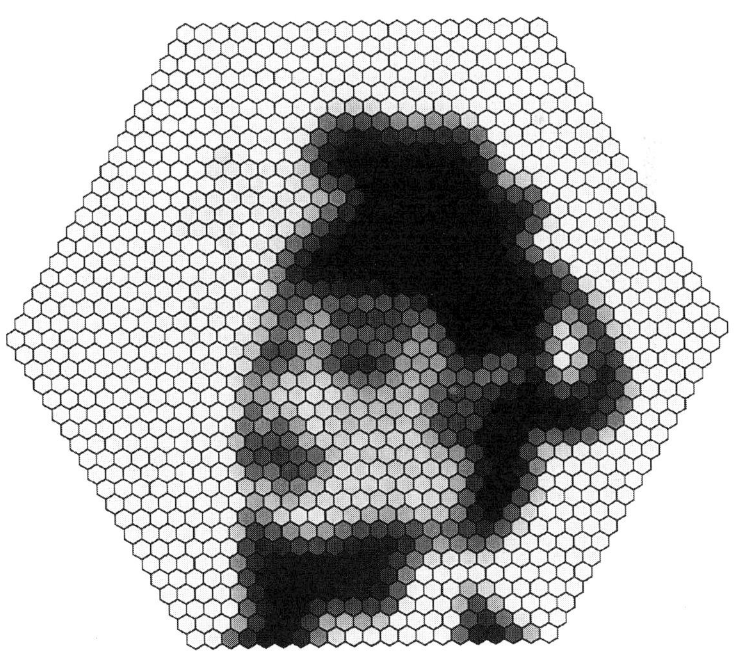

Fig. 7. Image sensed in the TV-camera-like operation mode of the hexagonal MSM-VSPD device.

connects to two scanning lines. In the hexagonal configuration, three groups of 41 scanning lines run through the cell array at angles of 0, 60 and 120 degrees, as shown in Fig. 6(c). Each cell connects to three scanning lines, and contains three MSM-VSPDs on a GaAs substrate. Each VSPD is connected to two scanning lines. Figure 6(b) shows the cell design in detail. The design is intentionally kept completely rotationally symmetric, in order to ensure that the characteristics of the three VSPDs in each cell are as similar as possible. The image in Fig. 7 is sensed in the TV-camera-like operation mode of the hexagonal device by setting one of the 0 degree lines at a time to a positive voltage, keeping all other lines at ground potential, and measuring the output currents of the 60 degree lines. This operation mode utilizes only one out of three detectors in each cell. The hexagonal device offers processing functions similar to those already discussed, but with the additional flexibilities of three-directional processing, such as multidirectional filtering or the concurrent extraction of three-directional projections.

UTILIZING MEMORY CURRENTS

Until now we have described spatial processing functions involving manipulation of the sensitivity of VSPDs. We will explain here how to perform temporal processing using another device property of metal-semiconductor-metal (MSM) type VSPDs. At a bias voltage of 0 V, we would not usually expect any photocurrent to flow through a MSM-VSPD. Experiments show, however, that there is indeed a small 'memory' current, though it is at least one to two orders of magnitude smaller than the photocurrents induced by bias voltages of around 5 V [22]. The memory current is roughly proportional to the charge transferred through the VSPD in the past, and to the current intensity of illumination. Photocurrents induce memory currents of opposite polarity. One could also say that each VSPD turns into a photovoltaic cell with a tiny, positive or negative conversion efficiency. The memory effect is quite stable – we typically observed memory current decay times of about 20 minutes, and the memory can last for days when no readout takes place. The physical mechanism behind the memory current is not entirely clear, but impurities at the GaAs and metal interface seem to be the most important factor allowing the build-up of an internal electric field that gives rise to the memory current.

Nitta *et al.* have shown that memory currents in MSM-VSPDs are very useful for storage of bipolar weights and on-chip learning in optical neurochips [22]. Here, the MSM-VSPDs are monolithically integrated on top of light emitting diodes (LEDs), as illustrated in Fig. 8. As useful as memory currents may be in optical neurochips, they have sometimes been a hindrance to image processing applications. But how about making good use of such interesting device characteristics and trying to devise schemes for image processing that take advantage of memory currents?

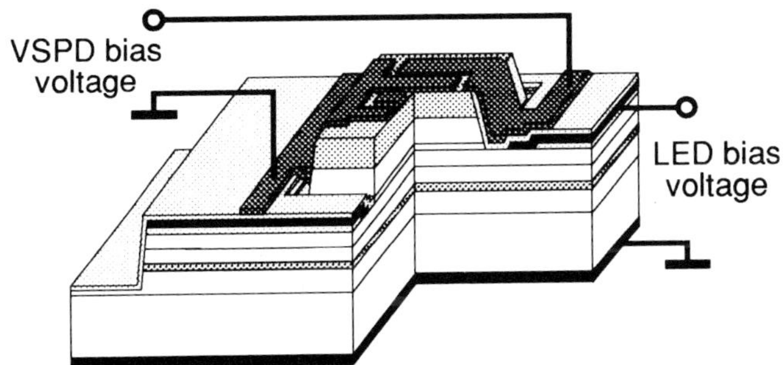

Fig. 8. Schematic diagram of the unit cell of the optical neurochip. A MSM-VSPD is monolithically integrated on top of a LED [22].

Three basic operations are important for such image processing operations. First of all, the memory current may be read out by setting the bias voltage across the VSPD to 0 V and switching the LED on. Secondly, the memory current may be increased by applying a negative bias voltage, switching the LED off, and then shining light on the VSPD. Finally, the memory current may be decreased by applying a positive bias voltage, switching the LED off, and shining light on the VSPD. Using these three operations and arrays of monolithically integrated LED/VSPD pairs, a variety of spatio-temporal image processing operations can be implemented. Nitta *et al.* have already demonstrated image subtraction and the recording of the movement traces of a light spot as a means of implementing motion detection [23]. For subtracting two images, we project the first image on to the chip while applying a positive bias voltage. Then we reverse the polarity of the bias voltage and project the second image for the same duration. The memory current distribution after this operation corresponds to the difference between the first and the second image, and may be read out as previously explained. For recording the traces of a moving light spot, the polarity of the bias voltage is repeatedly reversed. Thus the memory currents due to stable light spots and to the background light cancel out, while the moving light spots leave a trace of alternating, positive or negative memory currents. Resetting the memory effect quickly is difficult. The memory currents may be gradually erased, however, by applying the appropriate voltage patterns and LED intensities in a time sequential procedure.

CONCLUSION AND FUTURE WORK

Artificial retinas based on VSPD arrays perform, in conjunction with an optional neural network for postprocessing, a variety of useful image processing operations. They operate both in modes that filter and transform images and in modes that rapidly extract small numbers of features from images, with potential applications in the fields of image compression, robotics, fast machine vision, and others.

The advantages of the approach lie in its versatility and speed in conjunction with relatively low structural complexity. With regard to practical applications, a drawback is the relatively high cost of implementation in GaAs technology. Image processing employing memory currents in optical neurochips faces similar problems. In addition, the manufacturing process is quite sophisticated and it is difficult to reset the memory currents quickly.

Future work should focus on increasing sensitivity, resolution and frame rate, on integrating driver circuitry on-chip, and on the implementation in silicon technology. Increased sensitivity is required in processing modes which utilize only a small fraction of the detectors, for example in the edge enhancement mode of a high-resolution device. Currently two CMOS VLSI devices which address these issues are under evaluation. The core components of the first device are 256×256 imager cells with variable sensitivity and charge collection capability, augmented by on-chip scanning circuitry. The second device employs VSPDs which have a high internal gain, boosting output currents by two orders of magnitude [24]. Another interesting direction of research is implementing the functionality of the GaAs neurochips in silicon, eliminating the problem of fast memory erasure.

Several factors contribute to the trend towards implementation of CMOS image sensors. Using standard CMOS processes instead of special CCD processes reduces manufacturing costs significantly. CMOS technology also facilitates the integration of additional processing functions on the focal plane. Such on-chip processing derives benefit from the exponentially decreasing cost of computation, resulting in increasingly effective, compact, simple and affordable image processing systems.

Fossum's active pixel sensors are an example of the emerging CMOS sensor technology [6]. These sensors are several years away from matching the image quality and resolution of the best CCDs, but this might be of little importance for the target area: inexpensive camera systems for personal computers, video phones, and so on.

Rapid advances in integration technology will allow us to implement complex vision alogrithms in real time. Vision chips are likely to become ubiquitous and indispensible parts of personal computers, household appliances, and vehicles. Now may be the best time to reexamine the perceived competitive relationship between optical and elctronic processing, and move towards a synergy; to reevaluate existing optical processing techniques from this synergistic point of view, and to research exciting hybrid systems that combine optical elements with electronic processing.

ACKNOWLEDGEMENTS

The authors would like to thank E. Funatsu, K. Hara, Y. Miyake, S. Tai J. Ohta and T. Arbuckle for their technical support.

REFERENCES

1. M.A. Mahowald and C. Mead, 'Silicon retina,' in *Analog VLSI and Neural Systems*, Addison Wesley, Reading, Mass., 1989.
2. D.L. Standley, 'An object position and orientation IC with embedded imager,' *IEEE J. Solid-State Circuits*, **26**, No. 12, Dec. 1991, pp. 1852–1853.
3. C.B. Umminger and C.G. Sodini, 'Integrated analog sensor for automatic alignment,' *Technical Digest of IEEE ISSCC Int. Solid-State Circuits Conf.*, San Francisco, February 1995, pp. 156–157.
4. P. Seitz, D. Leipold, J. Kramer, and J.M. Raynor, 'Smart optical and image sensors fabricated with industrial CMOS/CCD semiconductor processes,' *Proc. SPIE*, **1900**, 1993, pp. 30–39.
5. M. Chevroulet, M. Pierre, B. Steenis, and J.-P. Bardyn, 'A battery operated optical spot intensity measurement system,' *Technical Digest of IEEE ISSCC Int. Solid-State Circuits Conf.*, San Francisco, February 1995, pp. 154–155.
6. A. Dickinson, B. Ackland, E.-S. Eid, D. Inglis, and E.R. Fossum, 'A 256×256 CMOS active pixel image sensor with motion detection,' *Technical Digest of IEEE ISSCC Int. Solid-State Circuits Conf.*, San Francisco, February 1995, pp. 226–227.
7. T.M. Bernard, B.Y. Zavidovique, and F.J. Devos, 'A programmable artificial retina,' *IEEE J. Solid-State Circuits*, **28**, No. 7, July 1993, pp. 789–798.
8. E.R. Fossum, 'Architectures for focal plane image processing,' *Opt. Eng.*, **28**, No. 8, Aug. 1989, pp. 865–871.
9. 'Vision Chips: Implementing Vision Alogorithms with Analog VLSI Circuits,' C. Koch and H. Li, eds., IEEE Computer Society Pess, 1994.
10. E. Lange, E. Funatsu, K. Hara, and K. Kyuma, 'Optical neurochips for direct image processing,' *Proc. of the IEICE Spring Conference*, Noda, Japan, March 1992, paper 6–83.
11. E. Lange, E. Funatsu, K. Hara, and K. Kyuma, 'A new artificial retina device – direct image processing in an array of variable sensitivity photodetectors,' *Proc. of the Autumn Meeting of the Japan Society of Applied Physics*, Suita, Japan, Sep. 1992, p. 810; K. Kyuma, E. Lange, and Y. Nitta, 'Optical neuro-devices,' *Optoelectronics – Devices and Technologies*, **8**, No. 1, March 1993, pp. 35–52.
12. E. Lange, E. Funatsu, K. Hara, and K. Kyuma, 'Artificial retina devices – fast front ends for neural image processing systems,' *Proc. Int Joint Conf. on Neural Networks*, Nagoya, Japan, Oct. 1993, pp. 801–804; E. Lange, E. Funatsu, K. Hara, and K. Kyuma, 'Stereo vision using artificial retina chips – a simple method for depth perception,' *Technical Digest of Topical Meeting of International Commission for Optics*, Kyoto, Japan, April 1994, p. 133.
13. B. Banish, E. Lange, and K. Kyuma, 'Kanji character recognition using an

artificial retina,' *Proc. JNNS Japanese Neural Network Society Meeting*, Tsukuba, Japan, November 1994, pp. 297–298.

14. K. Kyuma, E. Lange, J. Ohta, A. Hermanns, B. Banish, and M. Oita, 'Artificial retinas – fast, versatile image processors,' *Nature*, **372**, No. 6502, pp. 197–198.

15. E. Lange, Y. Nitta, and K. Kyuma, 'Optical Neural Chips,' *IEEE Micro*, **14**, No. 6, December 1994, pp. 29–41.

16. E. Lange, E. Funatsu, J. Ohta, and K. Kyuma, 'Direct image processing using arrays of variable sensitivity photodetectors,' *Technical Digest of IEEE ISSCC Int. Solid-State Circuits Conf.*, San Francisco, February 1995, pp. 228–229.

17. Y. Nitta, J. Ohta, S. Tai, and K. Kyuma, 'Variable sensitivity photodetector using metal-semiconductor-metal structure for optical neural networks,' *Opt. Lett.*, **16**, No. 8, April 1991, pp. 611–613; R.I. MacDonald and S.S. Lee, 'Photodetector sensitivity control for weight setting in optoelectronic neural networks,' *Appl. Opt.*, **30**, No. 2, Jan. 1991, pp. 176–179.

18. R.G. Stearns and R.L. Weisfield, 'Two-dimensional amorphous-silicon array for optical imaging,' *Appl. Opt.*, **31**, No. 32, Nov. 1992, pp. 6874–6881.

19. E. Funatsu, K. Hara, T. Toyoda, J. Ohta, and K. Kyuma, 'Variable-sensitivity photodetector of pn-np structure for optical neural networks,' *Jpn. J. Appl. Phys.*, **33**, No. 1B, 1994, pp. L113-L115.

20. M. Holler, S. Tam, H. Castro, and R. Benson, 'An electrically trainable artificial neural network (ETANN) with 10240 "floating gate" synapses,' *Proc. Int. Joint Conf. on Neural Networks*, Washington, DC, Vol. 2, June 1989, pp. 191–196.

21. Y. Kondo, Y. Koshiba, Y. Arima, M. Murasaki, T. Yamada, H. Amishiro, H. Shinohara, and H. Mori, 'A 1.2 GFLOPS neural network chip exhibiting fast convergence,' *Technical Digest of IEEE ISSCC Int. Solid-State Circuits Conf.*, San Francisco, February 1994, pp. 218–219.

22. Y. Nitta, J. Ohta, S. Tai, and K. Kyuma, 'Optical learning neurochip with internal analog memory,' *Appl. Opt.*, **32**, 1993, pp. 1264–1274.

23. Y. Nitta, J. Ohta, S. Tai, and K. Kyuma, 'Optical neurochip for image processing,' *Proc. Int. Joint Conf. on Neural Networks*, Nagoya, Japan, Oct. 1993, pp. 805–808.

24. A. Hermanns, J. Ohta, and K. Kyuma, 'Variable sensitivity Si MOS photodetectors for optical neurochip applications,' *Optoelectronics Conference*, Chiba, Japan, pp. 454–455, July 1994.

5 Principles and development of diffraction tomography

Emil Wolf
*Department of Physics and Astronomy and Rochester
Theory Center for Optical Science and Engineering,
University of Rochester, Rochester, NY 14627, USA*

INTRODUCTION

Inversion techniques which utilize X-rays, light, ultrasonic waves and electron beams have been remarkably successful in recent times and have led to major breakthroughs in fields such as condensed matter physics, biology and medicine. The best known of them is computerized axial tomography, often referred to as CAT. This technique has found many important applications and has revolutionized diagnostic medicine[1]. However, when the wavelength of the radiation is not appreciably smaller than the linear dimensions of the object, CAT does not provide adequate resolution. One must then use so-called diffraction tomography or other inversion techniques. Although the basic theory of diffraction tomography was formulated more than a quarter of a century ago it is only now that it is beginning to find applications. This is undoubtedly due to the fact that practical implementation of this technique had to await the development of fast computers.

In this chapter the basic mathematical principles of diffraction tomography are reformulated in a manner which provides a new physical insight into this technique and which brings into evidence its close relationship to the well-known technique of structure determination of crystals from X-ray diffraction measurements. The main contributions to diffraction tomography which have been made since publication of the first paper on this subject are briefly

---

[1] For reviews of this subject see, for example, Swindell and Barrett (1977), Gortion, Herman and Johnson (1975) and Mueller, Kaveh and Wade (1979). A more comprehensive treatment is given in an excellent book an Kak and Slaney (1988).

reviewed and some recent developments in this field are discussed. No attempt is made, however, to provide complete coverage of the rapidly expanding literature on this subject.

DIRECT SCATTERING AND THE FAR FIELD IN THE FIRST BORN APPROXIMATION

We begin by summarizing the main equations relating to the interaction of monochromatic scalar waves with weak scatterers. Let

$$V^{(i)}(\mathbf{r}, t) = e^{i(k\mathbf{s}_0 \cdot \mathbf{r} - \omega t)} \tag{1}$$

be a plane monochromatic wave of frequency ω incident on a scattering medium which occupies a finite domain $\mathfrak{R}$ (Fig. 1). In Eq. (1) $\mathbf{r}$ is the position vector of a typical point in space and t denotes the time. Further $\mathbf{s}_0$ is a unit vector along the direction of propagation of the wave and

$$k = \omega/c \tag{2}$$

is the wave number associated with the frequency ω, c being the speed of light in free space.

Let $n(\mathbf{r}, \omega)$ be the refractive index of the medium. If we assume that the scatterer is sufficiently weak we may represent the resulting field within the

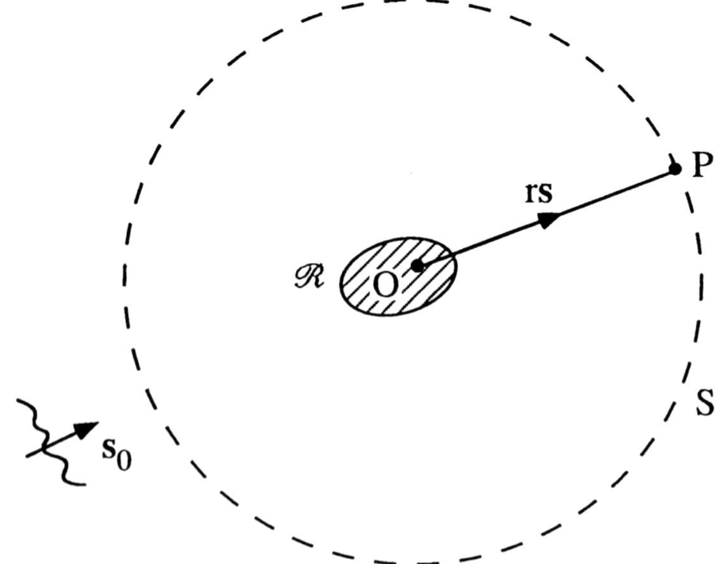

Fig. 1. Illustrating the notation. *S* is a sphere of very large radius, centred in the region of the scatterer.

accuracy of the first Born approximation as

$$V(\mathbf{r}, t) = [e^{iks_0 \cdot \mathbf{r}} + U^{(s)}(\mathbf{r}, \omega)]e^{-i\omega t}, \tag{3}$$

where the time-independent part $U^{(s)}(\mathbf{r}, \omega)$ of the scattered field is given by (Roman 1965)

$$U^{(s)}(\mathbf{r}, \omega) = \int_{\mathcal{R}} F(\mathbf{r}', \omega)e^{iks_0 \cdot \mathbf{r}'} \frac{e^{ik|\mathbf{r}-\mathbf{r}'|}}{|\mathbf{r}-\mathbf{r}'|} \, d^3\mathbf{r}', \tag{4}$$

with

$$F(\mathbf{r}, \omega) = \frac{k^2}{4\pi} [n^2(\mathbf{r}, \omega) - 1] \tag{5}$$

representing the *scattering potential.*

In most treatments one considers the scattered field only in the far zone. An expression for it can be readily obtained from Eq. (4) by setting $\mathbf{r} = r\mathbf{s}$ ($s^2 = 1$) and proceeding to the asymptotic limit as $kr \rightarrow \infty$, with $\mathbf{s}$ being kept fixed. With the help of the asymptotic approximation for the outgoing Green's function, viz.,

$$\frac{e^{ik|\mathbf{r}-\mathbf{r}'|}}{|\mathbf{r}-\mathbf{r}'|} \sim \frac{e^{ikr}}{r} e^{-iks.\mathbf{r}'}, \tag{6}$$

whose validity can be verified by elementary geometry (see Fig. 2), we find at once from Eq. (4) the required expression for the far field:

$$U^{(s)}(r\mathbf{s}, \omega) \sim A_\omega(\mathbf{s}, \mathbf{s}_0) \frac{e^{ikr}}{r}, \qquad (kr \rightarrow \infty, \mathbf{s} \text{ fixed}). \tag{7}$$

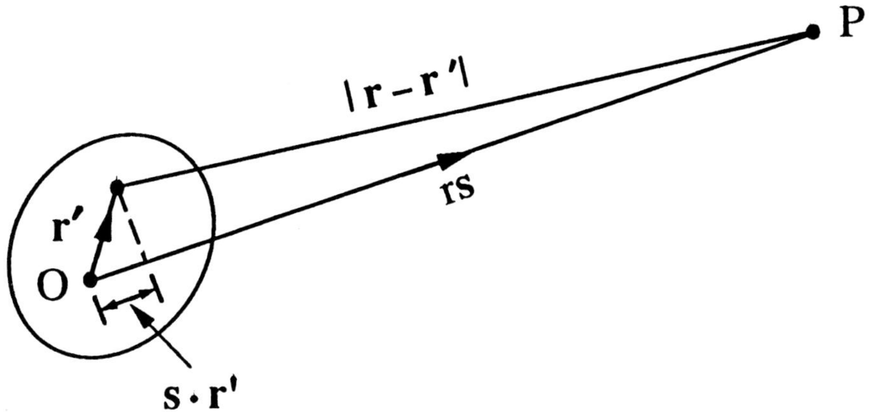

Fig. 2. Illustrating the derivation of the asymptotic formula (6), which is an immediate consequence of the approximation $|\mathbf{r} - \mathbf{r}'| \sim r - \mathbf{s} \cdot \mathbf{r}'$ for $r/r' \gg 1$.

Here $A_\omega(s, s_0)$, known as the *scattering amplitude*, is given by the simple expression

$$A_\omega(s, s_0) = \tilde{F}[k(s - s_0), \omega], \qquad (8)$$

where

$$\tilde{F}(K, \omega) = \int_{\mathfrak{R}} F(r', \omega)e^{-iK \cdot r'}d^3r' \qquad (9)$$

is the three-dimensional Fourier transform of the scattering potential.

The formulas (2.7) and (2.8) show that with a plane wave incident on the scatterer in the direction specified by the unit vector s_0, a measurement of the scattered field, in a direction specified by the unit vector s, provides information about one and only one Fourier component of the scattering potential, namely the component labelled by the vector[2]

$$K = k(s - s_0). \qquad (10)$$

Suppose now that the direction of incidence, s_0, is kept fixed and the complex amplitude of the scattered field in the far zone is measured for all possible directions s of scattering. According to Eqs. (8) and (10) one then obtains those three-dimensional Fourier components of the scattering potential that are labelled by K-vectors whose end points K lie on a sphere σ_1, of radius $k = \omega/c = 2\pi/\lambda$ centred at $-ks_0$, where λ is the wavelength associated with the frequency ω. In the theory of X-ray diffraction by crystals, where the potential $F(r, \omega)$ is a periodic function of position, this sphere is known as the *Ewald sphere of reflection* (James 1948), illustrated in Fig. 3(a).

Next let us suppose that the scatterer is illuminated in a different direction of incidence and the scattered field is again measured for all possible directions s of scattering. From such measurements one obtains those Fourier components of the scattering potential that are labelled by K-vectors whose end points lie on another Ewald sphere of reflection, σ_2, say. If this procedure is continually repeated, with different directions of incidence, one can determine all those Fourier components of the scattering potential that are labelled by K-vectors whose end points fill the domain covered by the Ewald spheres of reflection associated with all possible directions of incidence. This domain is the interior of a sphere Σ_L of radius $2k = 4\pi/\lambda$. In the theory of X-ray diffraction by crystals this sphere, shown in Fig. 3(b), is called the *Ewald limiting sphere* (James 1948).

---

[2] The formula (10) is the classical analogue of the well-known momentum transfer equation $\Delta p = p - p_0$ of the quantum-mechanical theory of elastic scattering. It follows formally at once from Eq. (10) by multiplying both sides by $\hbar = h/2\pi$, where h is Planck's constant, and using the de Broglie–Einstein relation $p_0 = \hbar ks_0$, $p = \hbar ks$ for the momentum of the incident and of the scattered particle, respectively. Then $\hbar K = \hbar k(s - s_0)$ is evidently the momentum transfer vector.

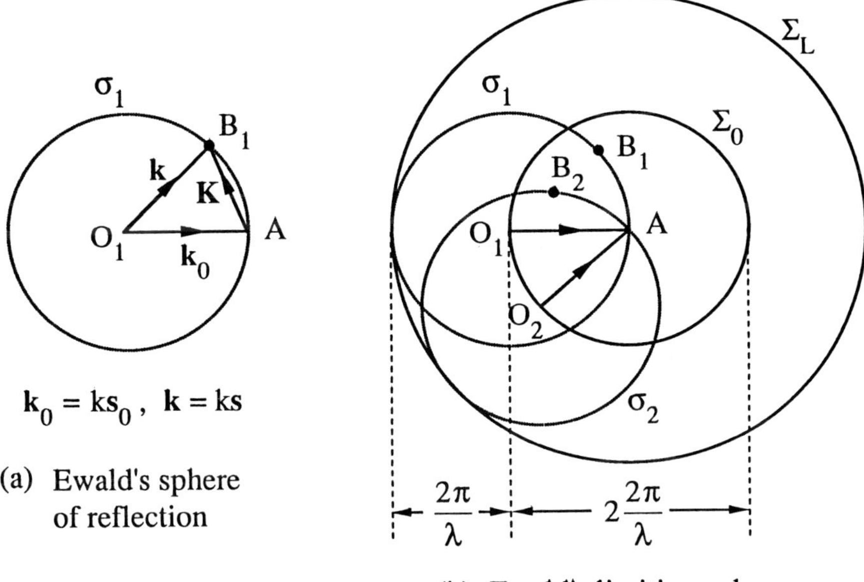

Fig. 3. (a) Ewald's sphere of reflection σ_1, associated with the incident wave vector $\mathbf{k}_0 = k\mathbf{s}_0 = \overrightarrow{O_1A}$. It is the focus of the end points of the vector $\mathbf{K} = \mathbf{k} - \mathbf{k}_0 = \overrightarrow{AB_1}$, with $\mathbf{k} = k\mathbf{s}$ representing the wave vectors of the field scattered in all possible directions $\mathbf{s}$.
(b) Ewald's limiting sphere Σ_L. It is the envelope of the spheres of reflections $\sigma_1, \sigma_2, ...,$ associated with all possible wave vectors $\overrightarrow{OA_1}, \overrightarrow{OA_2}, ...,$ of the incident fields. Σ_0 represents the sphere generated by the centres $O_1, O_2, ...,$ of the spheres of reflection associated with all possible directions of incidence.

We may illustrate the preceding formulas by a simple example. Consider a potential $F(\mathbf{r}, \omega)$ which is periodic along three mutually orthogonal directions x, y, z, with periods $\Delta x = a$, $\Delta y = b$, $\Delta z = c$. Let us take a Cartesian coordinate system along these directions and let us expand the potential into a Fourier series, viz.

$$F(x, y, z; \omega)$$

$$= \begin{cases} \sum_{h_1} \sum_{h_2} \sum_{h_3} g(h_1, h_2, h_3; \omega) \exp\left[2\pi i \left(\frac{h_1 x}{a} + \frac{h_2 y}{b} + \frac{h_3 z}{c} \right) \right] & \text{when } (x, y, z) \in \mathfrak{R} \\ 0 & \text{when } (x, y, z) \notin \mathfrak{R}, \end{cases}$$

$$(-\alpha < h_j < \alpha; \, j = 1, 2, 3).$$

$$(11)$$

Suppose that the domain $\mathfrak{R}$ occupied by the medium is a rectangular parallelepiped,

$$-\tfrac{1}{2}A \leqslant x \leqslant \tfrac{1}{2}A, \qquad -\tfrac{1}{2}B \leqslant y \leqslant \tfrac{1}{2}B, \qquad -\tfrac{1}{2}C \leqslant z \leqslant \tfrac{1}{2}C.$$

The Fourier transform, defined by Eq. (9) of the scattering potential (11), is then readily found to be

$$\tilde{F}(K_x, K_y, K_z; \omega) = ABC\pi^3 \sum_{h_1}\sum_{h_2}\sum_{h_3} g(h_1, h_2, h_3; \omega)\,\mathrm{sinc}\left[\frac{1}{2}\left(\frac{2h_1}{a} - \frac{K_x}{\pi}\right)A\right]$$

$$\times \mathrm{sinc}\left[\frac{1}{2}\left(\frac{2h_2}{b} - \frac{K_y}{\pi}\right)B\right]\mathrm{sinc}\left[\frac{1}{2}\left(\frac{2h_3}{c} - \frac{K_z}{\pi}\right)C\right], \qquad (12)$$

where

$$\mathrm{sinc}\,x = \frac{\sin \pi x}{\pi x}. \qquad (13)$$

If the periodic scattering medium is a sample of an orthorhombic crystal, the constants a, b, c (the lattice parameters) will be of the order of Angstrom units, whereas A, B and C will be of macroscopic size (millimetres or centimetres). Under these circumstances the ratios A/a, B/b, C/c will each be very large compared to unity and consequently the expression (12) will only have non-negligible values when

$$K_x \approx \frac{2\pi h_1}{a}, \qquad K_y \approx \frac{2\pi h_2}{b}, \qquad K_z \approx \frac{2\pi h_3}{c}. \qquad (14)$$

The 'momentum transfer' equation (10) then implies that the intensity of the scattered field will only have appreciable values in directions $s \equiv (s_x, s_y, s_z)$ such that

$$s_x - s_{0x} = h_1\,\frac{\lambda}{a}, \qquad s_y - s_{0y} = h_2\,\frac{\lambda}{b}, \qquad s_z - s_{0z} = h_3\,\frac{\lambda}{c}, \qquad (15)$$

where $s \equiv (s_x, s_y, s_z)$ and $s_0 \equiv (s_{0x}, s_{0y}, s_{0z})$.

Equations (15) are the well-known *von Laue's equations* (Sommerfeld 1954), which are the basic equations of the theory of X-ray diffraction by crystals. While the conditions for the validity of the first Born approximation on which our analysis is based are generally not satisfied for a real crystal, the directions at which the intensity maxima of the far-zone intensity of the diffracted field occur are usually very well approximated by the von Laue equations. This is due to the fact that when $A/a \gg 1$, $B/a \gg 1$ and $C/A \gg 1$ it is the periodicity of the potential rather than the detailed structure of the crystal which determines the location of the intensity maxima.

THE ANGULAR SPECTRUM REPRESENTATION OF THE SCATTERED FIELD

Before formulating the central theorem of diffraction tomography it is important to learn how information about the scatterer is encoded into the scattered field in planes at arbitrary distances from the scatterer.

Let R^+ and R^- be the two half-spaces $z < 0$ and $z > Z$, respectively, on the two sides of the scatterer, as shown in Fig. 4. We may represent the scattered field at any point $r \equiv (x, y, z)$ in the two half-spaces in the form of an angular spectrum of plane waves, i.e., as a superposition of plane waves each of which satisfies the same equation as the incident field, namely the Helmholtz equation $(\nabla^2 + k^2)U(r, \omega) = 0$. One may readily obtain this representation with the help of the following formula that is essentially due to Weyl (1919) [see also Mandel and Wolf (1995):

$$\frac{e^{ik|r-r'|}}{|r-r|} = \frac{ik}{2\pi} \iint\limits_{-\infty}^{\infty} \frac{1}{m} e^{ik[p(x-x') + q(y-y') + m|z-z'|]} \, dp \, dq, \tag{16}$$

where $r' \equiv (x', y', z')$ and

$$m = \begin{cases} +\sqrt{1 - p^2 - q^2} & \text{when } p^2 + q^2 \leqslant 1 \tag{17a} \\ +i\sqrt{p^2 + q^2 - 1} & \text{when } p^2 + q^2 > 1. \tag{17b} \end{cases}$$

If we substitute from Eq. (16) into Eq. (4) we obtain, after simple algebraic manipulation and interchanging of the order of the integrations, the following representation, known as the *angular spectrum* representation of the scattered field, valid at any point in the half-spaces $z > Z$ (upper sign) and $z < 0$ (lower

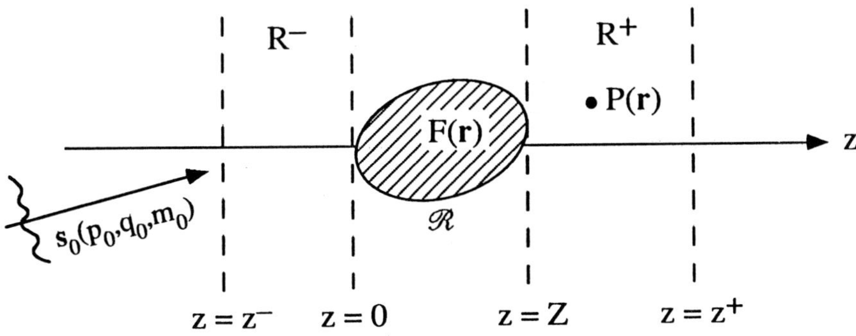

Fig. 4. Illustrating the notation relating to the angular spectrum representation (18) of the scattered field.

sign):

$$U^{(s)}(\mathbf{r}, \omega) = \iint\limits_{-\infty}^{\infty} a_\omega^{(\pm)}(p, q; s_{0x}, s_{0y}) e^{ik(px + qy \pm mz)}\, \mathrm{d}p\, \mathrm{d}q, \tag{18}$$

where

$$a_\omega^{(\pm)}(p, q; s_{0x}, s_{0y}) = \frac{ik}{2\pi m} \int_{\mathscr{R}} F(\mathbf{r}', \omega) e^{-ik[(p - s_{0x})x' + (q - s_{0y})y' + (\pm m - s_{0z})z']}\, \mathrm{d}^3 r'. \tag{19}$$

The formula (18) expresses the scattered field throughout the two half-spaces as a superposition of two kinds of plane-wave modes, viz.,

$$u_\omega(\mathbf{r}) = e^{ik(px + qy \pm mz)}, \tag{20}$$

waves for which $p^2 + q^2 \leq 1$ [m real, see Eq. (17a)], and waves for which $p^2 + q^2 > 1$ [m imaginary, see Eq. (17b)]. The former are ordinary homogeneous plane waves, propagating away from the scatterer. The latter are evanescent waves whose amplitudes decay exponentially with increasing distance from the scatterer (Fig. 5).

We note that for homogeneous waves the formula (19) may be expressed in the form

$$a_\omega^{(\pm)}(p, q; s_{0x}, s_{0y}) = \frac{ik}{2\pi m} \tilde{F}[k(p - s_{0x}), k(q - s_{0y}), k(\pm m - s_{0z}); \omega], \tag{21}$$

where $\tilde{F}$ is the Fourier transform, defined by Eq. (9), of the scattering potential. The formula (21) implies that each homogeneous wave carries information about one and only one three-dimensional spatial Fourier component of the scatterer, namely the one labelled by the Fourier component $\mathbf{K}^\pm \equiv (K_x, K_y, K_z^\pm)$ with

$$K_x = k(p - s_{0x}), \qquad K_y = k(q - s_{0y}), \qquad K_z^\pm = k(\pm m - s_{0z}). \tag{22}$$

However, the situation is quite different with evanescent waves, because for such waves m is purely imaginary and the integral on the right-hand side of

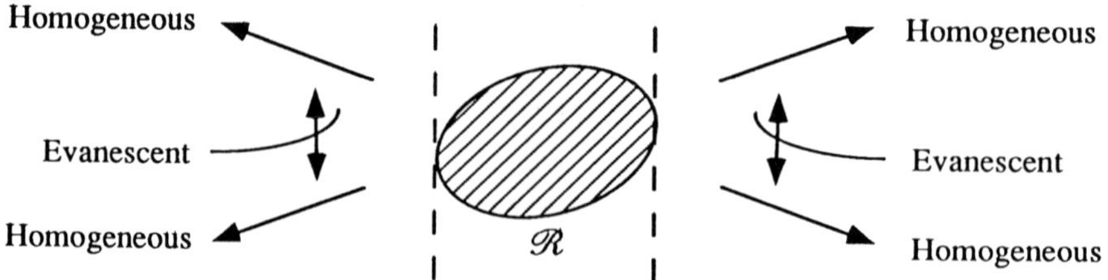

Fig. 5. Illustrating the different types of plane waves which contribute to the angular spectrum representation (18) of the scattered field.

Eq. (19) is no longer the Fourier transform of $F(\mathbf{r}, \omega)$ but rather is its analytic continuation.

On comparing Eq. (21) with Eq. (8) we see that

$$a_\omega^{(\pm)}(s_x, s_y; s_{0x}, s_{0y}) = \frac{ik}{2\pi m} A_\omega(s, s_0),\tag{23}$$

(with $s_x = p,\, s_y = q,\, s_z = \pm m = \pm\sqrt{1 - p^2 - q^2}$). The formula (23) has an important physical significance. It shows that for each pair of the (real) unit vectors s, s_0, the scattering amplitude $A_\omega(s, s_0)$, which characterizes the scattered field in the far zone, is proportional to the (generally complex) amplitude of one and only one homogeneous plane wave in the angular spectrum representation of the scattered field; namely to the plane wave which propagates in the s-direction, the positive or the negative sign being taken on the left-hand side of Eq. (23) according to $s_z \gtrless 0$.

When the scattering potential is periodic, the angular spectrum representation (18) of the scattered field will evidently be dominated by those homogeneous plane waves which propagate in directions specified by the direction cosines $(p, q, \pm m)$, whose values follow at once from Eqs. (22) and (14), i.e., they are given by the von Laue equations (15). Thus we see that the von Laue equations have a broader meaning than is usually attributed to them; for they also show which homogeneous plane-wave modes substantially contribute to the scattered field throughout the whole space surrounding the scatterer.

THE FUNDAMENTAL THEOREM OF DIFFRACTION TOMOGRAPHY

The formula (21) expresses the complex amplitudes of the homogeneous waves in the angular spectrum representation of the scattered field in terms of three-dimensional spatial Fourier components of the scattering potential. However, as we will now show, they may also be expressed in terms of the two-dimensional spatial Fourier components of the scattered field in planar cross-sections in the two half-spaces. Combination of these two results leads to the fundamental theorem of diffraction tomography.

Let us choose a plane $z = z^+$ in the half-space R^+ and a plane $z = z^-$ in the half-space R^-, both at arbitrary distances from the scatterer (see Fig. 4). According to Eq. (18) the scattered field in the two planes is given by the integrals

$$U^{(s)}(x, y, z^\pm; s_0; \omega) = \int\!\!\!\int_{-\infty}^{\infty} a_\omega^{(\pm)}(p, q; s_{0x}, s_{0y})e^{ik(px + qy \pm mz^\pm)}\, dp\, dq.\tag{24}$$

Let us take the two-dimensional Fourier transform of the scattered field in the two planes with respect to the x and y variables. This gives the following expression for the angular spectrum amplitudes of the scattered field:

$$a_\omega^{(\pm)}(p, q; s_{0x}, s_{0y}) = \frac{1}{(2\pi)^2} k^2 \tilde{U}^{(s)}(kp, kq; z^\pm; \mathbf{s}_0; \omega) e^{\mp ikmz^\pm}, \tag{25}$$

where

$$\tilde{U}^{(s)}(f_x, f_y; z^\pm; \mathbf{s}_0; \omega) = \iint\limits_{-\infty}^{\infty} U^{(s)}(x, y; z^\pm; \mathbf{s}_0; \omega) e^{-i(f_x x + f_y y)} \, dx \, dy. \tag{26}$$

The formula (25) holds for the spectral amplitudes of all the plane waves in the angular spectrum representation of the scattered field. On the other hand the formula (21) applies to the spectral amplitudes of the homogeneous waves only. Evidently when $p^2 + q^2 \leqslant 1$ we may equate the right-hand sides of the two equations, and we obtain the formula

$$\frac{ik}{2\pi m} \tilde{F}[k(p - s_{0x}), k(q - s_{0y}), k(\pm m - s_{0z}); \omega] = \left(\frac{k}{2\pi}\right)^2 \tilde{U}^{(s)}(kp, kq; z^\pm; \mathbf{s}_0; \omega) e^{\mp ikmz^\pm}. \tag{27}$$

If we set

$$f_x = kp, \qquad f_y = kq, \qquad f_z = km = \sqrt{k^2 - f_x^2 - f_y^2} \tag{28}$$

and

$$K_x = f_x - k s_{x0}, \qquad K_y = f_y - k s_{y0}, \qquad K_z^\pm = \pm f_z - k s_{z0}, \tag{29}$$

then the formula (26) may be expressed in the form

$$\tilde{F}(K_x, K_y, K_z^\pm; \omega) = -\frac{i f_z}{2\pi} \tilde{U}^{(s)}(f_x, f_y; z^\pm; \omega) e^{\mp i f_z z^\pm}, \tag{30}$$

which, in view of the inequality $p^2 + q^2 \leqslant 1$ (which holds only for homogeneous waves), is valid when

$$f_x^2 + f_y^2 \leqslant k^2, \tag{31}$$

m and f_z then being real.

The relation (30), together with the equations (29), is the mathematical formulation of the basic theorem of diffraction tomography, first derived by Wolf (1969). It shows that some of the three-dimensional spatial Fourier components $\tilde{F}(\mathbf{K}, \omega)$ of the scattering potential, namely those for which $|\mathbf{K}| \leqslant 2k = 4\pi/\lambda$ (representing points in the interior of the Ewald limiting sphere), can be determined from measurements of the scattered field in two arbitrary planes $z = z^+ > z$ and $z = z^- < 0$, one on each side of the scatterer. Stated differently, the formula (30) makes it possible to reconstruct a low-pass filtered version of the scattering potential from such measurements.

EARLY DEVELOPMENTS

Shortly after the formulation of the reconstruction procedure whose essence is the formula (30), Carter (1970), Carter and Ho (1974) and Ho and Carter (1976) tested it experimentally. It is clear from Eq. (30) that in order to determine the structure of a three-dimensional scatterer both the amplitude and the phase of the scattered field have to be determined. At optical wavelengths the phase of a field is generally difficult to measure. However, Wolf (1970) has shown that both the amplitude and the phase of a scattered field can be determined by holography. In the experimental papers that we just cited, holography was used for this purpose.

Because of the necessity of gathering and processing a large amount of data to perform the reconstructions, scattering objects of very simple kinds were used in the early experiments. The object used by Carter (1970) in the first experiment of this kind was a rectangular bar of fused silica (Fig. 6). The main results of Carter are shown in Figs 7(a) and 7(b). The width of the bar determined from these experiments was 2.016 mm, as compared with a directly measured value of 2.110 mm. In these experiments certain difficulties were encountered regarding the subtraction of the incident field from the total measured field, to determine the true scattered field. These difficulties are manifested by a 'drift' of the background seen in Figs 7(a) and 7(b).

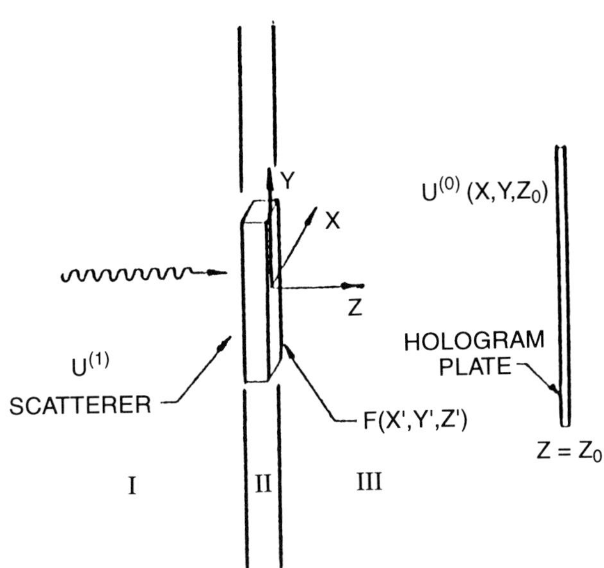

Fig. 6. Experimental set-up used in the first experiments to determine structural parameters of a simple scattering object by diffraction tomography (after Carter 1970).

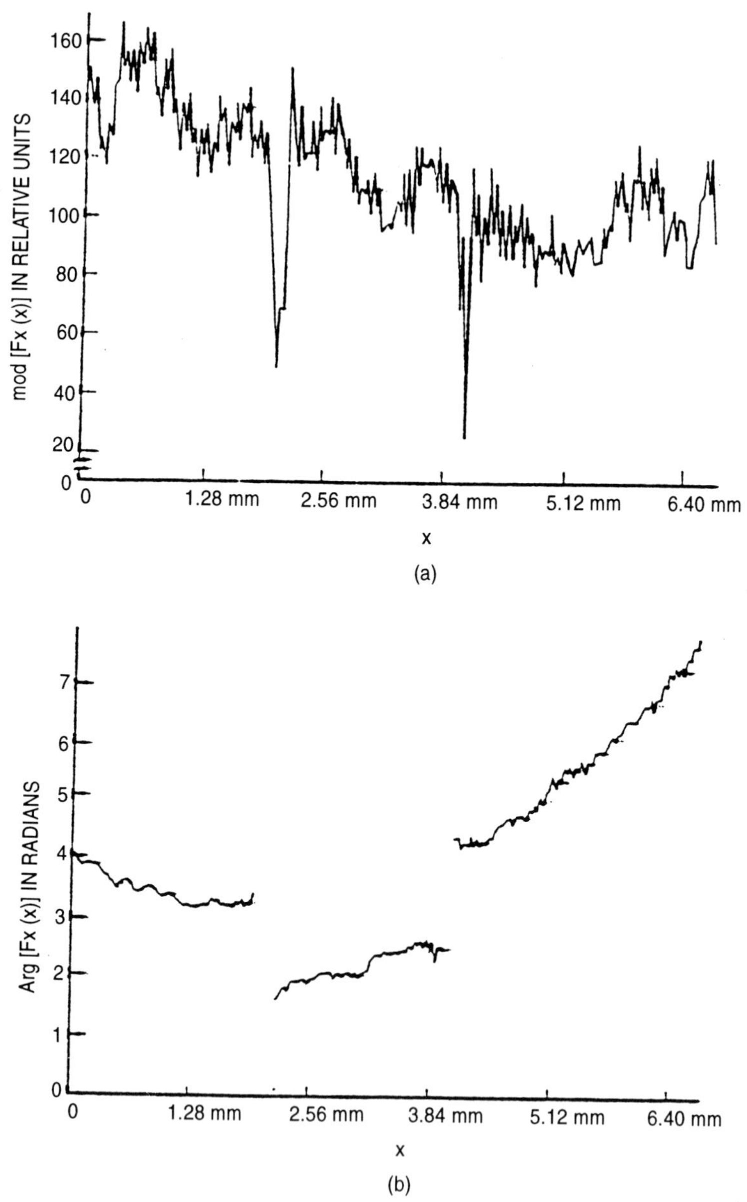

Fig. 7. Modulus (a) and phase (b) of the scattering potential $F(x)$ of the object shown in Fig. 6, reconstructed by diffraction tomography (after Carter 1970).

In a second series of experiments Carter and Ho (1974) reconstructed the structural parameters of a somewhat more complicated object, namely a bar which consisted of two right parallelepipeds of equal size, one made of fused silica and the other of extra dense flint glass (Fig. 8). The reconstructions are shown in Fig. 9. The width of the silica bar and of the extra dense flint portion deduced from these experiments agreed within 1.9% and 1.1% respectively with the directly measured values. Comparison of Figs 9 and 7 indicates that in the second series of experiments the troublesome 'drift' was largely eliminated. This was achieved by the use of a different computational procedure for removing the incident field.

In a third series of experiments Ho and Carter (1976) attempted to determine by this technique the cross-sectional diameter of a fused bar of circular cross-

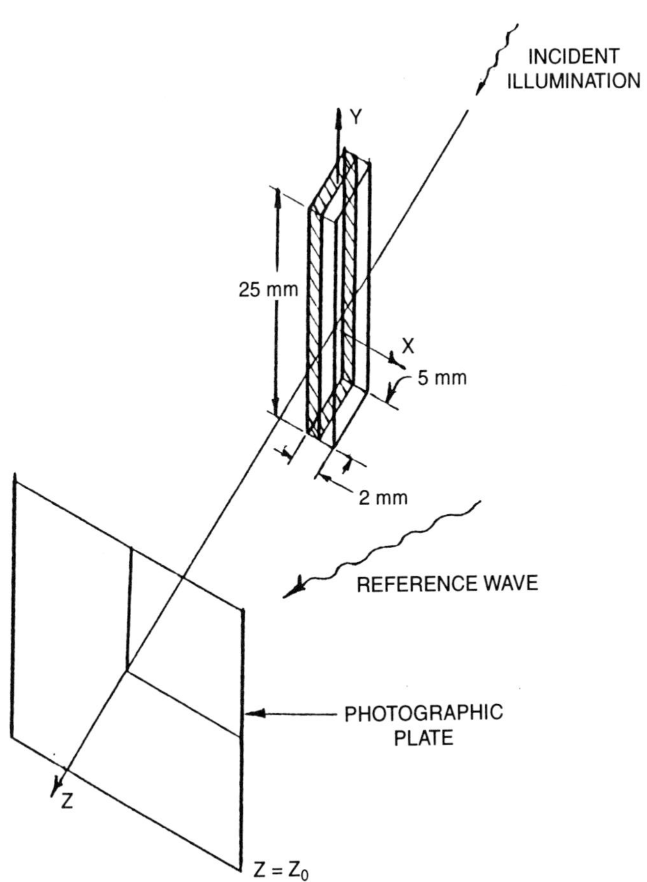

Fig. 8. Experimental set-up used to determine by diffraction tomography the structural parameters of a bar consisting of two parallelepipeds, one made of fused silica, the other of extra dense flint glass (after Carter and Ho 1974)

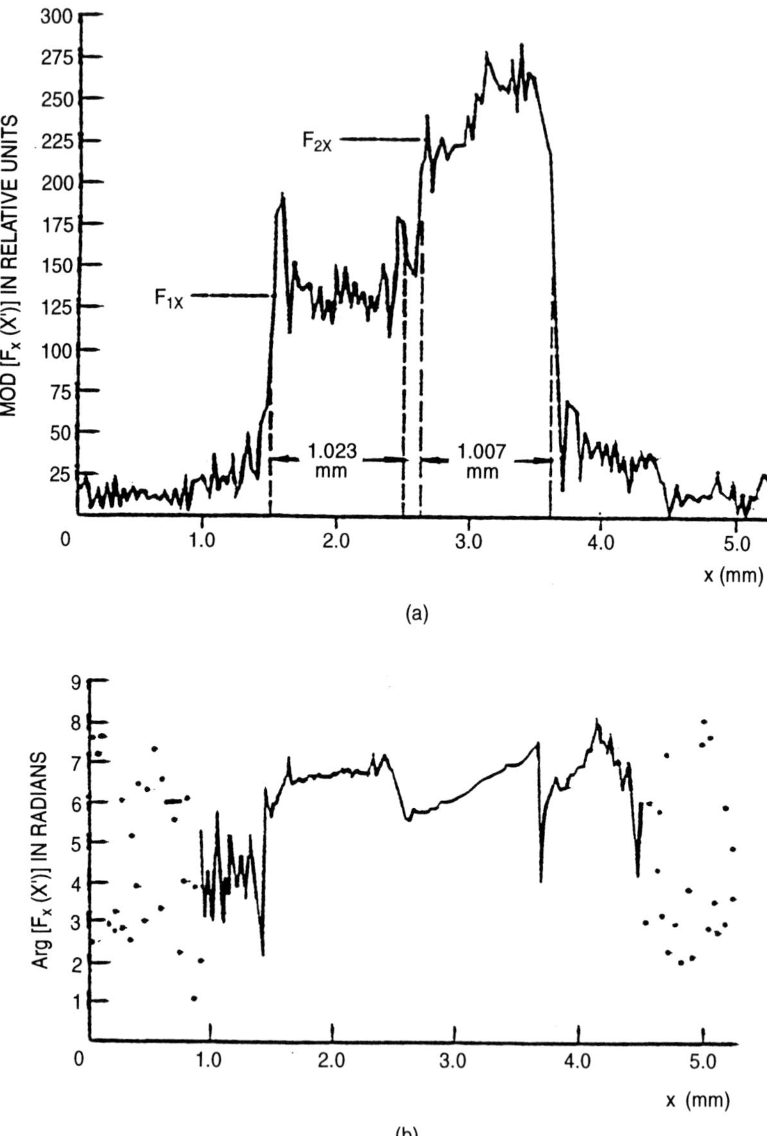

Fig. 9. Modulus (a) and phase (b) of the scattering potential $F(x)$ of the object shown in Fig. 8, reconstructed by diffraction tomography (after Carter and Ho 1974).

section. For such an object the first-order Born approximation cannot be expected to give a good approximation to the scattered field, unless the object is immersed in an index-matching liquid to suppress the effect of the abrupt change of the field across the curved boundary. Not surprisingly, therefore, this series of experiments was not as successful as those that we have just described.

Several theoretical papers followed the publication of the 1969 paper in which the principles of diffraction tomography were first outlined. Dändliker and Weiss (1970) suggested an analytic method to perform the reconstruction and discussed the accessible domain of spatial frequencies of the scatterer when only a limited set of incident plane waves is used. Evans (1970) compared some aspects of Kirchhoff's diffraction theory with diffraction tomography. A modification of the original theory for applications to microscopic objects using far-field measurements was discussed by Fercher, Bartelt, Becker and Wiltschko (1979). These authors also presented results of computer simulations and described experiments involving small spheres. In these experiments the scattered field was determined from interferometric measurements.

Iwata and Nagata (1975) have noted that when the scattering object is sufficiently small, the phase of the scattered field may vary rather rapidly over regions of the measurement plane over which the experimental data are being collected, producing practical difficulties in the computational reconstruction. They pointed out that in such cases measurement surfaces of other shapes may be more appropriate, to avoid this problem. They considered reconstruction of two-dimensional objects only, in which case the appropriate measurement surfaces are cylindrical and they presented a modification of the original theory for such situations.

The basic formula (30) of diffraction tomography is valid only within the accuracy of the first-order Born approximation. In some cases that approximation is not sufficiently accurate to describe the scattered field, even when it is weak compared with the incident field. This may be the case, for example, when the dimensions of the scattering object are so large that the cumulative effect of the small errors becomes significant. In such situations the use of the so-called first Rytov approximation may be more appropriate (Ishimaru, 1978). Iwata and Nagata (1975) described a modification, based on the first Rytov approximation, of the original theory of diffraction tomography for the reconstruction of two-dimensional scattering objects. In fact the first Rytov approximation has since found extensive use in diffraction tomo-graphy, particularly in geophysics (e.g. Devaney 1983, 1984). However many of the later investigations were only concerned with two-dimensional reconstructions.

MORE RECENT RESEARCHES

Although the theory of diffraction tomography was originally developed within the framework of optics, the technique was soon found to be particularly useful with other types of radiation. Mueller, Kaveh and Iverson (1978) and Mueller, Kaveh and Wade (1979) described a somewhat analogous reconstruction procedure in the context of ultrasonic 'imaging'. It makes use of a perturbation approximation to the solution of the wave equation that includes the (unknown) ultrasonic velocity field. The chief difference between this method and that outlined on pp. 91–92 is that it makes use of the Fourier decomposition rather than the angular spectrum representation of the wavefield. The paper also includes examples of computer simulated reconstructions based on this approach.

Computational reconstruction from measured data presents considerable practical difficulties. Several reconstruction methods have been developed. One of them is the so-called filtered back-propagation algorithm, developed by Devaney (1982, 1983) which may be regarded as a generalization of the back-projection algorithm of computerized axial tomography. Another inversion technique, which involves frequency domain interpolation of the accessible Fourier components, was developed by Pan and Kak (1983). Devaney (1984) also described a reconstruction algorithm specifically for use in seismic profiling and for well-to-well diffraction tomography with acoustical and electromagnetic waves (see Fig. 10). An inversion procedure based on generalized holography has been discussed by Porter (1989).

In addition to the early work of Carter and Ho already mentioned, laboratory tests of diffraction tomography were reported by Kaveh, Soumekh and Mueller (1980), Kaveh, *et al.* (1980), Kenue and Greenleaf (1982) and by Lo, Töksoz, Xu and Wu (1988). Field tests and the first applications soon followed, chiefly in geophysics (King, Witten and Reed 1989; Witten and King 1990a,c) to detect buried waste and to determine subsurface features using seismic waves. In particular Witten and King (1990b) located dinosaur bones by this technique. One of the chief applications envisioned for ultrasonic diffraction tomography is the detection of cancerous tumors in women's breasts. The first laboratory simulations of this important application were made by Sponheim and Johansen (1991), Sponheim, Johansen and Devaney (1991) and Ladas and Devaney (1993). A review of some of this work and of other researches on diffraction tomography has been given by Stamnes, Gelius, Johansen and Sponheim (1992) and by Stamnes and Wedberg (1995).

In conventional diffraction tomography both the amplitude and the phase of the scattered field are measured. Devaney (1989, 1991, 1992) (see, also Maleki, Devaney and Schatzberg 1992 has proposed a method which makes it possible, in principle, to reconstruct weakly scattering objects from measurements of the intensity of the total wave field (i.e., the sum of the incident and scattered field) in the far zone.

Diffraction tomography has also been adapted to one-dimensional length measurements by A. Fercher, Hitzenberger, Kamp and El-Zaiat (1995). In this technique the intensity of the scattered light is measured at various wavelengths with a fixed direction of illumination. The technique has been employed to determine the scattering potential of the human eye *in vivo* and to determine intraocular distances. The method is, in some respects, a generalization of a much older interferometric technique used for precise distance measurements, based on the use of so-called channelled spectra [see, for example, Born and Wolf (1980), p. 265].

Very recently Wedberg and Stamnes (1995a) compared three different methods for the retrieval of the phase of the scattered field for application in diffraction tomography and Wedberg and Wedberg (1995) and Wedberg and Stamnes (1995b) described diffraction tomography experiments by means of which cross-sections of semi-transparent cylindrical objects were reconstructed.

DIFFRACTION TOMOGRAPHY WITH RANDOM MEDIA

All the research which was described so far was concerned with the reconstruction of scattering potentials that were assumed to be deterministic functions of position. Only a small number of publications have been devoted to diffraction tomography with random media, i.e., media for which the scattering potential must be characterized statistically. Early publications dealing with this topic were papers by Rouseff and Porter (1991), who discussed the reconstruction of a two-dimensional anistropic, statistically homogeneous random medium, using generalized holograms, Howard (1991) who studied tomographic reconstruction of weakly scattering random media within the accuracy of the paraxial approximation, and Kunistyn and Tereshchenko (1992) who described applications to remote sensing of the ionosphere with radio waves.

We now briefly describe a recent generalization by Fischer and Wolf (1997) of the fundamental theorem of diffraction tomography, discussed earlier in this chapter, to a class of random media which are frequently encountered in practice.

When the medium is random, the scattering potential $F(r, \omega)$ and, consequently, the scattered field $U^{(S)}(r, \omega)$ are, for each frequency ω, random functions of position. It then follows from Eq. (30) that

$$\langle \tilde{F}^*(K_1^\pm, \omega)\tilde{F}(K_2^\pm, \omega)\rangle = \frac{1}{(2\pi)^2} f_{1z}f_{2z}\langle \tilde{U}^{(s)*}(f_{1\perp}; z^\pm; \omega)\tilde{U}^{(s)}(f_{2\perp}; z^\pm; \omega)\rangle e^{\mp i(f_{2z}-f_{1z})z^\pm}.$$

$$(32)$$

If we express $\tilde{F}(K, \omega)$ and $\tilde{U}^{(S)}(f_\perp; z; \omega)$ in terms of $F(r, \omega)$ and $U^{(S)}(\rho; z; \omega)$ respectively by the use of the Fourier representations (9) and (26) we find from Eq. (32) that

$$\tilde{C}_F(-K_1^\pm, K_2^\pm; \omega) = \frac{1}{(2\pi)^2} f_{1z}f_{2z}\tilde{W}^{(s)}(-f_{1\perp}, f_{2\perp}; z^\pm; \omega)e^{\mp i(f_{2z}-f_{1z})z^\pm}, \qquad (33)$$

where $\tilde{C}_F$ is the six-dimensional Fourier transform of the two-point spatial correlation function

$$C_F(\mathbf{r}_1, \mathbf{r}_2, \omega) = \langle F^*(\mathbf{r}_1, \omega) F(\mathbf{r}_2, \omega) \rangle \tag{34}$$

of the scatterer and $\tilde{W}^{(S)}$ is the four-dimensional Fourier transform of the cross-spectral density function

$$W^{(s)}(\boldsymbol{\rho}_1, \boldsymbol{\rho}_2; z^\pm; \omega) = \langle U^{(s)*}(\boldsymbol{\rho}_1; z^\pm; \omega) U^{(s)}(\boldsymbol{\rho}_2; z^\pm; \omega) \rangle \tag{35}$$

of the scattered field in the planes $z = z^\pm$. More explicitly

$$\tilde{C}_F(\mathbf{K}_1, \mathbf{K}_2, \omega) = \int_{\mathfrak{R}} \int_{\mathfrak{R}} C_F(\mathbf{r}_1, \mathbf{r}_2, \omega) e^{-i(\mathbf{K}_1 \cdot \mathbf{r}_1 + \mathbf{K}_2 \cdot \mathbf{r}_2)} \, d^3 r_1 \, d^3 r_2 \tag{36}$$

and

$$\tilde{W}^{(s)}(\mathbf{f}_{1\perp}, \mathbf{f}_{2\perp}; z^\pm; \omega) = \int_{z=z^\pm} \int_{z=z^\pm} W^{(s)}(\boldsymbol{\rho}_1, \boldsymbol{\rho}_2; z^\pm; \omega) e^{-i(\mathbf{f}_{1\perp} \cdot \boldsymbol{\rho}_1 + \mathbf{f}_{2\perp} \cdot \boldsymbol{\rho}_2)} \, d^2 \rho_1 \, d^2 \rho_2. \tag{37}$$

The formula (33) expresses some of the six-dimensional Fourier components of the (generally unknown) two-point spatial correlation function C_F of the scattering potential in terms of the four-dimensional Fourier components of the cross-spectral density $W^{(s)}$ of the scattered field in two arbitrary planes $z = z^+$ and $z = z^-$, one on each side of the scatterer. However, one can show that, in general, it is not possible to reconstruct the correlation function of the scattering potential from measurements of the cross-spectral density of the scattered field, not even its low spatial-frequency part. An exception is the reconstruction of scattering media which belong to an important class, the class of so-called quasi-homogeneous random media (Carter and Wolf 1988).

The two-point spatial correlation function of a quasi-homogeneous medium has the form

$$C_F(\mathbf{r}_1, \mathbf{r}_2; \omega) = I_F\left(\frac{\mathbf{r}_1 + \mathbf{r}_2}{2}, \omega\right) g_F(\mathbf{r}_2 - \mathbf{r}_1, \omega), \tag{38}$$

where

$$I_F(\mathbf{r}, \omega) = \langle F^*(\mathbf{r}, \omega) F(\mathbf{r}, \omega) \rangle \tag{39}$$

and

$$g_F(\mathbf{r}_2 - \mathbf{r}_1, \omega) \equiv \frac{C_F(\mathbf{r}_1, \mathbf{r}_2, \omega)}{\sqrt{I_F(\mathbf{r}_1, \omega)} \sqrt{I_F(\mathbf{r}_2, \omega)}}. \tag{40}$$

Moreover, for such media the second moment $I_F(\mathbf{r}, \omega)$ of the scattering potential varies much more slowly with $\mathbf{r}$ than the degree of spatial correlation $g_F(\mathbf{r}', \omega)$ of the scatterer varies with $\mathbf{r}' = \mathbf{r}_2 - \mathbf{r}_1$.

On taking the six-dimensional Fourier transform of Eq. (38), we readily find that

$$\tilde{C}_{\mathrm{F}}(\boldsymbol{K}_1, \boldsymbol{K}_2, \omega) = \tilde{I}_{\mathrm{F}}(\boldsymbol{K}_1 + \boldsymbol{K}_2, \omega)\tilde{g}_{\mathrm{F}}\!\left(\frac{\boldsymbol{K}_2 - \boldsymbol{K}_1}{2}, \omega\right), \tag{41}$$

where $\tilde{I}_{\mathrm{F}}$ and $\tilde{g}_{\mathrm{F}}$ are the three-dimensional spatial Fourier transforms [defined by formulas of the form (2.9)] of $I_{\mathrm{F}}(r, \omega)$ and $g_{\mathrm{F}}(r', \omega)$ respectively. On substituting from Eq. (41) into Eq. (33) and on setting

$$\boldsymbol{K}_2^{\pm} = \boldsymbol{K}_1^{\pm} = \boldsymbol{K}^{\pm}, \tag{42}$$

Eq. (33) gives

$$\tilde{g}_{\mathrm{F}}(\boldsymbol{K}^{\pm}; \omega) = \frac{1}{(2\pi)^2 \tilde{I}_{\mathrm{F}}(0, \omega)} f_z^2 \tilde{W}^{(\mathrm{s})}(-f_\perp, f_\perp; z^{\pm}; \omega), \tag{43}$$

where, as before, $\boldsymbol{K}^{\pm}$ are the vectors defined by Eqs. (28) and (29).

The formula (43) shows that all of the low spatial-frequency components of the degree of spatial correlation $g_{\mathrm{F}}(r', \omega)$ of the scattering potential of a quasi-homogeneous medium may be determined, up to a constant proportionality factor $\tilde{I}_{\mathrm{F}}(0, \omega)$, from measurements of the cross-spectral density function of the scattered field in two planes $z = z^+$ and $z = z^-$, one on each side of the scatterer. The proportionality factor $\tilde{I}_{\mathrm{F}}(0, \omega)$ can be determined by imposing the normalization condition $g_{\mathrm{F}}(0, \omega) = 1$.

RECONSTRUCTION FROM INCOMPLETE DATA AND SUPERRESOLUTION

In the theoretical analysis described earlier it was assumed that sufficient measured data is available to calculate all the Fourier components $\tilde{F}(\boldsymbol{K}, \omega)$ of the scattering potential $F(r, \omega)$ which lie in the interior of the Ewald limiting sphere, i.e. the sphere centred at the origin in $\boldsymbol{K}$-space of radius $|\boldsymbol{K}| = 2k = 4\pi/\lambda$. This, however, is never the situation in practice and, often, all the Fourier components cannot be determined, even in principle. For example in the so-called well-to-well tomography employed in geophysical remote sensing, sources of acoustical or electromagnetic waves are placed in a borehole and the waves are detected by receivers placed in an adjacent borehole, after the waves have passed through the intervening geological formation. In another configuration, known as offset vertical seismic profiling, a transmitter array is located along a line on the Earth's surface and the receivers are located in a borehole or vice versa (see Fig. 10). The limited regions of the Fourier space of the scattering potential accessible from such measurements were discussed by Devaney (1984) and are depicted in Figs. 11 and 12.

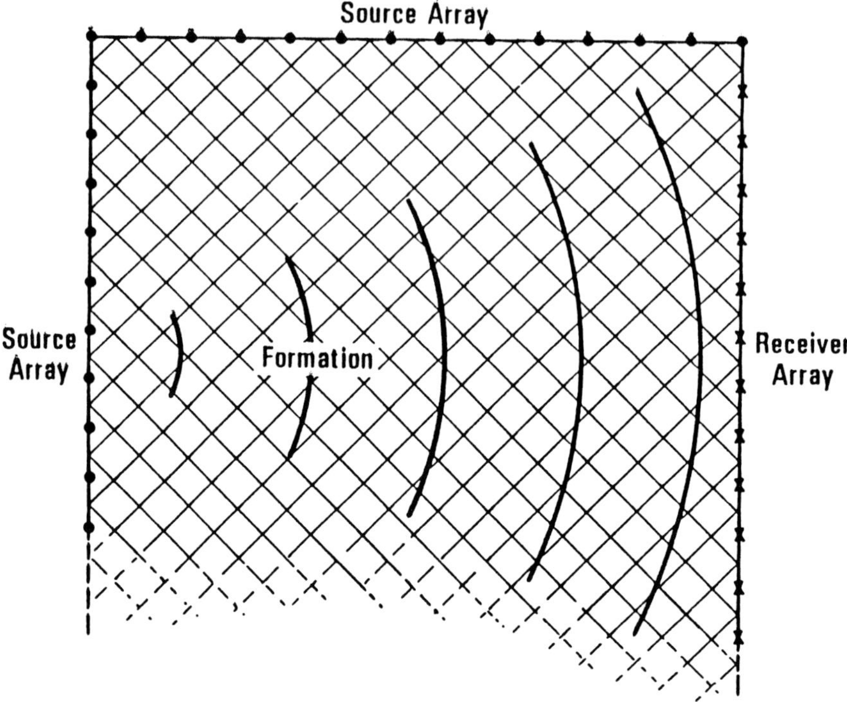

Fig. 10. Geometry for well-to-well tomography and for offset vertical seismic profiling (after Devaney 1984).

The preceding remarks indicate only one of the problems encountered in practical implementation of diffraction tomography. Others arise, for example, from the fact that only a discrete set of data and not a continuous one can be obtained in realizable experiments. Numerous publications are available which deal with questions such as the optimum utilization of the available data in the reconstruction process, and the best choice of sampling, effects of noise, etc. We will not review these topics in the present chapter, in spite of their practical importance. We will only present a theoretical result derived not long ago, concerning the information about the scatterer that can be obtained *in principle* from accurate knowledge of a continuous set of Fourier components whose representative K-vectors fill only a portion of the interior of the Ewald limiting sphere. More precisely we will briefly discuss the following theorem established by Habashy and Wolf (1994):

Suppose that the Fourier transform $\tilde{F}(K, \omega)$ [defined by Eq. (9)], of a scattering potential $F(r, \omega)$ is only known throughout some finite three-dimensional region $\mathcal{K}$ in K-space. If the scatterer occupies a finite three-dimensional domain $\mathcal{R}$ of r-space (i.e. if $F(r, \omega) = 0$) when $r \notin \mathcal{R}$ — see

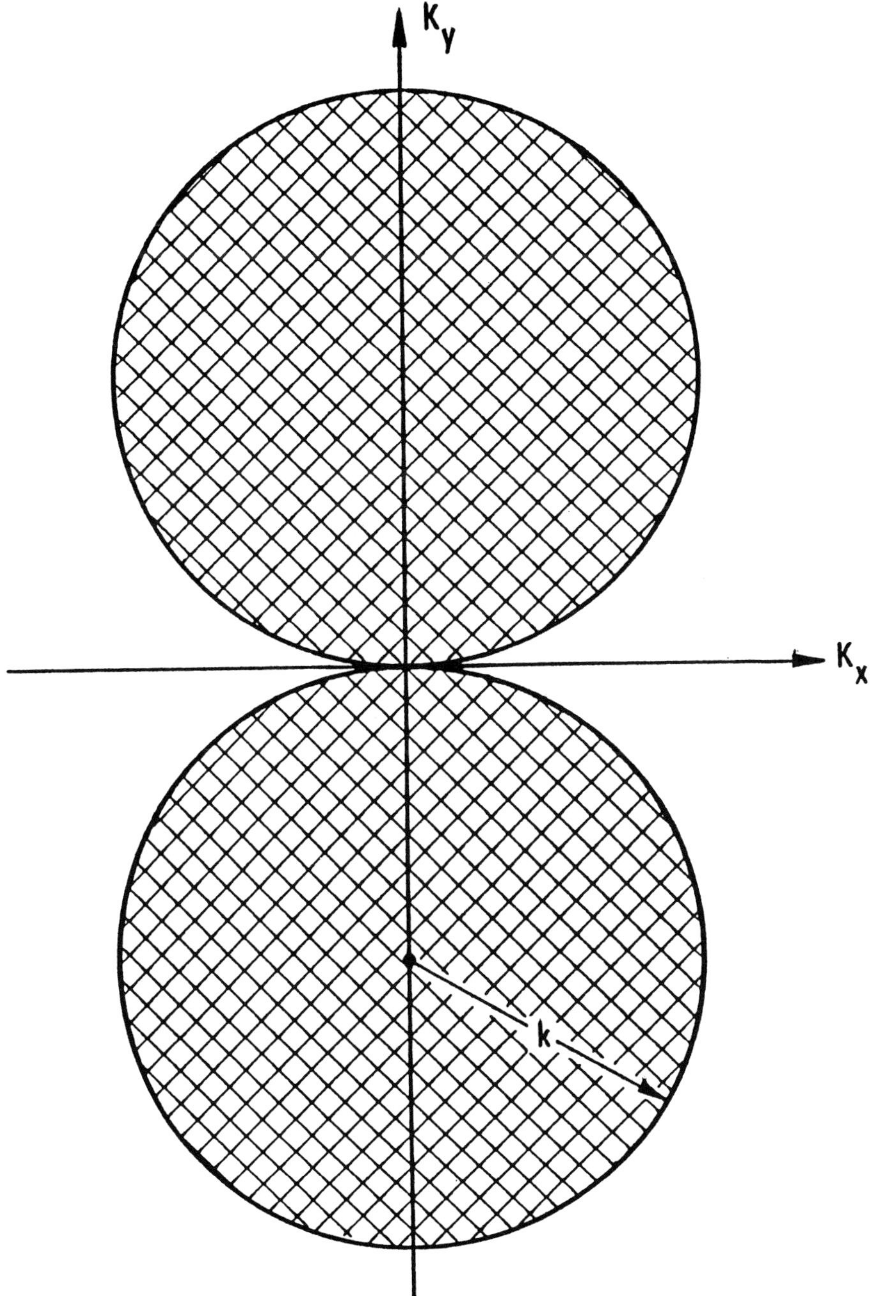

Fig. 11. Region of Fourier space throughout which the Fourier components of the scattering potential can be determined from a full set of well-to-well tomographic experiments (after Devaney 1984).

Fig. 13) and if the potential is square-integrable, then $F(r, \omega)$ may be determined uniquely from the known 'piece' of the Fourier transform $\tilde{F}(K, \omega)$ via the integral equation[3]

$$F_{\mathcal{H}}(r, \omega) = \int_{\mathcal{R}} F(r', \omega) C_{\mathcal{H}}(r - r')\, \mathrm{d}^3 r',$$
(44)

where

$$F_{\mathcal{H}}(r, \omega) = \int_{\mathcal{H}} \tilde{F}(K, \omega) e^{iK \cdot r}\, \mathrm{d}^3 K$$
(45)

and

$$C_{\mathcal{H}}(r) = \int_{\mathcal{H}} e^{iK \cdot r}\, \mathrm{d}^3 K.$$
(46)

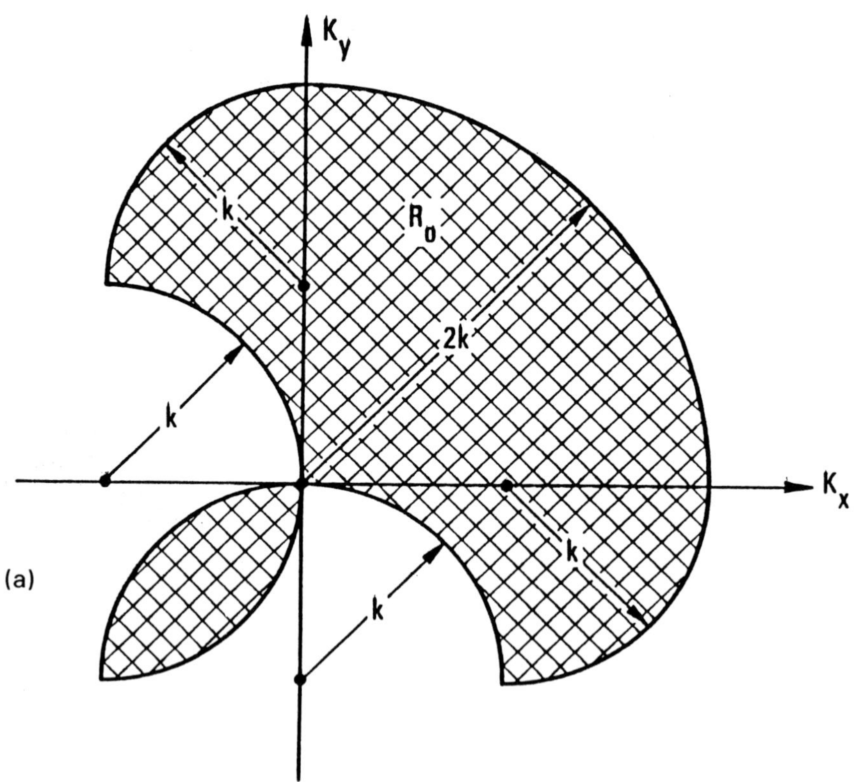

Fig. 12. (a) *Continued*

[3] A one-dimensional version of the integral equation (44) was obtained a long time ago by Toraldo di Francia (1969), in connection with determining the number of degrees of freedom of an optical image.

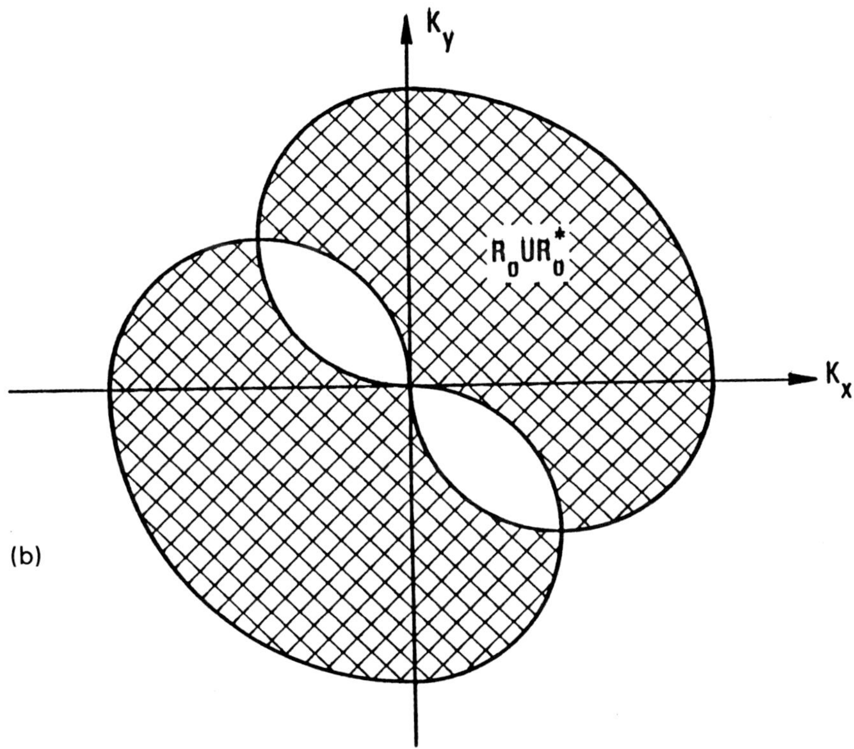

Fig. 12. Regions of Fourier space throughout which the Fourier components of the scattering potential can be determined from full sets of offset vertical seismic profiling experiments. (a) The region R_0 is obtained when the potential is complex, (b) the region $R_0 UR_0^*$ is obtained when the potential is real (after Devaney 1984).

This result seems to be rather surprising since it makes it possible, in principle, to determine the scattering potential $F(r, \omega)$ from the knowledge of some but not all its Fourier components. The theorem is rooted in a fundamental analytic property of Fourier transforms of functions of finite support (Fuks 1963) whose one-dimensional version is known as the Polya–Plancherel theorem.

The integral equation (44) may be solved by standard eigenfunction techniques that are briefly discussed in the paper by Habashy and Wolf (1944). The equation is closely related to an integral equation, introduced by Slepian and Pollak (1961) (see also Slepian 1964), in the theory of extrapolation of band-limited functions.

It is of interest to note that the integral equation (44) makes it possible, in principle, to extrapolate the low-pass filtered approximation to a scattering potential (based only on data contained within the interior of the Ewald

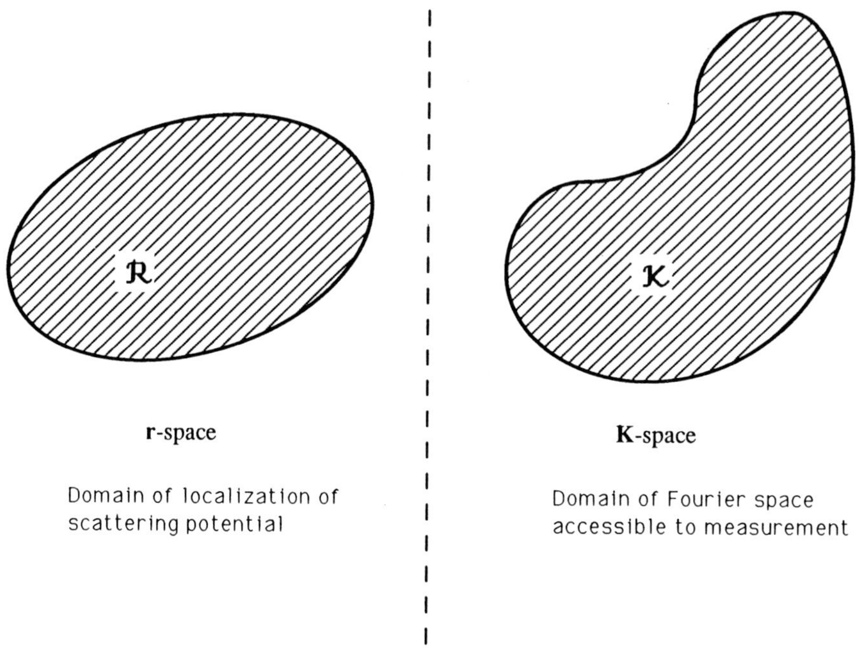

Fig. 13. Illustrating the notation relating to the integral equation (44), which makes it possible to determine the scattering potential $F(r, \omega)$ of a medium occupying a finite domain $\mathcal{R}$ of *r*-space from a 'piece' of its Fourier transform, i.e., from its Fourier components $\tilde{F}(K, \omega)$ known only throughout some finite region $\mathcal{K}$ of *K*-space.

limiting sphere) to obtain a better approximation which contains Fourier components associated with the exterior of that sphere. Consequently the integral equation (44) could be used to reconstruct details of a scattering potential beyond the usual resolution limit. However, this technique turns out to be useful only for very small scatterers, with linear dimensions of the order of, or smaller than, the wavelength of the incident radiation, and the available Fourier components must be known very accurately.

An example of a computational reconstruction of a model scattering potential beyond the traditional resolution limit is presented in Fig. 14.

In addition to the method just described there have been other methods developed to achieve superresolution. Among them is a method developed by Ramm (1985, Sec. 4.1), which is based on analytic continuation of the available Fourier components of the scattering potential and a method by Schatzberg and Devaney (1992) which directly incorporates the measured evanescent components of the scattered field in the reconstruction process.

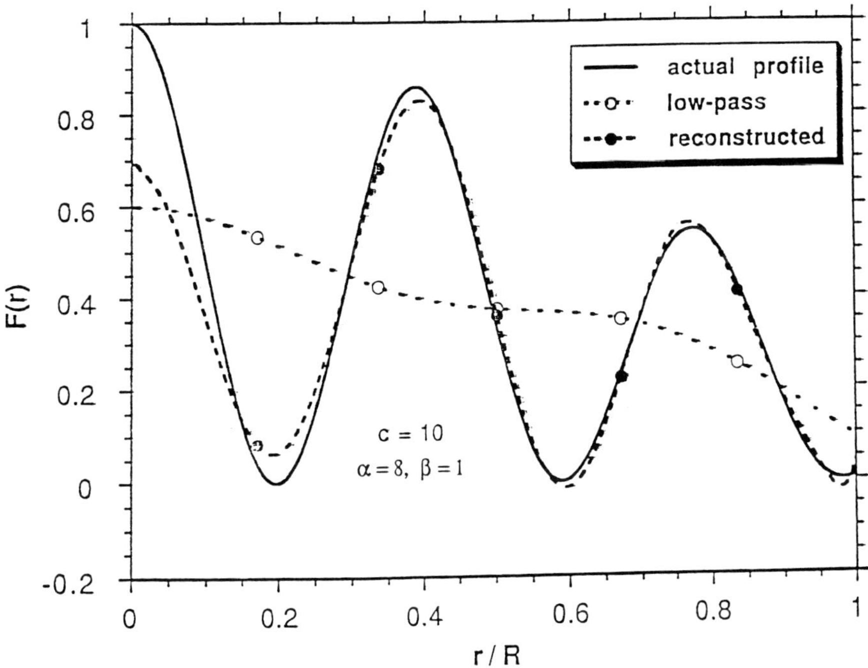

Fig. 14. Computer simulation of the reconstruction of the radially symmetric scattering potential

$$F(r) = \begin{cases} \cos^2(\alpha r/R)\exp(-\beta r^2/R^2) & \text{when } 0 \leqslant r \leqslant R \\ 0 & \text{when } r > R \end{cases}$$

from its low-pass filtered version, for selected values of the parameters α, β and $c = 2kR = 4\pi(R/\lambda)$ (after Habashy and Wolf 1994).

ACKNOWLEDGMENTS

I am much obliged to Dr. A. J. Devaney and to Mr. D. G. Fischer for careful reading of the manuscript of this article and for many useful suggestions.

Some of the research described in this chapter was supported by the US Department of Energy, the National Science Foundation, the New York State Science and Technology Foundation and the Army Research Office under the University Research Initiative Program.

REFERENCES

Born, M. and Wolf, E. (1980), *Principles of Optics* (Pergamon Press, Oxford), 6th ed.

Carter, W.H. (1970), 'Computational reconstruction of scattering objects from holograms', *J. Opt. Soc. Amer.* **60**, 306–314.

Carter, W.H. and Ho, Pin-Chin (1974), 'Reconstruction of inhomogeneous scattering objects from holograms', *Appl. Opt.* **13**, 162–172.

Carter, W.H. and Wolf, E. (1988), 'Scattering from quasi-homogeneous media', *Opt. Commun.* **67**, 85–90.

Dändliker, R. and Weiss, K. (1970), 'Reconstruction of the three-dimensional refractive index from scattered waves', *Opt. Commun.* **1**, 323–328.

Devaney, A.J. (1982), 'A filtered back propagation algorithm for diffraction tomography', *Ultrasonic Imag.* **4**, 336–350.

Devaney, A.J. (1983), 'A computer simulation study of diffraction tomography', *IEEE Trans. on Biomed. Eng.*, **BME-30**, 377–386.

Devaney, A.J. (1984), 'Geophysical diffraction tomography', *IEEE Trans. on Geosc. and Remote Sensing*, **GE-22**, 3–13.

Devaney, A.J. (1989), 'Structure determination from intensity measurements in scattering experiments', *Phys. Rev. Lett.* **62**, 2385–2388.

Devaney, A.J. (1991), 'Inverse scattering and diffraction tomography using intensity data;', in *Acoustical Imaging*, Vol. 18, ed. H. Lee and G. Wade (Plenum Press, New York), p. 97–103.

Devaney, A.J. (1992), 'Diffraction tomographic reconstruction from intensity data', *IEEE Trans. on Image Processing*, **1**, 221–228.

Evans, E. (1970), 'Comparison of the diffraction theory of image formation with the three-dimensional, first Born scattering approximation in lens systems', *Opt. Commun.* **2**, 317–320.

Fercher, A.F., Bartelt, H., Becker, H. and Wiltschko, E. (1979), 'Image formation by inversion of scattered field data: experiments and computational simulation', *Appl. Opt.* **18**, 2427–2439.

Fercher, A.F., Hitzenberger, C.K., Kamp, G. and El-Zaiat, S.Y. (1995), 'Measurement of intraocular distances by backscattering spectral interferometry', *Opt. Commun.* **117**, 43–48.

Fischer, D.G. and Wolf, E. (1996), 'Theory of diffraction tomography with quasi-homogeneous media', submitted to *Opt. Commun.* A preliminary account of this work was presented at the Annual Meeting of the Optical Society of America in Portland, OR, on Sept. 14, 1995, Abstract ThEE6.

Fuks, B.A. (1963), *Introduction to the Theory of Analytic Functions of Several Complex Variables* (Providence, RI: American Mathematical Society), p. 352.

Gordon, R., Herman, G.T. and Johnson, S.A. (1975), 'Image reconstruction from projections', *Sci. Amer.* **233**, 56–68.

Habashy, T. and Wolf, E. (1994), 'Reconstruction of scattering potentials from incomplete data', *J. Mod. Opt.* **41**, 1679–1685.

Ho, Pin Chin and Carter, W. H. (1976), 'Structural measurement by inverse scattering in the first Born approximation', *Appl. Opt.* **15**, 313–314.

Howard, J. (1991), 'Laser probing of random weakly scattering media', *J. Opt. Soc. Amer.* **A8**, 1955–1963.

Ishimaru, A. (1978), *Wave Propagation and Scattering in Random Media*, Vol. 2 (Academic Press, New York), Sec. 17–2.

Iwata, K. and Nagata, R. (1975), 'Calculation of refractive index distribution from interferograms using the Born and Rytov approximations', *Jap. J. Appl. Phys.* **14**, Suppl. 14–1, 379–383.

James, R.W. (1948), *The Optical Principles of the Diffraction of X-rays* (Bell and Sons, London), 14–15.

Kak, A. C. and Slaney, M. (1988), *Principles of Computerized Tomographic Imaging* (IEEE Press, New York).

Kaveh, M., Soumekh, M. and Mueller, R. K. (1980), 'Experimental results in ultrasonic diffraction tomography' *Acoustical Imaging* 9, 433–450.

Kaveh, M., Mueller, R. K., Rylander, R., Coulter, T. R. and Soumekh, M. (1980), 'Experimental results in ultrasonic diffraction tomography', in *Industrial Symposium on Acoustical Imaging* (Plenum, New York), 9, 433–450.

Kenue, S.K. and Greenleaf, J. (1982), 'Limited angle multi-frequency diffraction tomography', *IEEE Trans. on Sonics and Ultrasonics*, **SU-29**, 213–217.

King, W.C., Witten, A.J. and Reed, G.D. (1989), 'Detection and imaging of buried wastes using seismic wave propagation', *J. Env. Eng.* **115**, 527–540.

Kunitsyn, V.E. and Tereshchenko, E.D. (1992), 'Radio tomography of the ionosphere', *IEEE Antennas and Propagation Magazine*, **34**, 22–32.

Ladas, K.T. and Devaney, A.J. (1993), 'Application of ART algorithm in an experimental study of ultrasonic diffraction tomography', *Ultrasonic Imaging*, **15**, 48–58.

Lo, T., Toksöz, M. N., Xu, S. and Wu, R. (1988), 'Ultrasonic laboratory tests of geophysical tomographic reconstruction', *Geophysics*, **53**, 947–956.

Maleki, M.H., Devaney, A.J. and Schatzberg, A. (1992), 'Tomographic reconstruction from optical scattered intensities', *J. Opt. Soc. Amer.* **A9**, 1356–1363.

Mandel, L. and Wolf, E. (1995), *Optical Coherence and Quantum Optics* (Cambridge University Press, Cambridge and New York).

Mueller, R.K., Kaveh, M. and Iverson, R.D. (1978), 'A new approach to acoustic tomography using diffraction techniques', in *Acoustical Imaging*, Vol. 8, ed. A. Meterell (Plenum Press, New York), 615–628.

Mueller, R.K., Kaveh, M. and Wade, G. (1979), 'Reconstructive tomography and applications to ultrasonics', *Proc. IEEE* **67**, 567–587.

Pan, S.X. and Kak, A.C. (1983), 'A computational study of reconstruction algorithms for diffraction tomography: interpolation versus filtered back propagation', *IEEE Trans. on Acoustic, Speech and Signal Processing*, **AASP-31**, 1262–1275.

Porter, R.P. (1989), 'Generalized holography with application to inverse scattering and inverse source problems', in *Progress in Optics*, Vol. 27, ed. E. Wolf (North-Holland, Amsterdam), 315–397.

Ramm, A.G. (1985), 'Some inverse scattering problems of geophysics', *Inverse Problems*, **1**, 133–172.

Roman, P. (1965), *Advanced Quantum Theory* (Addison-Wesley, New York).

Rouseff, D. and Porter, R.P. (1991), 'Diffraction tomography and the stochastic inverse scattering problem', *J. Acoust. Soc. Amer.* **89**, 1599–1605.

Schatzberg, A. and Devaney, A.J. (1992), 'Super-resolution in diffraction tomography', *Inverse Problems*, **8**, 149–164.

Slepian, D. (1964), 'Prolate spheroidal wave functions, Fourier analysis and uncertainty – IV: Extension to many dimensions; generalized prolate spheroidal functions', *Bell Syst. Tech. J.* **43**, 3009–3057.

Slepian, D. and Pollak, H.O. (1961), 'Prolate spheroidal wave functions, Fourier analysis and uncertainty – I', *Bell Syst. Tech. J.* **40**, 43–63.

Sommerfeld, A. (1954), *Optics* (Academic Press, New York, 1954), p. 187.

Sponheim, N. and Johansen, I. (1991), 'Experimental results in ultrasonic tomography using a filtered back propagation algorithm', *Ultrasonic Imaging*, **13**, 56–70.

Sponheim, N., Johansen, I. and Devaney, A.J. (1991), 'Initial testing of a clinical ultrasound monograph', in *Acoustical Imaging*, Vol. 18, ed. H. Lee and G. Wade, 401–411.

Stamnes, J.J., Gelius, L.-J., Johansen, I. and Sponheim, N. (1992), 'Diffraction tomography applications in seismics and medicine' in *Inverse Problems in Scattering and Imaging*, ed. M. Bertero and E. R. Pike (Adam Hilger, Bristol), p. 268–292.

Stamnes, J.J. and Wedberg, T.C. (1995), 'Recent advances of diffraction tomography in geophysics, ultrasonics, and optics', *Particle and particle system characterization.* **12**, 95–104.

Swindell, W. and Barrett, H.H., (1977), 'Computerized tomography: taking sectional x-rays', *Phys. Today*, Dec. 1977, 32–41.

Toraldo di Francia, G. (1969), 'Degrees of freedom of an image', *J. Opt. Soc. Amer.* **59**, 799–804.

Wedberg, T.C. and Stamnes, J.J. (1995a), 'Comparison of phase retrieval methods for optical diffraction tomography', *Pure Appl. Opt.* **4**, 39–54.

Wedberg. T.C. and Stamnes, J.J. (1995b), 'Experimental examination of the quantitative imaging properties of optical diffraction tomography', *J. Opt. Soc. Amer.* **A12**, 493–500.

Wedberg, T.C. and Wedberg, W.C. (1995), 'Tomographic reconstruction of the cross-sectional refractive index distribution in semi-transparent, birifringent fibres', *J. Microscopy*, **177**, Part 1, 53–67.

Weyl, H. (1919), 'Ausbraitung elektromagnetischer Wellen über einem ebenen Leiter', *An. Physik*, **60**, 481–500.

Witten, A.J. and King, W.C. (1990a), 'Sounding of buried waste', *Civl. Eng.* **60**, 62–64.

Witten, A.J. and King, W.C. (1990b), 'Acoustic imaging of subsurface features', *J. Env. Eng.* **116**, 166–181.

Witten, A.J. and King, W.C. (1990c), 'Geophysical imaging with backpropagation and zeroth-order phase approximation', *Geophys. Res. Lett.* **17**, 673–676.

Wolf, E. (1969), 'Three-dimensional structure determination of semi-transparent objects from holographic data', *Opt. Commun.* **1**, 153–156.

Wolf, E. (1970), 'Determination of amplitude and phase of scattered fields by holography', *J. Opt. Soc. Amer.* **60**, 18–20.

# Diffractive optics: from promise to fruition

Jari Turunen
*Department of Physics, University of Joensuu, P.O. Box 111,
FIN-80101 Joensuu, Finland*

Frank Wyrowski
*Department of Diffractive Optics and Hybrid Systems,
Berlin Institute of Optics, Rudower Chaussee 5, D-12484 Berlin,
Germany*

INTRODUCTION

It has been understood for approximately three decades that a sophisticated utilization of diffraction by microstructured media can open up new avenues in optical research and technology. However, only in the past few years have the theoretical, numerical, and manufacturing techniques reached such a high level that this approach, known today as diffractive optics, has begun to fulfil its considerable promise on a broad front. In the spirit of the present volume, we aim to expand the understanding of the nature and the applicability of diffractive optics. We first describe its relation to established optical technology and some of its fundamental principles, then cover the state of the art in the synthesis of diffractive microstructures with the aid of some representative examples, and finally attempt to foresee some of the future directions in the research and application of diffractive optics.

DIFFRACTIVE OPTICS AS AN EXTENSION OF CLASSICAL OPTICS

The wide spectrum of methods to manipulate optical fields by material bodies may be divided into two main categories as illustrated in Fig. 1: (i) methods

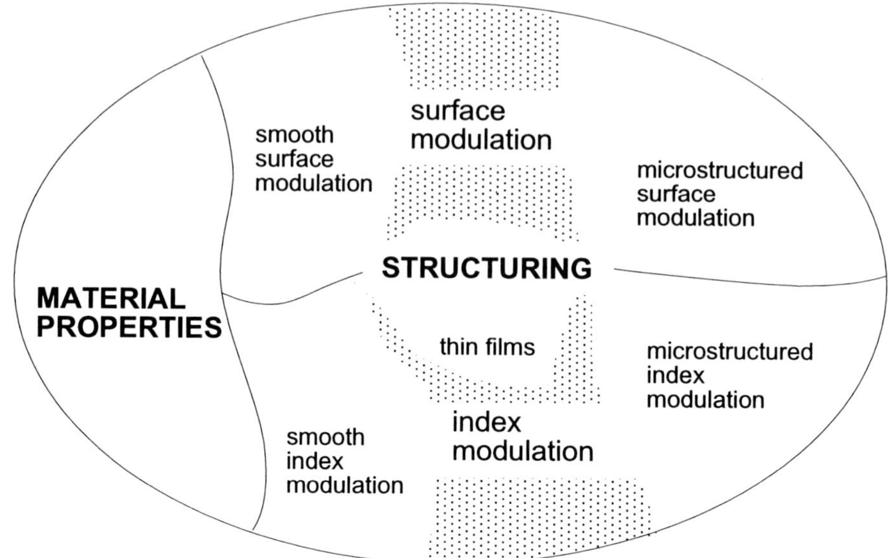

Fig. 1. Classification of methods to realize optical functions by interaction of light and matter.

that employ some inherent property of bulk matter, such as anisotropy or non-linearity, and (ii) methods based on structured media that may be generated by a variety of mechanical or photochemical processes such as grinding, polishing, diamond-turning, exposure and development, selective etching, and material deposition. The category of structured matter, which is our prime concern, may be further divided into the subcategories of surface-modulated and index-modulated media. In both cases the modulation can be coarse in units of the optical wavelength λ (spherical and gradient-index lenses can serve as examples) or the matter may be microstructured; ruled spectroscopic gratings and volume gratings are fine examples. The dashed area in the figure indicates that the transition between microstructured and macrostructured media is not a sharp one.

The local plane wave approximation, i.e., geometrical optics, is usually applicable to the analysis and synthesis of optical elements and systems that employ smoothly structured media. In contrast, the use of diffraction (or scattering) theory is often unavoidable in the case of microstructured media. It is, in fact, well known in grating theory [1] that a rigorous solution of Maxwell's equations is required whenever the grating period is comparable to the optical wavelength λ. With reference to Fig. 1, we may therefore characterize diffractive optics as that part of optical technology which relies on microstructured media. In diffractive optics the phenomenon of diffraction, traditionally seen as a limitation to the performance of optical instruments, is employed to advantage. Since the use of diffraction by microstructured media clearly adds to the choice of methods to realize optical functions, diffractive optics is an obvious extension of classical optical technology.

Diffractive elements, i.e., microstructured optical elements based on the use of diffraction, can substitute traditional refractive or reflective elements in a number of applications as illustrated in Fig. 2. Diffraction gratings are superior to prisms in spectroscopy and indeed remain the most important single application of diffractive optics, at least if the importance is measured in commercial terms [2]. Another well-known example of diffractive elements is the Fresnel zone plate, which is capable of focusing light. In practice, however, it is better to use diffractive surface-relief elements for this purpose [3], because then the absorption of incident energy is avoided. Such elements can be manufactured, e.g., by microlithographic methods adapted from the fabrication of integrated circuits. Alternatively, diffractive lenses can be fabricated in the form of index-modulated microstructures by means of off-axis holography [4], which has indeed had a tremendous impact on the general development of diffractive optics.

The importance of gratings in diffractive optics is enhanced by the fact that many important diffractive structures can be treated as gratings at least locally. For example, any diffractive lens may be viewed as a grating with a smoothly

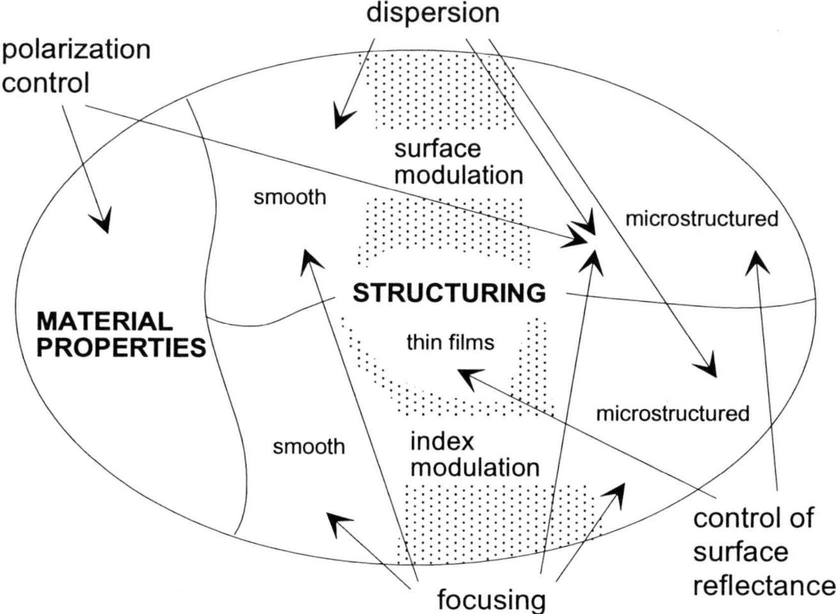

Fig. 2. Examples of optical functions that may be performed either by diffractive or conventional optical elements.

modulated period. Also subwavelength-period gratings, which do not diffract light in the classical sense but can nevertheless significantly influence the incident field, are rapidly gaining importance. Such gratings can act as antireflection layers [5], enhance the reflectance over a narrow wavelength band [6], or modulate the state of polarization of the incident field [7]. Therefore diffractive microstructured media can replace thin films and elements based on bulk matter, such as wave plates. In many cases it is useful to combine diffractive and refractive elements. For example, one can create a single-glass achromatic lens by fabricating a weak positive diffractive lens on the flat back surface of a refractive singlet [8].

One of the unique strengths of diffractive optics is that it can do more than just replace traditional elements or supplement them in a hybrid system. This aspect is largely due to the work of Lohmann [9], who introduced a revolutionary mathematical synthesis scheme of diffractive elements. Diffractive optics can now provide practical solutions to optical instrumentation problems in which conventional methods are unattractive because of weight, size, cost, or sheer complexity. A number of such examples can be readily found in spatial filtering, beam profile shaping [10], advanced resonator design [11], optical null testing [12], materials processing [13], and in many other applications. For example, it is not easy to construct a large two-dimensional array of refractive microlenses with no dead space between the elements, but with diffractive optics this is achieved as easily as the fabrication of a single microlens. In addition, each individual lens element can have its own focal length, aspheric coefficients, and optical axis.

FUNDAMENTALS OF DIFFRACTIVE OPTICS

To fully appreciate the flexibility of the mathematical synthesis scheme of diffractive elements it is beneficial to consider the task of an optical element on a somewhat abstract level as illustrated in Fig. 3. Let us assume that an electromagnetic wave of rather arbitrary form, described for simplicity by a scalar function U_i, illuminates an optical element that resides between the planes $z = -h$ and $z = 0$. The interaction of U_i with the modulated structure generates new fields U_d in the half-spaces $z > 0$ and $z < -h$. The field U_d should satisfy the specified optical function formally defined by means of a signal function s inside some signal window W, which may be a spatial point, a section of a plane, a bounded volume, or a spatial-frequency range. In a more general configuration there may, e.g., be several independent incident waves U_i, with s and W defined separately for each one of them.

We can formulate a synthesis problem in physical optics with the help of Fig. 3: given the incident field $U_i(x, y, -h)$, find a modulated structure in the region $-h < z < 0$ such that the field U_d satisfies the signal function s inside W.

It is a prerequisite for the solution of the synthesis problem that we are able

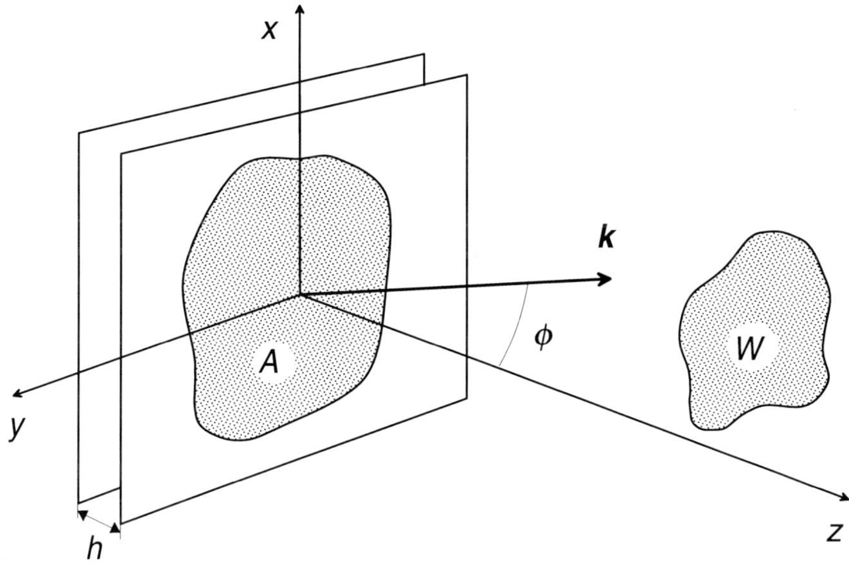

Fig. 3. A basic optical configuration for the transformation of an incident field U_i into a field U_d that satisfies a specified signal function s inside the signal window W. The transmission mode of operation is shown, and k represents the wavevector of one plane-wave components of U_d.

to determine U_d in W from the knowledge of U_i and the modulated structure. This problem may be divided into two parts: (i) prediction of U_d across the exit plane $z = 0$ (or $z = -h$) of the element, formally described by a transmission operator $\mathcal{T}$ (or reflection operator $\mathcal{R}$), and (ii) propagation of U_d into W, formally described by an operator $\mathcal{P}$. Very often W is a section of a plane $z = z_s$.

The propagation operator $\mathcal{P}$ may be expressed exactly by means of the angular spectrum representation of the electromagnetic field, which reduces to the approximate Fresnel and Fraunhoher propagation integrals when all plane waves of interest in U_i and U_d are paraxial ($\sin \theta \approx \phi$ in Fig. 3). The choice of the mathematical description for $\mathcal{T}$ (and $\mathcal{R}$) depends critically on the characteristic dimensions of the microstructure. In general, one has to find an exact solution of Maxwell's equations and the electromagnetic boundary conditions. In many cases, however, the transverse dimensions of all features are large compared to λ. We may then write a pointwise expression for $\mathcal{T}$ in the form $U_d(x, y, 0) = \mathcal{T} U_i(x, y, -h) = t(x, y)U_i(x, y, -h)$, where $t(x, y) = U_d(x, y, 0)/U_i(x, y, -h) = A(x, y)\exp[i\phi(x, y)]$ is the complex-amplitude transmission function associated with the element. The real functions $A(x, y)$ and $\phi(x, y)$ denote the amplitude transmittance and the phase delay, respectively.

The primary goal in the solution of the synthesis problem by any method is to approximate s in W as closely as necessary in view of the tolerances set by the application, with regard to an appropriately chosen merit function such as the signal-to-noise ratio. Typically it is not possible to realize s in exactly the specified form, because U_d must be an analytic function that satisfies the wave equation in free space. Hence, for example, the intensity $|U_d(x, y, z_s)|^2$ cannot be completely uniform inside a finite section W of a plane $z = z_s$ and simultaneously vanish everywhere outside W.

Let us denote by $U_s(x, y, z_s)$ an analytic field that satisfies the signal function s in W and call it the signal wave. We may then apply $\mathcal{P}^{-1}$ to determine $U_s(x, y, 0)$. In view of the complex-amplitude transmittance approach, we now need to fabricate an element with

$$t(x, y) = \alpha\,\frac{U_s(x, y, 0)}{U_i(x, y, -h)}, \tag{1}$$

where the constant α is chosen such that $|t(x, y)| \leqslant 1$ for all (x, y). The diffracted field would then be proportional to the signal wave, i.e., $U_d(x, y, 0) = \alpha U_s(x, y, 0)$, and an attenuated but otherwise perfect reconstruction of s is obtained in W: $U_d(x, y, z_s) = \alpha U_s(x, y, z_s)$.

The class of signals s that can be generated without error is limited by the finite size of the element (or the profile of U_i), which sets the resolution limit in W. It is therefore imperative to choose $U_s(x, y, z_s)$ in such a manner that $U_s(x, y, 0)$ is at least approximately confined within the aperture. In many instances only the intensity is fixed by s, which implies that we have a certain degree of freedom in the choice of the phase of $U_s(x, y, z_s)$, known as phase freedom. Should we choose a constant phase, $U_s(x, y, 0)$ will exhibit a high intensity near the optical axis and weak off-axis components, which nevertheless contain important information about $U_s(x, y, z_s)$. This situation is well known in holography, being highly unsatisfactory since a great deal of attenuation (a small value of α) is needed to obtain a linear recording of the important off-axis information. Diffuse illumination alleviates the problem and suggests the introduction of a random phase in $U_s(x, y, z_s)$. This, however, leads to pseudo-random speckle-like intensity variations in W since a significant proportion of $U_s(x, y, 0)$ will fall outside the finite aperture. While tolerable in, e.g., visual display applications of holography, such fluctuations are highly detrimental in many modern applications of diffractive elements. We therefore need a deterministic signal-dependent phase, which can be designed as described, e.g., in Ref. [*14*].

Let us suppose that U_s has been found and an element has been designed according to Eq. (1). If $\alpha = 1$, all the energy in the incident wave will pass through W and we therefore have an element with $\eta = 100\%$ diffraction efficiency. Few such ideal solutions of the synthesis problem are known. One is the trivial solution with no element in place and $U_s = U_d = U_i$, but some non-trivial solutions are known in the framework of approximate diffraction theories: in view of the complex-amplitude transmittance approach a triangular-groove grating can deflect a monochromatic plane wave with 100% efficiency, and the well-known two-wave coupled-wave theory predicts that a plane wave may be split into two arbitrarily weighted parts by means of a thick dielectric index-modulated grating.

The synthesis problem can have rather serious constraints. One normally wishes to avoid absorption losses using either dielectric microstructures or high-reflectance metallic surface-relief profiles. Then, at least in view of the complex-amplitude transmittance approach, we are unable to modulate $A(x, y)$. Fabrication considerations often call for a quantization of $\phi(x, y)$, with perhaps only two permitted levels. Such constraints can obviously violate the condition $U_d(x, y, 0) = \alpha U_s(x, y, 0)$ quite severely, leading to a field of the form

$$U_d(x, y, z_s) = \alpha U_s(x, y, z_s) + U_f(x, y, z_s) \tag{2}$$

in W. Here U_f represents noise, which disturbs the signal s unless it vanishes in W.

The synthesis problem has degrees of freedom such as the phase freedom, since s often specifies only some properties of U_s in W. The so-called scale factor freedom, i.e., the freedom to choose the value of α in Eq. (2), is always available. Since a small value of α implies a low diffraction efficiency η, one prefers to use the scale factor freedom by maximizing α in the synthesis procedure. In this sense the scale factor freedom is restricted. Perhaps the most fundamental degree of freedom, the so-called amplitude freedom, results from the fact that $U_d(x, y, z_s)$ is, by definition, free outside W. These degrees of freedom are instrumental in the design of diffractive elements, being more or less explicitly employed by modern synthesis algorithms.

It is possible to obtain quantitative information about the achievable diffraction efficiency η from the knowledge of U_s in W alone, without the need to actually synthesize the diffractive microstructure. An upper bound η_l for η can be obtained from the explicit formula [*15*]

$$\eta_l = \frac{\left[\iint |U_s(x, y, 0)| \, dx \, dy\right]^2}{\iint |U_s(x, y, 0)|^2 \, dx \, dy}, \tag{3}$$

given here for a phase-only $t(x, y)$ and uniform illumination (the integrations are carried out over the aperture). However, as soon as we step outside the paraxial domain of diffractive optics, we arrive at a quite hostile but interesting territory, where little can be said about the general features of the synthesis problem. The grating equation tells us that to deflect light by an angle greater than $\sim 15°$, we need structural features with dimensions not much larger than λ, which calls for rigorous diffraction analysis [*1*]. Numerical studies have revealed that it is possible to achieve 100% diffraction efficiency and perfect signal fidelity in a number of cases involving perfectly conducting gratings [*16, 17*], even if the upper bound η_l for a corresponding paraxial signal is significantly less than 100%. In such designs the amplitudes of evanescent waves play a significant role and, in fact, provide amplitude freedom without a reduction of η.

DESIGN SCHEMES IN DIFFRACTIVE OPTICS

The synthesis problem of diffractive optics may be approached by at least two distinctly different methods, which we call the direct approach and the inverse approach. These can be either non-iterative or iterative design schemes, which make use of the available degrees of freedom and also take into account the contraints of the synthesis problem [*18*].

The direct approach is the one a novice in the field of diffractive optics would probably try first. Its essence is that we define a type of microstructure that we are able to manufacture. This structure contains a set of parameters that may take at least two different values. Typically, we first choose a random parameter set and evaluate the merit function associated with that configuration. Then we begin an optimization process, in which the parameters are modified according to some coherent strategy until the merit function reaches a satisfactory value or the process stagnates. Any optimization scheme (gradient algorithms, simulated annealing, genetic algorithms, etc.) may be used. Of course, the numerical efficiency may depend critically on the choice of the optimization scheme, the initial distribution, and the skill of numerical implementation. Since the merit function is defined only inside W, the amplitude freedom is effectively used throughout the design procedure. The scale freedom can be accounted for by including in the merit function a variable related to the efficiency η and planning the optimization strategy in such a manner that it attempts to ultimately maximize η.

In the inverse approach we start from the signal s in W, construct $U_s(x, y, z_s)$, apply $\mathscr{P}^{-1}$ to obtain $U_s(x, y, 0)$, and construct $t(x, y)$, which generally does not satisfy the fabrication constraints. To avoid an excessive amount of noise U_f in W, we are forced to modify $t(x, y)$ by procedures known as coding.

The use of a carrier grating leads to efficient, non-iterative coding methods. The carrier divides $U_d(x, y, 0)$ into a number of waves (diffraction orders) that propagate in different directions. Each order is split into some pattern by the modulation of the carrier amplitude, phase, or frequency. One of the orders, e.g., the first, is chosen to generate the signal s, while the other orders carry the noise away from W, making use of amplitude and scale factor freedoms.

Iterative inverse methods can optimize the use of different freedoms and therefore give higher diffraction efficiencies and permit on-axis signals. Such schemes proceed as follows. Once $U_s(x, y, 0)$ has been calculated from $U_s(x, y, z_s)$ by $\mathscr{P}^{-1}$, we determine a transmittance $t(x, y)$ by demanding that the constraints at $z = 0$ be satisfied. Then we determine $U_d(x, y, 0) =$

$t(x, y)U_i(x, y, 0)$ and compute $U_d(x, y, z_s)$ by application of $\mathcal{P}$. If $U_d(x, y, z_s)$ satisfies s within the prescribed accuracy, the process is terminated. If not, we modify it such that s is satisfied within W, using some or all of the available freedoms (for example, if only the intensity in W is fixed, we leave the phase unchanged). Then we return to the plane $z = 0$, apply the constraints there, and continue the iteration until a satisfactory solution is found or the process stagnates. We stress that it is not necessary to demand that the constraints at $z = 0$ be fully satisfied in the beginning stages of the procedure. For example, if we wish to design an element with a binary-valued $t(x, y)$, it has proven beneficial to introduce the constraint gradually during the iteration.

If we compare the direct and the inverse approaches, the strength of the direct scheme is that it is available whenever we are able to determine the operators $\mathcal{T}$ (or $\mathcal{R}$) and $\mathcal{P}$ either analytically or numerically. The inverse approach, on the other hand, is applicable only if the complex-amplitude transmittance approach is valid. However, when available, the inverse approach is far superior in terms of numerical efficiency, not least because the fast Fourier transform (FFT) algorithm may be used to evaluate the Fresnel and Fraunhofer propagation integrals (operators $\mathcal{P}$ and $\mathcal{P}^{-1}$) in a highly efficient parallel fashion.

DESIGN EXAMPLES

We proceed to discuss the actual synthesis of diffractive elements by means of some representative examples, namely beam profile shaping and pattern projection.

We begin with an intuitive picture of a redistribution of optical energy that can be achieved by a conventional aspheric optical element in a non-imaging configuration. Let us assume that the incident ray density follows the intensity profile of the incident optical field. The rays are refracted by the aspheric surface according to Snell's law and redistributed upon propagation into W at $z = z_s$. If s specifies the ray density in W, the required aspheric profile remains to be synthesized. However, it appears difficult to fabricate the required aspheric surface even if we manage to design it. Diamond turning is feasible for rotationally symmetric profiles, but not if we need a general, rotationally non-symmetric element or an array of elements on a single substrate. Here diffractive optics provides the solution.

If the illumination is monochromatic of wavelength λ, the thick, smooth aspheric profile may be reduced to a thin microstructured profile by removing the extra multiples of 2π phase shift [3]. The local structure of the element is that of a triangular grating, and hence the grating equation gives the local propagation direction of the diffracted field. Application of the synthesis methods of diffractive optics then permits us to design the profile based on either geometrical optics or by applying direct or inverse design schemes together with a wave-optical description of $\mathcal{P}$. In the latter case, design freedoms permit us to replace the locally triangular structure by a quantized profile, which is more easily manufacturable by lithographic techniques adopted from microelectronics.

Let us consider the following task: the incident Gaussian field of $1/e^2$ half-width $w = 2.5$ mm and $\lambda = 514$ nm shown in Fig. 4(a) is to be converted into a square-shaped flat-top distribution of sides $a = 2.5$ mm at a distance $z_s = 300$ mm. The (square) aperture of the element has sides $D = 3w = 7.5$ mm. In Fig. 4(b) we show the profile designed using an object-dependent diffuser by an iterative Fourier-transform algorithm [*19*]. Figures 4(c) and (d) show the field intensity in the plane $z = z_s$ and its cross-section at $y = 0$.

As a second example we consider the problem of splitting an incident field into a pattern of spatially separated light spots. This is a problem of consider-

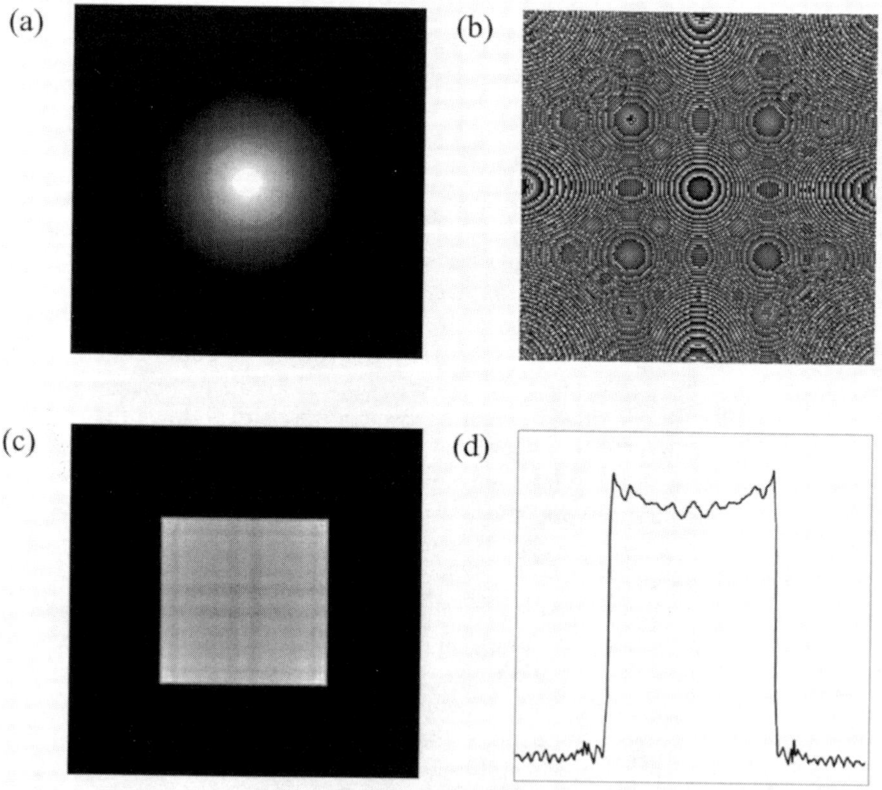

Fig. 4. Gaussian to flat-top transformation: (a) incident beam profile; (b) phase profile of the diffractive element; (c) the field intensity at $z = z_s$; (d) the cross-section at $y = 0$ (courtesy of H. Aagedal and S. Teiwes).

able practical importance in fields such as high-power laser materials processing, photonic switching, robot vision, etc. We are particularly tempted to use periodic structures here, because the periodicity of the element will itself provide the spatial separation of the spots provided that they are taken to represent various diffraction orders of the grating and focused into spots by a lens. The profile of each spot is the same at least in the paraxial domain, and the intensity of each spot is given by the efficiency of the particular diffraction order that gives rise to it. Since the signal is discrete, there is no speckle problem and we can make full use of the phase freedom. Hence we can achieve, in general, a higher diffraction efficiency for discrete than continuous signals.

In general, the use of amplitude freedom in combination with an optimization of the diffraction efficiency η results in the greatest concentration of noise in the immediate vicinity of the signal window, which may not always be desirable. We can, however, extend W such that the signal also contains a frame of vanishing orders on each side of the array. This is illustrated in Fig. 5. The efficiency of the element suffers a little, being reduced from 76.5% (signal without frame) to 73.3% (signal with frame). Use of multilevel structures improves η but adds to the complexity of fabrication.

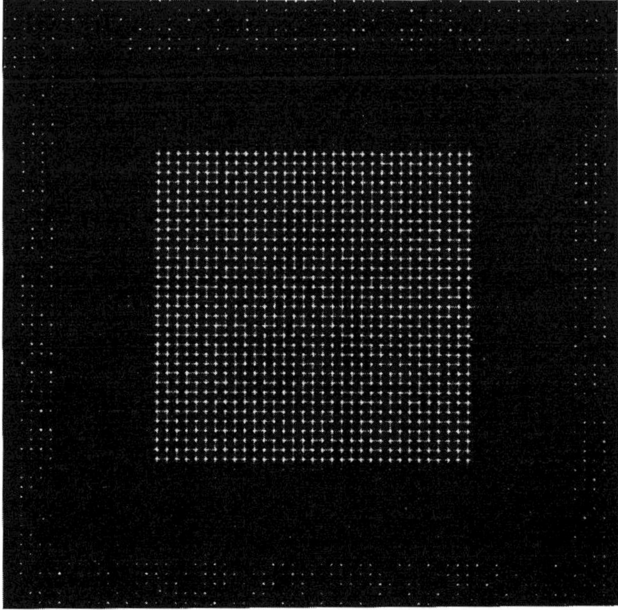

Fig. 5. The array of 33 × 33 spots of equal intensity, surrounded by a 10-spot-wide frame of orders with vanishing efficiency, generated by a binary diffractive phase element.

STATE OF THE ART AND FUTURE TRENDS

Research in the field of diffractive optics continues with an ever-expanding intensity. We are nevertheless in a position to state that this technology already lies on a solid basis. New analysis and synthesis methods will undoubtedly continue to emerge, but the selection already available is quite respectable. The rapidly increasing performance of affordable computers will automatically increase the complexity of microstructures that can be analysed and synthesized by existing methods. Similarly, the manufacturing technology of diffractive microstructures is well established and will continue to develop hand in hand with the lithographic fabrication of electronic circuits. We stress, however, that microlithography is not necessarily a solution to the problem of making diffractive optics an affordable optical technology, because the commercial value of diffractive microstructures per unit wafer area continues to be lower than that of electronic circuits. Fortunately, once a master diffractive element has been fabricated, perhaps at a considerable cost using direct-write electron-beam patterning, high-quality replicas of it can be produced cheaply by methods such as embossing and injection molding of plastics, which is not possible in electronics. Moreover, there are convenient methods to integrate diffractive elements with different optical functions [20], and the development of spatial light modulators is beginning to make real time reconfigurable diffractive elements feasible.

It is always an interesting proposition to attempt to foresee some future directions in a rapidly developing field of science, although such predictions can at best be regarded as educated guesses. We nevertheless wish to present some. As indicated in Fig. 1, the transition between macrostructured and microstructured elements is not sharp. This may open the way to the design of useful elements that lie somewhere between refractive and diffractive. Furthermore, it seems to us that the role of partial optical coherence in diffractive optics is worth detailed investigation. Diffractive optics is too often considered a technology suitable for the manipulation of fully coherent laser radiation. However, it can be useful even if the source is nearly incoherent and polychromatic. In fact, the reduced state of spatial coherence of, e.g., high-power multimode laser beams may actually enable us to design diffractive elements with improved performance since the speckle problems are reduced. Finally, with regard to Fig. 2 it appears possible to combine microstructured media with properties of bulk matter such as non-linearity. It is already known that periodic modulation by a grating can greatly enhance non-linear effects such as second-harmonic generation, and a continuation of this line of research is likely to yield unexpected discoveries.

In conclusion, we feel confident in predicting that diffractive optics will play a significant role in numerous fields of optical and optoelectronic technology in the forthcoming decade and beyond. The existing and foreseeable applications in optical communication, information technology, display technology, security applications, sensor technology, materials processing, laser beam focusing and profile shaping, laser resonator design, medical instrumentation and industrial process control will ensure the future blossoming of diffractive optics, which as-yet undetected developments and applications will undoubtedly strengthen.

REFERENCES

1. R. Petit, Ed., *Electromagnetic Theory of Gratings* (Springer, Berlin, 1980).
2. M.C. Hutley, *Diffraction Gratings* (Academic Press, London, 1982).
3. K. Miyamoto, 'The phase Fresnel lens,' *J. Opt. Soc. Amer.* **51**, 17–20 (1961).
4. E.N. Leith and J. Upatnieks, 'Reconstructed wavefronts and communication theory,' *J. Opt. Soc. Amer.* **A52**, 1123–1130 (1962).
5. R.C. Enger and S.K. Case, 'High-frequency holographic transmission gratings in photoresist,' *J. Opt. Soc. Amer.* **A73**, 1113–1118 (1983).
6. E. Popov and L. Mashev, 'Diffraction from planar corrugated waveguides at normal incidence,' *Opt. Commun.* **61**, 176–180 (1987).
7. D.C. Flanders, 'Submicrometer periodicity gratings as artificial anisotropic dielectrics,' *Appl. Phys. Lett.* **42**, 492–494 (1983).
8. T. Stone and N. George, 'Hybrid diffractive-refractive lenses and achromats,' *Appl. Opt.* **27**, 2968–2971 (1988).
9. A.W. Lohmann and D.P. Paris, 'Binary Fraunhofer holograms, generated by computer,' *Appl. Opt.* **6**, 1739–1748 (1967).
10. N. Streibl, 'Beam profile shaping with optical array generators,' *J. Mod. Opt.* **36**, 1559–1571 (1989).
11. J.R. Leger, D. Chen, and Z. Wang, 'Diffractive optical elements for mode shaping of a Nd:YAG laser,' *Opt. Lett.* **19**, 108–110 (1994).
12. J.C. Wyant and V.P. Bennett, 'Using computer-generated holograms to test aspheric wavefronts,' *Appl. Opt.* **12**, 2762–2765 (1974).
13. F. Wyrowski and R. Zuidema, 'Diffractive interconnection between high power Nd:YAG-laser and fibre bundle,' *Appl. Opt.* **33**, 6732–6740 (1994).
14. F. Wyrowski and O. Bryngdahl, 'Speckle-free reconstruction in digital holography,' *J. Opt. Soc. Amer.* **A6**, 1171–1174 (1989).
15. F. Wyrowski, 'Upper bound of the diffraction efficiency of diffractive phase elements,' *Opt. Lett.* **16**, 1915–1917 (1991).
16. E.V. Jull, J.W. Heath, and G.R. Ebbeson, 'Gratings that diffract all incident energy,' *J. Opt. Soc. Amer.* **67**, 557–560 (1977).
17. E. Noponen, J. Turunen, A. Vasara, M.R. Taghizadeh, and J.M. Miller, 'Synthetic diffractive optics in the resonance domain,' *J. Opt. Soc. Amer.* **A9**, 1206–1213 (1992).
18. F. Wyrowski and O. Bryngdahl, 'Digital holography as part of diffractive optics,' *Rep. Prog. Phys.* **54**, 1481–1571 (1991).
19. H. Aagedal, S. Teiwes, and F. Wyrowski, 'Consequence of illumination wave on optical function of non-periodic diffractive elements,' *Opt. Commun.* **109**, 22–28 (1994).
20. J. Jahns, 'Planar packaging of free-space optical interconnections,' *Proc. IEEE* **82**, 1623–1631 (1994).

7 Planar diffractive elements for compact optics

A. A. Friesem and Y. Amitai
Department of Physics of Complex Systems,
Weizmann Institute of Science, Rehovot, Israel

INTRODUCTION

There have been significant advances in replacing conventional optical elements with diffractive (or holographic) optical elements (DOEs) over the last thirty years. These include advances in design methods [1], in efficient recording materials and fabrication techniques [2], and in the incorporation of the elements into a variety of applications [3]. Yet, the usual configurations with diffractive elements have some drawbacks for many free-space applications. For example, when several elements are needed, the overall optical configuration is bulky and relatively heavy. Furthermore, these configurations do not lend themselves to easy modularization, and tolerances needed for aligning one element to another can be severe.

Some of these drawbacks can be overcome by supplementing the DOEs with planar optics [4], or, as it is sometimes called, substrate-mode holography [5]. This yields a planar diffractive configuration in which a multiplicity of DOEs can be recorded on a single substrate. The first, input, DOE diffracts the incident light from a source so it will be trapped inside the substrate by total internal reflection, while the last, output, DOE diffracts the light out from the substrate on to a detector (or the eye of a viewer). The path of going from conventional elements to diffractive elements to planar configurations can be viewed as analogous to that in the microelectronics industry where glass tubes were first replaced by transistors and then these were integrated into microcircuits.

The planar configurations have certain distinct advantages over those using free space. For example, the exact accuracy typically needed for alignment of the DOEs with respect to each other is achieved during the recording, where control is relatively easy. Also, since all the DOEs are integrated in a single substrate, the overall configuration is mechanically stable. For example, since light propagates through the substrate, rather than in free space, the volume of the configuration is relatively low. Finally, since the chromatic dispersion of one DOE can be corrected by other DOEs, the planar configuration is relatively insensitive to changes in the wavelength of the incident radiation [6].

The DOEs for planar optics configuration are mainly recorded as surface relief gratings or as volume gratings. Diffractive surface relief gratings typically exploit computer generated patterns that are transferred on to a substrate by lithographic and deposition techniques [7]. A wide range of patterns can be generated, so the DOEs can possess almost any desired function [8]. Moreover, the lithographic and deposition techniques are compatible with available microelectronics equipment and technology. Thus, it is possible to fabricate on one substrate both active elements such as surface emitting microlasers as well as passive elements such as beam splitters and microlenses leading to very compact overall systems [9].

Unfortunately, there are two major disadvantages associated with surface relief gratings. The first is their relatively low diffraction efficiency with the usual binary gratings. The efficiency can be significantly increased by resorting to multilevel gratings [10] – typically 8 or 16 levels. Unfortunately, this implies that the resolution of the computer generated masks must be substantially increased resulting in excessively high space-bandwidth products for the planar DOEs, beyond the capabilities of current plotting and lithographic equipment. The second disadvantage is the poor angular and wavelength discriminations of surface relief gratings [11]. For many applications, some including a multiplicity of DOEs on the same substrate, this lack of discrimination leads to excessive cross-talk. Specifically, as rays, trapped by total internal reflection inside the substrate, undergo several bounces, they will interact with DOEs for which no interaction is desired.

By recording the DOEs as volume gratings, it is possible to alleviate the problems of low efficiency and poor discrimination associated with surface relief gratings. The volume gratings are interferometically recorded in thick phase materials for obtaining high diffraction efficiencies, typically greater than 90 percent. Moreover, they have relatively high angular and wavelength discriminations in accordance with the Bragg relation [12].

When comparing DOEs recorded as surface relief gratings with those recorded as volume gratings, it is clear that it is easier to obtain arbitrary DOE functions through computer generation as well as more convenient replication with DOEs of the surface relief gratings. On the other hand, as a result of the high angular and wavelength discriminations, the cross-talk, which could be excessive with DOEs recorded as surface relief gratings, is essentially eliminated with DOEs recorded as volume gratings. Consequently, the DOEs of volume gratings can be incorporated into applications not possible otherwise. These include large-scale interconnects, wavelength division multiplexing, displays and optical data processing. We shall describe these in some detail in the following sections.

LARGE-SCALE OPTICAL INTERCONNECTS

Optical interconnects will undoubtedly play an important role in the development of advanced microelectronics and electro-optics systems. The conventional electrical wiring techniques have a limited transmission bandwidth and excessive cross-interference among the wires when the number of required interconnects is large. With optical interconnections [13, 14], on the other hand, the transmission bandwidth can be significantly increased and the cross-interference is negligible. A possible configuration for optical interconnects, between two integrated circuit boards, might exploit a DOE to selectively distribute the information from an array of input light sources in one board to arbitrary output detector locations at the other board. Since free-space optics are involved, there is no need for mechanical point-to-point contacts, and there is no interference between one optical path and another. Additional advantages can be gained by replacing the free-space interconnect configuration with that of planar optics.

The basic building block of the planar interconnect configuration, shown in Fig. 1, is composed of two DOEs that are recorded on the same substrate [15]. The first DOE collimates the light from an input light source into a plane wave, that is trapped inside the substrate by total internal reflection, whereas the second DOE focuses the collimated wave on to an output detector. A large number of DOE pairs can be recorded on the same substrate to provide optical interconnections for a large number of source–detector pairs [16]. Each diffractive doublet can transmit more than one channel simultaneously [17], although there are strict limitations on the positions of each source and detector. Since the substrate with the planar configuration can be directly adjacent to the planes of the sources and detectors, the overall interconnect system can be very compact and modularized. In a slightly modified version of this interconnect building-block the surfaces of the substrate are coated with a

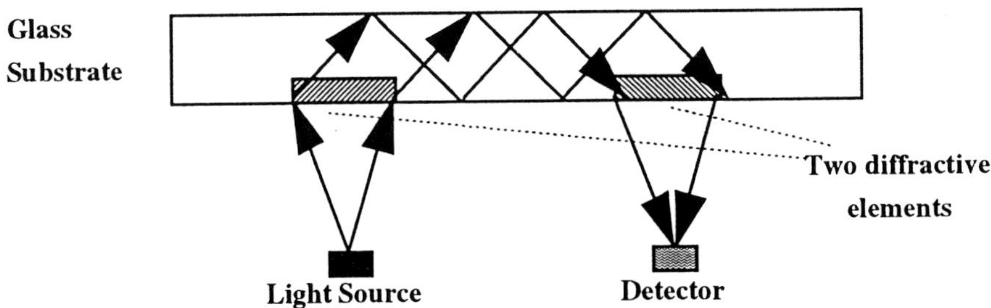

Fig. 1. The interconnect basic building block.

reflective coating [18]. With this approach, the bouncing angles inside the substrate can be reduced, but it is not appropriate for applications where a large number of building blocks and/or a transparent substrate are required.

Figure 2 illustrates schematically a possible combination of a large number of building blocks for performing large-scale optical interconnects. Here two parallel boards are essentially adjacent to each other. One contains a multiplicity of transducers (e.g., sources and detectors), and the other a multiplicity of DOEs. We assume that $n/2$ pairwise interconnections are to be established among a collection of $n \gg 1$ points, where $n/2$ source–detector pairs in the electronic transducers board are to be connected. For each source–detector pair, a doublet DOE is recorded in the optical board. This architecture is rather general, and it can be modified and adapted to various applications such as optical switching [19], optical buses for multiprocessor systems [20], board-to-board [21] and chip-to-chip [22] communications, neural networks based on optoelectronic neurons [23], and optical clock distribution [24].

The interconnect architecture in Fig. 2 can be very compact and convenient to use. Indeed, it can potentially approach the least possible size of any two-dimensional [25] or three-dimensional [26] optical architecture. In addition, a fan-out capability where a single source simultaneously transmits information to several detectors [27–29] can be obtained with multiplexed DOEs, where each channel is associated with a separate DOE. On the other hand, fan-in from several sources on to a single detector could be achieved with low insertion loss and negligible cross-talk if each channel is operating at a slightly different wavelength [30].

The DOEs must be recorded so as to operate with high efficiency and low aberrations. Usually, there is a difference between the recording and readout wavelengths. For example, the readout wavelength is typically in the near-infrared domain, whereas the recording wavelength must be shorter than 530 nm for most phase materials. Such wavelength shifts generally result in excessive aberrations and lower diffraction efficiencies. Thus, the DOEs must

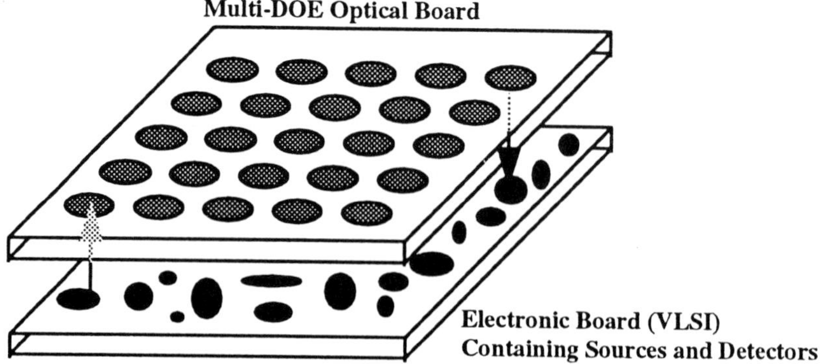

Fig. 2. Schematic geometry of large-scale interconnect architecture.

be recorded with predistorted wavefronts. A possible way to obtain such predistorted wavefronts is by resorting to a recursive design technique [*15*], where the needed wavefronts are derived from interim holograms whose readout geometries differ from those used during recording. Specifically, the predistorted object and reference waves are derived from two separate intermediate holograms.

The recording of a DOE is shown schematically in Fig. 3. The intermediate holograms are illuminated simultaneously by a plane wave, and the reconstructed output waves with the desired predistortions are used to record the DOE. The paraxial parameters of these predistorted output waves are calculated so as to satisfy the Bragg condition for obtaining high diffraction efficiency as well as to compensate for the aberrations that result from the recording-readout wavelength shift. Exploiting such recursive techniques, a basic interconnect building block with high diffraction efficiencies and nearly diffraction limited performance was successfully recorded even in the presence of a wavelength shift and DOEs with comparatively small f-number [*15*].

The recording of the DOEs for the general interconnect configuration can be done with the arrangement shown in Fig. 4. Here, the reconstructing plane wave is inserted through a glass prism. Because of the inherent high obliquity, only the desired first diffracted orders emerge from the intermediate holograms toward the final DOE, whereas the zero order is coupled out of the prism by total internal reflections and the other orders are evanescent waves. Since the intermediate holograms are typically adjacent to the final DOE, there will be little, if any, overlap from one intermediate hologram to the other.

The intermediate holograms can be recorded interferometically by exploiting the recursive technique, or can be recorded directly as (Computer generated holograms) CGHs. Since the efficiency of the intermediate holograms need not be high, they can be simply recorded as surface relief binary gratings. If the

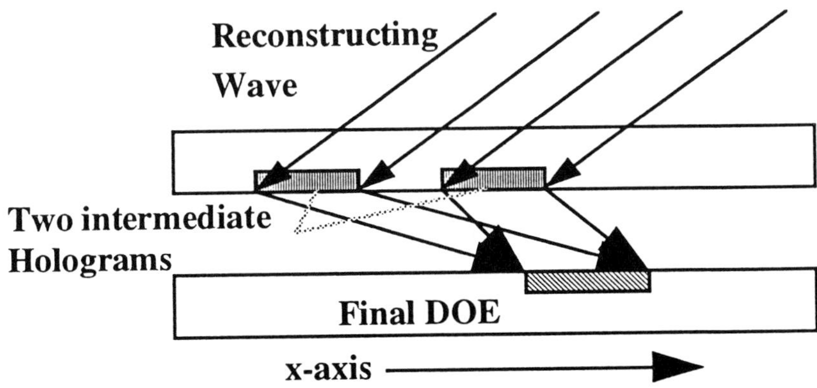

Fig. 3. Schematic arrangement of recording the DOE for the basic interconnect building block by means of a recursive technique.

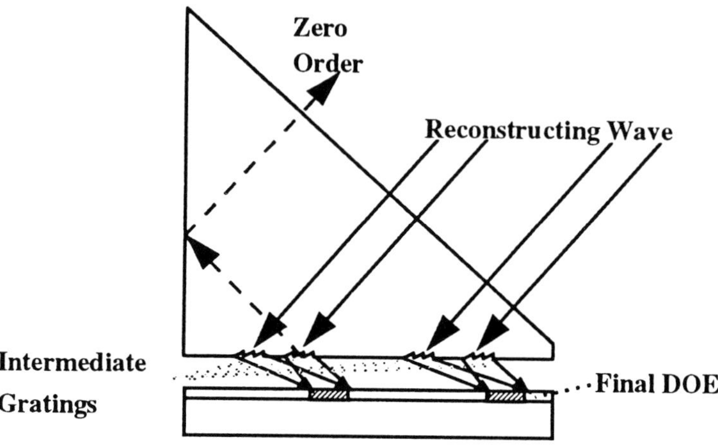

Fig. 4. The recording of DOEs for the general interconnect configuration.

intermediate holograms are recorded as CGHs, their grating function must be calculated explicitly. This can be done either numerically with ray tracing methods [*31*], or analytically with Fourier expansion techniques [*32*].

WAVELENGTH DIVISION MULTIPLEXING

Wavelength division multiplexing/demultiplexing (WDM/WDDM) techniques can effectively increase the information capacity that can be transmitted through optical fibres. With these techniques, a large number of communication channels can be transmitted simultaneously over a single fibre. During the last decade, various devices for implementing WDM and WDDM have been proposed. These are based on birefringent materials, surface relief gratings, Mach–Zender interferometry, and waveguides. Unfortunately, the devices have either low efficiencies or cannot handle a large number of channels.

A novel device for WDM is based on a combination of planar diffractive and conventional refractive elements and takes advantage of the dispersive characteristics of the diffractive elements [*33–35*]. Such a device is shown schematically in Fig. 5. It is basically a modified version of the arrangement presented in Fig. 1. An input fibre source, containing n different communication channels $C_1, ..., C_n$, with the wavelengths $\lambda_1, ..., \lambda_n$, is brought close to the substrate. The waves emerging from the fibre source are collimated by a conventional refractive lens and coupled into the substrate by the first DOE. The incoming waves, each of a different wavelength, are diffracted into guided waves with different bouncing angles. After travelling a certain distance, so the various channels are laterally separated, the guided waves are coupled out by a second DOE and focused by another refractive lens on to the receiving fibres.

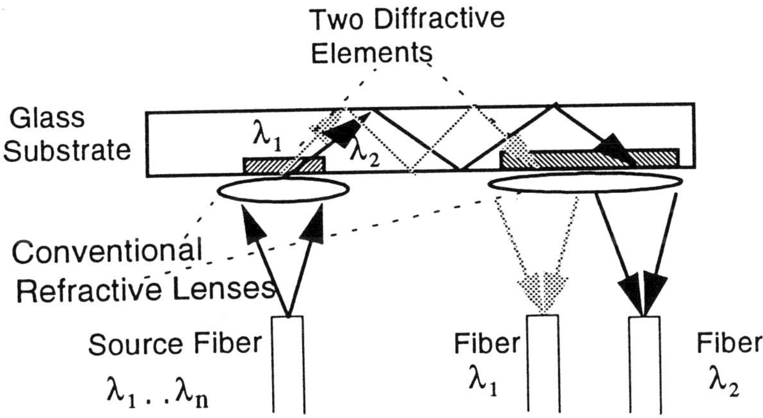

Fig. 5. WDM device based on planar optics and refractive lenses.

Unfortunately, in order to completely separate the various channels, the distance between the two DOEs must be relatively large. In addition, the substrate must be thick enough to prevent cross-talk between the channels, and the collimation and the focusing must be done with refractive lenses to reduce chromatic aberrations, so the overall configuration is rather large.

A more promising WDM device, based entirely on planar DOEs, is presented in Fig. 6. Here, a number of building blocks are combined to form the WDDM/WDM device [36]. For WDDM, the central element comprises n different DOEs, $\mathcal{H}_1^s, ..., \mathcal{H}_n^s$, which both collimate the incoming waves and diffract them in different directions inside the substrate in accordance with their wavelengths. Each wave is then diffracted out from the substrate and focused by a single DOE, $\mathcal{H}_i^r$, on to its receiving fibre. Alternatively, the propagation direction of the waves can be inverted to yield a WDM device which multiplexes a number of channels from their separated fibre sources on to one receiving fibre.

In order to obtain high efficiency and negligible cross-talk between the channels, each $\mathcal{H}_i^s$ DOE must have high diffraction efficiency for a specific wavelength λ_i, and very low efficiencies for the other wavelengths λ_j, $j \neq i$. To obtain the desired high diffraction efficiency, the DOEs must be recorded so as to fulfil the Bragg condition for the appropriate wavelength [15]. To ensure that the diffraction efficiencies are low at other wavelengths, the spectral separation $\Delta\lambda_i$ between two adjacent wavelengths, λ_i, λ_{i+1}, must fulfil the relation [37]

$$\Delta\lambda_i \geq \frac{\sqrt{3}}{2} \frac{\lambda_i^2}{Dv \sin \dfrac{\beta_{in}}{2} \tan \beta_{in}}, \tag{1}$$

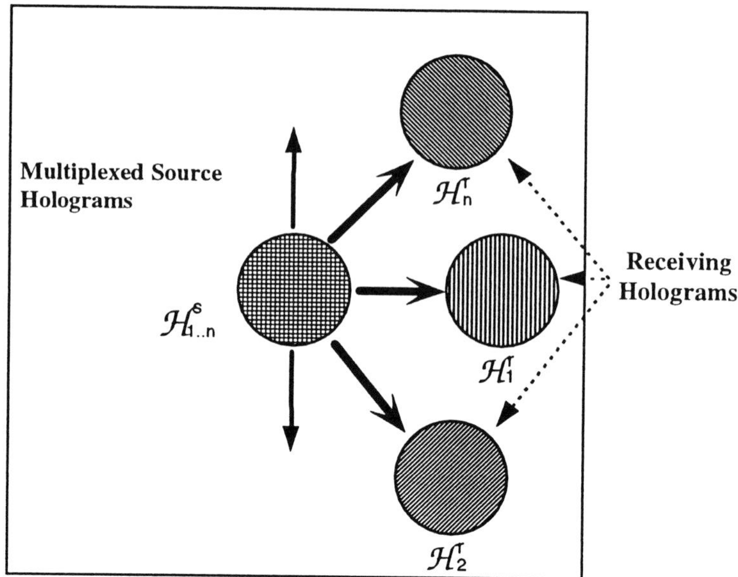

(a) Top View

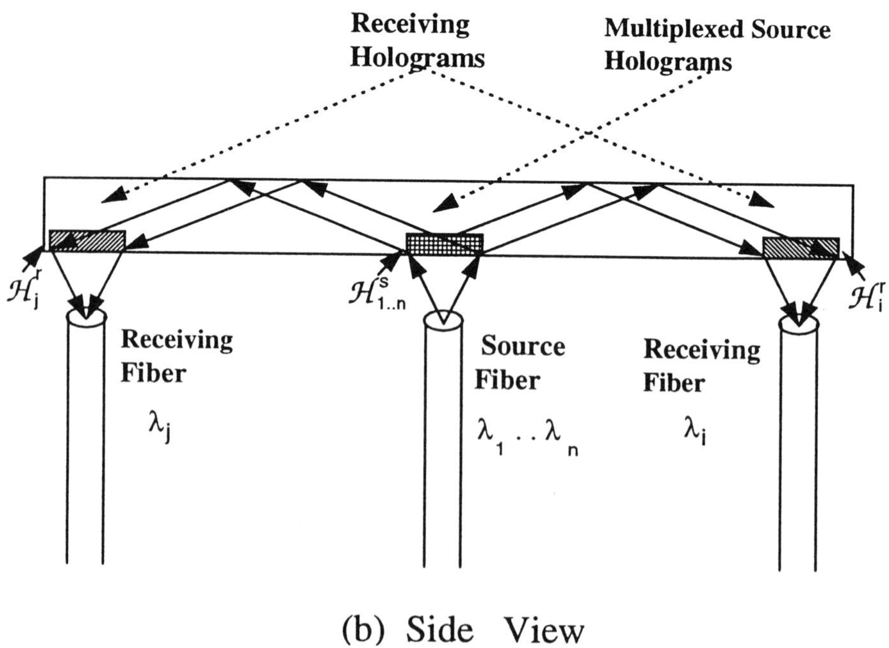

(b) Side View

Fig. 6. WDDM/WDM devices based wholly on planar optics.

where ν is the refractive index of the recording material, D is the emulsion thickness and β_{in} is the bouncing angle inside the substrate.

Another important parameter is the number of channels that such a WDM device can handle simultaneously. This number depends on the number of different DOEs that can be superimposed on the same location without exceeding the dynamic range of the refractive index of the recording material. Namely, the total sum of the desired refractive index modulations η_i for all the multiplexed DOEs must be less than the maximum possible refractive index modulation η_{max} of the recording material. Hence, the maximum number of channels can be written [37], as

$$N = \frac{\eta_{\text{max}}}{\eta i} = \frac{2\eta_{\text{max}}D}{\lambda\sqrt{\cos\beta_{\text{in}}}}.$$

(2)

Consider, for example, a recording material with a thickness of $D = 100$ microns, refractive index $\nu = 1.5$ and illumination wavelength $\lambda = 1.3$ microns. Then, in accordance with Eqs. (1) and (2), the required spectral separation must be $\Delta\lambda_i = 7$ nm and the maximum number of channels with no cross-talk is $N = 20$.

THREE-DIMENSIONAL HOLOGRAPHIC DISPLAYS

Three-dimensional holographic displays are being successfully incorporated into advertising, packaging, educational illustrations and even art. Yet, the current form of display holograms has some severe drawbacks. For example, the readout light sources must be located at some distance from the hologram, resulting in bulky, space consuming, and sometimes inconvenient display arrangements. In addition, the transmitted part of the readout wave, which is not diffracted by the holograms, can bother the observer.

Recently, there have been several proposals [38–40], based on edge-illuminated holograms, for constructing compact displays that overcome the drawbacks mentioned above. The salient feature of these is to reconstruct the holograms with a readout wave which enters the substrate through a polished edge so as to reach the hologram at a large angle of incidence. For these, the substrate must be rather thick and a large offset angle must be used in the recording.

An alternative and more convenient method for realizing compact and efficient white-light holographic displays [41], based on planar optics, is shown in Fig. 7. The display arrangement is composed of two holograms, $\mathcal{H}^s$ and $\mathcal{H}^d$, on one substrate. The readout wave is a spherical wave at a distance R_c from the centre of the first hologram $\mathcal{H}^s$. The output wave from $\mathcal{H}^s$ is a diverging beam, diffracted at an angle β^{in} inside the substrate. The second element $\mathcal{H}^d$ diffracts the wave from $\mathcal{H}^s$ outward so as to form a virtual image of

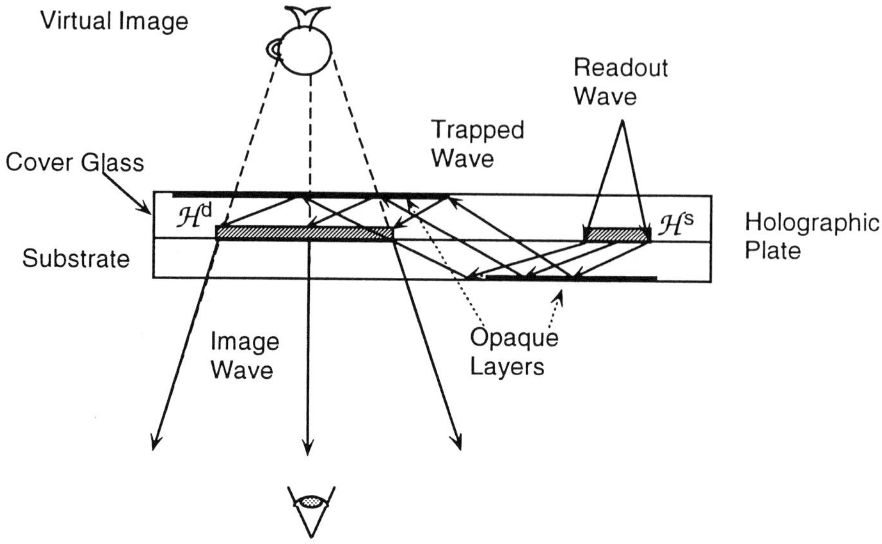

Fig. 7. Three-dimensional holographic display based on planar optics.

a three-dimensional object. Since $\mathcal{H}^d$ is illuminated with a diverging beam that is spreading inside the substrate, the actual size of $\mathcal{H}^s$ can be much smaller than that of $\mathcal{H}^d$. Thus, the readout wave can be located very close to the holographic display resulting in a very compact and convenient arrangement.

Several design issues must be addressed during practical implementation of this planar holographic display arrangement. First, in order to avoid extraneous light reflections from obscuring the desired three-dimensional display, parts of the substrate should be covered with opaque layers. Also, since each ray from $\mathcal{H}^s$ undergoes several bounces inside the holographic plate and impinges on the surface of $\mathcal{H}^d$ at several locations, we must ensure that diffraction occurs only at the desired locations, in order to avoid extraneous diffraction and thereby the formation of ghost images. We found [41] that undesired diffractions can be avoided if

$$T \geq \frac{\sqrt{3}}{4} \frac{\lambda l}{Dv \sin \dfrac{\beta_{in}}{2} \tan^2 \beta_{in}}, \qquad (3)$$

where T is the thickness of the substrate and l is the distance between the centres of the two holograms.

Representative three-dimensional images from a planar display that was reconstructed with white light are shown in Fig. 8. Here, in order to illustrate the three-dimensionality of the reconstructed image, two different images are shown, each viewed at a different location. Since our holographic display

 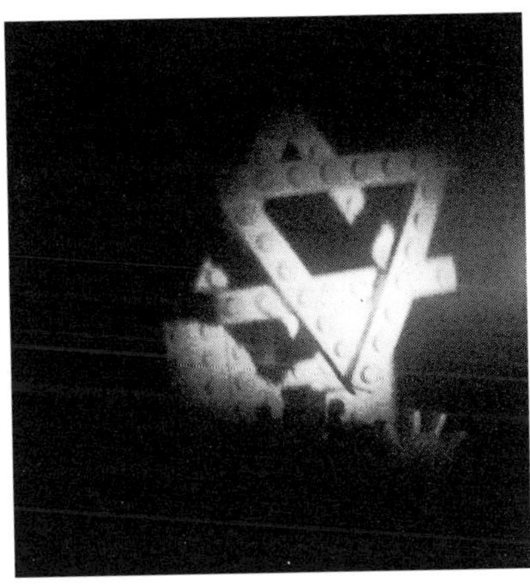

Fig. 8. Virtual image from planar holographic display, from two different viewing points, when it is reconstructed with a white light source.

arrangement is very discriminating to readout wavelength, only a relatively narrow spectral bandwidth of the source was diffracted; in this case it was reddish reconstruction. In this arrangement, the chromatic dispersion of $\mathcal{H}^s$ was compensated almost completely by $\mathcal{H}^d$ to yield sharp images.

VISOR DISPLAYS

One of the important applications for DOEs is in visor displays. In these a DOE serves as an imaging lens and a combiner, where a two-dimensional quasi-monochromatic display is imaged to infinity and reflected into an observer's eyes. The display can be derived either directly from a CRT or indirectly by means of a relay. Typically, the display comprises an array of points whose geometry at readout differs from that at recording. As a result, the imaged array contains aberrations that decrease the image quality. Another problem, which is usually common to all types of diffractive optical elements, is their relatively high chromatic dispersion. This is a major drawback in applications where the light source is a CRT which is not purely monochromatic.

Several planar holographic designs, based on either edge-illuminated holograms [42] or double-DOE configurations [43,44], were proposed for improving the performance of holographic sights and visor displays. An example of a holographic visor display, based on planar holographic doublet configuration, is schematically presented in Fig. 9. The doublet comprises two DOEs. One is a collimating lens $\mathcal{H}^d$ and the other a simple linear grating $\mathcal{H}^g$, both of which are recorded on the same substrate. A two-dimensional display is located at a distance R_d from the centre of $\mathcal{H}^d$, where R_d is the focal length of

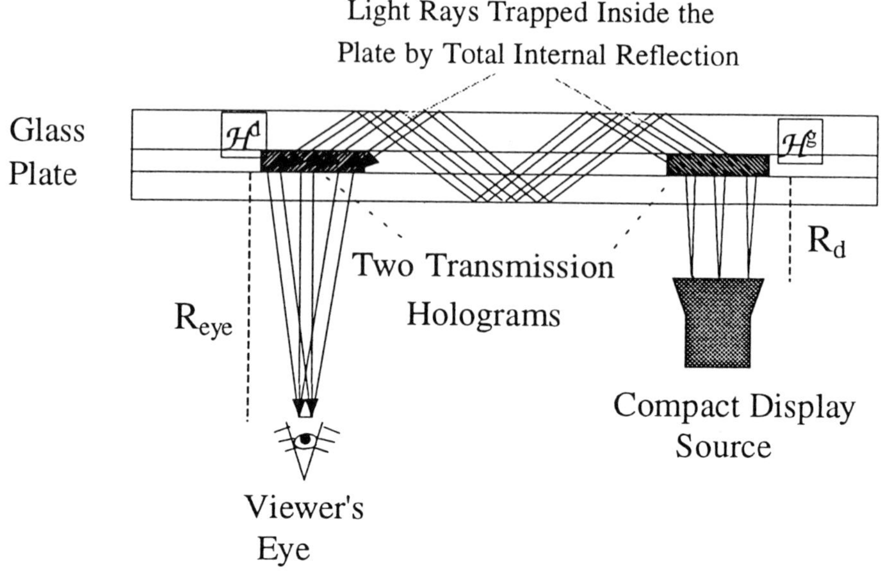

Fig. 9. Holographic visor display based on planar doublet configuration.

$\mathcal{H}^d$. The light from the display is thus transformed into an angular spectrum of plane wavefronts by $\mathcal{H}^d$. Specifically, each viewing angle of the input is diffracted into a plane wave at an angle $\beta_{in}(x)$ inside the substrate, where x is the lateral coordinate of $\mathcal{H}^d$. The linear grating $\mathcal{H}^g$ diffracts the trapped wavefronts outward. An observer, located at a distance R_{eye} thus sees an image of the display, located at infinity.

The design of the linear grating $\mathcal{H}^g$ is straightforward. It has a grating function $\mathcal{H}^g = (2\pi/\lambda)(v \sin \beta_{in})\xi$, where ξ is the lateral coordinate of $\mathcal{H}^g$. The design of the collimating lens $\mathcal{H}^d$ is much more complicated because a simple holographic lens, recorded with only spherical waves, has, in general, very large aberrations over a large field of view. In order to compensate for the large aberrations, it is necessary to record the holographic lens with two aspherical object and reference waves. These aspheric waves can be derived from CGHs or from intermediate holograms, in accordance with a recursive design technique. It has been demonstrated that with a proper use of the recursive technique, a planar doublet visor display, having diffraction limited performance [44], relatively low chromatic dispersion and uniform high diffraction efficiency over a wide field of view [45], can be obtained. Figure 10 shows representative output imagery from a planar visor display that was illuminated with a green light having a bandwidth of ±4 nm. The field of view of this image was ±8 degrees.

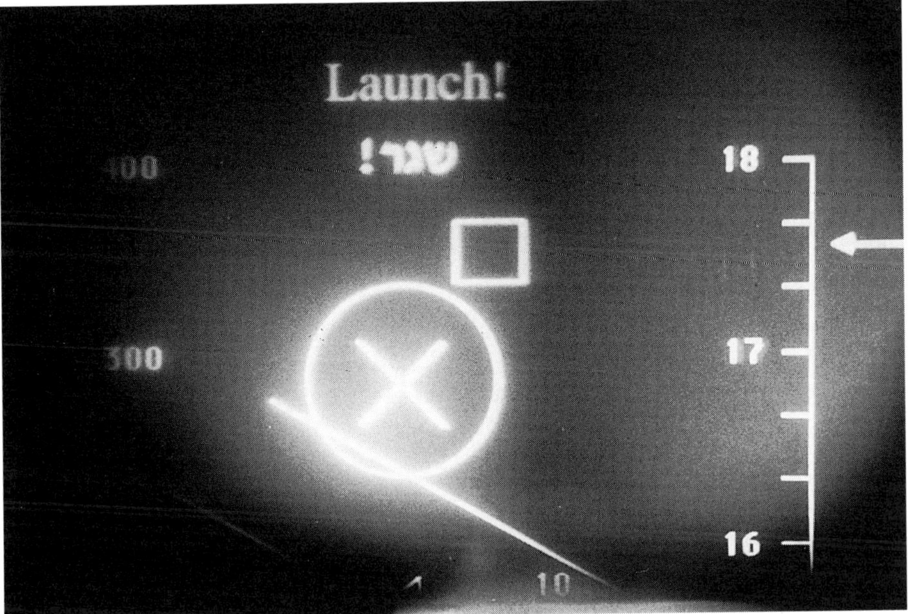

Fig. 10. Output imagery with a planar holographic visor display configuration.

OPTICAL DATA PROCESSING

Planar optics configurations should be extremely useful in developing compact correlators and convolvers for optical data processing applications. In these, the basic building block is a Fourier transform lens. Most optimization methods for designing holographic Fourier transform lenses that operate in off-axis configurations involve a single holographic lens [46,47]. The lens transforms off-axis incident plane waves into spherical wavefronts that converge to diffraction limited spots at the Fourier plane. Unfortunately, a single lens introduces phase errors that cannot be tolerated when the Fourier lenses are incorporated into optical data processing. Moreover, such lenses are not symmetrical elements; that is, exchanging the places of the input and output planes does not yield the same result. Therefore, it is difficult to cascade such elements, as is needed for optical data processing applications.

A planar arrangement of two off-axis holographic lenses can perform Fourier transformation with no phase errors [48]. When the two lenses are separated by their focal length f, they yield a Fourier transformation whose intensity as well as phase can be accurately controlled. In such a configuration, shown in Fig. 11, the first lens performs the desired intensity transformation at its back focal plane while the second lens corrects the phase error introduced by the first

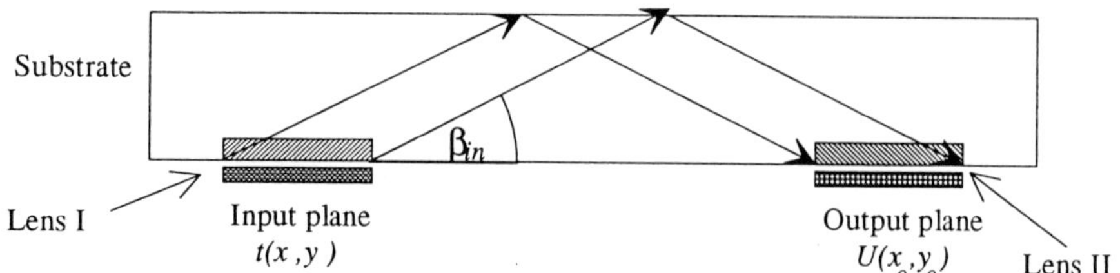

Fig. 11. Planar holographic doublet configuration for Fourier transformation.

lens. The required grating function for each element can be derived from the second-order approximation of the interference pattern of an on-axis plane wave and an off-axis spherical wave, to yield the quadratic grating function of the form

$$\phi_{\text{h}} = \frac{-2\pi}{\lambda}\left(\frac{x^2 \cos^2 \beta_{\text{in}} + y^2}{2f} - x \sin \beta_{\text{in}}\right), \tag{4}$$

where λ is the wavelength *inside* the substrate.

The field distribution at the output of the Fourier transform doublet has the form

$$U(x_0, y_0) = C \iint t(x, y)\exp\left[-i2\pi\left(\frac{x \cdot x_0 \cos^2 \beta_{\text{in}} + y \cdot y_0}{2f}\right)\right] \mathrm{d}x\, \mathrm{d}y, \tag{5}$$

where x_0 and y_0 are the coordinates of the output plane and C is a constant. Equation (5) indicates that the field behind the second lens in the doublet is indeed the complete Fourier transformation of input transparency $t(x, y)$ with no phase errors. The spatial frequencies in this doublet configuration are scaled in accordance with

$$\tilde{\omega}_x = \frac{x_0 \cos^2 \beta_{\text{in}}}{\lambda f}, \qquad \tilde{\omega}_y = \frac{y_0}{\lambda f}. \tag{6}$$

The $\cos \beta_{\text{in}}$ term in $\tilde{\omega}_x$ is due to the off-axis geometry and introduces a scale change between the x and y coordinates. Such a scale change is deterministic and must be taken into account when incorporating the doublet into data processing systems.

The doublet configuration of Fig. 11 was experimentally evaluated for Fourier transformation by recording two holographic lenses, with $f = 250$ mm and $\beta_{\text{in}} = 45°$. A transparency of lines forming concentric squares was placed at the input and was illuminated with a plane wave of wavelength $\lambda = 514.5$ nm. The inverse Fourier transformation of the output imagery was performed with a conventional refractive lens, and the final image was detected with a CCD camera. Photographs of the input as well as the output are shown in Fig. 12.

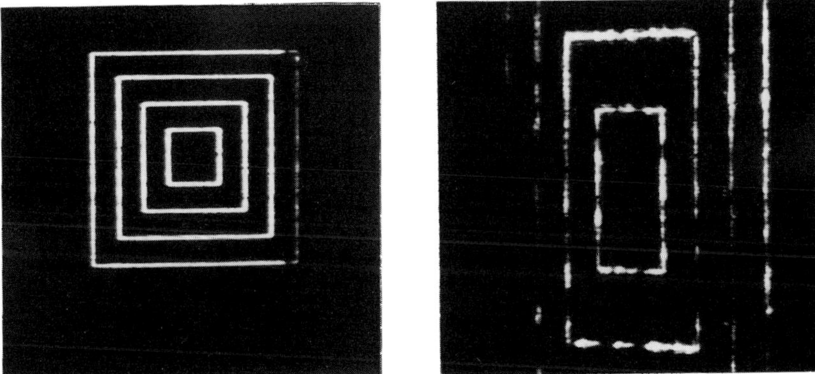

Fig. 12. Experimental results: (a) input, (b) output from the planar configuration.

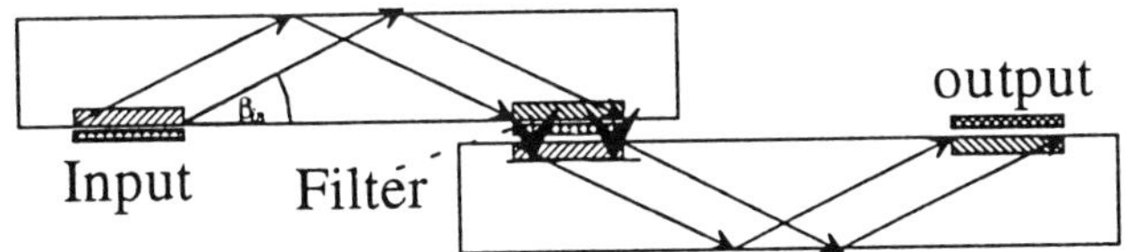

Fig. 13. Planar optical correlator arrangement.

The results indicate that the doublet performs the desired Fourier transformation with the expected scaling factor of $\cos^2(45°) = 0.5$. The simultaneous high sharpness of the lines in both the x and y directions clearly indicates that there is little, if any, observable phase error.

Finally, Fig. 13 illustrates how two doublets and a filter would be combined to form an optical correlator. For such a correlator, the scaling factor should be taken into account when the required optical filters are designed. The overall configuration is very compact, and alignment between the individual elements can be set conveniently during the recording.

PLANAR DEVICES FOR OTHER APPLICATIONS

There are many specialized planar diffractive elements that could be incorporated into applications other than those described in the preceding sections. Here we briefly describe two such elements.

Polarization Selective Devices

As can be shown from the coupled wave theory [12], the coupling coefficient, which controls the diffraction efficiency for thick phase holograms, is different for the S and P polarization states of the incident beam. This difference depends strongly on the offset angle between the incident and the output beams inside the substrate; that is, by increasing the diffraction angle there will be higher discrimination between the polarization states. Since the off-axis angle in planar diffractive configurations is usually very high, this phenomenon might be exploited to record diffractive elements which have very high diffraction efficiency for one of the linear polarization states and be transparent for the orthogonal polarization state. As a result, a wide variety of devices which are based on such polarization-selective elements might be designed. These include beam splitters [49], multistage optical buses [50], switching elements [51], and polarization-division multiplexers for optical communication [52].

Galilean Telescopes

Beam expanders for magnifying a narrow collimated beam into a beam of larger diameter and smaller divergence angle are usually composed of a Galilean telescopic configuration of two lenses along a common axis and with a common focal point. With monochromatic light sources it may be advantageous to exploit a planar diffractive configuration [53,54]. A possible geometry for a planar Galilean telescope is shown in Fig. 14. The negative lens converts a narrow input beam into a diverging spherical wave, which

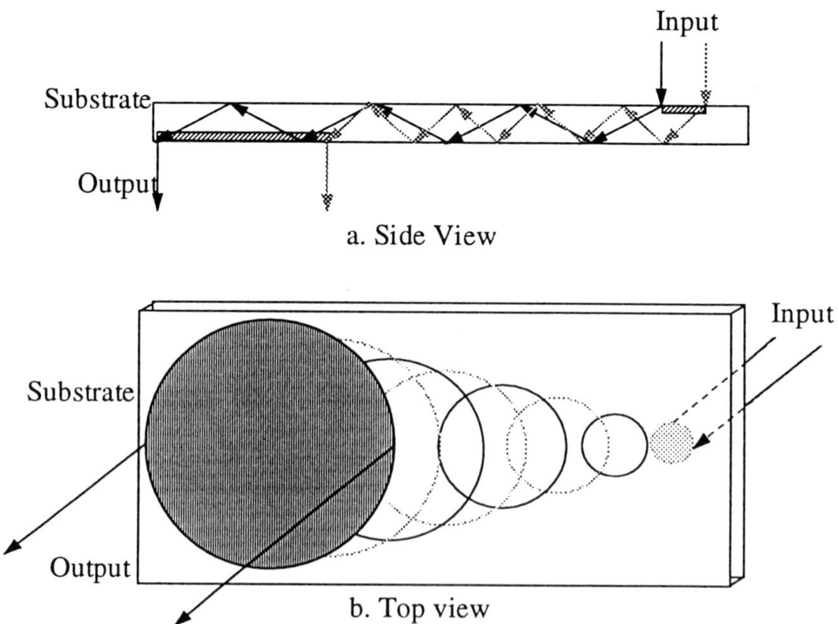

Fig. 14. Holographic beam expander based on planar doublet configuration.

propagates toward the positive lens, while undergoing total internal reflections within the substrate. Finally, the wave is collimated by the positive lens as a magnified plane wave. Obviously, the compact configuration should be useful in many applications.

CONCLUDING REMARKS

Planar optics is a new technology that should lead to more compact and modularized optical and electro-optical systems. We described some specific applications where this technology can be incorporated. These included large-scale optical interconnects for advanced microelectronics, wavelength division multiplexing for optical communication, novel configurations for displays, and compact architectures for optical data processing. These are but examples for a technology that has a bright future. Despite some difficulties that must still be overcome, we expect that planar optics technology could be incorporated into advanced integrated optics and microelectronics systems that are yet to be developed.

ACKNOWLEDGEMENTS

We would like to thank S. Reinhorn, R. Shechter, I. Shariv, S. Gorodeisky and V. Weiss for their contributions. Some of the work described in this chapter was supported by the MINERVA Foundation.

REFERENCES

1. H.P. Herzig and R. Dandliker, 'Holographic optical elements for use with semiconductor lasers,' in *International Trends in Optics* (J.W. Goodman ed.), Academic Press, San Diego 1991.
2. B.J. Chang, 'Dichromated gelatin holograms and their applications,' *Opt. Eng.* **19**, 642–648 (1980).
3. D.H. Close, 'Holographic optical elements,' *Opt. Eng.* **14**, 408–419 (1975).
4. J. Jahns and B.A. Brumback, 'Integrated-optical split-and-shift module based on planar optics,' *Opt. Comm.* **76**, 318–20 (1990).
5. R.K. Kostuk, M. Kato, and Y.-T. Huang, 'Substrate mode holograms for optical interconnects' in *Optical Computing 1989 Technical Digest Series*, **9**. *Opt. Soc. Amer.* 168–171 (1989).
6. R.K. Kostuk, Y.T. Huang, D. Hetherington, and M. Kato, 'Reducing alignment and chromatic sensitivity of holographic optical interconnects with substrate-mode holograms,' *Applied Optics* **28**, 4939–4944 (1989).
7. M.W. Farn, M.B. Stern, and W. Veldkamp, 'The making of binary optics,' *Opt. & Phot. News* **2**, 20–22 (1991).

8. G.J. Swanson and W.B. Veldkamp, 'Diffractive optical elements for use in infrared systems,' *Opt. Eng.* **28**, 605–608 (1989).
9. J. Jahns, F. Sauer, B. Tell, K.F. Brown-Goebeler, A.Y. Feldbaum, C.R. Nijander, and W.P. Townsend, 'Parallel optical interconnections using surface-emitting microlasers and a hybrid imaging system,' *Opt. Comm.* **109**, 328–337 (1994).
10. E. Hasman, N. Davidson, and A.A. Friesem, 'Efficient multilevel phase hologram for CO_2 lasers,' *Opt. Lett.* **16**, 423–424 (1991).
11. N. Davidson, R. Duer, A.A. Friesem, and H. Hasman, 'Blazed holographic gratings for polychromatic and multidirectional incidence light,' *J. Opt. Soc. Amer.* **A9**, 1196–1199 (1992).
12. H. Kogelnik, 'Coupled wave theory for thick holograms and their applications,' *Bell Syst. Tech. J.* **48**, 2909–2947 (1969).
13. K.-H. Brener and F. Sauer, 'Diffractive-reflective optical interconnects,' *Appl. Opt.* **27**, 4251–4254 (1988).
14. J.W. Goodman, F.J. Leonberger, S.Y. Kung, and R. Athale, 'Optical interconnections for VLSI systems,' *Proc IEEE* **72**, 850–866 (1984).
15. Y. Amitai and J.W. Goodman, 'Design of substrate-mode holographic interconnects with different recording and readout wavelengths,' *Applied Optics* **30**, 2376–2381 (1991).
16. Y. Amitai, 'Large scale optical interconnects based on planar optics,' *Proceedings of the SPIE* **1972**, 217–222 (1993).
17. J. Jahns and J. Walker, 'Imaging with planar optical systems,' *Opt. Comm.* **76**, 313–317 (1990).
18. S.J. Walker and J. Jahns, 'Optical clock distribution using integrated free-space optics,' *Opt. Comm.* **90**, 359–371 (1992).
19. Y.-T. Huang and Y.-H. Chen, 'Optical switches with a substrate-mode grating structure,' *Optik* **98**, 1–4 (1994).
20. J.-H. Yeh, R.K. Kostuk, and K.-Y. Ta, 'High-speed optical bus for multiprocessor systems with substrate mode holograms,' *Proceedings of the SPIE* **2153**, 69–77 (1994).
21. Y.-T. Huang and R.K. Kostuk, 'Substrate-mode holograms for board-to-board two-way communications,' *Proceedings of the SPIE* **1474**, 39–44 (1991).
22. J. Jahns, Y.H. Lee, C.A. Burrus, and J.L. Jewell, 'Optical interconnects using top-surface-emitting microlasers and planar optics,' *Appl. Opt.* **31**, 592–597 (1992).
23. P.E. Keller and A.F. Gmitro, 'Design and analysis of fixed planar holographic interconnects for optical neural networks,' *Appl. Opt.* **31**, 5517–5526 (1992).
24. F. Lin and E. Strzelecki, 'Self-aligned holographic clock distribution network,' *Proceedings of the SPIE* **1154**, 227–230 (1989).
25. H.M. Ozaktas, Y. Amitai, and J.W. Goodman, 'Comparison of system size for some optical interconnection architectures and the folded multi-facet architecture,' *Opt. Comm.* **82**, 225–228 (1991).
26. H.M. Ozaktas, Y. Amitai, and J.W. Goodman, 'A three dimensional optical interconnection architecture with minimal growth rate of system size,' *Opt. Comm.* **85**, 1–4 (1991).
27. R.T. Chen, H. Lu, D. Robinson, M. Wang, G. Savant, and T. Jannson, 'Guided-wave planar optical interconnects using highly multiplexed polymer waveguide holograms,' *Journal of Lightwave Technology* **10**, no. 7, 888–897 (1992).
28. V. Minier, A. Kervorkian, and J.M Xu, 'Diffraction characteristics of superim-

posed holographic gratings in planar optical waveguides,' *IEEE Photonics Technology Letters* **4**, 1115–1118 (1992).

29. J.-H. Yeh and R.K. Kostuk, 'Design issues for substrate mode holograms used in optical interconnects,' *Proceedings of the SPIE* **2176**, 207–217 (1994).

30. Y. Amitai, 'Multiplexed substrate-mode holographic interconnects,' *Optical Design for Photonics Technical Digest* **9**, 84–87 (Optical Society of America, Washington, DC, 1993).

31. M.J. Hayford, 'Holographic optical design using CODE V™,' *Proceedings of the SPIE* **883**, 2–7 (1988).

32. E. Socol, Y. Amitai, and A.A. Friesem, 'Design of planar optical interconnects,' *Proceedings of the SPIE* **2426**, 433–442 (1995).

33. Y.T. Huang, D.-C. Su, and Y.-K. Tsai, 'Wavelength-division-multiplexing and demultiplexing by using a substrate-mode grating pair,' *Opt. Lett.* **17**, 1629–1631 (1992).

34. V. Minier, A. Kevorkian, and J.M. Xu, 'Superimposed phase gratings in planar optical waveguides for wavelength demultiplexing applications,' *IEEE Photonics Technology Letters.* **5**, 330–333 (1993).

35. M.M. Li and R.T. Chen, 'Five-channel surface-normal wavelength-division demultiplexer using substrate-guided waves in conjunction with polymer-based Littrow hologram,' *Opt. Lett.* **20**, 797–799 (1995).

36. Y. Amitai, 'Design of wavelength-division multiplexing/demultiplexing using substrate mode holographic elements,' *Opt. Comm.* **98**, 24–28 (1993).

37. L. Solymar and D.J. Cooke, *Volume Holography and Volume Gratings* (Academic Press, London, 1981) Ch. 4.

38. S.A. Benton, S.M. Birner, and A. Shikakura, 'Edge-lit rainbow holograms,' *Proceedings of the SPIE* **1212**, 149–157 (1990).

39. A.N. Putilin, V.N. Morozov, Q. Huang, and J.H. Caufield, 'Waveguide holograms with white light illumination,' *Opt. Eng.* **30**, 1615–1619 (1991).

40. T. Kubota, K. Fujioka, and M. Kitagawa, 'Method for reconstructing a hologram using a compact device,' *Appl. Opt.* **31**, 4734 (1992).

41. Y. Amitai, I. Shariv, A.A. Friesem, M. Kroch, and S. Reinhorn, 'White-light holographic display based on planar optics,' *Optics Letters* **18**, 1265–1267 (1993).

42. J. Upatnieks, 'Edge illuminated holograms,' *Appl. Opt.* **31**, 1048–1052 (1992).

43. J. Upatnieks, 'Compact holographic sight,' *Proceeding of the SPIE* **883**, 171–176 (1988).

44. Y. Amitai, S. Reinhorn, and A.A. Friesem, 'Visor-display design based on planar holographic optics,' *Applied Optics* **34**, 1352–1356 (1995).

45. Revital Shechter, 'Planar holographic elements with uniform diffraction efficiency,' M.Sc. dissertation (Weizmann Inst. Sci., Rehovot, 1995).

46. J. Kedmi and A.A. Friesem, 'Optimal holographic Fourier-transform lens,' *Appl. Opt.* **23**, 4015–4019 (1984).

47. Y. Amitai and J.W. Goodman, 'Design of Fourier transform holographic lenses in the presence of recording-readout wavelength shift,' *Optics Letters* **16**, 952–954 (1991).

48. S. Reinhorn, S. Gorodeisky, Y. Amitai, and A.A. Friesem, 'Fourier transformation with planar holographic doublet,' *Opt. Lett.* **20**, 495–497 (1995).

49. M. Kato, Y.-T. Huang, and R.K. Kostuk, 'Multiplexed substrate-mode holograms,' *Jour. Opt. Soc. Amer.* **A7**, 1441–1447 (1990).

50. R.K. Kostuk, M. Kato, and Y.-T. Huang, 'Polarization properties of substrate-mode holographic interconnects,' *Appl. Opt.* **26**, 3848–3854 (1990).
51. Y.-T. Huang and Y.-H. Chen, 'Polarization-selective elements with s substrate-mode grating pair structure,' *Opt. Lett.* **18**, 921–923 (1993).
52. J.-T. Chang, D.-C. Su, and Y.-T. Huang, 'Substrate-mode polarization-division-multi/demultiplexer for optical communication,' *Appl. Opt.* **33**, 8143–8145 (1994).
53. I. Shariv, Y. Amitai, and A.A. Friesem, 'Compact holographic beam expander,' *Opt. Lett.* **18**, 1268–1270 (1993).
54. D.-C. Su, Y.-K. Tsai, and Y.-T. Huang, 'Galilean telescope arrays in infrared region using substrate-mode holograms,' *Jour. Opt.* **24**, 113–116 (1993).

8 Resonant light scattering from weakly rough metal surfaces

Kevin A. O'Donnell
The School of Physics, Georgia Institute of Technology, Atlanta, Georgia, USA

INTRODUCTION

We are surrounded by surfaces with different types of roughness; indeed we perceive the world around us largely through the light diffusely reflected by the rough surfaces of objects. However, is it fair to say that the scattering of light by rough surfaces is well understood? At first glance it is perhaps surprising that, as a physical problem, the scattering of light by rough surfaces is not only poorly understood but is a highly active and developing field. A broad development of new theoretical and experimental techniques has been stimulated by a variety of interesting and even remarkable phenomena noted in this field.

A summary of the wide range of works in surface scattering is beyond the scope of this chapter and may be found elsewhere [1–3]. Instead, our purpose here is to consider cases in which roughness plays an essential role in the resonant excitation of the surface electromagnetic waves known as surface plasmon polaritons. The effects of interest arise when light waves are scattered from metal surfaces with weak random roughness. Further, our discussions reflect the bias of the author and thus concern mostly recent experimental work in this field. The relevant theoretical methods are themselves a substantial subject and, while they are not developed here, we do discuss such works particularly when they provide a physical understanding of the effects noted.

The type of surface scattering problem considered here is deceptively simple to describe. As shown in Fig. 1(A), consider a plane light wave, propagating in free space, incident on the rough surface of a metal. We assume that the metal is otherwise homogeneous and isotropic, is linear in its response to the light

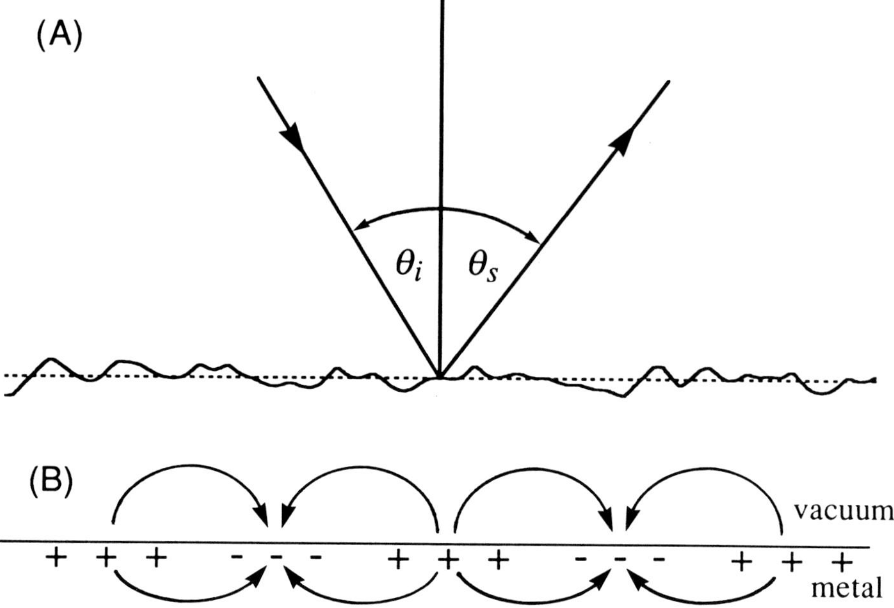

Fig. 1. Top: scattering of light by a rough metal surface. The incident and scattering angles θ_i and θ_s are positive as shown. Bottom: a surface plasmon polariton on a flat metal surface. The curved lines denote the electric field lines arising from the charge distributions shown.

wave, and that both media extend an infinite distance from the interface. The surface corrugation $h(x)$ is taken to be either purely random or else a superposition of a random and a periodic function.

A perfectly flat surface would produce, of course, only a specularly reflected wave of amplitude consistent with the Fresnel reflection coefficient. The surface corrugation complicates the issue considerably. Far from the surface, it is typically found that the specular reflection may appear with reduced magnitude, and that scattered light is now directed to other angles in the upper half-space. How are these effects related to the properties of $h(x)$? Indeed, there have never been any general, closed-form expressions derived from rigorous theory that predict the scattering behaviour; theoretical works instead commonly employ approximate perturbational or numerical techniques. This situation implies that there is no shortage of partially or even poorly understood phenomena in surface scattering.

The organization of this chapter is as follows. First, the properties of a surface plasmon polariton are discussed and we briefly consider the related effect of near-total absorption by an otherwise highly reflective diffraction grating. We then present the first observations of backscattering enhancement due to polariton excitation on a rough metal surface; the case considered has random roughness with a relatively narrow spectrum. Finally, observations of the diffuse light bands are presented. The surface studied there could be termed a roughened diffraction grating, in which a strong surface Fourier component as well as a continuous range of roughness wavenumbers allow interaction with surface polaritons.

SURFACE PLASMON POLARITONS

A surface plasmon polariton is a type of electromagnetic wave that may be created in the scattering situation of Fig. 1(A). The polariton is unusual in that the field takes on a maximum at the surface and decays exponentially in both media as one moves away from the surface. Other terms that have been used equivalently include surface plasmon and surface electromagnetic wave; here we employ the full phrase surface plasmon polariton (often abbreviated to surface polariton or simply polariton in discussions) for the surface wave that occurs for metals at optical and infrared frequencies.

A surface plasmon polariton corresponds to the longitudinal oscillations of the conduction electrons of a metal. The situation is depicted in Fig. 1(B) for a polariton on a planar metal surface; at optical frequencies the electron plasma wave decays rapidly as a function of distance into the metal. The properties of the electromagnetic field associated with the plasma wave may be derived from Maxwell's equations [4, 5]. The most important result for our purposes here is that, in the direction along the interface, this field has the wavenumber

$$k_{\mathrm{sp}} = \frac{\omega}{c}\,\mathrm{Re}\!\left(\sqrt{\frac{\varepsilon}{\varepsilon + 1}}\right), \tag{1}$$

where ω is the frequency and ε is the complex dielectric constant of the metal. Upon evaluating Eq. (1) for typical metals it is generally found that k_{sp} is slightly greater than ω/c; on a flat interface a surface polariton thus cannot couple to an outgoing light wave, nor can an incident light wave excite a surface polariton. More verbosely, for two waves to couple the periodicities of the waves along the flat interface must match one another; with $k_{\mathrm{sp}} > \omega/c$, k_{sp} remains greater than the parallel wavevector component $(\omega/c)\sin\theta$ of a wave propagating at angle θ in free space.

The techniques of prism coupling and diffraction grating coupling have been widely used to allow light waves to excite surface plasmon polaritons [4, 5]; the latter method is of direct relevance to the rough surfaces of interest here. A metallic diffraction grating with period p_0 would generate a set of diffracted orders consistent with the grating equation

$$k' = k + nk_{\mathrm{g}}, \tag{2}$$

where $k = (\omega/c)\sin(\theta_i)$ and k' are, respectively, wavevector components parallel to the mean surface of the incident and diffracted waves, $k_g = 2\pi/p_0$ is the wavenumber of the grating, and $n = 0, \pm 1, \pm 2, \ldots$ is the order of diffraction. With appropriately chosen parameters, it is a simple matter to excite a surface polariton by producing an evanescent diffracted order with k' identical to $\pm k_{sp}$ of Eq. (1) (the plus and minus signs denote, respectively, polaritons travelling to the right or left in Fig. 1(B)).

The consequences of such an excitation can be remarkable and produce the resonant absorption anomalies of diffraction gratings. We briefly consider the case of a gold sinusoidal grating with $p_0 = 555$ nm as was first discussed by Hutley and Maystre [6]; the results shown here are from a later work [7]. Experimental data for the power contained in the zero order with illumination wavelength $\lambda = 633$ nm is shown in Fig. 2 for p polarization (electric field in the plane of incidence) and s polarization (the orthogonal case). While the result in s polarization is relatively flat with a high reflectivity, in p polarization there is a deep minimum for $\theta_i \cong 4.8°$. With $n = -1$ in Eq. (2), the minimum is consistent with polariton excitation for $k' = -k_{sp} = -1.057(\omega/c)$. The minimum corresponds only to absorption, as there are no non-zero propagating orders present until the $n = -1$ order emerges as a propagating wave at $\theta_s = -90°$ for $\theta_i \cong 8.1°$. It was demonstrated in Ref. 6 that if the grating has a slightly greater amplitude than in Fig. 2, a *total* absorption of an incident p-polarized wave occurs due to the excitation of the surface plasmon polariton.

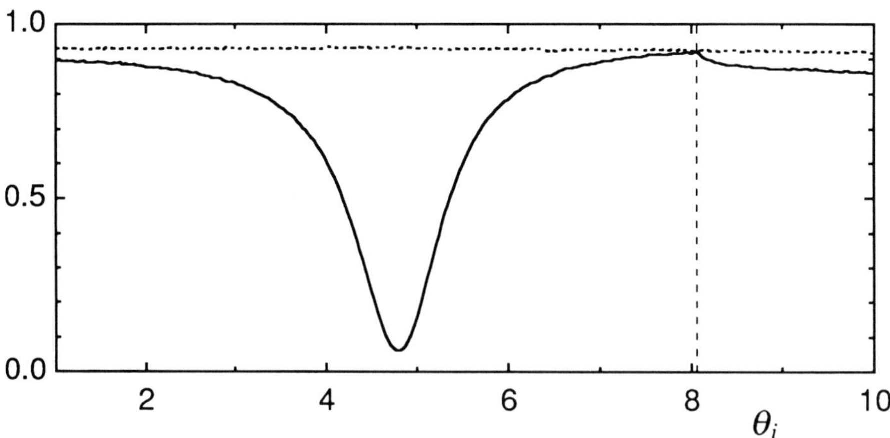

Fig. 2. Experimental results for the powers of the zeroth diffracted order as a function of incidence angle θ_i in p and s polarization (solid and dashed curves, respectively). The grating is of sinusoidal profile in gold with period 555 nm and amplitude 16.6 nm; the illumination wavelength is 633 nm. The dashed vertical line denotes the emergence of the first diffracted order.

There are a number of observations to be made at this point. First, the resonant absorption anomalies occur only for an incident light wave with p polarization; this is consistent with coupling to the p-polarized field of the polariton as created by the charge distribution shown in Fig. 1(B). Also, a surprisingly weak grating amplitude (approximately $\lambda/40$ here) is all that is required to provide strong coupling to the resonance of the metal with easily observable consequences. We also discuss briefly two further considerations that are necessary in the following discussions concerning randomly rough surfaces. We consider here only one-dimensional roughness, for which $h(x)$ remains a one-dimensional function as would be the case for a diffraction grating. Illuminating such a randomly corrugated surface with an incident wave whose wavevector is perpendicular to the surface's grooves produces scatter only in the plane of incidence; all waves remain in the plane of Fig. 1(A). Secondly, the amplitude or intensity scattered by such a surface is itself a random process and averaging is required to obtain a meaningful result. The angular dependence of the mean diffusely scattered intensity is thus of primary importance, and averaging is required to remove speckle noise in both theoretical and experimental works.

POLARITON-RELATED BACKSCATTERING ENHANCEMENT

Backscattering enhancement has attracted much attention in surface scattering in recent years. Mechanisms such as multiple scattering within surface valleys or geometrical optics have been found to produce some of the effects. However, a remarkable polariton-related backscattering effect was predicted in 1985 in two pioneering theoretical papers [8, 9] that preceded much of the later interest; subsequent theoretical works have also made similar predictions [10, 11]. The effect has remained elusive in experimental work and was only observed recently [12] under circumstances that were significantly different from the theoretical cases. In this section we present some of the theoretical results, describe the physical mechanisms responsible, and then discuss the approach employed in the experimental observations.

The theoretical predictions [8, 9] were developed from perturbation theory for $h(x)$ being a Gaussian random process with correlation function

$$C(\Delta x) = \langle h(x)h(x + \Delta x) \rangle = \sigma^2 e^{-(\Delta x/a)^2}, \tag{3}$$

where $\langle \ \rangle$ denotes an ensemble average, σ is the surface standard deviation, and a is the correlation length of $h(x)$. Figure 3 shows the p-polarized mean diffuse intensity I_p with $\sigma \cong \lambda/90$ as taken from Ref. 8. An unusual narrow peak appears in I_p in the backscattering direction; the peak persists there with reduced height as θ_i is increased. No such peak is seen in the s-polarized diffuse intensity I_s [13]. These and other results of this section represent mean scattered power per unit angle for unit identically polarized incident power; the

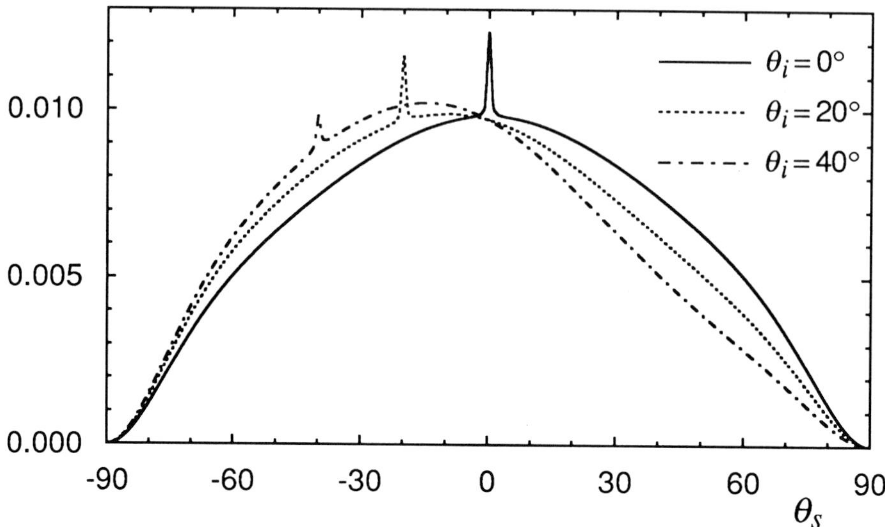

Fig. 3. Top: from Ref. 8, the diffuse intensity I_p from a silver surface with one-dimensional roughness for the incidence angles indicated. Parameters are: $a = 100$ nm, $\sigma = 5.0$ nm, $\lambda = 457.9$ nm, and $\varepsilon = -7.5 + 0.24i$ (silver). The peak occurs in the backscattering direction ($\theta_s = -\theta_i$). An error of a factor of 2π in Fig. 2 of Ref. 8 has been corrected.

area of I_p in Fig. 3 corresponds to slightly less than 2 percent of the incident power and much of the scattered power escapes in the specular reflection.

The physical mechanism responsible for the peak in I_p was described in a later work [13] and again requires an interaction of a p-polarized incident wave with the surface plasmon polariton. The argument proceeds as follows. The incident wave scatters from a point on the rough surface and launches a surface polariton. The polariton then reaches a second point, where it is scattered to create a propagating wave escaping from the surface. The time-reversed version of this process also will occur, in which the incident wave interacts with the second point, excites a counterpropagating surface polariton, and then couples to an outgoing wave at the first point. These two sequences produce diffuse scattering contributions that interfere constructively in the direction of retroreflection. Further, this argument holds for all pairs of points on a random surface and, upon averaging, a narrow backscattering peak is produced of diffraction width inversely proportional to the mean free path of the polariton.

However, a key question still remains: what property of the rough surface is truly essential in producing the effect? A satisfactory answer involves an extension of the discussion of grating diffraction (above) and is consistent with the surface wavenumber formalisms developed in Refs. *8–11*. First, the frequency content of the random process $h(x)$ is specified by its roughness spectrum $S(k)$, which is the Fourier transform of $C(\Delta x)$. The arguments of the previous paragraph require that an incident wave must couple simultaneously to $\pm k_{sp}$. As was the case for the grating anomaly in (above), this inward coupling may be expressed using Eq. (2) (but in the ± 1st orders to excite $\pm k_{sp}$) as

$$+k_{sp} = \frac{\omega}{c} \sin \theta_i + k_1 \qquad \text{and} \qquad -k_{sp} = \frac{\omega}{c} \sin \theta_i - k_2, \tag{4}$$

where k_1 and k_2 are wavenumbers present in the rough surface (i.e., wavenumbers for which $S(k)$ is non-zero). The methods of Refs. *8–11* include multiple scattering, and predict that a polariton that has thus been excited may now couple to a propagating wave through yet another grating equation. The outward coupling of the excited polaritons may be expressed by more versions of Eq. (2) using the ∓ 1st orders (so as to produce propagating waves), with $k = \pm k_{sp}$ and $k' = (\omega/c)\sin \theta_s$ as in

$$\frac{\omega}{c} \sin \theta_s = +k_{sp} - k_3 \qquad \text{and} \qquad \frac{\omega}{c} \sin \theta_s = -k_{sp} + k_4, \tag{5}$$

where k_3 and k_4 are again wavenumbers that are available in $S(k)$. Eqs. (4–5) thus express the scattering of an incident wave to an outgoing wave via a polariton channel; the net effect involves a difference of two surface wavenumbers.

We now make an important observation: for a modest range of θ_i between $\pm \theta_{max}$, Eqs. (4–5) imply an interaction with a limited region of $S(k)$. At the limiting angle $\theta_i = \theta_{max}$, the wavenumbers of the surface roughness that are necessary to excite both polaritons follow from Eqs. (4) as

$$k_1 = k_{min} = k_{sp} - \frac{\omega}{c} \sin \theta_{max} \qquad \text{and} \qquad k_2 = k_{max} = k_{sp} + \frac{\omega}{c} \sin \theta_{max}. \tag{6}$$

If $S(k)$ is non-vanishing between k_{min} and k_{max}, it is readily verified that Eqs. (4) predict that the intermediate values of k allow polariton excitation to occur for all θ_i between $\pm \theta_{max}$. Similarly, Eqs. (5) predict that the excited polaritons will be coupled to scattered waves directed to a continuous range of θ_s also lying between $\pm \theta_{max}$. Thus the essential property to produce backscattering enhancement is quite simply stated: $S(k)$ need only be significantly non-zero between k_{min} and k_{max}.

Fabrication Techniques

It is not unreasonable to state that the single most difficult task of an experimentalist in this field involves the fabrication of suitable surfaces. The author has no idea how to fabricate surfaces consistent with the parameters of Fig. 3; we have also made unpublished (and quite unsuccessful) attempts to observe the effect in the infrared with $\lambda = 3392$ nm, $a/\lambda \cong 0.12$, and $a/\lambda \cong 0.007$. There have been many experimental observations of backscattering enhancement and numerous 'unexplained' effects have been noted; however, no effects reasonably consistent with the polariton-related mechanism have been observed. For example, the simple demonstration of a backscattering effect that occurs in p but not in s polarization has not, to the author's knowledge, appeared in the literature.

The closing discussion of the previous section in combination with the method with which the grating of Fig. 2 was fabricated have led to a fresh experimental approach [12]. The grating of Fig. 2 was created with holographic fabrication techniques [14], in which a pair of overlapping laser beams produced a sinusoidal intensity pattern that was recorded in a photoresist-coated glass plate. The development of the plate then produced a sinusoidal surface relief. It would appear to be possible to expose a single photoresist plate to a series of N fringe patterns with differing periodicities, develop the plate, and obtain a surface with a range of wavenumbers that satisfy Eqs. (4–5).

These vague statements must be mathematically justified; the problem amounts to constructing a realization of a random process $h(x)$ from its Fourier components. The full details appear in Ref. *12*, but the essential requirements are that the exposing fringe patterns must be randomly phased, and that N must be large to obtain both a satisfactory spectrum and convergence of $h(x)$ to convenient Gaussian statistics. For extreme simplicity, we have also chosen to produce a spectrum that is essentially flat between k_{min} and k_{max}, with negligible spectral power elsewhere. The parameters for the surface to be discussed here are $N = 500$ exposures, an illumination wavelength of 612 nm, $\theta_{max} = 13.5°$, and, with k_{sp} estimated to fall at $1.059(\omega/c)$ for gold, it follows from Eqs. (6) that $k_{min} = 2\pi/(741 \text{ nm})$ and $k_{max} = 2\pi/(474 \text{ nm})$. An optically thick layer (200 nm) of pure gold was deposited on the photoresist in the final step of the fabrication process.

A segment of $h(x)$ from the resulting surface as obtained with a Talystep stylus profilometer is shown in Fig. 4. The surface was found to have an rms roughness of $\sigma = 10.9$ nm and is obviously a narrow-band random process. Figure 4 also shows $S(k)$ as estimated from 23 profilometer scans of length 80 μm. The sharp spectral limits are clearly visible at k_{min} and k_{max}, and quite low levels of spectral power are present for smaller k.

Experimental Observations

The scattering instrument employed is quite simple in principle and is described elsewhere [12]. The averaging over speckle fluctuations mentioned earlier is achieved by integrating over a 0.5° scattering angle, as well as moving the sample over a 15×30 mm area to provide each data point.

The mean diffusely scattered intensities I_p and I_s at $\lambda = 612$ nm are shown in

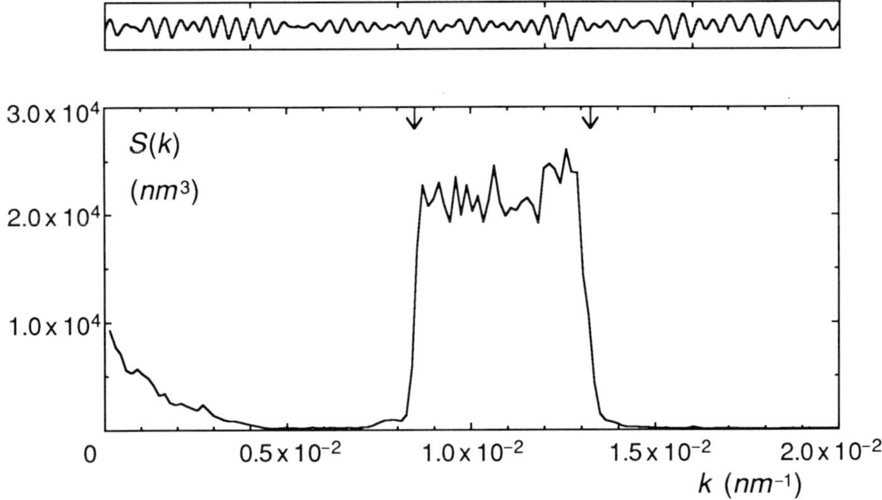

Fig. 4. Results from surface profilometry. Uppermost plot: a segment of unprocessed profilometer data from the rough surface. The vertical scale is ±50 nm and ticks on the horizontal axis are spaced by 5000 nm; these scales exaggerate surface slopes by a factor of 15. Lower plot: roughness spectrum $S(k)$; normalization is such that $\sigma^2 = \int_0^\infty S(k)\,dk$. The two arrows denote the desired spectral limits k_{min} and k_{max}.

Fig. 5 for three values of θ_i. First, in these results there is a considerable amount of diffuse scatter directed to large $|\theta_s|$. We attribute this to single scatter from the surface roughness consistent with the grating equation

$$\frac{\omega}{c}\sin\theta_s = \frac{\omega}{c}\sin\theta_i + nk_r, \tag{7}$$

where $n = \pm 1$, and k_r is a roughness wavenumber falling between k_{min} and k_{max}. With $n = 1$, Eq. (7) is consistent with diffuse scatter for $\theta_s \geqslant 56°$ when $\theta_i = 0°$, but these scattered modes become evanescent for $\theta_i \cong 10°$ and remain so for larger θ_i. With $n = -1$, Eq. (7) predicts scattering for $\theta_s \geqslant -56°$ when $\theta_i = 0°$, $\theta_s \leqslant -41°$ when $\theta_i = 10°$, and for θ_s within the range $(-79°,\ -31°)$ when $\theta_i = 18°$. All of these considerations are closely consistent with the behavior of I_p and I_s at large $|\theta_s|$ in Fig. 5.

More unusual behaviour is apparent in Fig. 5 at small $|\theta_s|$; plots of I_p showing greater detail in this region are also presented in Fig. 6. For $\theta_i = 0°$, 5°, and 10°, I_p presents a compact shape with angular limits that remain fixed at $\theta_s \cong \pm 13°$, and a narrow peak of nearly twice the height of the remainder of this distribution appears at the backscattering position. However, at $\theta_i = 18°$ this scattering distribution has disappeared in I_p and a much lower level of scatter is seen near the specular angle that is only slightly higher than I_s. In

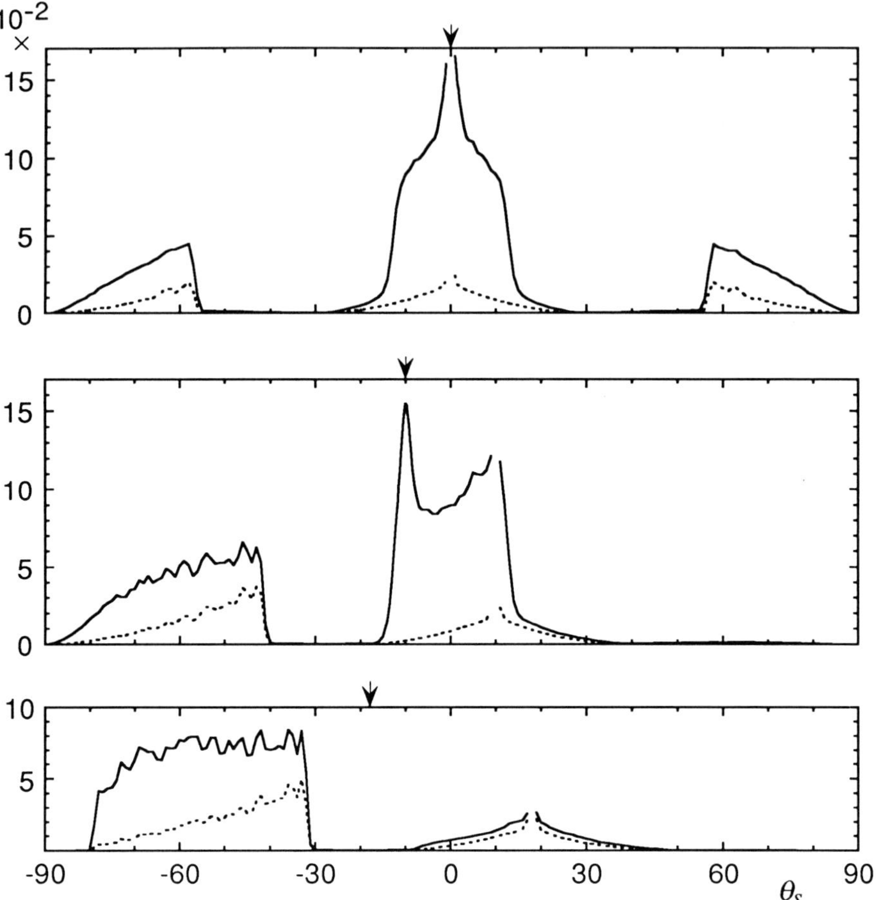

Fig. 5. The diffuse intensities I_p (solid lines) and I_s (dashed lines) plotted as a function of scattering angle θ_s. Incidence angles are 0° (top), 10° (centre), and 18° (bottom). The wavelength is 612 nm and the backscattering direction ($\theta_s = -\theta_i$) is indicated by the arrow.

fact, I_s always presents a low curve that does not change form significantly, and only proceeds into the direction of forward-scattering with increasing θ_i.

All of these observations at small $|\theta_s|$ are consistent with backscattering enhancement in I_p involving the surface plasmon polaritons. As discussed, $S(k)$ has been constructed to allow the polaritons to outwardly couple to θ_s within angular limits at ±13.5°, for all θ_i allowing polariton excitation. The resulting distribution should remain within these fixed angular limits as θ_i is varied and, as seen in Figs 5 and 6, may thus be easily distinguished from the motion of the single scatter consistent with Eq. (7). It appears likely that the

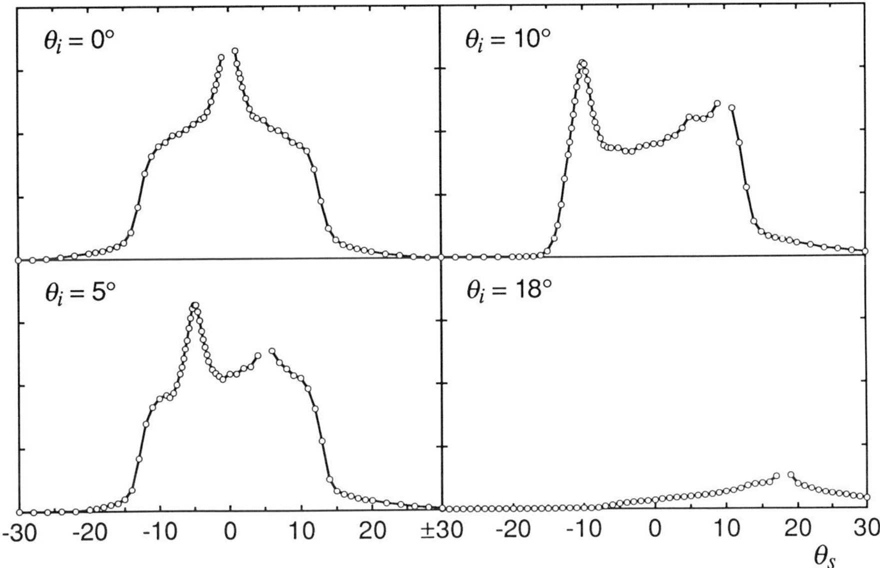

Fig. 6. For wavelength 612 nm, detail of the diffuse intensities I_p for small θ_s. Ticks on the y-axis are spaced by 0.05. For the range of θ_s shown, the powers contained in I_p are 0.052, 0.053, 0.050, and 0.0085 for $\theta_i = 0°$, 5°, 10°, and 18°, respectively.

low scattering levels seen in I_s near the specular direction, as well as in I_p for $\theta_i = 18°$ where the inward polariton coupling has broken down, correspond to single scattering (according to Eq. (7)) from the small amount of low frequencies seen in $S(k)$ in Fig. 4.

It is also of interest to consider the wavelength dependence of the scattering properties of this surface. As the wavelength is varied, there will be angular shifts of the scattering distributions created by $+k_{sp}$ and $-k_{sp}$. This may be seen by determining the scattering angle coupling limits of $+k_{sp}$ from the first of Eqs. (5) as

$$\frac{\omega}{c} \sin \theta_{max} = +k_{sp} - k_{min} \quad \text{and} \quad \frac{\omega}{c} \sin \theta_{min} = +k_{sp} - k_{max}. \quad (8)$$

From the second of Eqs. (5), it follows that the outward coupling of $-k_{sp}$ falls within the range $(-\theta_{max}, -\theta_{min})$. The intensity I_p for $\theta_i = 4°$ with these coupling limits superposed is shown in Fig. 7 for wavelengths 543, 594, 612, 633, and 674 nm. It is seen that the scattering distribution for small $|\theta_s|$ is most compact at $\lambda = 612$ nm, and broadens to span the coupling limits at 594, 633, and 674 nm. Backscattering enhancement is apparent for all λ except for 543 nm, where an unusual right-skewed distribution is observed. Indeed, if one determines the *inward* coupling limits imposed by Eqs. (4) at 543 nm, it is found that $+k_{sp}$

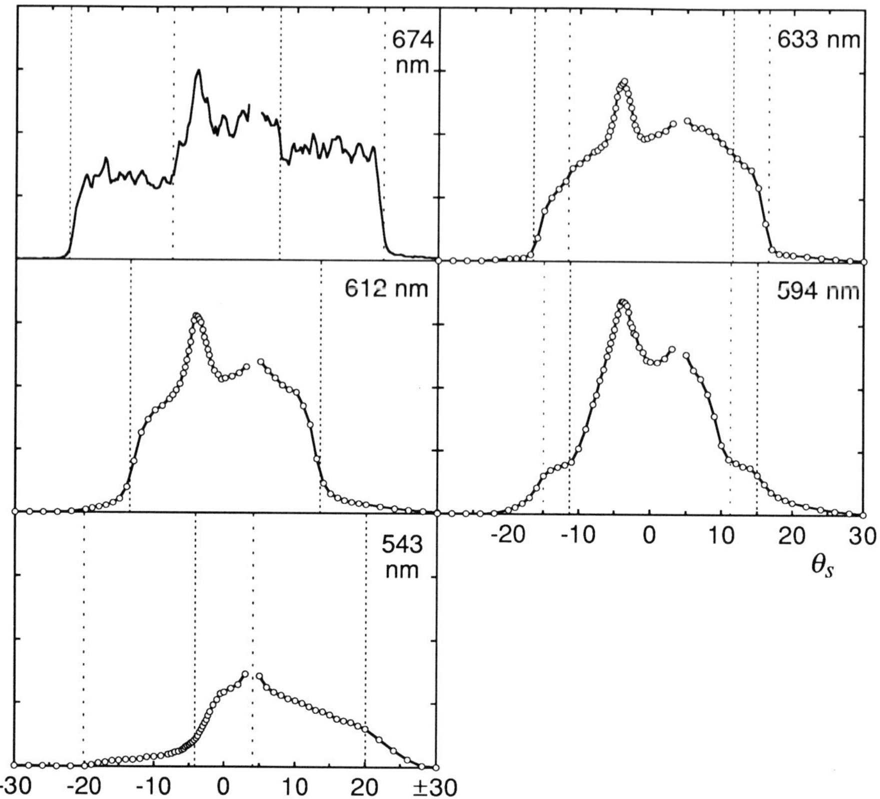

Fig. 7. For $\theta_i = 4°$, the diffuse intensity I_p for wavelengths indicated. The dashed vertical lines denote the outward coupling limits of $+k_{sp}$ (dense lines) and $-k_{sp}$ (sparse lines); these limits overlap at $\lambda = 612$ nm. Ticks on the y-axis are spaced by 0.05. Results at $\lambda = 674$ nm have not been averaged or properly normalized and the fluctuations are reproducible speckle noise.

should be excited for θ_i on the interval $(-3.9°, 20.1°)$, while $-k_{sp}$ should be excited for θ_i within $(-20.1°, 3.9°)$. Thus $+k_{sp}$ should be strongly excited at $\theta_i = 4°$, but the inward coupling to $-k_{sp}$ should be breaking down. Hence these considerations are consistent with the far stronger scattering levels within the $+k_{sp}$ angular limits and, for that matter, the lack of backscattering enhancement seen at 543 nm in Fig. 7.

A more fundamental effect notable in Fig. 7 is that, for the four cases where $+k_{sp}$ and $-k_{sp}$ are strongly excited and have scattering distributions that overlap well at $\theta_s = -\theta_i$, the width of the backscattering peak increases slightly as λ decreases. The observed trend is indeed just the opposite of that seen for another type of backscattering enhancement when $a > \lambda$ and $\sigma > \lambda$, where the peak width increases linearly with λ [15]. The broadening of the peak with decreasing λ is consistent with the numerical calculations for a Gaussian spectrum due to Michel [*11*], who noted a change in peak width for a particular surface by a factor of seven over a considerably wider range of λ. In an analytical theory [*8*], this effect arises from both the broadening of the polariton resonance and the increase in radiative damping of the polariton as λ decreases.

In summary, the results of Figs 5–7 are quite convincing in demonstrating that the backscattering enhancement is due to surface plasmon polariton excitation. The effect arises for incident and scattering angles consistent with polariton excitation, it occurs in I_p but not in I_s, and it has the appropriate wavelength dependence. Indeed, it has been quite recently shown that these data compare well with the perturbation methods of Refs. *8* and *10* as applied to the experimental $S(k)$, with no free parameters [*16*]. The differences with the original theoretical predictions of Fig. 3 are entirely due to the differing forms of $S(k)$. Without the narrow spectral limits in $S(k)$, it would be expected that the single scattering seen only at high angles in Fig. 5 would expand to encompass all θ_s, and that the outward coupling of the polaritons would similarly cover all θ_s. For the broad Gaussian spectra assumed in the theoretical works, this produces the smooth distributions of Fig. 3.

THE DIFFUSE LIGHT BANDS AND RELATED EFFECTS

The second polariton-related scattering phenomenon considered here concerns the diffuse light bands arising from roughened diffraction gratings. The diffuse bands had been noted in visual observations over twenty years ago by Hutley and Bird [*17*] and by Cowan [*18*] using holographic diffraction gratings with incidental two-dimensional roughness. For a p-polarized incident wave, unusual bands appeared in the diffusely scattered light that were attributed to the excitation of surface plasmon polaritons. There have been subsequent theoretical studies of the diffuse bands using approximate surface impedance methods [*19*], perturbation theory [*20*], and further experimental observations have been presented [*21*].

In all experimental observations cited above the random roughness of the grating arose through uncontrolled imperfections introduced during fabrication procedures; certainly none of the experimental papers have measured the angular scattering distributions for a well-characterized surface. This comment is not merely a technical issue concerning the subtleties of experimental technique for, as will be seen below, previously unnoticed effects associated with the bands (including backscattering enhancement) are apparent only when controlled experiments are conducted. As was the case earlier, the difficulties to be surmounted are largely those associated with the fabrication of suitable surfaces.

In the following, we describe experimental studies conducted with a highly one-dimensional surface with profile $h(x) = A \sin(k_g x) + \zeta(x)$ consisting of both a deterministic sinusoidal part and of a random part $\zeta(x)$ that is consistent with Gaussian statistics. The surface was fabricated on a photoresist-coated plate by superposing two exposing patterns. In the first, the plate was exposed to a fine-scaled speckle pattern of correlation length approximately 500 nm as the plate was translated by 1 mm. The resulting net exposure was thus highly one-dimensional with a narrow correlation length in one dimension and a correlation length comparable to the scan length in the second dimension. Moreover, any point of the plate received contributions from many statistically independent speckles during exposure, so that the statistics of this exposure should again follow convenient Gaussian statistics. The second exposure was to a sinusoidal intensity pattern of period $p_0 = 555$ nm in a standard holographic fabrication geometry [14]. The period is identical to that of the grating employed in Fig. 2. In fact, the grating of Fig. 2 as well as the roughened grating of interest here were fabricated at different places on the same plate; they are patches A and E, respectively, of Ref. 7.

After development, the plate was coated with an optically thick layer (200 nm) of pure gold. A segment of a typical profilometer scan is shown in Fig. 8; a random surface structure as well as a periodic behaviour is apparent. The two surface components are more obvious in the spectrum $S(k)$ (also shown in Fig. 8), for which the random roughness produces a smoothly decaying shape, and the periodic component produces a narrow peak for wavenumber k_g. From the area of the peak, the grating amplitude A is determined to be 13.1 nm and, from the area of the remainder of the spectrum, the standard deviation of $\zeta(x)$ is estimated as 13.0 nm.

Observations

The diffuse scatter from the surface appeared as a narrow speckled band following the plane of incidence. Distinct diffraction orders consistent with the deterministic grating were also noted. Further, it was found that a resonant absorption anomaly due to the periodic surface component (again appearing as a minimum of the zero order) was still present at $\theta_i = 4.8°$ with $\lambda = 633$ nm, and was only slightly shallower than that of Fig. 2. Also, for consistency with the conventions of Ref. 7, in this section I_p and I_s are each normalized such that an area of $\frac{1}{2}$ would imply diffuse reflection of all incident power.

We consider here the effects seen as θ_i passes through the resonant absorption anomaly. The diffuse intensities I_p and I_s are thus shown for $\theta_i = 0°$, 4.8°, and 10° in Fig. 9. First, it is clear that I_s remains reasonably symmetric and is similar in shape for the three incidence angles; the peak of I_s simply follows the specular angle as θ_i increases. However, dramatic changes are seen in I_p; for $\theta_i = 4.8°$ there is a broad increase in I_p for negative θ_s (I_p rises by as much as a factor of 4 for $\theta_s \cong -50°$), and the area of I_p increases by a factor of approximately 1.5 upon comparison with the result at $\theta_i = 0°$. The rise of I_p

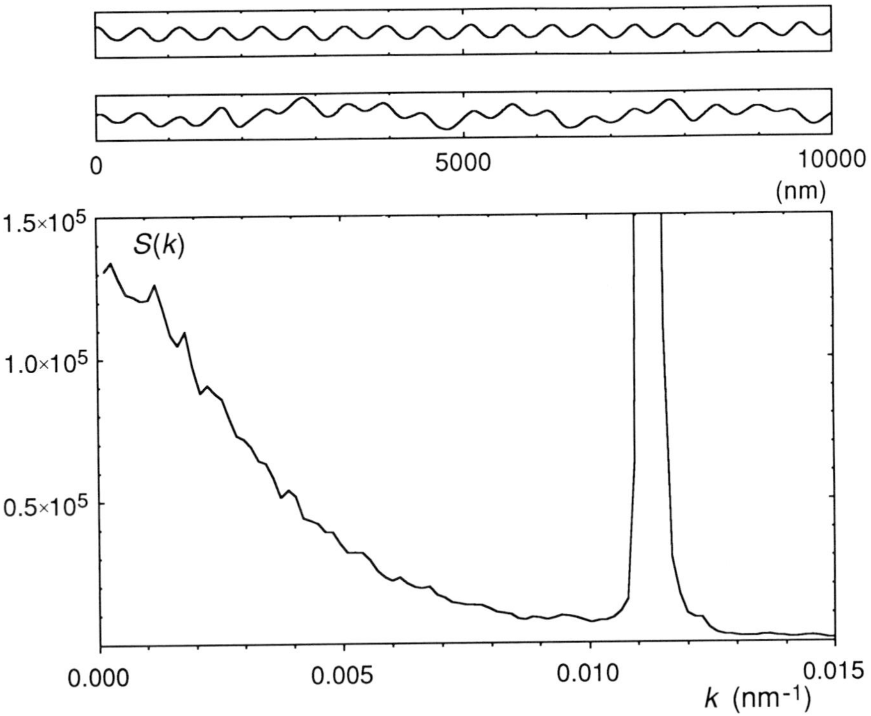

Fig. 8. Top: profilometer scan of the unroughened grating employed in Fig. 2. Middle: profilometer scan of the intentionally roughened grating. The vertical scale of each scan is ±60 nm; this scale exaggerates the slope by a factor of five. Bottom: positive frequency parts of the spectrum $S(k)$ of the roughened grating. Normalization is such that σ^2 in nm^2 is $\frac{1}{2\pi}\int_{-\infty}^{\infty} S(k)\,dk$. The peak at $k_g \cong 0.0113$ nm^{-1} rises by a factor of five above the scale.

occurs only for θ_i where the grating's resonant absorption is significant and, for $\theta_i = 10°$, the area of I_p has returned to levels similar to those of normal incidence. There is also other unusual behaviour in I_p at $\theta_s \cong \pm 4.8°$. For $\theta_i = 0°$ there is a steep slope in I_p at $\theta_s \cong \pm 4.8°$, at $\theta_i = 4.8°$ there is a remarkable maximum at $\theta_s \cong -4.8°$ and a possible minimum for $\theta_s \cong 4.8°$ (the latter is partly obscured by the zero order), and at $\theta_i = 10°$ there is a less distinct maximum and a modest minimum at $\theta_s \cong -4.8°$ and $\theta_s \cong 4.8°$, respectively.

All of the unusual behaviour seen in I_p may be attributed to polariton excitation. We start by considering the observed broad increase in I_p for $\theta_i = 4.8°$. Here the grating is providing a direct excitation of $-k_{sp}$. For the case of Fig. 2 this produces absorption, but for the roughened grating the excited surface wave can also couple to propagating waves via the random roughness.

This may be expressed from Eq. (2) for scattering from a roughness wavenumber with $k = -k_{sp}$, $k' = (\omega/c)\sin\theta_s$, and $n = 1$ as

$$\frac{\omega}{c}\sin\theta_s = -k_{sp} + k_r, \tag{9}$$

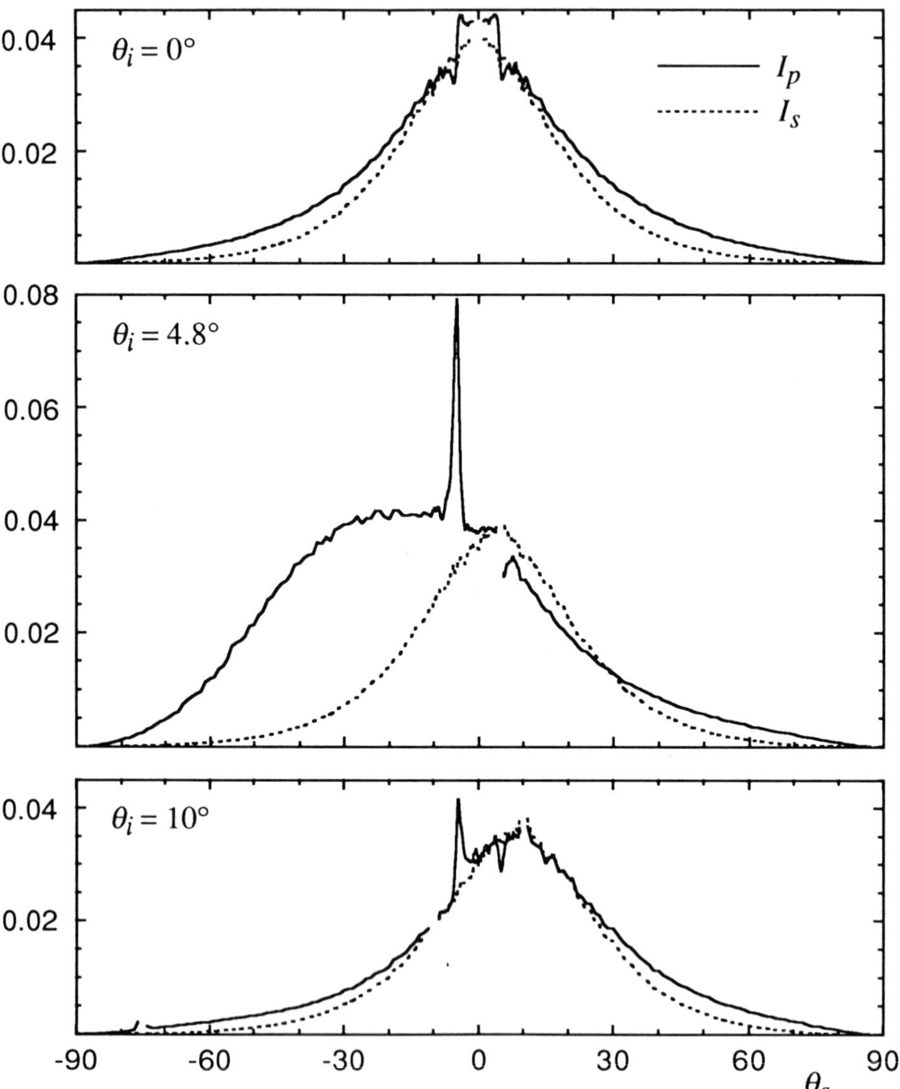

Fig. 9. For wavelength 633 nm, the mean diffuse intensities I_p and I_s for the roughened grating. Diffracted orders are not shown.

where k_r is any of the roughness wavenumbers available in $S(k)$ in Fig. 8. Equation (9) is consistent with scattering contributions for negative θ_s as these are the couplings for which the height of $S(k)$ is greatest.

The narrow structures seen in Fig. 9 in I_p for $\theta_s \cong \pm 4.8°$ are indeed one-dimensional versions of the diffuse light bands observed in the earlier works. However, what physical mechanism gives rise to the bands? The coupling of interest must produce a scattering peak for θ_s equal to the θ_i producing resonant absorption. With some thought, the bands can then only arise from processes that are time-reversed analogues of resonant absorption: the bands represent the strong coupling of the excited polaritons to outgoing waves through the periodic surface component. This be may expressed by versions of Eq. (2) with $k = \mp k_{sp}$, $n = \pm 1$, as in

$$k' = -k_{sp} + k_g \qquad \text{and} \qquad k' = +k_{sp} - k_g, \tag{10}$$

where $k' = (\omega/c)\sin \theta_s$. With $k_{sp} = 1.057(\omega/c)$ as in Fig. 2, the first of Eqs. (10) is responsible for the band at $\theta_s \cong 4.8°$, and the second of Eqs. (10) provides the band at $\theta_s \cong -4.8°$. We have sidestepped an important question: what provides the original polariton excitation? Certainly it cannot be only the deterministic grating, as the bands persist for θ_i far from the resonant absorption anomaly. Thus the original polariton excitation necessarily involves the wavenumbers of the random surface roughness. However, the precise manner in which this excitation occurs is a delicate issue [7].

There are a number of other subtle points that we will not consider in detail here. For example, it is seen in Fig. 9 that the bands take on a variety of forms including a maximum, a minimum, or a steeply sloped region. An s-shaped band is also sometimes seen [7]. Such effects can be interpreted as interference between various scattering contributions; the interference is seen not only in I_p but in other Mueller matrix elements [7]. One effect that is certainly worthy of discussion here is the pronounced increase in height of the backscattered band in the middle plot of Fig. 9. Is this backscattering enhancement?

It is possible to develop a perturbation theory applicable to the roughened grating [7]. Specifically, the scattered amplitude is obtained to first order in the random roughness, but in such a way that it reduces to an exact treatment of the periodic grating when the roughness is removed. Upon calculating the mean diffuse intensity many terms are created and they have been collected as follows: Σ_1 is a single term which involves interaction with the zero grating order, Σ_2 contains the intensities of contributions involving non-zero grating orders, and Σ_3 contains only the interference terms between scattering processes that become time-reversed versions of one another. Thus the role of Σ_3 in producing the strong backscattered band is of key interest; other terms that do not belong to the three classes considered make small contributions and are thus neglected.

In Fig. 10, Σ_1, Σ_2, and Σ_3 are shown for parameters similar to those of the experimental grating for $\theta_i = 4.8°$. The term Σ_1 produces a scattering distri-

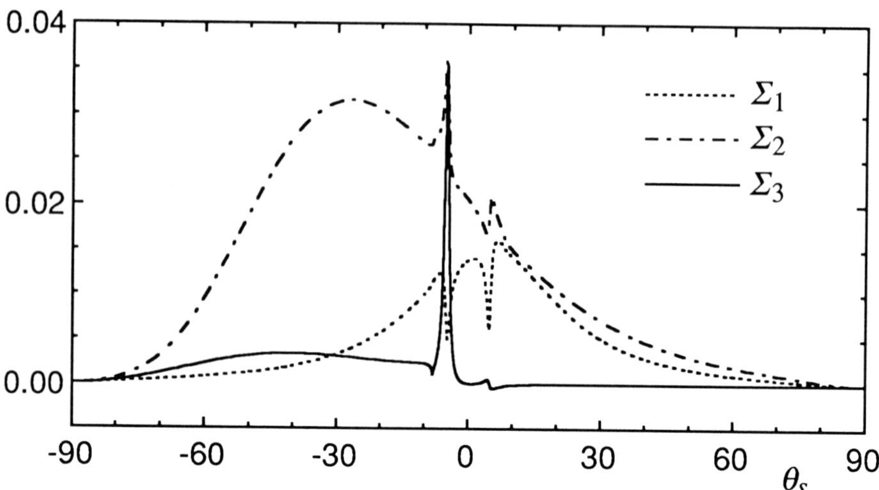

Fig. 10. For $\theta_i = 4.8°$, contributions to the diffuse intensity I_p from the perturbation method.

bution that is skewed toward positive θ_s and is relatively consistent with single scattering, while Σ_2 produces a more unusual distribution skewed toward negative θ_s. Examination of Σ_2 reveals that it is dominated by one term that expresses, as was discussed earlier, the excitation of the surface polariton by the grating, with the subsequent scattering of the polariton by the roughness as in Eq. (9). Most importantly, the results also indicate that Σ_3 produces a large peak in the backscattered band; for other θ_i it is found that Σ_3 is almost insignificant. Thus the enhanced band in the backscattering configuration indeed arises from coherent interference of pairs of time-reversed scattering sequences, as had been the case for the purely randomly rough surface discussed above. The situation here turns out to be slightly different from there because a complete description requires the random surface roughness as well as couplings of higher than first order in the wavenumber of the strong grating component [7]. Moreover, the polariton coupling provided by the grating is essential in producing the effect, and the backscattering enhancement (as well as Σ_3) is thus significant only for θ_i producing resonant absorption. The spectrum of the random roughness is by itself incapable of producing the persistent backscattering effect discussed above, apparently because $S(k)$ is of insufficient height for wavenumbers near k_{sp}.

CONCLUSIONS

It may seem quite surprising that weakly corrugated metal surfaces can produce the strong scattering effects discussed here. In particular, if the structure of a rough surface allows incident and outgoing waves to interact with surface plasmon polaritons, backscattering enhancement, bands, and other unusual effects may be seen in the diffusely scattered light. All of these phenomena represent multiple scattering in the sense that they require an interaction with at least two surface wavenumbers, with the surface plasmon polariton providing an intermediate resonant scattering channel. The relevant theoretical methods are still under development and, with a wide range of opportunities also available for experimental investigation, surface scattering should remain a highly active field for many years to come.

ACKNOWLEDGEMENTS

The author thanks his former postdoc T. R. Michel for pointing out the interesting possibilities in resonant light scattering, as well as graduate students C. S. West and M. E. Knotts for much assistance with the experimental work. The author is also grateful to A. R. McGurn, A. A. Maradudin, and E. R. Méndez for providing the data shown in Fig. 3.

REFERENCES

1. A.G. Voronovich, *Wave Scattering from Rough Surfaces*, Springer-Verlag, Berlin (1994).
2. J.A. Ogilvy, *Theory of Wave Scattering from Random Rough Surfaces*, Adam Hilger, Bristol (1991).
3. J.A. DeSanto and G.S. Brown, in *Progress in Optics*, Vol. 23, (ed. E. Wolf), Elsevier, New York (1986).
4. H. Raether, *Surface Plasmons on Smooth and Rough Surfaces and on Gratings*, Springer-Verlag, Berlin (1988).
5. V.M. Agranovich and D.L. Mills, eds., *Surface Polaritons*, North-Holland, Amsterdam (1982).
6. M.C. Hutley and D. Maystre, *Opt. Commun.* **19**, 431–436 (1976).
7. T.R. Michel, M.E. Knotts and K.A. O'Donnell, *J. Opt. Soc. Amer. A* **12**, 548–559 (1995).
8. A.R. McGurn, A.A. Maradudin, and V. Celli, *Phys. Rev. B* **31**, 4866–4871 (1985).
9. V. Celli, A.A. Maradudin, A.M. Marvin, and A.R. McGurn, *J. Opt. Soc. Amer. A* **2**, 2225–2239 (1985).
10. A.A. Maradudin and E.R. Méndez, *Appl. Opt.* **32**, 3335–3343 (1993).
11. T.R. Michel, *J. Opt. Soc. Amer. A* **11**, 1874–1885 (1994).
12. C.S. West and K.A. O'Donnell, *J. Opt. Soc. Amer. A* **12**, 390–397 (1995).
13. A.A. Maradudin, T.R. Michel, A.R. McGurn, and E.R. Méndez, *Ann. Phys. (N.Y.)* **203**, 255–307 (1990).
14. M.C. Hutley, *Diffraction Gratings*, Academic Press, London (1982), Chapter 4.
15. K.A. O'Donnell and E.R. Méndez, *J. Opt. Soc. Amer. A* **4**, 1194–1205 (1987).

16. A.A. Maradudin, A.R. McGurn, and E.R. Méndez, *J. Opt. Soc. Amer. A* **12**, 2500–2506 (1995).
17. M.C. Hutley and V.M. Bird, *Optica Acta* **20**, 771–782 (1973).
18. J.J. Cowan, *Opt. Commun.* **12**, 373–378 (1974).
19. V.L. Brudny and R.A. Depine, *J. Mod. Optics* **40**, 427–439 (1993) and references therein.
20. A.D. Arsenieva, A.A. Maradudin, J.Q. Lu, and A.R. McGurn, *Opt. Lett.* **18**, 1588–1590 (1993).
21. J.M. Simon and S.A. Ledesma, *Optik* **89**, 145–150 (1992) and references therein.

9 Femtosecond time-and-space-domain holography

Alexander Rebane
*Physical Chemistry Laboratory, Swiss Federal Institute of
Technology, ETH-Zentrum, CH-8092 Zürich, Switzerland*

INTRODUCTION

Time-and-space-domain holography [1–3] is a new method in coherent optics
which allows one to manipulate in addition to the spatial waveforms also the
temporal characteristics of optical wave amplitude. The holographic recording
in time-and-space domain uses the *frequency selectivity* of certain light-
sensitive materials which allows one to capture instead of the conventional
spatial interference, the *frequency-and-space-domain interference* of the object
with the reference wave pulse amplitude. The play-back of the object wave in
the time domain is accomplished by the coherent optical response or photon
echo stimulated by the read pulse.

The high spectral selectivity of the storage media is based upon the effect of
persistent spectral hole burning (PSHB) [4, 5] which in turn is feasible because
at liquid-helium temperatures ($T < 4$ K) the *homogeneous* spectrum of a single
chromophore (molecule, atom, colour center, etc.) in a solid matrix comprises
a narrow purely electronic zero-phonon line (ZPL) [6]. The width of the ZPL
is typically $\Gamma_{ZPL} = 1 - 10^{-2}$ GHz for organic molecules (Γ_{ZPL} in the kHz range
are observed for rare-earth ions in crystals) while the overall *inhomogeneous*
band width of the absorption spectrum Γ_{inh} exceeds the homogeneous line
width by a factor as large as $\Gamma_{inh}/\Gamma_{ZPL} \sim 10^4$. PSHB occurs when the resonant
excitation via the intense and narrow ZPL forces the chromophores either to
undergo a photochemical reaction or shelves them in a long-living metastable
state. As a result, a hole is created in the initially smooth inhomogeneous
frequency distribution of ZPLs which is an imprint of the intensity spectrum of
the illuminating light.

The physical background of PSHB, the basic principles of time-and-space-domain holography, as well as the potential application for optical storage, were discussed in ICO Book II [7] (for a detailed review on PSHB see [8–11]). PSHB significantly enriches the property of photon echo to reproduce the spatial wave front [12, 13] and the temporal envelope [14, 15] of the excitation pulses (for a review on photon echoes see e.g. [16, 17] and also [18] in ICO Book I) by granting direct control over the optical characteristics of the media such as the index of absorption, and correlated to it via dispersion relations index of refraction, with high selectivity in the frequency dimension. We can think about PSHB actually as a bank of slightly damped classical oscillators where the frequency distribution of the resonances can be tailored with the laser beam. The photon echo is then the sum of the linear coherent responses of all oscillators and is called photochemically accumulated stimulated photon echo (PASPE). The time-domain amplitude of PASPE is simply connected to the distribution of the oscillators in the frequency dimension via Fourier and Hilbert transformation (for details about the linear treatment of photon echoes see [19]).

By using a wavelength-tunable cw monochromatic laser, thousands of holograms can be recorded and played back *in series* at different frequencies within the PSHB material [20–22] (see also reviews [10, 23] and references therein). By combining the cw technique with the Stark effect it is possible to construct a 'molecular computer' which carries out logical operations with the holographic images controlled via an external electric field [24].

In the time-domain approach the frequencies are addressed *in parallel* which has the advantage of potentially very fast data transfer rates basically limited only by the bandwidth Γ_{inh}. In this chapter we make use of the fact that in many organic PSHB materials consisting of polymer films doped with porphyrin-type molecules at liquid-helium temperature the width Γ_{inh} is on the order of several teraherz, and we apply this to record femtosecond time-and-space-domain holograms. We first summarize the basic hole-burning features of these materials and then describe the holographic storage with a femtosecond laser. We proceed by discussing some interesting diffraction properties which arise from the circumstance that the coherence length of the femtosecond pulses is much less than the dimensions of the hologram. In the next experiment we combine the cw technique mentioned above with the femtosecond time-domain readout to synthesize arbitrary teraherz pulse trains. Finally we present experiments on associative holographic recall in the time-and-space domain and show a new way of coherent optical image processing with short pulses based on interference in the frequency dimension.

PERSISTENT SPECTRAL HOLE BURNING IN ORGANIC DYE-DOPED POLYMERS

Here we consider as a typical example of photochemical hole-burning of the molecules of chlorin (2,3-dihydroporphyrin) in an organic polymer matrix at liquid-helium temperature. This PSHB system has been studied extensively [10,25–27] and it possesses key features that are common to the porphyrin- and naphthalocyanine-type molecules used in our holographic experiments.

Figure 1 presents a schematic view of the homogeneous and of the inhomogeneous absorption lineshape of chlorin in polyvinylbutyral (PVB) film in the wavelength range of the transition from the ground S_0 electronic state to the first singlet excited electronic state S_1. The homogeneous spectrum of a single chromophore consists of the narrow ZPL accompanied by a broad phonon sideband. The ZPL has a width at $T = 1.7$ K of about 150 MHz which decreases with $T \to 0$ towards the value $\Gamma_{ZPL} = (2\pi\tau_1)^{-1}$, where $\tau_1 \sim 8$ ns is the excited electronic state's decay time (energy relaxation time). The homogeneous linewidth defines the highest possible resolution of the storage in the frequency dimension, whereas the optical dephasing time $T_2 = (\pi\Gamma_{ZPL})^{-1}$ gives the maximum span of the coherent response (echo) in the time domain. The zero-phonon lines corresponding to the transitions to the vibrational levels of S_1 are also present but they are much broader than the purely electronic ZPL and contribute to the phonon sideband. At temperatures $T < 2$ K the ratio between the integrated intensity of the ZPL and the total integrated intensity of the

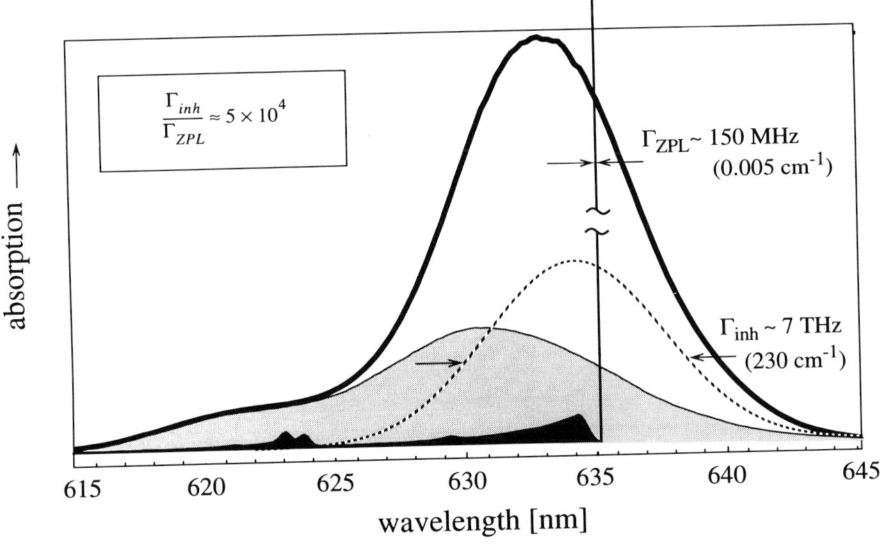

Fig. 1. Absorption spectrum of chlorin in PVB at liquid-helium temperature. The homogeneous spectrum of a single chromophore consists of the ZPL (narrow vertical line) and of the phonon sideband (black area) which also includes the vibrational lines. The inhomogeneous absorption band profile (bold curve) is given by the convolution of the homogeneous spectrum with the inhomogeneous distribution function (dashed curve). The shaded area indicates that part of the inhomogeneous band where the absorption is due to the phonon sideband and the vibrational lines. The remaining (unshaded) area under the bold curve belongs to the absorption by ZPLs only.

homogeneous spectrum is about 0.6–0.7. Note that because of the largely different widths the ZPL has much higher peak absorption than the sideband. The inhomogeneous frequency distribution of the ZPLs is described by a nearly Gaussian function with a width of 7 THz. The inhomogeneous distribution function $g(\nu)$ is defined usually as the concentration of chromophores absorbing via ZPL in the frequency interval $(\nu, \nu + d\nu)$. The observed inhomogeneous absorption band profile is given by the convolution of the inhomogeneous distribution function with the homogeneous absorption spectrum. Note that the relative contribution of ZPLs to the total absorption is largest at the longer-wavelength flank of the inhomogeneous band. The phonon sideband, together with the vibrational zero-phonon lines, contribute more to the shorter-wavelength side of the absorption band. The ratio between the width of the inhomogeneous distribution and the ZPL linewidth is in the present case about 5×10^4.

The photochemical reaction in chlorin leading to the formation of persistent holes is shown in Fig. 2. At room temperature the product form of this

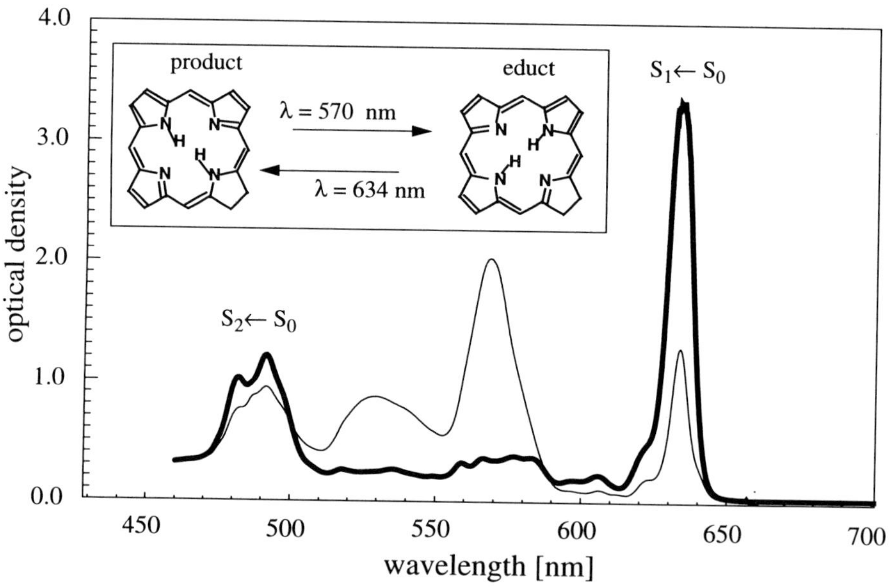

Fig. 2. Photochemical hole burning of chlorin in PVB. Bold curve – absorption of the sample at liquid-helium temperature before illumination. The $S_1 \leftarrow S_0$ band at 634 nm corresponds to the (educt) spectrum shown Fig 1. The band at 490 nm corresponds to the transition from the ground state to a higher excited singlet state and does not contribute to the PSHB. Thin curve – absorption after illumination with spectrally broad light in the interval of 625–645 nm. The illumination bleaches the educt absorbing at a wavelength of 634 nm and creates product absorption at 570 nm. The inset shows the structure of the educt and of the product form of chlorin.

molecule is unstable; however, at low temperature light-induced switching between the educt form and the product form is possible. When the sample is cooled from room temperature to liquid-helium temperature, then most of the molecules are still in the educt form. Illumination in the wavelength interval of 625–645 nm decreases the concentration of the educt and creates a new absorption band centred around 570 nm which is due to the $S_1 \leftarrow S_0$ transition of the photoproduct. The quantum yield of this photochemical reaction in the S_1 state is estimated to be about 0.1% or less. The reverse process bringing the product back to the educt state occurs either by illumination at 570 nm or by heating the sample to $T > 50$ K.

For femtosecond hologram storage it is important that the initial inhomogeneous distribution is a smooth bell-shaped function of sufficient width to accommodate the whole breadth of the frequency spectrum of the ultrashort pulses (hundreds of cm^{-1}). It is also important that the photoproduct absorption does not overlap significantly with the educt absorption band and that the relative contribution of the phonon sideband is not too large. Then the profile of the hole created by PSHB in the educt absorption is proportional to the intensity spectrum of the illuminating light.

Because of the relatively low quantum yield of the photochemical reaction the chromophores have to absorb many photons before the PSHB occurs. In the case of chlorin the average dose of the absorbed energy required to record a hologram in the frequency interval of 100 cm^{-1} is about 10–100 mJ cm^{-2}. The energy of the femtosecond pulses in the experiments described below is only

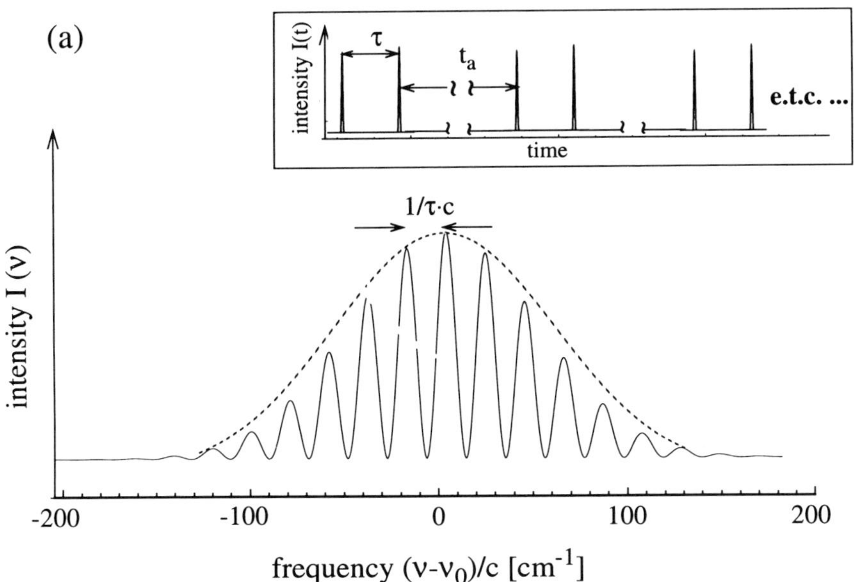

Fig. 3. *Continued*

about 10^{-9}–10^{-10} J and therefore we use repeated illumination and accumulation of the PSHB effect. Figure 3 shows schematically the frequency distribution of the PSHB in the inhomogeneous band of the educt at one spatial location of the sample after different doses of illumination with pairs of femtosecond pulses. The time delay between the two pulses is $\tau < T_2$ and the hologram has the form of a frequency-dimension grating with the period $1/\tau c$. Note that to avoid interference between successive illumination cycles the interval t_a between the repeated illuminations should be larger than the optical dephasing time T_2.

The large inhomogeneous bandwidth providing the theoretical time resolution of $(\Gamma_{inh})^{-1} < 100$ fs is observed in a broad variety of polymers and glasses as solid hosts for organic dye molecules. The inhomogeneous broadening results from the high degree of structural disorder of these

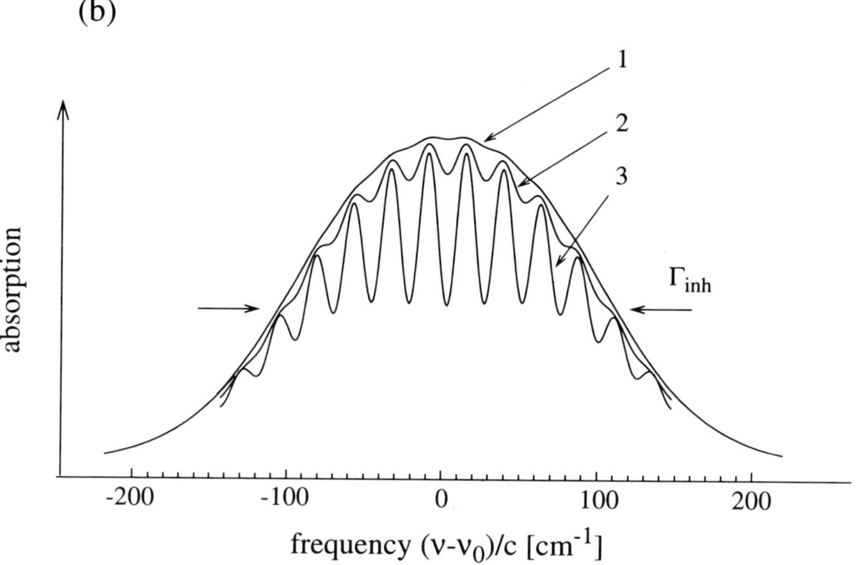

Fig. 3. (a) Mutual intensity spectrum of two temporally non-overlapping fs-duration pulses separated in time by $\tau < T_2$ and interfering with each other in the PSHB material. The result of the interference is the grating in the frequency dimension with the period equal to $(tc)^{-1}$; the dashed curve is the spectral envelope of the pulses; the carrier frequency ν_0 coincides with the maximum of the educt inhomogeneous band. The inset depicts the sequence of repeated illumination with pulse pairs separated by the interval $t_a \gg T_2$. (b) Three different stages of accumulation in the hole structure in the inhomogeneous band. After illumination with only a few pairs of low-energy fs pulses the change of the inhomogeneous band is very small (1). A detectable grating structure appears only after illumination with about 10^7 pulse pairs (2) and a high-contrast hologram is recorded with about 10^9 pulse pairs (3).

materials. Artificially it is possible to further increase Γ_{inh} by doping the polymer simultaneously with different PSHB molecules absorbing at nearby wavelengths.

HOLOGRAM STORAGE WITH FEMTOSECOND PULSES

Consider the hologram storage experiment shown schematically in Fig. 4. The light source is a 100-MHz repetition rate 300 mW average power self-modelocked Ti:sapphire laser generating at the wavelength of 750 nm bandwidth-limited pulses of duration 70 fs (spectral width $\Delta\nu_L/c \sim 140$ cm^{-1}). The PSHB material is a 100 μm-thick and 2×2 cm cross-section polyethylene film doped with naphthalocyanine-type molecules at concentration 10^{-4} mol/l. The hole-burning properties are qualitatively the same as in chlorin, although the educt $S_1 \leftarrow S_0$ transition occurs here at 750 nm and matches the wavelength of the Ti:sapphire laser. The PSHB film is positioned in an optical cryostat at temperature $T = 2$ K. The laser output beam is expanded and divided into two parts. One of the expanded beams serves as the plane reference pulse wave propagating along the z-axis direction, the other illuminates the object (a coin). The light scattered from the coin propagates along the direction, $\vec{Fh} = (0, -\sin\theta, \cos\theta)$, where the angle is $\theta \sim 10°$. The reference wave illuminates the hologram at time $t = 0$, while the object pulse arrives with the delay $\tau = \tau_0 - y\,(\sin\theta)/c$, where $\tau_0 = 5$ ps is the delay at the origin of coordinates. Note that the delay is positive, i.e. the object pulse stays always behind the reference pulse. Because the profile of the coin surface which scatters the light has a depth comparable to the coherence length of the 70-fs-duration pulses the time-domain shape of the object wave arriving at the hologram may actually vary from one spatial point to another.

The hologram writing procedure consists in illuminating the film with the object and reference beams for about ten seconds. Considering the 100 MHz repetition rate of the laser the resulting space-and-frequency-domain structure of the hologram is accumulated from about 10^9 identical writing pulse pairs. The phase of the interference grating is given by $\phi = \tau\nu$, where ν is the frequency and where τ contains the information about the delay and the angle between the beams.

For the readout the object beam is blocked with the shutter and the hologram is illuminated with the attenuated reference beam. In this experiment the thickness of the PSHB film is small and fulfils the condition for a thin hologram. Geometrically the diffraction can take place in both, in the positive and negative diffraction orders ($+1$ diffraction order is in the direction of the former object beam and the -1 diffraction order is in the complementary direction, $\vec{Fh} = (0, \sin\theta, \cos\theta)$), but the intrinsic causality related asymmetry of the scattering of the spectrally selective hologram restricts the diffraction to the positive direction (for a discussion of causality and asymmetric diffraction

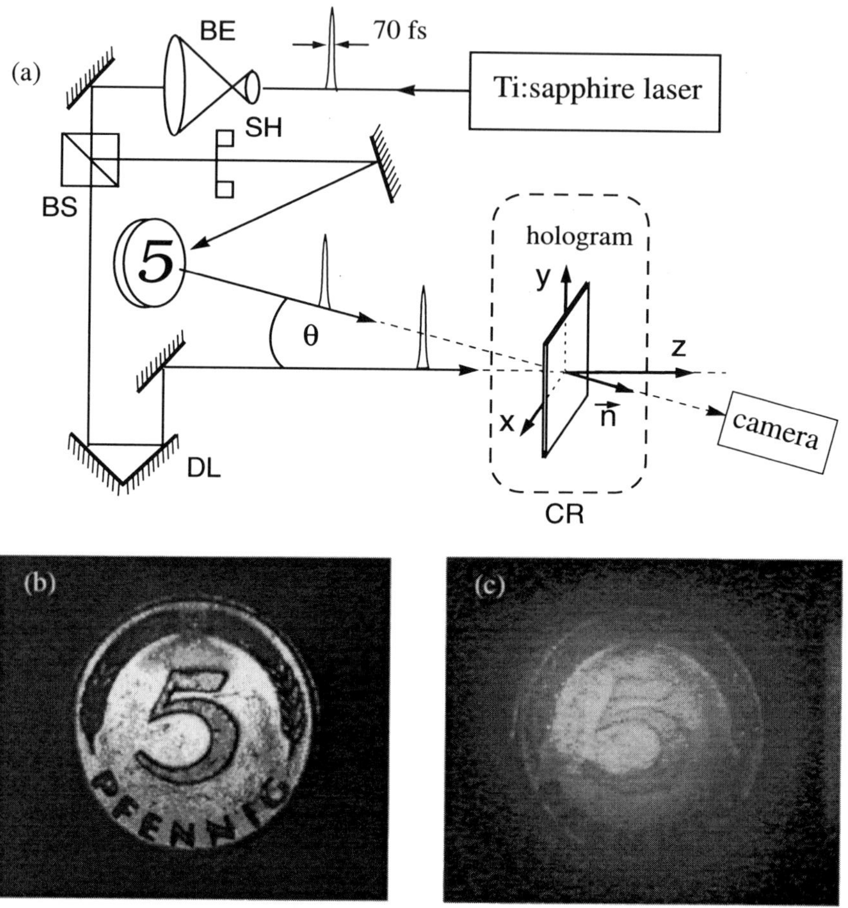

Fig. 4. (a) Experimental arrangement to write PSHB holograms with 70-fs-duration pulses. The reference beam propagates along the z-axis, whereas the object beam is scattered from the coin in the direction of the unit vector $\vec{n} = (0, -\sin\,\theta, \cos\,\theta)$. BE is the beam expander; BS is the beam splitter; SH is the shutter to block the object beam during readout; DL is the optical delay to set the time τ_0; and CR is the optical cryostat containing the PSHB film immersed in liquid helium. (b) The original image of the coin illuminated with the Ti:sapphire laser beam. (c) Image reconstructed from the hologram.

see [*3, 7, 19, 28*]). Figure 4 shows the hologram image detected with a CCD camera placed in the object beam direction. In the display of the result of this experiment the camera can capture only the spatial image although the linear coherent response of the hologram plays back in fact the image and the time-domain dependence, where the latter contains the delay τ as well as the profile depth modulation of the coin.

To explicitly determine the time-domain profile of the PASPE signal one can apply a non-linear cross-correlation technique based on sum-frequency or second-harmonic generation in a non-linear crystal. Figure 5 shows the result of such a cross-correlation measurement [*29*] where the femtosecond time profile is obtained by focusing the hologram signal together with a 67-fs-duration reference pulse in a thin $LiIO_3$ crystal. In this experiment the light source to write and read out holograms is a colliding-pulse-modelocked ring dye laser operating at the wavelength of 620 nm, the PSHB material is polystyrene doped with protoporphyrin molecules whose $S_1 \leftarrow S_0$ transition coincides with the wavelength of the laser (for further details see [*29*]). The cross-correlation trace shows that the hologram signal reproduces the time profile and the delay of the femtosecond object pulse. Deconvolution of the cross-correlation curve

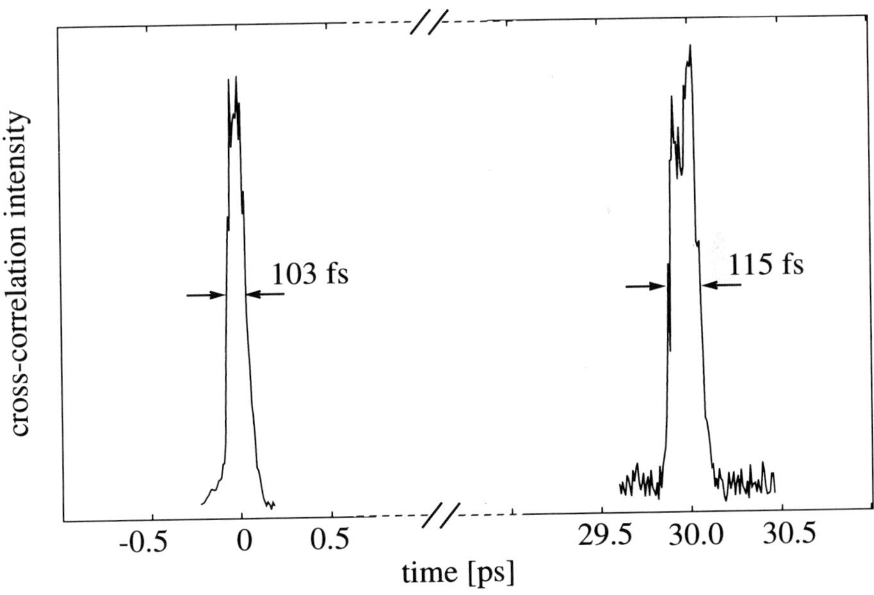

Fig. 5. Cross-correlation of the PSHB hologram signal with the 66-fs-duration laser pulse. The hologram is recorded and read out using the colliding-pulse modelocked dye laser. The holographic PASPE signal is at delay 30 ps after the directly transmitted hologram read pulse (the pulse at zero delay). The PASPE signal width after deconvolution is 75 fs [29].

assuming a sech^2 pulse shape gives a photon echo pulse duration of 75 fs which is close to the theoretical limit defined by the inverse value of the inhomogeneous bandwidth of this PSHB material.

Combined with the image storage experiment described above this demonstrates the feasibility of playing back the time-and-space-domain images with femtosecond resolution and practically without distortions of the original waveform. In this way femtosecond pulses can be used for ultrafast data storage and, as we will show below, for coherent optical processing without the usual limitation on the coherence time of the light waves. Another possible application of this technique can be for time-and-space-domain holographic interferometry to study fast-varying images of transient processes suggested in [*30*].

DIFFRACTION ON THE 'TIME EDGE'

In the experiment shown in Fig. 4 the coherence length of the pulses generated by the femtosecond Ti:sapphire laser is much shorter than the dimension of the PSHB polymer film. We can use this circumstance to write time-and-space-domain holograms where the diffraction direction changes abruptly as function of the y-coordinate. Let us set the delay $\tau_0 = 0$ and replace the object coin with a mirror. Then both the object and the reference beams illuminating the hologram are plane wave pulses which overlap in time at the origin of coordinates. Geometrically the 70-fs-duration pulses resemble two thin sheets of light of thickness ~20 μm which cross in a narrow stripe parallel to the x-axis. In the half-plane $y > 0$ the time delay is $\tau < 0$, while in the half-plane $y < 0$ the delay is $\tau > 0$ (for details of the experiment see [*31*]). In the plane $z = 0$ the holographic information is coded in the frequency-and-space-domain pattern of intensity distribution shown schematically in Fig. 6. If we take a vertical cut of this pattern parallel to the y-coordinate axis we will obtain the spatial fringe pattern at the given frequency ν, while a cut taken in the horizontal direction parallel to the frequency axis gives us an idea about the spectral structure recorded at a given spatial location.

After the writing exposure we illuminate the hologram with either of the two attenuated beams (the other beam is blocked) and observe the images diffracted in the complementary directions (see Fig. 7). In this experiment the causality related asymmetry of diffraction leads to the situation where the lower half-plane of the hologram scatters light in the $+1$ direction, while the upper half-plane scatters in the -1 direction. By comparing this picture to the previously shown frequency-and-space-domain fringe pattern (Fig. 6) we see that the asymmetry is caused by the different direction of the time arrow (telling which of the writing pulses was applied first) coded in the *phase change* of the fringes: in the region $y > 0$ the time arrow is pointing in the negative direction, while for $y < 0$ the time arrow is pointing in the positive direction.

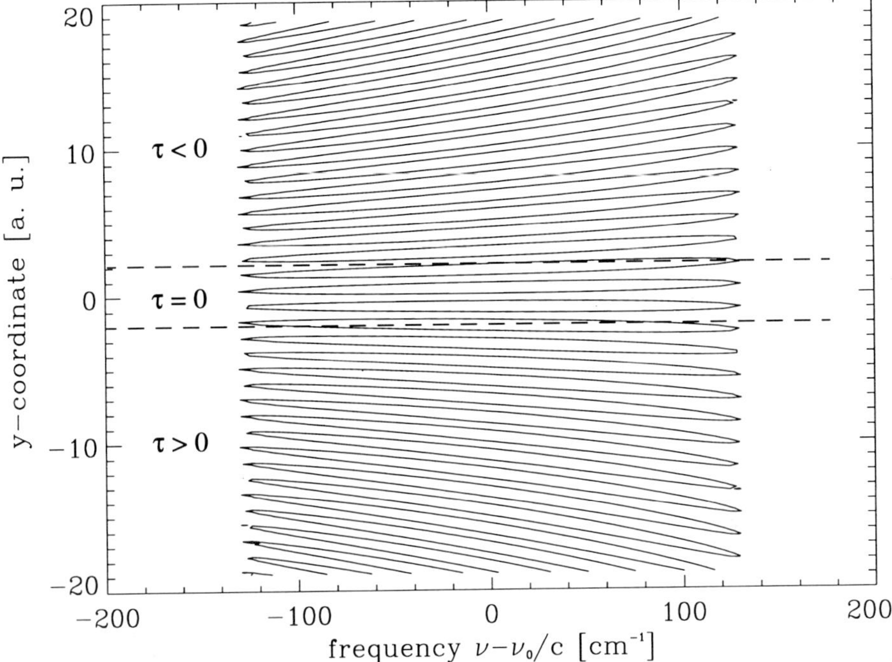

Fig. 6. Schematic view of the frequency-and-space-domain fringe structure of the 'time edge' hologram. The inclination direction of the fringes in the plot is given by the sign of the delay. In the time overlap region ($\tau = 0$) the fringes are parallel to the frequency axis.

Let us now turn our attention to the virtual 'time edge' which occurs in the 0.1 mm-wide region in the centre of the hologram where the two writing pulses actually overlap in time. In this narrow region the time arrow is not defined and the hologram scatters in both diffraction directions. At the same time, at small delay values the period $1/\tau c$ of the interference fringe pattern in the frequency dimension is comparable or larger than the width of the phonon sideband. As a result, not only the narrow ZPL, but also the much broader phonon sideband contributes to the PASPE signal. The additional intensity originating from the phonon sideband gives rise to the bright stripe seen in the images in Fig. 7 in the vicinity of the 'time edge'. It is worthwhile to note that in this way the femtosecond time-and-space-domain hologram image carries spectral information about the PSHB material. But we can observe even more interesting effects if we notice that the sharp 'time edge' causes diffraction of the PASPE signal itself and that the resulting diffraction pattern also depends on the homogeneous spectrum of the PSHB material. The 'time edge' can be seen as a virtual opaque half-plane which gives rise to a spatial diffraction pattern which depends on the propagation distance behind the hologram and on

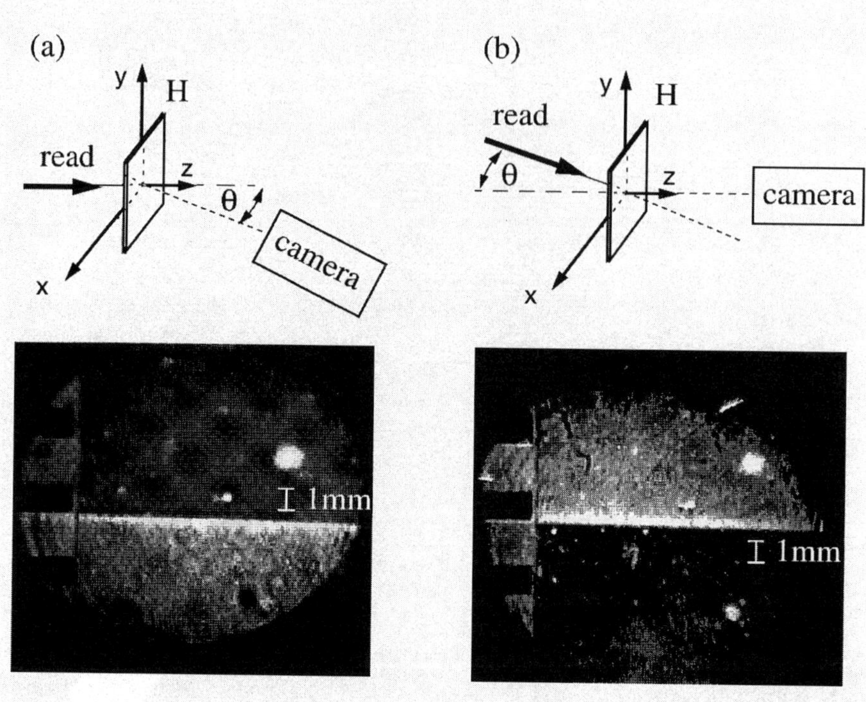

Fig. 7. Image reconstructed from the hologram in the two complementary diffraction directions. The CCD camera is focused in the plane H of the hologram. (a) The hologram is illuminated with the reference beam and the camera is positioned in the object beam direction. (b) The hologram is illuminated with the object beam and the camera is positioned in the reference beam direction. The bars show the actual scale of the hologram image [32].

the amplitude distribution in the plane of the hologram. As shown recently in [*32*] the intensity distribution of the light scattered from the 'time edge' is proportional in the far (Fourier) diffraction zone to the homogeneous hole-burning spectrum of the PSHB molecules.

This surprising conclusion can be used to construct a simple 'mono-chromator' which allows us to observe the 'spectrum' of the ZPL and of the phonon sideband of the PSHB molecules. Figure 8(a) shows the corresponding optical arrangement which consists of a cylindrical lens with $f = 300$ mm positioned behind the plane H of the hologram and performing the one-dimensional spatial Fourier transform of the amplitude scattered in the $+1$ direction. The 'spectrum' is then simply proportional to the spatial intensity

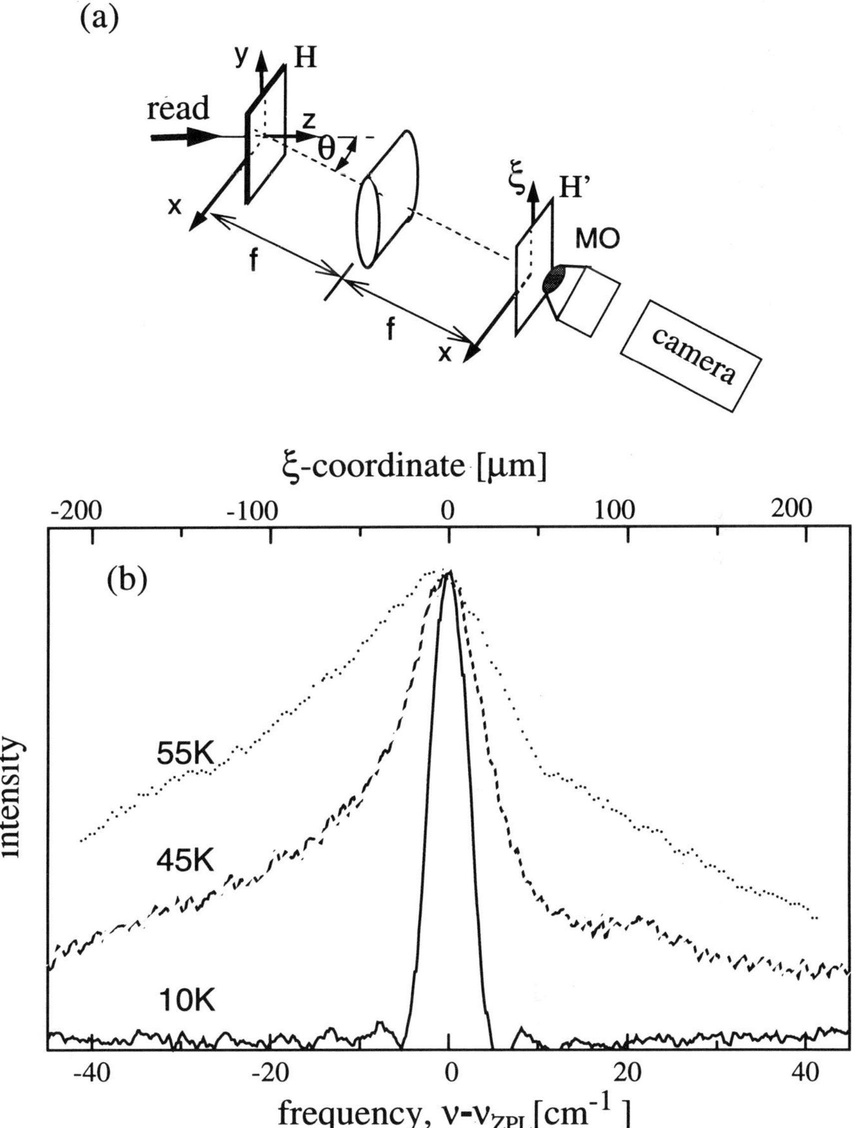

Fig. 8. Simple 'monochromator' based on the spatial Fourier transformation of the 'time edge' image. (a) Scheme of the optical arrangement; H is the hologram plane; H' is the Fourier transform plane; *f* is the focal length of the cylindrical lens; ξ is the Fourier transform coordinate; and MO is the microscope objective. (b) Image intensity in the H' plane integrated over the *x*-direction. The three curves correspond to the 'homogeneous hole-burning spectrum' at different temperatures. The curves are normalized to the maximum value. The ξ-axis is calibrated directly in units of optical frequency, $\xi = 200$ cm^{-1} mm^{-1} [32].

distribution in the ξ-direction in the conjugated plane H$'$ of the lens. Indeed, the signal from the phonon sideband appears in the 'time edge' as a narrow stripe and therefore in the Fourier plane gives an intensity that is spread out along the ξ-direction. The signal from the ZPL will contribute, in contrast, to a δ-like peak because it is nearly constant in the hologram plane. Figure 8(b) shows the 'spectra' measured at different sample temperatures. The resolution of the 'monochromator' is in this first experiment too low to resolve the real structure of the homogeneous spectrum; however, qualitatively the results agree with the expected behaviour that at low temperatures (10 K) the phonon sideband is small and the 'spectrum' consists mostly of the ZPL signal (the 7 cm^{-1} wide peak), while at elevated temperatures (above 40 K) the relative contribution of the sideband is large with decreasing ZPL (for a mathematical analysis of the problem see [32]).

The diffraction effects associated with the 'time edge' become more pronounced as the coherence length of pulses gets shorter and we believe that especially interesting will be to study the situation when the pulse duration is approaching only a few optical oscillation periods.

SPECTRAL PROGRAMMING OF ULTRASHORT TIME-DOMAIN PULSE TRAINS

In this section we show that it is possible to produce arbitrary time-domain pulse shapes from a PSHB hologram synthesized in the frequency dimension. For this purpose we first record in the PSHB material a set of gratings at different frequencies with variable predefined relative amplitudes and phases and then apply a short subpicosecond laser pulse to generate the diffracted PASPE response in the time domain. To write the gratings we use essentially the cw technique employing a tunable monochromatic laser mentioned in the Introduction [20]. Such spectrally selective gratings diffract light only when the illumination wavelength coincides with the frequency of the burnt spectral hole. Given the narrow ZPL width thousands of narrow-band holograms can be recorded at different frequencies within the broad inhomogeneous band of an organic PSHB material [21].

The procedure of this experiment is described in detail in [33, 34]. As the tunable narrow-band light source we utilize a dye laser with a linewidth of 0.5 cm^{-1}. The PSHB material is chlorin in PVB at temperature 2 K with the hole-burning properties discussed earlier in this chapter. The maximum diffraction efficiency of the recorded narrow-band gratings is on the order of 1%. The readout is carried out by use of 200-fs-duration pulses at a wavelength of 634 nm cut out from the white-light continuum generated by means of an amplified femtosecond Ti:sapphire laser system. The time profile of the hologram signal is measured with the non-linear cross-correlation technique in a thin LiIO$_3$ crystal.

Figure 9(a) shows the time response obtained when one grating is recorded in the PSHB sample. In this most elementary case the grating diffracts only one spectral component out of the broad spectrum of the illuminating fs-pulse. The intensity in the time domain appears as a smooth quasi-continuous profile with the duration approximately given by the inverse value of the spectral width of the dye laser used to write the grating. Figure 9(b) shows the time response from two gratings recorded at close frequencies. As expected, the diffracted intensity contains a coherent beat with the period equal to the inverse value of the frequency separation between the gratings. By varying the relative phase between the gratings we can observe the shift of the beat structure in the time domain. Figure 10 shows a complicated teraherz pulse train which is produced

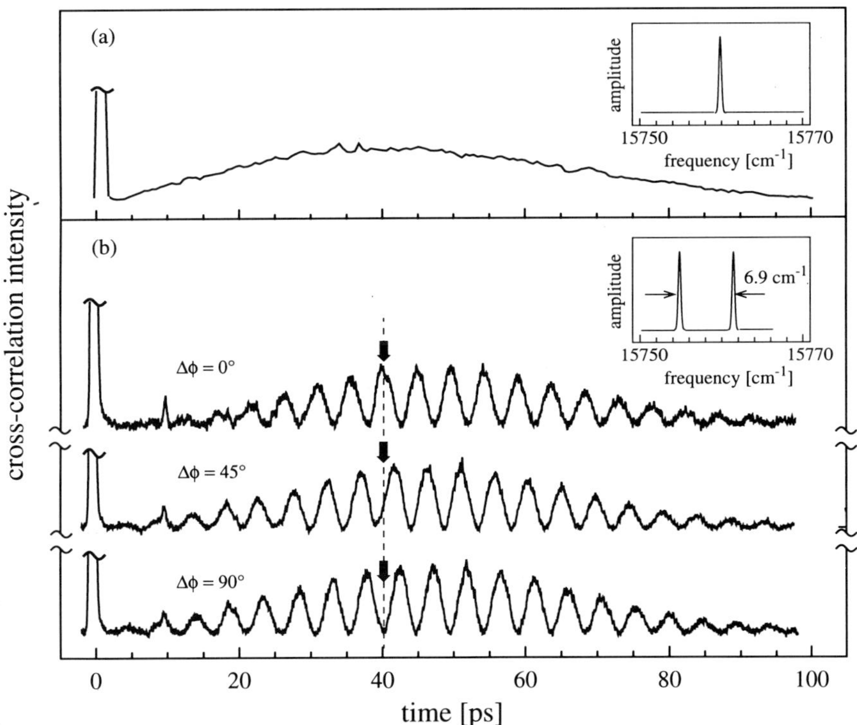

Fig. 9. Time response of simple spectrally synthesized holograms. (a) PASPE signal obtained from one narrow-band grating. The peak at the zero delay occurs due to residual scattering of the readout pulse on the surface of our sample. (b) The beat signal obtained when two gratings are recorded at adjacent frequencies. The three curves show the time response at three different relative phase values between the gratings, $\Delta\phi = 0°$, 45°, 90°. The arrows and the dashed line guide the eye to notice the shift of the beat signal. Insets show the amplitude of the gratings in the frequency dimension [34].

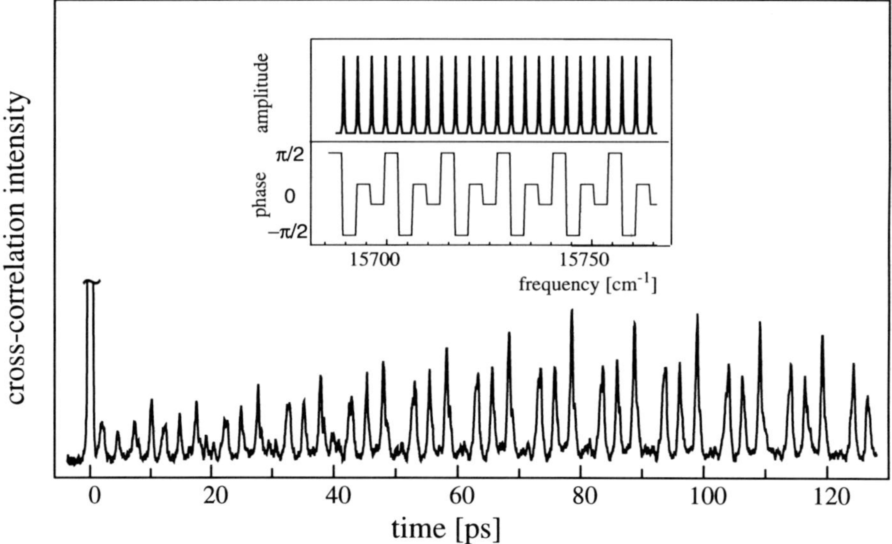

Fig. 10. The response of a more complicated spectrally programmed hologram which produces a series of subpicosecond pulses with given amplitudes and phases. The inset shows the corresponding amplitude and phase functions of the gratings recorded in the frequency dimension [34].

when the cw gratings are recorded according to the given amplitude and phase algorithm (see the inset of the figure). The maximum time span of synthesized pulse trains is in this experiment about 0.5 ns and is basically limited by the finite homogeneous ZPL linewidth. The measured duration of the individual pulses in the train is about 350 fs.

The spectral synthesis of time-domain pulse shapes can be realized also by dispersing the spectrum of the short light pulse in space with a combination of diffraction grating or prisms [35]. The amplitudes and phases of the frequency components are modified by use of spatial modulators, holograms or acousto-optic cells [36]. Another symmetrically placed grating is used to recombine the spatially separated frequencies into one time-modulated beam. This technique has become standard for fs laser pulse shaping for various applications in spectroscopy. The potential advantage of PSHB is that the resolution of ZPLs is much higher than that usually achieved with diffraction gratings or prisms and allows more information to be coded in the frequency dimension. The principal utility of the PSHB method consists, however, in the fact that the spectral resolution is an intrinsic property of the material. In our experiment, for example, the spatial coordinates are still free for additional manipulations such as, e.g., generating different pulse sequences in different directions.

ASSOCIATIVE RECALL IN TIME-AND-SPACE DOMAIN

In this section we show that time-and-space-domain holograms possess another unusual property which is based on the intrinsic time arrow and which is best displayed when the storage is performed with short pulses. It turns out that when the hologram is recorded without an explicit reference wave pulse (in the previous examples the reference pulse defines the delay of the object) in an associative manner, then the recall performed with a time-and-space-domain fragment of the object waveform still distinguishes between the 'future' and the 'past' [37].

Figure 11(a) shows an expanded beam of a modelocked picosecond dye laser which passes through an optical delay composed of a stack of glass plates with different thicknesses. The transmitted wave consists of four parallel-propagating picosecond pulses with delays of 0, 34, 68 and 102 ps, respectively. These pulses pass through a glass plate with a ground surface and are scattered in the forward direction so that they uniformly illuminate the PSHB hologram film inside the optical cryostat. The distance from the ground glass surface to the hologram is chosen large enough so that the light from each fragment illuminates the whole area of the PSHB sample. Since every pulse scatters from a different region of the ground surface, the waveforms reaching the PSHB film are all different and not correlated to each other. We call the whole beam the object and the four wave pulses the fragments. The optical dephasing time of the PSHB material (polystyrene doped with octaethylporphyrin molecules) is several hundreds of picoseconds.

To record the associative hologram we illuminate the PSHB film with the object beam. Because the relative delay of the four fragments is less than T_2, interference takes place and the hologram is recorded even though no explicit reference beam is present. In this experiment each fragment serves simultaneously as the object and as the reference wave pulse. The time arrow remembers the relative time ordering of the fragments (so long as the fragments do not overlap in time).

The readout is accomplished by illuminating the hologram with the attenuated object beam where part of the fragments are blocked with the shutter. The fragments used for the readout serve as the 'key' for the associative recall of the hologram. Figure 11(b) shows the original image of the object observed by the camera before recording of the hologram. We can replace the photocamera with a picosecond streak camera and measure the time-resolved intensity of the object wave. The arrows indicate which pulse in the time domain corresponds to which part of the spatial image. Figure 11(c) shows the spatial image and the corresponding time response reproduced from the hologram when the 'key' consists of three first pulses. In this case the associative recall plays back the image of the fourth fragment as well as its delay in the time domain. However, if we use as the 'key' the last of the four fragments, the hologram does not reproduce any part of the waveform. Thus the associative recall is selective with respect to the time arrow: it can reproduce only these parts of the scene which are later (or simultaneous) with respect to the 'key' fragment.

This experiment demonstrates especially that spectrally selective storage can be used to model associative recall of information depending on the occurrence

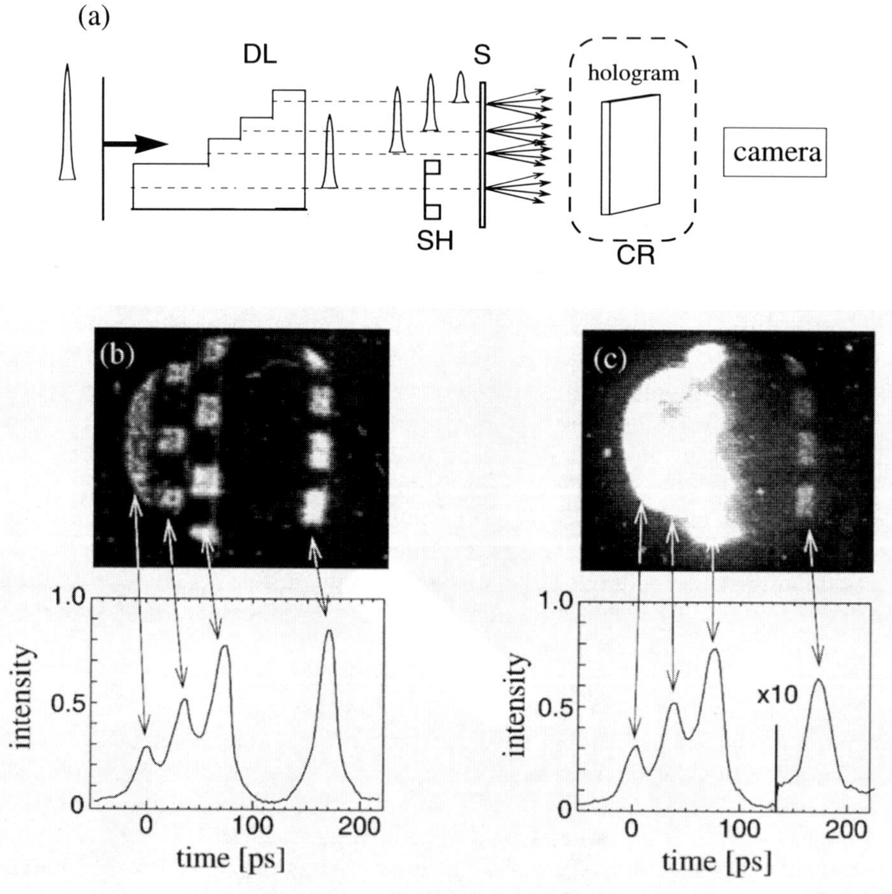

Fig. 11. Associative recall of time-and-space-domain holograms. (a) Experimental arrangement; DL is the optical delay; S is the scattering screen. (b) Original spatial image and the time-resolved intensity of the object wave. Arrows indicate the correspondence between the spatial image parts and time-domain pulses. (c) Associatively reconstructed signal when the fragment consisting of three first pulses (the brightly illuminated part on the left side of the image) is used to read out the hologram. The reconstructed signal is seen on the right side of the image [37].

in time. This property is strikingly similar to how we ourselves remember events in time and in space. A version of time-domain associative holography was discussed earlier by Gabor [38] who also pointed out the analogy to the human memory.

A digital version of associative frequency-selective memory is studied in [39]. In this case the holograms serve as programmable multidimensional (in the spatial, frequency and time dimension) interconnecting optical elements between digital input and output signals in a Hopfield-type error-corrective optical neural network scheme.

FREQUENCY-DOMAIN INTERFERENCE OF HOLOGRAPHIC IMAGES

Interference between two holograms with a given relative phase can be applied to perform coherent addition and subtraction of images useful as the basic building block of an optical computer. In spectrally selective materials logical operations can be performed with pulses in the time domain [40]. In addition, destructive interference can be used for selective erasure of time-domain data [41]. As already mentioned, the Stark effect in PSHB materials can be used to control the interference between holograms by an external electric field [23, 24].

The experiment described below demonstrates a novel interference principle of time-and-space-domain holograms where the phase is essentially fixed, while the constructive and destructive interference is achieved by manipulating the spectral envelopes of the holograms. Figure 12 shows the intensity spectrum of the writing laser beam measured before the beam is split into the object and reference arm. The laser spectrum can be switched between two functions, $S_1(\nu)$ and $S_2(\nu)$ which have the same envelope width $\Delta\nu_L = 9$ cm^{-1}. In both cases the laser spectrum has a structure characterized by the modulation period ν_{mod}. The inverse value of the modulation period gives approximately the maximum duration of the laser pulses (the pulses are not transform-limited), $t_{pulse} \sim (\nu_{mod})^{-1} = 20$ ps. The spectral shapes are taken such that their sum, $S(\nu) = S_1(\nu) + S_2(\nu)$, gives the smooth envelope profile. Using this specially arranged light source we record the holograms in a setup similar to that shown in Fig. 4 where the reference is a plane wave and the object beam carries different spatial images. The PSHB sample is polystyrene doped with octaethylporphyrin molecules (for experimental details see [42]).

The delay τ (and also the optical phase difference) between the writing beams is fixed and not changed during the whole experiment. The value of the delay is set at $\tau < t_{pulse}$, so that the object wave and the reference wave partially overlap in time. On the other hand, the delay is chosen longer than the inverse value of the spectral envelope, $(\Delta\nu_L)^{-1} < \tau$.

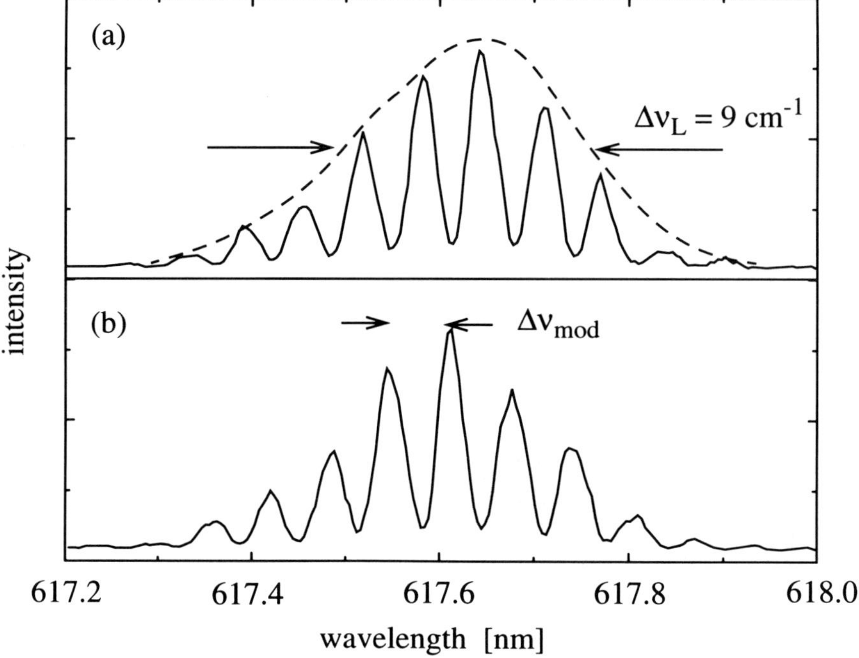

Fig. 12. Two different spectra of the laser beam used to record the PSHB holograms. The spectrum is measured before the beam is split into object and reference arms (see the text) [42].

The main idea of this method is that the causality related asymmetry of the combined hologram changes depending on the overlap of the images. For this we first record a hologram with the object image shown in Fig. 13(a), and after that superimpose the second hologram with the different object image shown in Fig. 13(b). Taken separately, in each of the two holograms the writing beams overlap in time and therefore causality does not cancel any of the two basic ($+1$ and -1) diffraction directions. However, if in the first exposure the laser spectrum is given by $S_1(\nu)$ and in the other exposure the spectrum is $S_2(\nu)$, then the two functions add together and merge into one smooth spectral envelope profile. As a result, the effective pulse duration t'_{pulse} is given by the smooth envelope function, $t'_{pulse} \sim (\Delta\nu_L)^{-1} < \tau$. Consequently, at the spatial locations where the two images have equal intensity the scattering occurs now only in the $+1$ direction, while in the -1 direction the signal is cancelled. Figures 13(c) and (d) show the interference images observed in the -1 and in the $+1$ directions, respectively. We emphasize

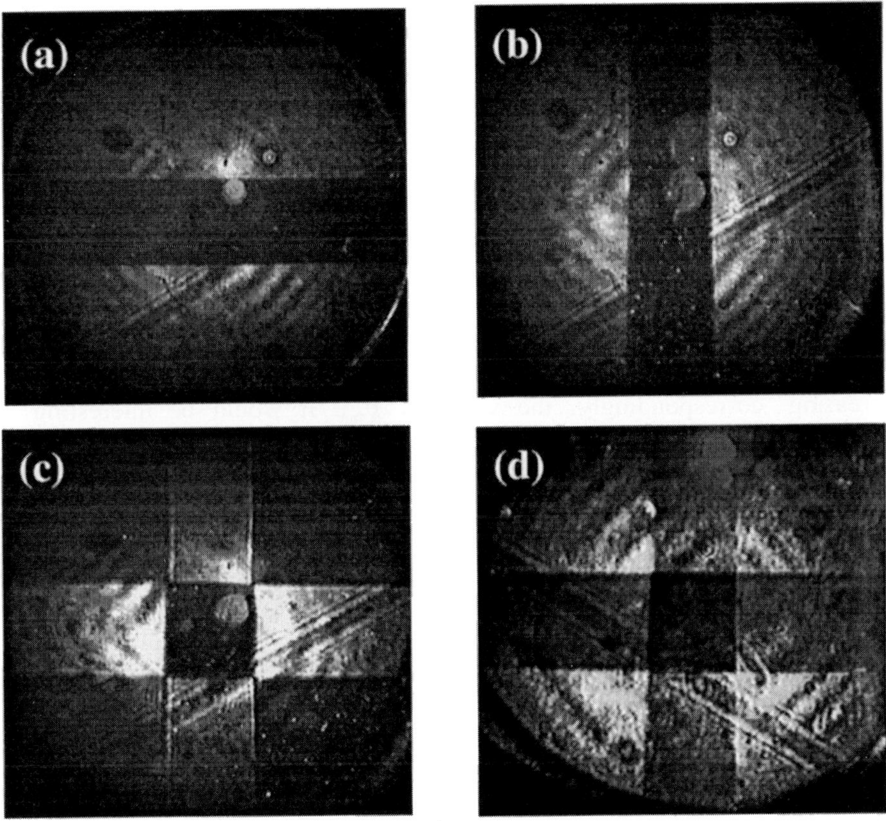

Fig. 13. Images reconstructed from the hologram. (a) and (b) – the first and the second images recorded and read out separately; (c) – destructive interference observed from the double-exposure hologram when the detection is in the −1 diffraction direction; (d) – constructive interference obtained from the same double-exposure hologram when the detection is in the +1 diffraction direction [42].

here that the destructive *and* the constructive interference between the two holograms occurs simultaneously from essentially one double-exposure hologram.

Let us note also that the interference experiment discussed above involving the complementary spectral intensities is in certain respects similar to the known Babinet principle which describes the spatial diffraction properties of complementary opaque screens. Following this similarity may be useful in designing time-and-space-domain logical processors using interference in spectrally selective media.

SUMMARY

The experiments performed so far have been designed to investigate the principally new ways of recording, manipulating and synthesizing broad-band time-and-space-domain optical signals. The fact that frequency selectivity is an intrinsic property of the PSHB material is of principal importance especially for multidimensional (spatial coordinates plus frequency and time) coherent optics. Practical applications in the future will imply, however, that the materials not only have a large ratio between the inhomogeneous and homogeneous linewidths at low temperature, but also maintain this property at higher temperatures (up to room temperature). It may be easier to achieve this goal if we allow the homogeneous linewidth to be on the order of 1 cm^{-1} by increasing, correspondingly, the value of Γ_{inh}. It would be interesting to observe whether the holograms recorded in such ultrabroad bandwidth materials can reach higher resolution in the time domain than presently achieved in organic materials and what the diffraction properties of such holograms will then be.

ACKNOWLEDGEMENTS

The author would like to thank Karl K. Rebane and Urs P. Wild for valuable discussion about this chapter. Part of our research has been supported by the Swiss Priority Program of Optical Sciences, Applications, and Technologies.

REFERENCES

1. T.W. Mossberg, *Opt. Lett.* **7**, 77 (1982).
2. A. Rebane, R. Kaarli, P. Saari, A. Anijalg, and K. Timpmann, *Opt. Commun.* **47**, 173 (1983); A. Rebane and R. Kaarli, *Chem. Phys. Lett.* **101**, 279 (1983).
3. P. Saari, R. Kaarli, and A. Rebane, *J. Opt. Soc. Am. B* **3** (4), 527 (1986); P. Saari and A. Rebane, *Proc. Acad. Sci. Estonian SSR Phys. Math.* **33**, 322 (1984); A. Rebane and J. Feinberg, *Nature* **351**, 378 (1991).
4. A.A. Gorokhovskii, R.K. Kaarli, and L.A. Rebane, *JETP Lett.* **20**, 216 (1974) (in Russian).
5. B.M. Kharlamov, R.I. Personov, and L.A. Bykovskaya, *Optics Commun.* **12**, 191 (1974).
6. K.K. Rebane, *Impurity Spectra of Solids*, Plenum Press, New York (1970).
7. K.K. Rebane, in *Current Trends in Optics*, J.C. Dainty (ed.), Academic Press (1994), p. 177.
8. L.A. Rebane, A.A. Gorokhovskii, and J.V. Kikas, *Appl. Phys. B* **29**, 235 (1982).
9. J. Friedrich and D. Haarer, *Ang. Chemie Int. Ed.* **23**, 113 (1984).
10. W.E. Moerner, (ed.), *Persistent Spectral Hole Burning: Science and Applications*, Springer-Verlag, Berlin (1988); W.E. Moerner, *J. of Mol. Electron.* **1**, 55 (1985).

11. K. Holliday and U.P. Wild, in *Molecular Luminescence Spectroscopy: Methods and Applications Part III*, S. Schulman (ed.), Wiley, New York (1993) Chap. 5.

12. E.I. Styrkov and V.V. Samartsev, *Opt. Spectrosc.* **40**, 224 (1976); C.V. Heer and P.F. McManamon, *Opt. Commun.* **23**, 49 (1977).

13. N.W. Carlson, W.R. Babbitt, and T.W. Mossberg, *Opt. Lett.* **8**, 623 (1983); M.K. Kim and R. Kachru, *J. Opt. Soc. Am. B* **4**, 305 (1987).

14. S.O. Elyutin, S.M. Zakharov, and E.A. Manykin, *Sov. Phys. JETP* **49**, 421 (1979); V.A. Zuikov, V.V. Samartsev, and R.G. Usmanov, *JETP Lett.* **32**, 270 (1980).

15. N.W. Carlson, L.J. Rothberg, A.G. Yodh, W.R. Babbitt, and T.W. Mossberg, *Opt. Lett.* **8**, 483 (1983); W.R. Babbitt and T.W. Mossberg, *Appl. Opt.* **25**, 962 (1986).

16. M. Mitsunaga, *Opt. Quant. Electron.* **24**, 1137 (1992).

17. L.A. Nefed'jev and V.V. Samartsev, *J. Appl. Spectrosc.* **57**, 386 (1992).

18. Y.N. Denisyuk, in *International Trends in Optics*, J.W. Goodman (ed.) Academic Press (1991) p. 279.

19. P.M. Saari, A.K. Rebane, and R.K. Kaarli, in *Optical Holography in Three-Dimensional Media*, Yu. N. Denisyuk (ed.), Nauka, Leningrad (1986) p. 30 (in Russian).

20. A. Renn, A.J. Meixner, U.P. Wild, and F.A. Burkhalter, *Chem. Phys.* **93**, 157 (1985); A. Renn, A.J. Meixner, and U.P. Wild, *J. Chem. Phys.* **92**, 2748 (1990).

21. B. Kohler, S. Bernet, A. Renn, and U.P. Wild, *Opt. Lett.* **18**, 2144 (1993); E. Maniloff, S. Altner, S. Bernet, F. Graf, A. Renn, and U.P. Wild, *Appl. Opt.* **34**, 4140 (1995).

22. M. Mitsunaga, N. Uesugi, H. Sasaki, and K. Karaki, *Opt. Lett.* **19**, 752 (1994).

23. U.P. Wild and A. Renn, *J. Mol. Electron.* **7**, 1 (1991).

24. U.P. Wild, A. Renn, C. De Caro, and S. Bernet, *Appl. Opt.* **29**, 4329 (1990).

25. F.A. Burkhalter, G.W. Suter, and U.P. Wild, *Chem. Phys. Lett.* **94**, 483 (1983).

26. W.-Y. Huang, Dissertation Nr. 10461, Swiss Federal Institute of Technology, Zürich (1994).

27. E.S. Maniloff, F.R. Graf, H. Gygax, S.B. Altner, S. Bernet, A. Renn, and U.P. Wild, *Chem. Phys.* **193**, 173 (1995).

28. A. Rebane, S. Bernet, A. Renn, and U.P. Wild, *Opt. Commun.* **86**, 7 (1991).

29. A. Rebane, J. Aaviksoo, and J. Kuhl, *Appl. Phys. Lett.* **54**, 93 (1989).

30. A. Rebane, J. Aaviksoo, *Opt. Lett.* **13**, 993 (1988); A. Rebane, O. Ollikainen, *Opt. Commun.* **78**, 327 (1990).

31. A. Rebane, O. Ollikainen, H. Schwoerer, and U.P. Wild, Conference on Spectral Hole-Burning and Related Spectroscopies, 24–26 August, 1994, Tokyo, Japan, Washington D.C., OSA Technical Digest Series, 15 (1994).

32. A. Rebane, O. Ollikainen, H. Schwoerer, D. Erni, and U.P. Wild, *J. Lumin.* **64**, 283 (1995).

33. H. Schwoerer, D. Erni, A. Rebane, and U.P. Wild, *Opt. Commun.* **107**, 123 (1994).

34. H. Schwoerer, D. Erni, and A. Rebane, *J. Opt. Soc. Am. B* **12**, 1083 (1995).

35. C. Froehly, B. Colombeau, and M. Vampouille, in *Progress in Optics XX*, E. Wolf (ed.), North-Holland, p. 65, 1983.

36. Yu.T. Mazurenko, *Appl. Phys. B* **50**, 101 (1990); A.M. Weiner, D.E. Leaird, D.H. Reitze, and E.G. Paek, *IEEE J. Quantum Electron.* **28**, 2251 (1992); C.W. Hillegas, J.X. Tull, D. Goswami, D. Strickland, and W.S. Warren, *Opt. Lett.* **19**, 737 (1994).

37. A. Rebane, *Opt. Comm.* **65**, 175 (1988).
38. D. Gabor, *Nature* **217**, 1288 (1968).
39. A. Rebane and O. Ollikainen, *Opt. Commun.* **83**, 246 (1991); O. Ollikainen, A. Rebane, and K. Rebane, *Opt. Quant. Electron.* **25**, 569 (1993).
40. M. Arend, E. Block, and S.R. Hartmann, *Opt. Lett.* **18**, 1789 (1993).
41. N.N. Akhmediev, *Opt. Lett.* **15**, 1035 (1990); R. Kaarli, P. Saari, R. Sarapuu, and H. Sõnajalg, *Opt. Commun.* **86**, 211 (1991).
42. D. Erni, A. Rebane, and U.P. Wild, *Opt. Lett.* **20**, 1065 (1995).

10 Holographic 3D disks using shift multiplexing

Demetri Psaltis, George Barbastathis
Department of Electrical Engineering

Michael Levene
*Department of Computation and Neural Systems,
Mail-Stop 136–93, California Institute of Technology, Pasadena,
CA 91125*

INTRODUCTION

The idea of using holograms to store information emerged in the 1960s, when van Heerden (1963) pointed out that information can be stored with high density in three-dimensional media. This thread was picked up by other researchers (Leith *et al.* 1966; Gabor 1969; Kogelnik 1969), culminating in the experimental storage of multiple holograms in $LiNbO_3$ by Staebler *et al.* (1972) at RCA and d'Auria *et al.* (1974) at Thompson-CSF. Holographic storage offers high capacity combined with parallelism in data access. Recently, optoelectronic components such as Spatial Light Modulators (SLMs) and CCD cameras became available that allow the recording and retrieval of binary data pages containing 1 Mbit each. This has made it possible to demonstrate the storage of a large number of holograms in a single crystal (Mok 1993; Mok *et al.* 1994) and in photopolymer films (Pu *et al.* 1994).

In this chapter we will describe briefly general aspects of holographic storage: techniques, materials, and capacity considerations. Then we will describe in more detail the shift multiplexing method (Psaltis *et al.* 1995), which is well suited for the realization of holographic 3D disks.

HOLOGRAPHIC STORAGE

Volume holograms are stored as a result of interference between two mutually coherent light beams, the signal and the reference (Fig. 1). The signal carries the information, typically in the form of amplitude modulation imprinted on the wavefront. The reference is usually a plane wave. In a thick medium one needs to reproduce the reference used for recording as accurately as possible in order to get diffraction from the hologram. If instead the reference deviates in angle or wavelength, then diffraction contributions from different parts of the hologram become phase mismatched causing the diffraction efficiency to drop (Bragg mismatch). The amount by which the angle or wavelength need to change before the reconstructed power drops to zero is called Bragg selectivity and depends on the geometry and the thickness of the material.

As an example, consider the transmission geometry of Fig. 1, a very common setup for holographic storage. The plane wave reference is incident at angle θ_R, the signal at θ_S, and they are both at wavelength λ_R. The respective angle and wavelength Bragg selectivities are:

$$\Delta\theta_R = m \; \frac{\lambda \cos \theta_S}{L \sin(\theta_R + \theta_S)}, \qquad m = 1, 2, \ldots \tag{1}$$

$$\Delta\lambda_R = m \; \frac{\lambda^2 \cos \theta_S}{2L \sin^2 \frac{1}{2}(\theta_R + \theta_S)}, \qquad m = 1, 2, \ldots \tag{2}$$

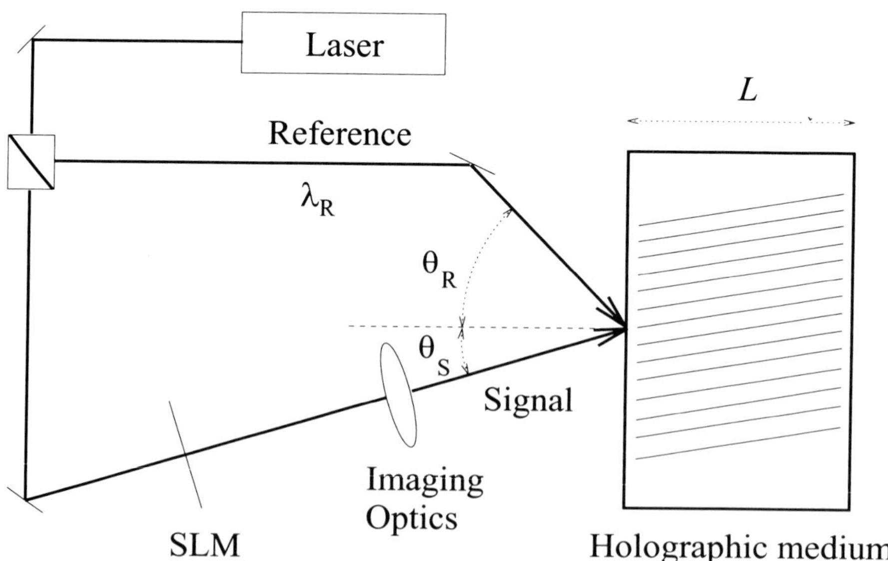

Fig. 1. Volume holographic setup.

Multiplexing is performed by changing the angle (Staebler *et al.* 1972) or the wavelength (Yin *et al.* 1993; Yariv, 1993) of the reference beam by an amount equal to the respective selectivity. For typical parameters $\theta_R = \theta_S = 30°$, $L = 5$ mm, $\lambda = 488$ nm we obtain $\Delta\theta_R = 5.6° \times 10^{-3}$, and $\Delta\lambda_R = 8.2 \times 10^{-2}$ nm. For example, if the angular range yielded by typical lenses is 20°, or if the range of a tunable laser source is 300 nm, we obtain $M_\theta \approx 3500$, and $M_\lambda \approx 3600$ for the number of holograms that can be stored, respectively. Assuming that each hologram contains approximately 1 Mbit of information, and that the area occupied by the holograms is 4 mm^2 (2 μm pixel size), we obtain the volume density of the holographic memory as ≈ 18 Gbits cm^{-3}.

Phase-code multiplexing (Denz *et al.* 1991; Trisnadi & Redfield 1992) is directly related to angle multiplexing in the sense that instead of using one plane wave reference at a time, one uses all of them at once, observing the Bragg-limited angular separation. The phases of the reference components implement some set of orthogonal functions, e.g., Walsh–Hadamard codes. Upon reconstruction, each member of the orthogonal reference set reconstructs its own hologram; orthogonality serves to eliminate all other reconstructions, yielding minimal cross-talk (Curtis & Psaltis 1993). The maximum number of superimposed holograms using this method is equal to the order M_ϕ of the system of orthogonal functions. Bragg mismatch is employed to eliminate multiple reconstructions due to the multiple reference components. Therefore M_ϕ is limited by the number of beams angularly separated by $\Delta\theta_R$ (Eq. 1) that fit in the aperture of the reference imaging system. In that sense, phase code multiplexing offers the same capacity as angle multiplexing ($M_\phi = M_\theta$) for the same optics. In practice, the space-bandwidth product of the available SLMs currently limits M_ϕ to $\leqslant 1000$.

The capacity offered by holographic storage can be further augmented by combining angle and wavelength multiplexing (Campbell *et al.* 1994), or either one of the two techniques with methods that do not utilize Bragg mismatch (Lee *et al.* 1989). These methods are based on the property of reconstructed holograms to follow the motion of the reference beam when Bragg mismatch is not present. For example, rotating the reference beam around the optical axis causes the reconstruction to rotate similarly until either it becomes Bragg-mismatched or moves out of the detector plane (which one of the two occurs first is determined by the spatial signal bandwidth). This effect is utilized in the method of peristrophic multiplexing (Curtis *et al.* 1994) (in Greek, 'peristrophic' means rotational). Recently, surface storage density of 10 bits μm^{-2} without any observed errors in the reconstructions was demonstrated using a combination of angle and peristrophic multiplexing (Psaltis & Pu 1995). The thickness of the material used for this experiment was 100 μm, and a volume density of 100 Gbits cm^{-3} was achieved.

Recording Materials

In order to build a high-capacity parallel holographic memory, the material, the format and the multiplexing method must be selected. The most commonly used volume holographic materials are photorefractive crystals, e.g. $LiNbO_3$ (Chen *et al.* 1968), SBN (Thaxter & Kestigian 1974; Thaxter 1968; Ford *et al.* 1992), and $BaTiO_3$ (Valley & Klein 1983); typical thicknesses range from fractions of a millimetre to a few centimetres) and photopolymers (Curtis & Psaltis 1994a) (commercially available thicknesses are hundreds of micrometres). Both materials form phase (index) gratings when illuminated by an interference pattern. In photorefractives, the refractive index is perturbed by the space charge field formed when photoexcited electrons diffuse and drift into the dark regions (Kukhtarev *et al.* 1979). The electronic grating thus formed is erasable by photoexcitation when the hologram is uniformly illuminated, therefore photorefractives can be used for optically rewritable memory applications. For long-term storage applications, photorefractive holograms can be fixed thermally (Amodei & Staebler 1971; Staebler *et al.* 1975; von der Linde & Glass 1975; Kewitsch *et al.* 1993) or electrically (Micheron & Bismuth 1972; Qiao *et al.* 1993; Horowitz *et al.* 1993; Orlov *et al.* 1993) so that they do not decay during readout and in the dark. Both techniques are reversible if new information must be stored in the material.

In photopolymers, holograms are formed by polymerization of the original monomer material in the illuminated regions. Monomers from the dark regions diffuse into the illuminated regions due to the density gradient that is formed, thus causing local change in the refractive index. After recording, curing by ultraviolet incoherent illumination fixes the recorded holograms which are then used as read-only (Write Once–Read Many, WORM) memory (Curtis & Psaltis, 1994a, 1992). Du Pont's HRF-150-100 photopolymer was used to demonstrate holographic WORMs with 1000 holograms (Pu *et al.* 1994), and storage density of 10 bits μm^{-2} (Psaltis & Pu, 1995). Recently, photorefractive polymers (Chandross *et al.* 1974) have received a lot of attention (Ducharme *et al.* 1991; Verkhovskaya *et al.* 1992; Meerholz *et al.* 1994); in these materials, recording is based on a photochemical reaction induced by the interference grating. Photorefractive polymers yield high diffraction efficiencies, but they require high applied fields during recording and readout.

Storage Capacity

The maximum capacity $\mathscr{C}$ in bits (resolvable gratings) of any 3D holographic medium of volume V at wavelength λ can be derived (van Heerden 1963) as $\mathscr{C} = V/\lambda^3$. For visible wavelengths $\mathscr{C}$ is ~10 Tbits cm^{-3}, well in excess of the numbers practically achieved. This theoretical storage capacity is very difficult to reach in practice, because of the finite dynamic range of holographic materials and the geometric constraints of optical systems. Depending on the material, the limitation comes either from the finite allowable range of refractive index modulation or from erasure sustained by holograms already stored when new holograms are superimposed over them, depending on the material. If we assume that the saturation diffraction efficiency η_0 is to be equally shared by M holograms, then the diffraction efficiencies η of the individual holograms depend on M as

$$\eta = \left(\frac{M/\#}{M}\right)^2.$$
(3)

By using appropriately calculated exposure times, recording of each new hologram affects previously recorded holograms just enough to equalize all diffraction efficiencies; the set of recording times is referred to as the 'exposure schedule' (Bløtekjaer 1979; Psaltis *et al.* 1988). The constant $M/\#$ (*M*-number) is characteristic of the material and the experimental set-up. For large M, and assuming exponential recording and erasure curves with time constants τ_w and τ_e respectively [these conditions are satisfied in diffusion dominated photorefractive crystals (Kukhtarev *et al.* 1979; Hall *et al.* 1985), commonly used for holographic storage experiments], the $M/\#$ is given by

$$M/\# = \sqrt{\eta_0}\,\frac{\tau_e}{\tau_w}.$$
(4)

The finite dynamic range limits the number of holograms that can be stored, since the diffracted power cannot be allowed to become lower than the noise level. On the other hand, from (3) we infer that, in order to maximize the overall capacity while keeping the diffraction efficiency sufficiently high, it is desirable to store as many bits per hologram as possible. The limitation in this case comes from the resolution of the holographic system (i.e. the imaging optics used to relay the information from the SLM to the hologram and then relay the reconstruction to the CCD).

Holographic Disks

The design of a holographic disk (Psaltis 1992; Li & Psaltis 1994) is shown Fig. 2. According to the previous discussion, one or two multiplexing methods combined (e.g., angle and peristrophic) are used to superimpose holograms on the same location (in Fig. 2 the angle multiplexing mechanism alone is shown). After the total number of holograms allowed by the geometry and the medium dynamic range is used up, rotational motion of the disk is utilized to access different locations on the disk surface where the process is repeated (spatial multiplexing). Typically, the size of each location containing multiple holograms is a few square millimetres. This poses a challenge in the implementation, since disk motion cannot be continuous but rather it occurs in the form of 'jumps' from one location to the other during both recording and readout.

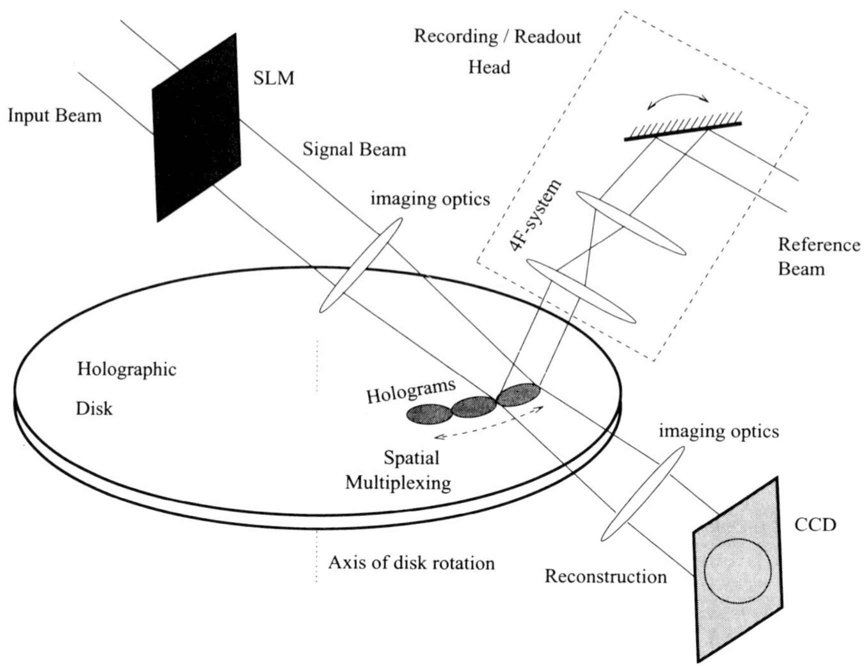

Fig. 2. Design of a 3D holographic disk.

Alternatively, a continuously spinning disk can be used, but the light source needs then to be pulsed. Moreover, all methods described so far require an additional structure inside the disk head in order to implement selective readout. Angle and peristrophic multiplexing require a beam steering mechanism; wavelength multiplexing requires an accurate, high-resolution tunable laser source; phase-code multiplexing requires a second SLM to implement the orthogonal phase code used for reference. In the next two sections we describe a new storage method, shift multiplexing, that is very simple to implement and thus eliminates most of these restrictions.

SHIFT MULTIPLEXING

Storage methods described so far require the use of either a plane wave as reference (angle, wavelength, peristrophic) or a collection of plane waves with orthogonal phases (phase-code). A different option for the reference beam is to use a spherical wave (Wagner & Psaltis 1987; Solymar & Cooke 1991). Since this is the Fourier transform of a plane wave, translation of the spherical reference is equivalent to a change in the angle of incidence of a plane wave reference. Therefore shifting the hologram relative to the reference produces

Bragg mismatch, causing the diffraction efficiency to drop to zero after a total translation equal to the shift selectivity. This property of spherical holograms suggests a new holographic multiplexing method, called shift multiplexing (Psaltis *et al.* 1995).

Shift multiplexing is particularly suitable for holographic 3D disks, as shown in Fig. 3. Successive holograms are separated by the shift selectivity (typically a few microns). Since each hologram is a few millimetres wide, neighbouring holograms stored on the disk overlap almost completely, but they are still resolvable due to the shift selectivity effect. To selectively reconstruct holograms belonging to the same track, the disk is rotated relative to the stationary head. The head needs to move only in the radial direction to access different tracks on the disk. No additional multiplexing mechanism is needed. Since both disk rotation and radial head motion are an integral part of the optical disk configuration, the implementation of a shift multiplexed disk is very simple. Moreover, the small displacement needed to access successive holograms allows better utilization of the uniform rotational disk motion.

In order to analyse diffraction from spherical holograms, we will consider the simple model of Fig. 4. The reference is a spherical wave of infinite numerical aperture originating at $z = -z_0$, incident normally on the holographic

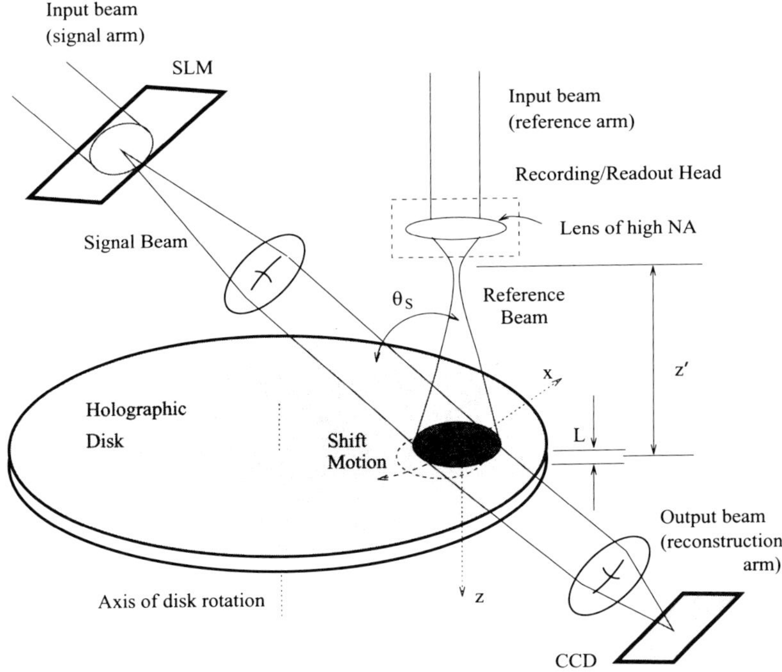

Fig. 3. Design of a 3D holographic disk using shift multiplexing.

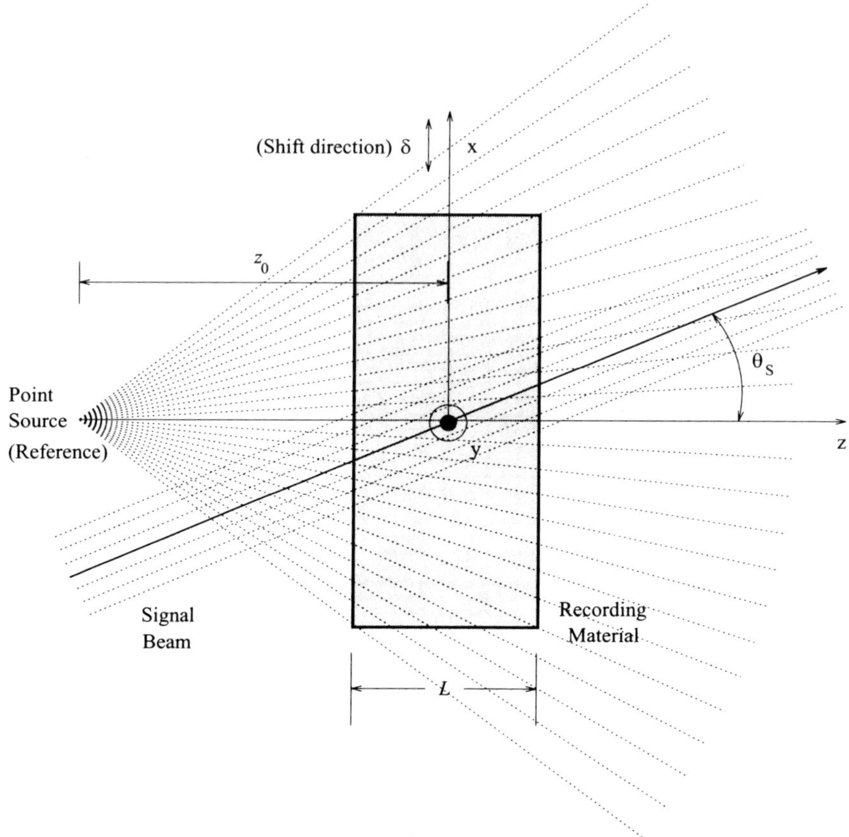

Fig. 4. Geometry for the theoretical calculation of the shift selectivity using a spherical wave as reference and a plane wave as signal.

material. The signal is assumed to be a single plane wave, incident at angle θ_S. The hologram is recorded in the region $|z| < L/2$, and is assumed to be infinite in the transverse directions x, y. In the paraxial approximation we can express the reference as

$$R(x,y,z) = \frac{1}{i\lambda(z + z_0)} \exp\left\{i2\pi \frac{z + z_0}{\lambda}\right\} \exp\left\{i\pi \frac{x^2 + y^2}{\lambda(z + z_0)}\right\}, \tag{5}$$

and the signal as

$$S(x,z) = \exp\left\{i2\pi u_S \frac{x}{\lambda} + i2\pi\left(1 - \frac{u_S^2}{2}\right)\frac{z}{\lambda}\right\}, \tag{6}$$

where $u_S \equiv \sin \theta_S \approx \theta_S \ll 1$. If we neglect the variable modulation depth due to the intensity variation of the spherical wave, the phase hologram recorded by the two beams is described by modulation in the refractive index:

$$\Delta\varepsilon(x, y, z) \propto |R(x, y, z) + S(x, z)|^2$$
$$= |R(x, y, z)|^2 + |S(x, z)|^2 + R(x, y, z)S^*(x, z)$$
$$+ R^*(x, y, z)S(x, z). \tag{7}$$

The hologram is reconstructed by the beam $R(x - \delta, y, z)$, i.e. the spherical reference displaced by the shift δ. From the four terms of (7) only the fourth one contributes significantly (the first and second are DC terms and the third term is Bragg-mismatched). Keeping the fourth term only, the field diffracted from a thin layer of the hologram is given by

$$\mathcal{E}_d(x, y, z; \delta) = R(x - \delta, y, z)R^*(x, y, z)S(x, z)$$
$$= \exp\left\{i\pi\,\frac{\delta^2}{\lambda(z + z_0)}\right\}\exp\left\{-i\pi\,\frac{2\delta x}{\lambda(z + z_0)}\right\}$$
$$\times \exp\left\{i2\pi u_S\,\frac{x}{\lambda} + i2\pi\left(1 - \frac{u_S^2}{2}\right)\frac{z}{\lambda}\right\}. \tag{8}$$

If $\delta = 0$ the reconstruction is proportional to $S(x, z)$ as expected. However for $\delta \neq 0$ the direction of propagation deviates from θ_S by an amount equal to

$$\Delta\theta_S \approx \frac{\delta}{(z + z_0)\cos\theta_S}. \tag{9}$$

This deviation from the direction of the original volume grating causes reconstructions from successive slices of the hologram to add up out of phase, i.e. Bragg mismatch occurs. A long but straightforward calculation (under the Born and paraxial approximations) yields the diffraction efficiency η as a function of δ:

$$\eta(\delta) = \frac{(\text{diffracted power})}{(\text{read-out power})} \propto \text{sinc}^2\left(\frac{L\,\delta u_S}{\lambda z_0}\right), \tag{10}$$

where $\text{sinc}(\xi) = \sin(\pi\xi)/(\pi\xi)$. The diffraction efficiency becomes zero at the Bragg nulls

$$\delta_{m,\text{Bragg}} = m\,\frac{\lambda z_0}{L u_S}, \qquad m = 1, 2, \ldots \tag{11}$$

The finite spot size $\delta_{\text{NA}} = \lambda/2(\text{NA})$ of a truncated spherical wave introduces ambiguity in the location of the point source with respect to the hologram. This ambiguity is added to the shift selectivity, giving the final expression

$$\delta_m = \delta_{m,\text{Bragg}} + \delta_{m,\text{NA}}$$

$$= m\left(\frac{\lambda z_0}{Lu_S} + \frac{\lambda}{2(\text{NA})}\right). \tag{12}$$

Finally, we need to account for refraction, i.e., the change in the spherical wavefront induced by the index of refraction n of the holographic medium. If we let z' denote the distance measured in air of the point source from the centre of the holographic material, then z_0 is obtained from z' from the relation

$$z_0 - \frac{L}{2} = n\left(z' - \frac{L}{2}\right). \tag{13}$$

Then the modified selectivity formula is

$$\delta'm = m\left\{\frac{\lambda\left[z' - \left(1 - \frac{1}{n}\right)\frac{L}{2}\right]}{Lu_S^{\text{in}}} + \frac{\lambda}{2(\text{NA})}\right\}, \tag{14}$$

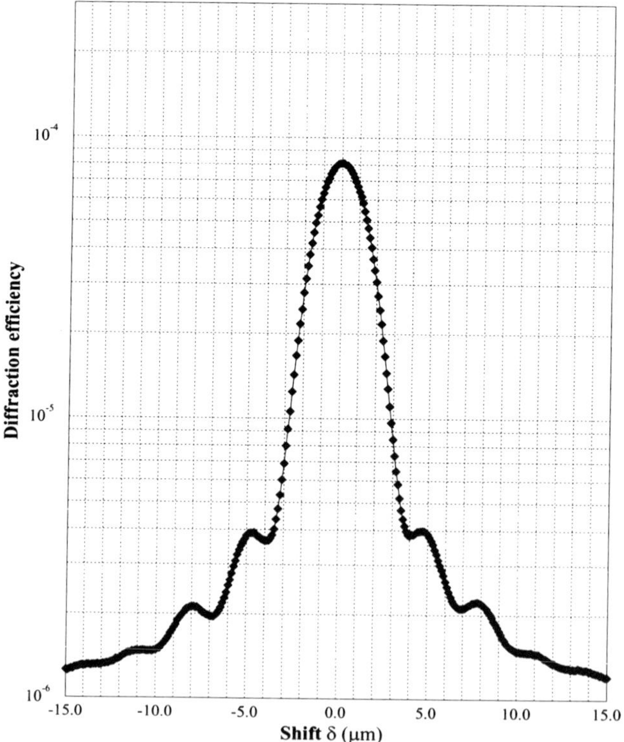

Fig. 5. Experimental selectivity curve.

where λ denotes the wavelength of light in vacuum, and $u_S^{in} \approx \theta_S^{in}$ is the angle of incidence of the signal beam inside the material, as obtained from Snell's law.

A typical experimental selectivity curve is shown in Fig. 5. The recording material (iron-doped $LiNbO_3$) had index of refraction $n \approx 2.24$, thickness $L = 4.5$ mm, and was located $z' = 1$ cm from the point source. The signal was a chessboard pattern, recorded as a Fresnel region hologram, with angle of incidence $\theta_S = 400$ (measured outside the crystal). For the parameters used in the experiment, Eq. (14) yields $\delta = 3.57$ μm. In the experiment the first null occurred at approximately 3.7 ± 0.2 μm (the margin of error is mainly due to stage inaccuracy and backlash), deviating by 3.6% from the theoretical prediction.

DESIGN CONSIDERATIONS

Issues important for the design of holographic storage systems are cross-talk, exposure schedule, and storage density. In this section we will briefly examine these issues in the context of shift multiplexing.

Cross-talk

In the theoretical calculation of the previous section we considered a hologram recorded with plane wave signal beam and showed that its diffraction efficiency goes to zero at $\delta = \delta_{m,\,Bragg}$ (in the limit of high numerical aperture). On the other hand, information-bearing signal beams contain a finite spectrum of plane waves. Only one component of the spectrum at a time can become exactly Bragg-mismatched (depending on the amount of shift), whereas the remaining components are still reconstructed, albeit weakly. When storing multiple holograms, these weak contributions build up to cause cross-talk (Gu *et al.* 1992; Yariv 1993; Curtis *et al.* 1993; Curtis & Psaltis 1993; Curtis & Psaltis 1994b; Bashaw *et al.* 1994; Yi *et al.* 1994; Yi *et al.* 1995).

Typically one chooses the shift separation between successive holograms equal to δ_m for some order m using the spatial frequency u_S of the carrier in (11). When storing Fourier plane holograms, this practice guarantees that the central pixel of every page will be noise free. As one moves towards the page edges, however, the angular deviation from the carrier component increases and hence cross-talk builds up. Similar effects occur in angle (Gu *et al.* 1992), phase-code (Curtis & Psaltis 1993), and wavelength (Yariv 1993; Curtis *et al.* 1993) multiplexing. The severity of the cross-talk effects decreases with m, therefore in a practical system one is interested in choosing the smallest m that will yield an acceptable cross-talk noise level.

Experimentally, we can use selectivity curves, such as the one shown in Fig. 5, to measure the cross-talk at the theoretical nulls. Alternatively, we can store a few holograms, intentionally leave one of them blank, and then measure the diffraction efficiency η_X at the blank hologram. Results from two such experiments are shown in Fig. 6. We observe that η_X is higher for multiple holograms than for a single hologram, as expected, and that the theoretical prediction of η_X decreasing with m is verified. Cross-talk is very high at the first null in the shift multiplexing geometry (in fact it is theoretically equal to cross-talk in the case of angle multiplexing with a plane wave reference incident on-axis). However, for $m \geq 2$ the cross-talk level is acceptable for most cases (SNR > 10).

Exposure Schedule

The problem of exposure schedule is related to the dynamics of holographic recording and comes up in other storage methods as well. In shift multiplexing, however, the exposure schedule is very simple: constant exposure suffices to equalize the average diffraction efficiencies of the holograms, due to the translation between successive exposures. For diffusion dominated photorefractive materials, the common exposure time is given by

$$t_0 = \tau_w \ln\left(1 + \frac{\tau_e}{(M-1)\tau_w}\right) \approx \frac{\tau_e}{M},\tag{15}$$

where M is the number of overlapping holograms ($M \gg 1$), and τ_w, τ_e are the recording and erasure time constants, respectively.

We used the constant exposure schedule technique to store 200 holograms in

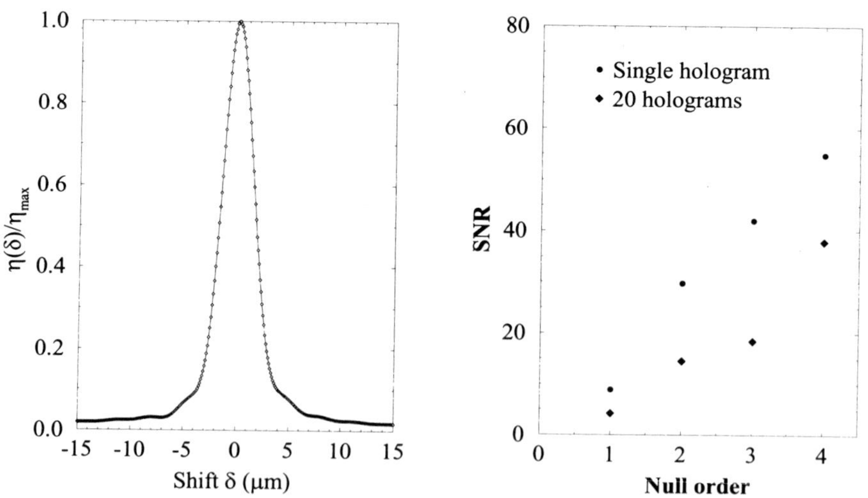

Fig. 6. Experimental characterization of the cross-talk. Left: selectivity curve for a single hologram (the signal bandwidth in this case was higher than that of Fig. 5). Right: signal-to-noise ratio ($\eta_{max}/\eta X$) versus null order.

LiNbO$_3$. The parameters used in this experiment were $\delta_4 = 14$ μm, $t_0 = 10$ sec. The reconstruction intensities for all holograms as well as sample reconstructions are shown in Fig. 7.

Because holograms are shifted relative to each other between successive exposures, the constant exposure method causes non-uniform erasure along the shift direction. In certain cases (most notably in Fourier plane storage) this non-uniformity cannot be tolerated, hence a different exposure schedule should be used: instead of storing holograms by successively translating by an amount δ_m at a time, one needs to shift completely out of the first hologram before recording the next one (i.e. spatial multiplexing precedes shift multiplexing in

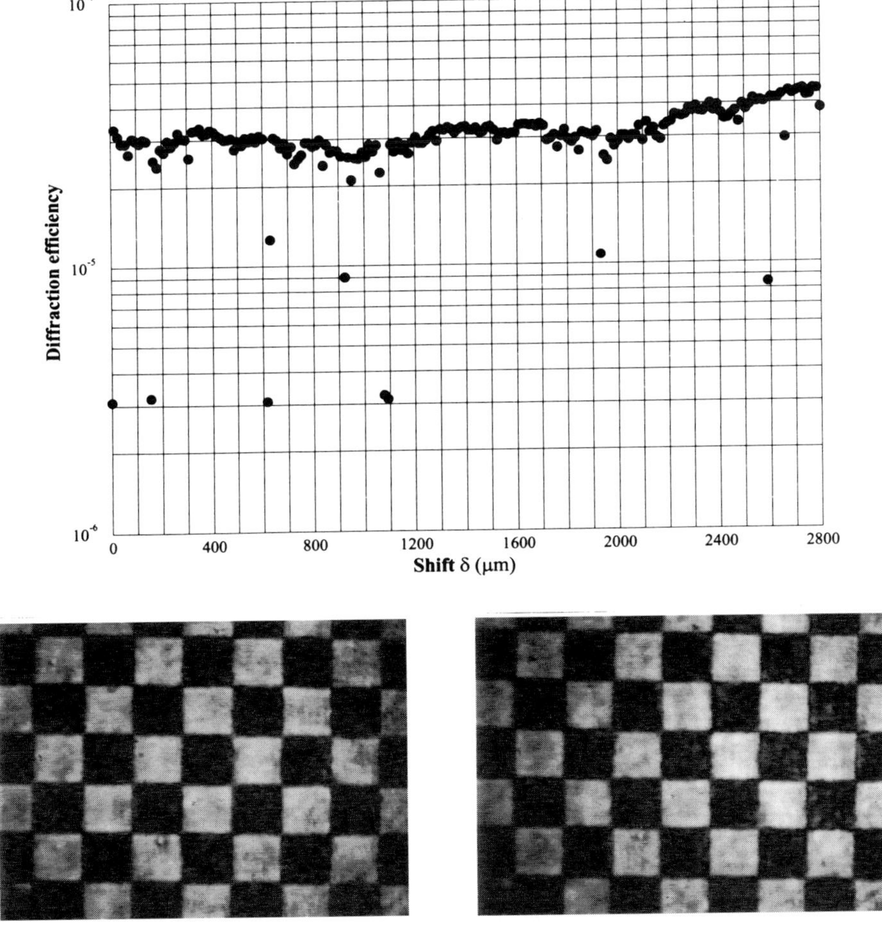

Fig. 7. Experimental storage of 200 shift multiplexed holograms: diffraction efficiency versus hologram location measured using an intensity integrating detector (top). Reconstructions of holograms #9 (bottom left) and 199.

this case). Since the holograms do not overlap, no erasure is caused in this first pass. In the holographic disk geometry, the beginning of the first hologram is eventually reached; then more holograms are recorded after an additional shift by δ_m, and so on. This method eliminates the non-uniformity (since erasure is the same for all holograms belonging to the same pass) and leads to the familiar exposure schedule derived for angle multiplexing by Psaltis *et al.* (1988), hence Eq. (3) holds. It can be shown that the expected average diffraction efficiency given by Eq. (3) is higher by a factor of 2.31 compared to the case of constant exposure. The price to pay for eliminating the non-uniformity is considerable complication in the recording procedure.

Surface Storage Density

Once the geometry has been chosen, and the cross-talk properties (i.e. the optimal m) and exposure schedule have been determined, the actual capacity of the holographic disk can be calculated. It is preferable to express the capacity in terms of the surface storage density $\mathcal{D}$, i.e. bits of information stored per μm^2 on the disk. Thus (Li & Psaltis 1994)

$$\mathcal{D} = \frac{N_p^2 M}{\mathcal{A}},\qquad(16)$$

where N_p^2 is the number of pixels per page, M is the number of overlapping holograms, and $\mathcal{A}$ is the area (in μm^2) on the disk taken up by a single hologram.

Conventional non-holographic storage methods (such as those used in compact disks) are diffraction limited to $\mathcal{D} \approx 1$ bit μm^{-2}. This is equivalent to having $N_p^2/\mathcal{A} = 1$ bit μm^{-2}. Holographic storage in volume materials allows $M \gg 1$, hence $\mathcal{D}$ can attain very high values.

The parameters N_p, $\mathcal{A}$, and M, and therefore the density of a shift multiplexed holographic disk, are determined from the material thickness L, the focal distance z_0, the angle of incidence of the signal θ_S, the numerical aperture NA of the reference, and the imaging systems used to store and retrieve the data. These parameters may depend on fundamental limits, such as the diffraction-limited resolution, or the currently available technology, e.g., the space-bandwidth product of the available SLMs, lenses and CCD cameras. In addition, various interrelations are introduced by the particular aspects of the chosen geometry, the material dynamic range, the desired error rate, etc. Examples of density calculations for other multiplexing techniques taking these parameters into account are given by Li & Psaltis, (1994), Yi *et al.* (1994) and Yi *et al.* (1995). In this section we will outline the surface density calculation for a shift multiplexed holographic disk (Fig. 3) using results developed in previous paragraphs.

We will assume that the disk is made of photopolymer, since this material has already been used successfully to demonstrate high density storage in practice (Psaltis & Pu 1995). Du Pont's commercially available photopolymer has thickness $L = 100$ μm and refractive index $n = 1.525$. We will consider image plane holograms, recorded with the signal incident at $\theta_S = 40°$ outside the polymer, and use $b = 2$ μm for the pixel size. A reasonable value for the $F/\#$ of the imaging optics used in the signal arm is 1.4 with $F = 5$ cm, when the maximum number of allowed pixels is $N_p = 5600$. Currently, SLMs with such high space-bandwidth product are not available, so we will limit ourselves to $N_p = 1000$. The size of the focused image is $s = N_p b = 2$ mm. We will assume that a lens with focal length $F_r = 3$ mm and NA $= 0.6$ is used to produce the spherical wave reference. We will use $z' = 2$ mm (focal distance measured in air) and $\delta_2 = 42.4$ μm (2nd null for the shift separation) as a compromise between low cross-talk and low null order m.

Therefore the number of holograms per area is:

$$\frac{M}{\mathcal{A}} = \frac{1}{\delta_2 s} = \frac{1}{84.8 \times 10^3 \, \mu m^2}$$

yielding density $\mathcal{D} = 11.2$ bits μm^{-2} for the 100 μm thick film. This number can be doubled by using the symmetric locations $\pm\theta_S$ around the optical axis (this complicates the implementation, by requiring either use of two hologram capturing mechanisms or motion of the same mechanism to capture either one of the two reconstructions) so that the maximal density for shift multiplexing becomes 22.4 bits μm^{-2}. The number of overlapping holograms required to reach this density is $M = 123$. For the polymer used in this example, $M/\# = 6.5$ was measured (Pu *et al.* 1994), hence the expected diffraction efficiency (using the recording scheme with spatial multiplexing first, as described previously) is $\eta = 2.8 \times 10^{-3}$, well within the range of comfortable detection.

Alternatively, we can use a photorefractive material such as $LiNbO_3$ (refractive index $n = 2.24$) which is available at higher thicknesses. As an example, we take $L = 1$ mm, $z' = 3$ mm to obtain $\mathcal{D} = 76.9$ bits μm^{-2}, which again can be doubled to yield 153.8 bits μm^{-2} with $M = 803$ overlapping holograms. Typical $M/\#$'s for photorefractive crystals are in the range $0.5 \sim 2$; using $M/\# = 1$ yields $\eta = 1.5 \times 10^{-6}$, in which case detection is more challenging. Increasing the thickness L further will yield higher densities and also somewhat higher $M/\#$, but will also cause M to increase rapidly, until the material dynamic range will not suffice for successful detection of the reconstructions. With the currently available materials and imaging devices, it is reasonable to expect that 100 bits μm^{-2} can be achieved using shift multiplexing.

CONCLUSIONS

In conclusion, we have presented the principles and the design considerations for holographic 3D disks using shift multiplexing. We have shown that shift selectivity is a result of Bragg mismatch caused by the translation of the spherical reference wave. We have given experimental results for the cross-talk and exposure schedule characteristics of shift-multiplexed memories, and demonstrated how the noise–dynamic range–density trade-offs can be accounted for in order to design and characterize practical systems. The shift multiplexing method is particularly suitable for convenient and robust implementation of holographic 3D disk architectures, and can yield surface storage densities in the order of 100 bits μm^{-2}.

ACKNOWLEDGEMENTS

This research was supported by the Air Force Office of Scientific Research under Grant No. F49620-92-J-0400, and by Rome Labs Grant No. F30602-94-C-0182. We are grateful to Allen Pu for his invaluable assistance and to Yayun Liu for technical support. Part of this work was conducted while George Barbastathis was supported by a Charles Lee Powell Foundation graduate fellowship. Michael Levene acknowledges the support of a National Defence Science and Engineering Graduate fellowship.

REFERENCES

Amodei, J. J., & Staebler, D.L. 1971. Holographic pattern fixing in electrooptic crystals. *Appl. Phys. Lett.*, **18**(12), 540–542.

Bashaw, M.C., Heanue, J.F., Aharoni, A., Walkup, J.F., & Hesselink, L. 1994. Cross-talk considerations for angular and phase-encoded multiplexing in volume holography. *J. Opt. Soc. Am. B*, **11**(9), 1820–1836.

Bløtekjaer, K. 1979. Limitations on holographic storage capacity of photochromic and photorefractive media. *Appl. Opt.*, **18**(1), 57–67.

Campbell, S., Yi, X. M., & Yeh, P. 1994. Hybrid sparse-wavelength angle multiplexed optical data storage system. *Opt. Lett.*, **19**(24), 2161–2163.

Chandross, E.A., Pryde, C.A., Tomlinson, W.J., & Weber, H.P. 1974. 'Photolocking – a new technique for fabricating optical waveguide circuits'. *Appl. Phys. Lett.*, **24**(2), 72–74.

Chen, F.S., LaMacchia, J.T., & Fraser, D.B. 1968. Holographic storage in lithium niobate. *Appl. Phys. Lett.*, **15**(7), 223–225.

Curtis, K., & Psaltis, D. 1992. Recording of multiple holograms in photopolymer films. *Appl. Opt.*, **31**(35), 7425–7428.

Curtis, K., & Psaltis, D. 1993. Cross talk in phase-coded holographic memories. *J. Opt. Soc. Am. A*, **10**(12), 2547–2550.

Curtis, K., & Psaltis, D. 1994a. Characterization of the Du-Pont photo-polymer for 3-dimensional holographic storage. *Appl. Opt.*, **33**(23), 5396–5399.

Curtis, K., & Psaltis, D. 1994b. Cross-talk for angle-multiplexed and wavelength-multiplexed image plane holograms. *Opt. Lett.*, **19**(21), 1774–1776.

Curtis, K., Gu, C., & Psaltis, D. 1993. Cross-talk in wavelength-multiplexed holographic memories. *Opt. Lett.*, **18**(12), 1001–1003.

Curtis, K., Pu, A., & Psaltis, D. 1994. Method for holographic storage using peristrophic multiplexing. *Opt. Lett.*, **19**(13), 993–994.

d' Auria, L., Huignard, J.P., Slezak, C., & Spitz, E. 1974. Experimental holographic read-write memory using 3-D storage. *Appl. Opt.*, **13**(4), 808–818.

Denz, C., Pauliat, G., & Roosen, G. 1991. Volume hologram multiplexing using a deterministic phase encoding method. *Opt. Commun.*, **85**, 171–176.

Ducharme, S., Scott, J.C., Twieg, R.J., & Moerner, W.E. 1991. 'Observation of the photorefractive effect in a polymer'. *Physical Review Letters*, **66**(4), 1846–1849.

Ford, J.E., Ma, J., Fainman, Y., Lee, S.H., Taketomi, Y., Bize, D., & Neurgaonkar, R.R. 1992. Multiplex holography in strontium barium niobate with applied electric field. *J. Opt. Soc. Am. A*, **9**(7), 1183–1192.

Gabor, D. 1969. Associative holographic memories. *IBM Journal of Research and Development*, March, 156–169.

Gu, C., Hong, J., McMichael, I., Saxena, R., & Mok, F. 1992. Cross-talk-limited storage capacity of volume holographic memory. *J. Opt. Soc. Am. A*, **9**(11), 1978–1983.

Hall, T.J., Jaura, R., Connors, L.M., & Foote, P.D. 1985. 'The photorefractive effect – a review'. *Progress in Quantum Electronics*, **10**(2), 77–145.

Horowitz, M., Bekker, A., & Fischer, B. 1993. Image and hologram fixing method with $(Sr_xBa_{1-x})Nb_2O_6$ crystals. *Opt. Lett.*, **18**(22), 1964–1966.

Kewitsch, A., Segev, M., Yariv, A., & Neurgaonkar, R.R. 1993. Selective page-addressable fixing of volume holograms in $(Sr_{0.75}Ba_{0.25})Nb_2O_6$ crystals. *Opt. Lett.*, **18**(15), 1262–1264.

Kogelnik, H. 1969. Coupled wave theory for thick hologram gratings. *Bell Syst. Tech. J.*, **48**(9), 2909–2947.

Kukhtarev, N.V., Markov, V.B., Odulov, S.G., Soskin, M.S., & Vinetskii, V.L. 1979. 'Holographic storage in electrooptic crystals, I. Steady state'. *Ferroelectrics*, **22**, 949–960.

Lee, H., Gu, X.-G., & Psaltis, D. 1989. Volume holographic interconnections with maximal capacity and minimal cross talk. *J. Appl. Phys.*, **65**(6), 2191–2194.

Leith, E.N., Kozma, A., Upatnieks, J., Marks, J., & Massey, N. 1966. Holographic data storage in three-dimensional media. *Appl. Opt.*, **5**(8), 1303–1311.

Li, H.-Y.S., & Psaltis, D. 1994. Three dimensional holographic disks. *Appl. Opt.*, **33**(17), 3764–3774.

Meerholz, K., Volodin, B.L., Kippelen, B. Sandalphon, & Peyghambarian, N. 1994. 'A photorefractive polymer with high optical gain and diffraction efficiency near 100 percent'. *Nature*, **371**(6497), 497–500.

Micheron, F., & Bismuth, G. 1972. Electrical control of fixation and erasure of holographic patterns in ferroelectric materials. *Appl. Phys. Lett.*, **20**(2), 79–81.

Mok, F.H. 1993. Angle-multiplexed storage of 5000 holograms in lithium niobate. *Opt. Lett.*, **18**(11), 915–917.

Mok, F.H., Burr, G.W., & Psaltis, D. 1994. Angle and space multiplexed random access memory (HRAM). *Optical Memory and Neural Networks*, **3**(2), 119–127.

Orlov, S., Psaltis, D., & Neurgaonkar, R.R. 1993. Dynamic electronic compensation of fixed gratings in photorefractive media. *Appl. Phys. Lett.*, **63**(18), 2466–2468.

Psaltis, D. 1992. Parallel optical memories. *Byte*, **17**(9), 179.

Psaltis, D., & Pu, A. 1995. Holographic 3D disks. *The International Journal of Optoelectronics-Devices and Technologies*, **10**(3), 333–342.

Psaltis, D., Brady, D., & Wagner, K. 1988. Adaptive optical networks using photorefractive crystals. *Appl. Opt.*, **27**(9), 1752–1759.

Psaltis, D., Levene, M., Pu, A., Barbastathis, G., & Curtis, K. 1995. Holographic storage using shift multiplexing. *Opt. Lett.*, **20**(7), 782–784.

Pu, A., Curtis, K., & Psaltis, D. 1994. A new method for holographic data storage in photopolymer films. In: *IEEE Nonlinear Optics: Materials, Fundamentals and Applications*, p. 433–435. Conference Proceedings, Waikoloa, Hawaii, 1994

Qiao, Y., Orlov, S., Psaltis, D., & Neurgaonkar, R. R. 1993. Electrical fixing of photorefractive holograms in $(Sr_{0.75}Ba_{0.25})Nb_2O_6$. *Opt. Lett.*, **18**(12), 1004–1006.

Solymar, L., & Cooke, D.J. 1991. *Volume Holography and Volume Gratings*. Academic Press.

Staebler, D.L., Amodei, J.J., & Philips, W. 1972 (May). Multiple storage of thick holograms in $LiNbO_3$. In: *VII International Quantum Electronics Conference*.

Staebler, D.L., Burke, W.J., Phillips, W., & Amodei, J.J. 1975. Multiple storage and erasure of fixed holograms in Fe-doped $LiNbO_3$. *Appl. Phys. Lett.*, **26**(4), 182–184.

Thaxter, J.B. 1968. Electrical control of holographic storage in strontium barium niobate. *Appl. Phys. Lett.*, **15**(7), 210–212.

Thaxter, J.B., & Kestigian, M. 1974. Unique properties of SBN and their use in a layered optical memory. *Appl. Opt.*, **13**(4), 913–924.

Trisnadi, J., & Redfield, S. 1992. Practical verification of hologram multiplexing without beam movement. *Photonic Neural Networks, Proc. SPIE*, **1773**, 362–371.

Valley, G.C., & Klein, M.B. 1983. Optimal properties of photorefractive materials for optical data processing. *Opt. Eng.*, **22**(6), 704–711.

van Heerden, P.J. 1963. Theory of optical information storage in solids. *Appl. Opt.*, **2**(4), 393–400.

Verkhovskaya, K.A., Fridkin, V.M., Bune, A.V., & Legrand, J.F. 1992. 'Photoconducting ferroelectric polymers'. *Ferroelectrics*, **13**(1–4), 7–15.

von der Linde, D., & Glass, A.M. 1975. Photorefractive effects for reversible holographic storage of information. *Appl. Phys.*, **8**, 85–100.

Wagner, K., & Psaltis, D. 1987. Multilayer optical learning networks. *Appl. Opt.*, **26**(23), 5061–5076.

Yariv, A. 1993. Interpage and interpixel crosstalk in orthogonal (wavelength multiplexed) holograms. *Opt. Lett.*, **18**(8), 652–654.

Yi, X.M., Yeh, P., & Gu, C. 1994. Statistical analysis of cross-talk noise and storage capacity in volume holographic memory. *Opt. Lett.*, **19**(9), 1580–1582.

Yi, X.M., Campbell, S., Yeh, P., & Gu, C. 1995. Statistical analysis of cross-talk noise and storage capacity in volume holographic memory – image plane holograms. *Opt. Lett.*, **20**(7), 779–781.

Yin, S., Zhou, H., Zhao, F., Wen, M., Zang, Y., Zhang, J., & Yu, F.T.S. 1993. Wavelength-multiplexed holographic storage in a sensitive photorefractive crystal using a visible-light tunable diode-laser. *Opt. Commun.*, **101**(5–6), 317–321.

11 Dense optical interconnections for silicon electronics

David A. B. Miller
*AT&T Bell Laboratories,
Holmdel, New Jersey, USA*

INTRODUCTION

No-one can doubt the impact of semiconductor electronics over the last 30 years. Silicon integrated circuits have, for example, led to a revolution in our ability to process information. Semiconductor optoelectronics is also having a major impact, an impact particularly clear now in long-distance fibre optic communications, and in many consumer devices, such as compact disk players, and their descendants, the compact disk read-only memory (CD-ROM). Optoelectronics also seems set to be a key enabling technology in bringing high-bandwidth information services into the home through optical networks.

Optoelectronics and electronics are technologies that would seem to complement one another very well for handling information. Electronics, with its ability to make sophisticated and flexible systems at low cost, dominates computing. On the other hand, optics is very good at communicating information, and has won the race for long-distance, high-bandwidth information transmission. It is also offering useful new options for information storage.

Despite this complementarity, optics and electronics are perhaps not as closely intermixed as we might have imagined they would become. One problem is that optoelectronic systems have remained stubbornly expensive. Packaging individual optoelectronic devices to fibres is costly. Optoelectronic devices themselves have remained at levels of integration that are comparable to the complexity of vacuum tube circuits. The economy of scale that has driven and continues to drive silicon integrated circuits has not happened in optoelectronics. Integrated optics, for example, which integrates waveguide optics with active electrooptic or, more recently, optoelectronic devices, while still likely to be a useful technology, shows no prospect of approaching integration levels common in electronics.

Without some qualitative change in optoelectronics, we are left with a rather depressing metaphor for future information processing machines. The electronic circuits become ever denser (though increasingly sophisticated) cities, with roadways ever more crowded with information trying to get from one part of the city to the other. For transport between cities the information travels out on crowded multilane highways to 'optoelectronic' airports with gigantic jets.

SMART PIXELS

There is, however, a new class of technologies emerging in optoelectronics. This class attempts to use large numbers of optoelectronic devices, usually in two-dimensional arrays, and integrate them with electronics; it is most often referred to as 'smart pixels' [1] (see Fig. 1). It has become possible because of the emergence of two developments: first, high-yield optoelectronic devices capable of operating with light beams perpendicular to the surface of the device; and, second, integration technologies that allow these optoelectronic devices to be integrated very efficiently with electronics.

The most difficult device problem has arguably always been in providing optical outputs (either modulators or emitters). One modulator solution is the quantum well modulator, which is the core of the class of smart pixel devices known as SEEDs (self-electrooptic effect devices) [2]. SEEDs allow arrays of thousands of fully functional optical inputs and outputs capable of operating at very high speeds. Another class is liquid crystal devices that may allow even larger numbers of somewhat slower devices [3]. Liquid crystal devices have

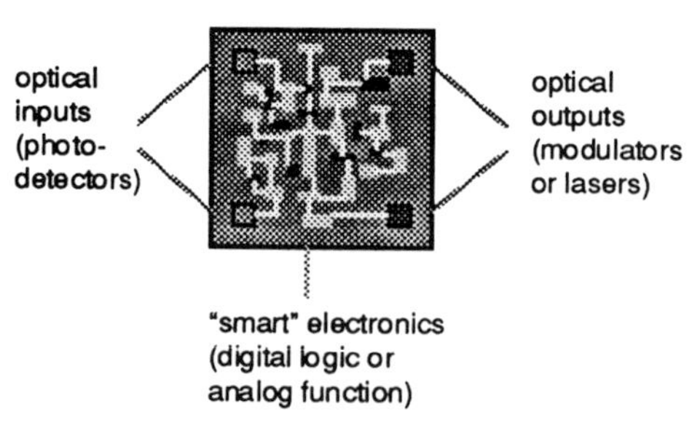

Fig. 1. Concept of 'smart pixels'. Individual smart pixels will usually have optical inputs and outputs, and they will typically be arranged in a two-dimensional array.

adapted integration techniques used previously for liquid crystal displays. We may yet see large arrays of vertical-cavity surface-emitting lasers (VCSELs) also integrated. SEEDs have used both monolithic integration with GaAs electronics [4], and, very recently, hybrid integration with silicon electronics using solder-bonding techniques [5].

The underlying proposition of these smart pixel devices is twofold: (1) take advantage of the many communications advantages of optics (including many currently not exploited in fibre optic systems) together with all of the processing abilities of electronics to allow higher performance systems; (2) avoid the high current cost of each optoelectronic 'connection' by working with large arrays of such connections at once. To exploit such smart pixels will require advances in two areas other than the device and integration technologies. First, we need optical technologies that can work with arrays of light beams. Second, we need systems architecture concepts that can exploit such array parallelism. Some switching architectures have been investigated and demonstrated already [6, 7]. Below we will look more closely at some of the key features of silicon technology and of optics, and discuss one recent technology, the hybrid SEED [5], in more detail.

SILICON TECHNOLOGY

The number of transistors on a silicon circuit has approximately doubled every 18 months for the past two decades, and it currently looks likely that this trend will continue [8], perhaps even for 15 years or more. Current electronic logic chips can have millions of transistors, and memory chips can have even more. It may even be that this progress has now become self-fulfilling, with different manufacturers using these predictions to know where they should be at a given time in the future to be competitive. The investment in delivering such future technology is certainly large, and no other technology seems likely to compete with electronics in the immediate future as far as complexity is concerned (though biological systems are an existing proof that systems much more complex and sophisticated than silicon are possible). Certainly optics has no way of matching such complexity, and no current prospect of an investment large enough to allow it to catch up even if it had a viable approach.

With all this advance of silicon, is there any role for optics? Will it not always be behind? Will optical machines not always be relegated to special purposes, to be overtaken by the advancing general purpose silicon? We could, of course, argue that optical devices may be able to operate with performance silicon cannot beat, such as ultrafast non-linear optical logic gates, but it is difficult to imagine how we could make systems with sufficient complexity to rival silicon for performing mainstream information processing tasks, and so optics appears to be relegated again only to special purpose tasks. It is also likely that silicon devices as we know them will run into physical limits as miniaturization continues, but it is by no means clear that optics offers the generation beyond the tiniest silicon device.

Silicon electronics has its weaknesses, however, and there are many related to the difficulty of interconnections. Electrical interconnections are dominated by problems of classical electromagnetics. Such problems do not necessarily scale well as silicon devices improve. They are connected with classical resistance and capacitance of lines, and with Maxwell's equations, phenomena that have little to do with silicon devices themselves. Furthermore, smaller silicon devices also tend to be able to run faster, and trying to exploit this increased speed only makes the problems of interconnections worse.

Suppose, for example, that, as we made the silicon devices smaller, we also scaled down the electrical wiring on the chip, scaling down both the widths and the thicknesses of the lines. On a silicon chip, the time taken to send information is essentially limited by the time taken to charge up the electrical line on the chip. Halving the width and thickness of the metal lines would multiply the resistance by a factor of 4 for a given length; the capacitance would remain the same (the line would be half as wide, but the dielectric would be half as thick), and hence the resistance capacitance (RC) product for a given length of line would have gone up by a factor of 4. At best, we might argue that the line only needs to be half as long since the transistors are half the size also, but even then the RC product merely stays the same. Not only is the line not any faster than it was before, but the smaller transistors can probably run about twice as fast, so the issue of the delay of the interconnect line is becoming even more of an impediment to exploiting the capability of the silicon. Of course, silicon technologists are well aware of such problems [8], and are doing their best to minimize them. For example, metal line thicknesses will likely not scale down with the width of the lines, and attempts will be made to reduce dielectric constants and metallic resistances. There can be little doubt that the interconnections will more and more limit the capability of silicon systems as silicon itself becomes more capable. Indeed, circuit architecture innovations to compensate for long interconnects and higher clock speeds are seen as one of the highest priority research needs in silicon technology [8].

Yet another set of problems arises as the signals leave the chip. For low speeds (e.g., less than 10 MHz) we can often think of interconnections as simple wires, with signals travelling 'instantaneously' from one end to the other. But at high speeds (e.g., >100 MHz), we must treat the signals as waves on transmission lines. Impedances must be matched to avoid reflections, terminating resistors (usually about 50 ohms) must be used, and associated low-impedance line drivers (with correspondingly high power dissipations) must be constructed. Cross-talk between lines becomes a more significant issue because the lines become progressively more efficient broadcasting and receiving antennas at higher frequencies. Signals become distorted because of the finite frequency response of the lines. The lines also become lossier at higher frequencies because of the skin effect (which forces electrical conduction into an ever thinner layer near the surface of the metallic conductors at higher frequencies). Synchronization of the arrival of signals also becomes progressively harder. Again, though engineers become very adept at minimizing the effects of these various issues, they are problems that only become progressively worse as the silicon circuits become larger and faster.

REASONS FOR OPTICAL INTERCONNECTIONS

There are many reasons why optics can help with interconnecting information, and only a few of these are exploited at the moment. All of the potential advantages of optics in this regard come from the same fundamental difference between optical and 'base-band' (i.e., radio frequency) electromagnetic waves. This difference can be expressed either in the higher frequency, the shorter wavelength, or the larger photon energy of optics, all three of which are different manifestations of the same difference (Fig. 2).

High Frequency of Light

The high frequency of light has several consequences.

(a) The carrier frequency of light is so much higher than any frequency at which we can modulate that such modulation makes essentially no difference to the propagation of light, at least over the distances within machines. Hence there is no frequency-dependent loss or frequency-dependent cross-talk when using light as the information carrier (whether in fibres or free-space) within machines. This is in strong contrast with wires, where such frequency-dependent loss and cross-talk become rapidly worse at high frequencies. There may well, of course, be loss and cross-talk with optics, but they get no worse at high speeds.

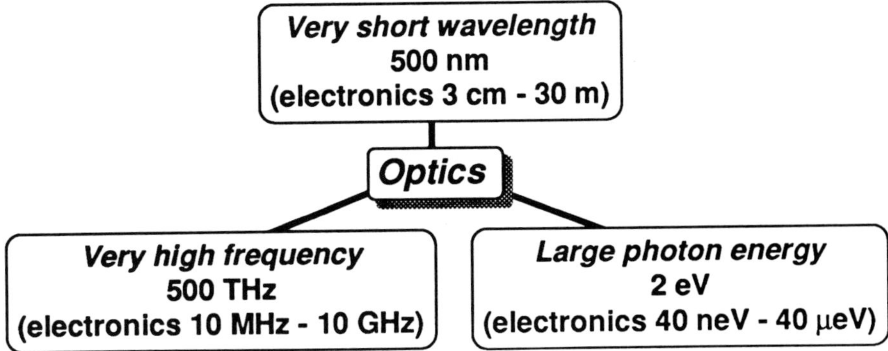

Fig. 2. Fundamental difference between optics and electronics for communication, expressed in terms of wavelength, frequency and photon energy.

(b) There are situations in optics where we must match impedances, as for example when we transition from propagating in air or glass into a semiconductor material. Because the modulation bandwidth we use is so narrow compared to the carrier, we can use very simple 'resonant impedance transformers' – dielectric anti-reflection coatings. There is no comparable simple impedance transformer we can use for broad-band radio-frequency waves.

Short Wavelength of Light

The short wavelength of light leads to the following benefits.

(c) Because the wavelength of light is so small, we can make waveguides that are much larger in cross-sectional dimensions than the optical wavelength (e.g., a 10 micron diameter core in an optical fibre is much larger than the approximately 1 micron wavelength of light). As a result, we can guide the waves entirely with dielectrics, and hence have very low loss propagation. In purely electrical systems, the radio-frequency wavelengths are often comparable to or larger than the entire system, and certainly larger than most current integrated circuits. To guide the information to where we want it to go, we need to use the very large effective dielectric constants of metallic conductors. Since these conductors are not perfect, we sustain losses. (It is tempting to think that it is electrons that carry information in wires, but this is not really correct. The proof is simple – the information does not travel at the electron velocity, but rather at some fraction of the velocity of light in low-loss transmission lines or at an effective velocity set by line resistance and capacitance in the dissipative propagation found on, for example, connections within integrated circuits.)

(d) The small wavelength allows us to contemplate information flowing on and off chips without the use of waveguides. We can instead image many light beams from one chip to another (Fig. 3). Such imaging is only possible because the wavelength is so much smaller than the systems we are interested in. The smallest region we can image in practice is about one wavelength in size, and so attempting to image 'base-band' electromagnetic waves would typically only allow one interconnection out of a system – essentially a radio transmitter. With optics, however, it is routine to image thousands of 'outputs' from one plane on to thousands of 'inputs' on another with a single lens. Despite the fact that all of the beams may overlap at the lens, the lens separates out all of the information again to image the multiple beams at the receiving 'inputs'. Optics therefore allows very large numbers of connections from one plane to another through 'free space'. An incidental benefit is that all of the connections in such imaging-based systems are automatically very well synchronized with one another, hence avoiding another practical problem of 'signal skew' common in electrical systems. Another related consequence is that it is possible to make very global interconnect topologies (such as so-called 'perfect shuffles') in which many of the 'beams' cross through one

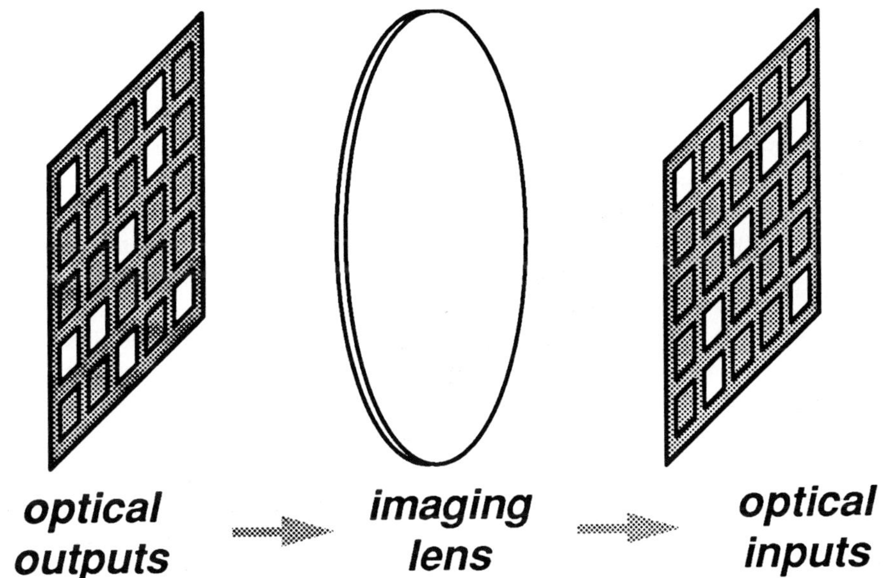

Fig. 3. Concept of imaging arrays of outputs on one plane to arrays of inputs on another.

another. Such patterns can be useful for operations such as telecommunications switching fabrics and fast Fourier transforms.

Large Photon Energy of Light

The fact that the photon energy of light is so large has two consequences, one already extensively exploited, and another that has only relatively recently been understood. A key physical point about light's large photon energy is that light is both generated and detected quantum mechanically, not classically – for example, detection of light in practice involves counting photons, not measuring electric field amplitudes.

(e) Detecting photons allows us to generate 'd.c.' currents and voltages without any direct electrical connection with the light source. In other words, optical interconnects give us perfect electrical isolation between two circuits, and make the absolute voltage levels in the different circuits irrelevant. This solves an important problem in electrical systems, and is exploited extensively in so-called 'opto-isolators', which usually contain a light-emitting diode (connected to the 'transmitting' circuit) and a photodiode (connected to the 'receiving' circuit).

(f) Use of optical emitters or modulators and photodiodes also fundamentally enables us to avoid the problems of the low impedance of electrical transmission lines [9]. The low impedance of electrical lines (typically

50 ohms) is a direct consequence of the low impedance of free space (377 ohms) and the fact that transmission line impedance only scales logarithmically with the conductor sizes. A related problem is that all electrical lines have about 1 pF cm^{-1} or more capacitance, and there is little that can be done to reduce that either. A key consequence of such low impedance and high capacitance is that line driver circuits have to be made, and these unavoidably consume relatively large amounts of power (e.g., tens to hundreds of mW per high-speed electrical line). The use of transmission lines becomes unavoidable at high clock rates (e.g., >100 MHz or so) off chip. The resulting power dissipation and use of chip area for drivers is a significant issue in chip design when high data throughput is required. The reason why optics can avoid the low impedance problem is that the voltage generated in a photodetector bears no particular relation to the classical 'voltage' in the light beam. It is quite possible, for example, to generate 1 V in a photodetector from a light beam with 600 microvolts of classical voltage [9] – a consequence, fundamentally, of the photoelectric effect. This particular benefit has not historically been exploited in practice because the optical sources have always consumed more power than a 50 ohm resistor anyway, so there was no power savings, and the driver problem was, if anything, worse than that of an electrical line. The emergence of quantum well modulator technology has, however, led to quite practical low-power optical output devices that can demonstrably send digital signals from chip to chip with substantially less power (e.g., <10 mW total dissipation at 150 Mb s^{-1}) than electrical connections [10]. Other approaches, such as low threshold lasers, may be able to offer similar benefits. This feature of optics is likely to be particularly important for large arrays of optical inputs and outputs, as in smart pixels, and may allow much larger amounts of information to be sent on and off chips optically than is practical electrically.

The various features of optics for interconnections discussed above clearly address many of the problems that are being encountered more and more in making electronic information processing systems. These features come directly from the differences in the physics between optics and electronics. The fact that the beneficial features of optics are so firmly rooted in the physics gives us reason to believe that optics will eventually be used extensively in interconnection in electronic systems.

HYBRID SEED TECHNOLOGY

A technology that allows easy conversion between electronics and optics and vice versa, and that allows large numbers of devices, is crucial for the more radical concepts like 'smart pixels'. Such a technology should ideally fit well with silicon electronics as it exists, without requiring any substantial changes to that technology, and should be capable of operating at the speeds and voltages of silicon circuits. Arguably the first such technology to emerge, and the most advanced at this time, is the hybrid SEED (self-electrooptic effect device) [5].

The hybrid SEED relies on quantum well diode modulators [11]. Quantum wells, made of alternating very thin (e.g., 10 nm) layers of two different semiconductor materials, show a large change of optical absorption when an electric field is applied perpendicular to the layers [12]. If we put many (e.g., 50–100) such layers inside a diode, reverse biasing the diode gives us a convenient and low-power way to apply the necessary field. Such diodes need only be a few microns thick overall yet still have sufficient change in optical transmission to make useful modulators. This allows two-dimensional arrays of such diode modulators to be made using standard semiconductor growth and processing techniques. The modulators are very fast, limited in practice only by the electrical drive circuits and parasitics, and can operate with moderate

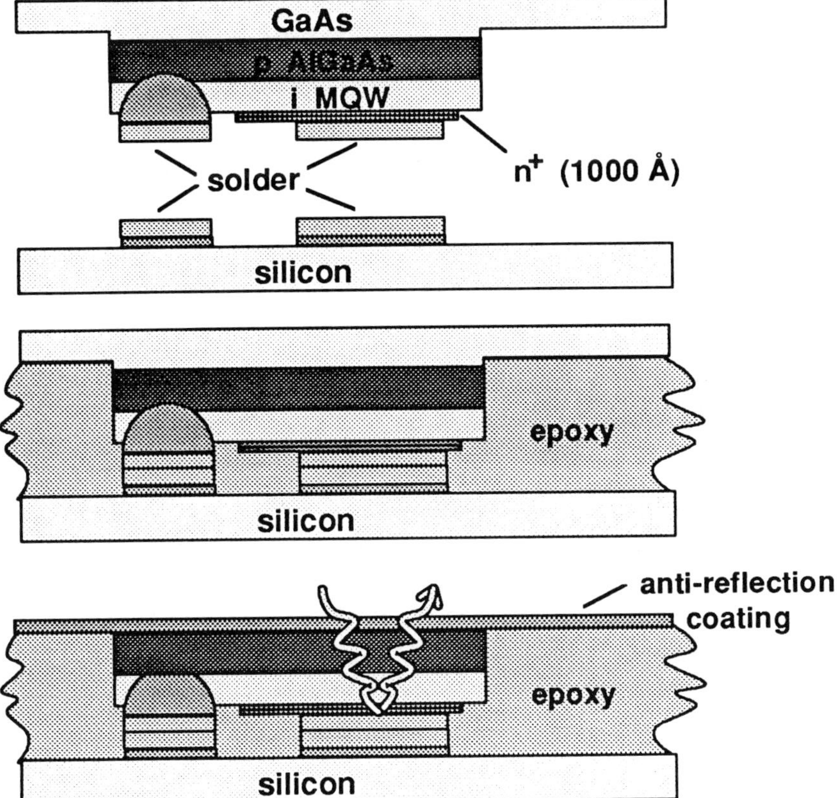

Fig. 4. Illustration of the hybrid solder bonding process used to bond arrays of quantum well diode modulators and detectors to silicon CMOS circuitry in the hybrid SEED process. With the final removal of the epoxy, individual modulators are left connected to the silicon circuitry in a two-dimensional array.

Fig. 5. Picture of part of a hybrid SEED chip. The quantum well diodes are the regular array of rectangles, and are approximately 15 × 45 microns in area and a few microns thick. Underneath is active silicon circuitry.

voltages (e.g., 5 V). The modulator diodes also function as good photodiodes, so the same device can be used for optical inputs and outputs.

In the hybrid SEED, arrays of such quantum well diodes are made, with reflecting metal on the top. Then the array is turned over and solder-bonded to the silicon circuit, and, as shown in Fig. 4, the entire substrate material is removed chemically from the quantum well diodes to leave isolated diodes bonded to the silicon circuit. The resulting array can be used as reflection modulators or photodetectors, depending on the silicon circuits to which they are connected. The use of reflection modulators is convenient since it means that light does not have to pass through the silicon circuits, so conventional silicon circuit mounting can be used, and the double pass of the light beam through the modulator increases the amount of modulation of the light beam.

The only additional processing needed of the silicon circuit is to deposit some metals and solder. This can be done after the usual fabrication of the silicon wafer, and no change is required in the usual fabrication process. As a result, this technique can be used with silicon circuits from any fabrication process.

Figure 5 shows a picture of a silicon circuit with attached quantum well diodes. In this case, the diodes are about 15×45 microns in area, and the solder bonds are about 15×15 microns in size. It is interesting to note that these optical input and output 'pads' are much smaller than the electrical bonding pads used on chips to carry signals electrically on and off chips.

More than 1000 of these quantum well diodes have been successfully bonded to a silicon chip [5]. Operating circuits have been demonstrated with this approach, including sophisticated and highly integrated receivers and transmitters [13] and large complex circuits [14] running as fast as 700 Mb s^{-1}. These 700 Mb s^{-1} results were obtained using 0.9 micron feature size silicon, which is remarkable because such silicon with electrical inputs and outputs is normally run at clock rates of about 100 MHz; the optics may be allowing much faster operation than would normally be contemplated in purely electrical systems.

An important point about this hybrid approach is that, as silicon electronics gets better, the optoelectronic systems based on this approach get better also. We expect still to be able to hybridize to future generations of silicon technology with smaller feature sizes and faster transistor performance. It is also reasonable to expect that in the future even larger numbers of optoelectronic devices can be integrated.

CHALLENGES

There are many challenges in trying to take advantage of optics in information processing machines. In the evolutionary approach of gradually substituting optical fibre for copper wire down to ever shorter distances, cost is perhaps the major issue. Beating electrical wiring on cost is hard; though it has its technical limitations, it does what it does for low cost, and it is difficult to compete when trying to substitute directly for it with similar performance.

The approach we have discussed here is, however, more revolutionary. Rather than trying to substitute for copper or aluminium wiring in systems already optimized around the constraints of such a technology, the 'smart pixels' approach would attempt to generate new kinds of systems that can directly exploit the many abilities of optics. This approach raises many challenges.

The first kind of challenge is the more obvious but real technical one of delivering the devices and systems components at low enough cost and high enough performance. Device technologies, such as the hybrid SEED, are emerging, and appear to have good chances of delivering impressive performance. Specific areas of work for the hybrid SEED will be in continuing to make larger numbers of devices in arrays, and to progressively reduce the operating voltage to track the expected reduction in the voltages in electronics.

An area that is still in need of innovation and more practical technology is that of optics and optomechanics for arrays of light beams. Substantial systems have been built, proving that array optics and optomechanics is possible – for example, one system operated successfully with more than 60 000 light beams [6]. Such systems need to evolve from a 'brass-board' level that allows systems prototypes, towards practical, low-cost, manufacturable systems.

The SEED technology, being based on modulators, also needs external laser power sources. It has usually been run using diode laser sources, and these are likely to continue to improve, though SEED systems usually want good wavelength stability and output power. Recent developments in diode-pumped solid state lasers are also promising here, including techniques that allow short pulse generation [15]. The use of external lasers has the significant system advantage of allowing the entire system to be 'clocked' globally by clocking one laser, and the use of short pulses has several additional advantages, including very well-defined 'clock' pulses, and also ways of reducing the operating energy of systems [16]. There is even a possibility of using such short pulses to make convenient ways of working with multiple wavelengths at once [17].

The second kind of challenge is a more subtle one – we need to devise the classes of systems that can best exploit what optics can do. We should not underestimate the extent to which current electronic systems are optimized around the particular problems and benefits of electrical connections. For example, it is not unusual for a high-performance electronic chip to have its area and power dissipation devoted approximately equally to each of the three functions of (i) input/output, (ii) clock distribution, and (iii) logic and memory circuitry; such a balance is likely not accidental, and reflects the intrinsic optimization done by the chip designer. Optics can likely substantially impact both the clock distribution and the chip input/output, hopefully allowing a higher system performance overall. At the chip level, therefore, and also likely at other levels in the system (such as the board-to-board level), there are many opportunities for optics to allow a re-optimization of the system design for higher performance overall. It is, however, important to be doing the research now on how we might do this optimization to take advantage of optical technology as it becomes available.

Another important aspect of this second challenge is that the optically enhanced systems must interface to the existing electronic world; they must make this interface in two ways, both physically and in a more subtle way that we can refer to as 'hierarchically'. The physical interface difficulty is what we can call a 'fire-hose' problem – a reference to the metaphor of attempting to drink from a fire hose pumping out vast quantities of water at high pressure. Optical systems may allow us to communicate vast amounts of data in and out of a chip, but this is of little use if we have no way of feeding this 'fire-hose'' of data into the chip or of handling a similar 'fire-hose' of data out of it. There are several kinds of systems that do solve this problem. One is the kind of system that we might encounter in telecommunications switching, where we can make up the large 'fire-hose' of data by bringing together a large number of smaller pipes of data, one pipe from each input line to the switching system (with a similar inverse arrangement at the output), with each pipe being carried on an optical fibre into an array of fibres that might be imaged on to the chip. Another kind of system would be one in which the internal data rate might be higher than the overall input and output rates. One such example would be a processor for multiplying vectors by a relatively fixed matrix – the input and output would only be vectors (one-dimensional parallelism), but internally the parallelism could be two-dimensional. Examining the architectural possibilities here should be a subject of research now.

Information processing machines are built and operated using a very sophisticated hierarchy of technologies and levels of abstraction. At one extreme, we have device physics and device technology. At the other extreme is the human interface with perhaps a computer screen or a telephone. In between, we find algorithms, logic design, assembler language, high-level computer languages, backplane architectures, bus protocols, and many other concepts. Here we come to the 'hierarchical' challenge for the use of optics. The development of each of these levels in the hierarchy has required substantial investment. If we come along with a concept for a processor that requires that we change many levels in this hierarchy, it will be very difficult to implement because it will be expensive to replace all of those many levels.

It is already clear that, even at the hardware level, the use of optics requires that we change several levels of the hierarchy; we must try in our architectural concepts to avoid changing many more if at all possible, at least initially. Technologies such as the hybrid SEED help in that they are intrinsically very well linked to conventional electronic technologies. At higher levels, this difficulty of changing levels in the hierarchy suggests that we might look at optically assisted processors as 'co-processors', processors that are addressed and controlled at a higher level by conventional processors, but that perform specific tasks or 'instructions' particularly efficiently. Examples of this approach in the electronic world are math co-processor chips as used in personal computers, and digital signal processor chips. Optically assisted co-processors might include signal processing functions (including, for example, Fourier transforms and correlations), switching functions (such as a high-performance switching fabric for telecommunications switching or interconnection multiple processors), or matrix math functions (such as inversion and multiplication). A key difference of the emerging smart pixel approach compared to many previous optical approaches is that the smart pixels can operate digitally, thus avoiding problems of limited analogue accuracy, and in general their digital format and use of existing silicon technology helps a lot in reuse of existing levels of the hierarchy.

PROGNOSIS

When will we see such two-dimensional array optics being used in real machines? Of course, that is a difficult question to answer, and depends on many things, some technical and some not. One possible driver is machines that must handle very large throughputs of data. We could speculate that by the year 2000 we may need to have telecommunications switching machines with capacities as large as 1 Tb s^{-1} [*18*]. It may also be that large database machines will approach similar capacities at this time, and perhaps also specialized image processing machines. Certainly, if one machine of this class is produced it is likely that machines for other purposes will try to exploit the technology. Machines with capacities of about 100–200 Gb s^{-1} are being researched now based on electrical technology. Such machines typically consist of many cabinets, with very large numbers of electrical cables connecting the cabinets. It is probably possible to make machines with 1 Tb s^{-1} capacity this way, but the intercabinet wiring problem is likely to be substantial, and it is not clear that this approach can continue scaling as problems such as switching may require [*18*]. Hence there may be a real opportunity for optics.

If two-dimensional optics for interconnection comes into first serious use by about the year 2000, then by the year 2010 we might expect to see volume use, with cheap optical power, and cost-reduced optomechanics. By that time systems would likely be designed so that optics is essential to get the architectures and performance customers need.

Of course, we could argue that such predictions, in addition to being speculative, are so distant as to be useless. In fact, however, if we are to meet such 'deadlines' of 2000 and 2010, history suggests that we had better be doing the research now – 15 years is a short time to turn a new paradigm into a practical reality with a mass market.

Too often in the past we have seen promising schemes for 'optical computing' defeated because of the continuing improvement in silicon technology. In the case discussed here, improvement of silicon technology only helps the capabilities of these optoelectronic systems. Optics here is not competing with silicon – instead it is competing with aluminium and copper, which are much more slowly moving targets. Indeed it may be the optoelectronics, through its ability to provide dense, high-speed and low-power interconnection, that allows us to exploit the real capabilities of silicon electronics.

REFERENCES

1. See for example, Special Issue on Smart Pixels (ed. S.R. Forrest and H.S. Hinton), *IEEE J. Quantum Electron.* **29**, No. 2, February 1993.
2. A.L. Lentine and D.A.B. Miller, Evolution of the SEED technology: bistable logic gates to optoelectronic smart pixels, *IEEE J. Quantum Electron.* **29**, 655–669 (1993); D.A.B. Miller, Quantum well self-electrooptic-effect devices, *Opt. Quantum Electron.* **22**, S61–S98 (1990).
3. D.J. McKnight, K.M. Johnson, and R.A. Serati, 256 × 256 liquid-crystal-on-silicon spatial light modulator, *Appl. Optics* **33**, 2775–2784 (1994).
4. L.A. D'Asaro, L.M.F. Chirovsky, E.J. Laskowski, S.S. Pei, T.K. Woodward, A.L. Lentine, R.E. Leibenguth, M.W. Focht, J.M. Freund, G.G. Guth, and L.E. Smith, Batch fabrication and operation of GaAs-AlGaAs field-effect transistor-self-electrooptic effect device (FET-SEED) smart pixel arrays, *IEEE J. Quantum Electron.* **29**, 670–677 (1993); D.A.B. Miller, M.D. Feuer, T.Y. Chang, S.C. Shunk, J.E. Henry, D.J. Burrows, and D.S. Chemla, Field-effect transistor self-electrooptic effect device: integrated photodiode, quantum well modulator and transistor, *IEEE Phot. Tech. Lett.* **1**, 61–64 (1989).
5. K.W. Goossen, J.A. Walker, L.A. D'Asaro, S.P. Hui, B. Tseng, R. Leibenguth, D. Kossives, D.D. Bacon, D. Dahringer, L.M.F. Chirovsky, A.L. Lentine, and D.A.B. Miller, GaAs MQW modulators integrated with silicon CMOS, *IEEE Phot. Tech. Lett.* **7**, 360–362 (1995).
6. H.S. Hinton, T.J. Cloonan, F.B. McCormick, A.L. Lentine, and F.A.P. Tooley, Free-space digital optical systems, *Proc. IEEE* **82**, 1632–1649 (1994).
7. F.B. McCormick, T.J. Cloonan, A.L. Lentine, J.M. Sasian, R.L. Morrison, M.G. Beckman, S.L. Walker, M.J. Wojcik, S.J. Hinterlong, R.J. Crisci, R.A. Novotny, and H.S. Hinton, Five-stage free-space optical switching network with field-effect transistor self-electrooptic-effect-device smart-pixel arrays, *Appl. Opt.* **33**, 1601–1618 (1994).
8. *The National Technology Roadmap for Semiconductors* (Semiconductor Industry Association, San Jose, 1994).
9. D.A.B. Miller, Optics for low-energy communication inside digital processors: quantum detectors, sources, and modulators as efficient impedance converters, *Optics Lett.* **14**, 146–148 (1989).
10. A.L. Lentine, R.A. Novotny, T.J. Cloonan, L.M.F. Chirovsky, L.A. D'Asaro, G. Livescu, S. Hui, M.W. Focht, J.M. Freund, G.D. Guth, R.E. Leibenguth, K.G. Glogovsky, and T.K. Woodward, 4 × 4 arrays of FET-SEED embedded control

2 × 1 optoelectronic switching nodes with electrical fan-out, *IEEE Phot. Tech. Lett.* **6**, 1126–1129 (1994).

11. D.A.B. Miller, Quantum well optoelectronic switching devices, *Int. J. High Speed Electronics* **1**, 19–46 (1990).

12. D.A.B. Miller, D.S. Chemla, T.C. Damen, A.C. Gossard, W. Wiegmann, T.H. Wood, and C.A. Burrus, Electric field dependence of optical absorption near the bandgap of quantum well structures, *Phys. Rev. B* **32**, 1043–1060 (1985).

13. T.K. Woodward, A.V. Krishnamoorthy, A.L. Lentine, K.W. Goossen, J.A. Walker, J.E. Cunningham, W.Y. Jan, L.A. D'Asaro, L.M.F. Chirovsky, S.P. Hui, B. Tseng, D. Kossives, D. Dahringer, and R.E. Leibenguth, *T*-Gb/s two-beam transimpedance smart-pixel optical receivers made from hybrid GaAs MQW modulators bonded to 0.8 µm silicon CMOS, *IEEE Phot. Tech. Lett.* **8**, 422–424 (1996).

14. A.L. Lentine, K.W. Goossen, J.F. Walker, L.M.F. Chirovsky, L.A. D'Asaro, S.P. Hui, B. Tseng, R.E. Leibenguth, D. Kossives, D. Dahringer, D.D. Bacon, T.K. Woodward, and D.A.B. Miller, Arrays of optoelectronic switching nodes comprised of flip-clip bonded MQW modulators and detectors on silicon CMOS circuitry, *IEEE Phot. Tech. Lett.* **8**, 221–223 (1996).

15. For a recent example, see S. Tsuda, W.H. Knox, E.A. de Souza, W.Y. Jan, and J.E. Cunningham, Low-loss intracavity AlAs/AlGaAs saturable Bragg reflector for femtosecond mode locking in solid-state lasers, *Optics Lett.* **20**, 1406–1408 (1995).

16. A.L. Lentine, L.M.F. Chirovsky, and T.K. Woodward, Optical energy considerations for diode-clamped smart pixel optical receivers, *IEEE J. Quantum Electron.* **30**, 1167–1171 (1994).

17. E.A. De Souza, M.C. Nuss, T.C. Damen, W.H. Knox, and D.A.B. Miller, Wavelength division multiplexing with femtosecond pulses, *Optics Lett.* **20**, 1166–1168 (1995).

18. T. Egawa, K. Yukimatsu, and K. Yamasaki, Recent research trends and issues in photonic switching technologies, *NTT Review* **5**, 30–37 (1993).

12 Fan-in loss for electrical and optical interconnections

Joseph W. Goodman and Jane C. Lam
*Department of Electrical Engineering,
Stanford University, Stanford, CA 94305*

INTRODUCTION

Power loss that occurs when an optical signal is split (or 'fans out') to N separate beams is easy to understand – the optical power is divided N ways, and as a result only $1/N$th of the total power is available in any one outgoing optical signal.

Power loss that occurs when N optical signals are combined (i.e. 'fan in') to a single signal is a bit more subtle. It occurs for some geometries and it doesn't occur for others. This is often particularly mysterious for individuals who do not have an optics background. Since the field of optical interconnections is interdisciplinary, involving computer architects, electrical systems designers, algorithm experts, in addition to those with an optics background, it seems useful to discuss fan-in losses from a general point of view. A special emphasis is made here to draw analogies between phenomena encountered in electrical circuits and phenomena encountered with optical circuits. For an earlier discussion of fan-in loss in optical interconnections, see Goodman (1985).

SPATIAL AND TEMPORAL MODES

As background, it is important to understand that both temporal and spatial signals can be thought of as consisting of a superposition of a multitude of separate *modes*. These concepts are briefly reviewed in what follows.

Temporal Modes

A time waveform $g(t)$ of duration T seconds can be represented by means of an expansion of orthonormal waveforms

$$g(t) = \sum_{k=-\infty}^{\infty} c_k \psi_k(t). \tag{1}$$

To a good approximation, the energy of this signal is contained in a finite number $K+1$ of these elementary waveforms, i.e.

$$g(t) = \sum_{k=-K/2}^{K/2} c_k \psi_k(t). \tag{2}$$

As an example, in the frequency-domain version of the Nyquist sampling theorem, a time waveform $g(t)$ of duration T seconds which has no significant frequency components outside of the frequency band $(-B/2,\ B/2)$ can be represented by a discrete set of $K+1 = BT+1$ separate time-harmonic signals with frequencies separated by $1/T$. Thus

$$g(t) = \sum_{k=-K/2}^{K/2} c_k \exp\left(j2\pi \, \frac{k}{K} \, Bt\right) \mathrm{rect}\left(\frac{t-T/2}{T}\right), \tag{3}$$

where $\mathrm{rect}(t)$ is unity for $-\frac{1}{2} < t < \frac{1}{2}$, zero elsewhere. The number of separate frequency modes $K+1$ is the same as the number of Nyquist time samples that exist in the T-second interval occupied by the signal. We can refer to these modes as *temporal* modes of $g(t)$.

For the above expansion, the coefficients c_k are the complex Fourier coefficients of the periodic extension of the given $g(t)$ to the infinite time axis. If the time waveform $g(t)$ arises from a noise source, then these coefficients are complex-valued random variables. Strictly speaking they are correlated, and therefore the average powers carried by the different modes cannot simply be added to find the total average power. However, they are asymptotically uncorrelated as $BT \to \infty$. We shall assume that the time-bandwidth product BT is sufficiently large for us to treat them as uncorrelated, and to add the powers of individual modes to find the total power.[1]

A generalization to bandpass signals occupying the frequency band $(f_0 - B/2,\ f_0 + B/2)$ is straightforward and yields equivalent results.

If the above signal is passed through a linear, time-invariant system, it may or may not be possible to selectively pass certain particular modes. If the impulse response duration τ is shorter than $1/B$, then the modes cannot be selectively passed. If the impulse response duration is T or longer, then the

---

[1] A more sophisticated argument can be constructed for the case when this is not true, using the Karhunen–Loéve expansion.

system can selectively pass any of the modes. Intermediate lengths of the impulse response result in the ability to select groups of modes.

Spatial Modes

A two-dimensional version of the expansion into orthonormal functions can be applied to a monochromatic source with a field that is zero outside the region $-X/2 < x < X/2$, $-Y/2 < y < Y/2$. The (n, m)th expansion function is now $\psi_{nm}(x, y)$ and the expansion takes the form

$$g(x, y) = \sum_{n=-N/2}^{N/2} \sum_{m=-M/2}^{M/2} c_{nm} \psi_{nm}(x, y), \tag{4}$$

where it has been assumed that a finite number $(N + 1)(M + 1)$ of modes will suffice. If in addition the propagation of the wave is confined to the finite angular region (θ_X, θ_Y) limited to $-\Theta_X/2 < \theta_X < \Theta_X/2$, $-\Theta_Y/2 < \theta_Y < \Theta_Y/2$, then in accordance with the Nyquist sampling theorem the orthonormal set can be taken to be complex exponentials and the modal expansion is given by

$$g(x, y) = \sum_{n=-N/2}^{N/2} \sum_{m=-M/2}^{M/2} c_{nm} \exp\left[j \frac{2\pi}{\lambda} \left(\frac{n}{N} \Theta_X x + \frac{m}{M} \Theta_Y y\right)\right]$$
$$\times \operatorname{rect}\left(\frac{x}{X}\right) \operatorname{rect}\left(\frac{y}{Y}\right), \tag{5}$$

where

$$N = \frac{X}{\lambda} \Theta_X, \qquad M = \frac{Y}{\lambda} \Theta_Y,$$

and λ is the optical wavelength. The total number of spatial modes is $(N + 1)(M + 1)$, which we call the space-bandwidth product of the wave.

If the optical waves are confined to a waveguide, rather than existing in free space, then a similar expansion into orthogonal modes exists, but the modes are no longer plane waves travelling in different directions.

If the source is narrow-band, rather than monochromatic, and we consider a finite length of time T, each spatial mode can be thought of as containing K temporal modes, by direct analogy with Eq. (1). If the source is incoherent spatially, and if its space-bandwidth product is large, then the various spatial modes are approximately uncorrelated. In this case the total power received can be expressed as the sum of the powers carried by the captured spatial modes.

A linear optical instrument with an angular aperture (Φ_X, Φ_Y), measured at the source, satisfying $\Phi_X > \Theta_X, \Phi_Y > \Theta_Y$ can selectively pass any one of the spatial modes. However, if the angular aperture satisfies $\Phi_X < \lambda/X, \Phi_Y < \lambda/Y$, none of the spatial modes can be selectively passed. If that same instrument has a temporal impulse response of duration τ satisfying $\tau < 1/B$, then no one of the temporal modes can be selectively passed. If the impulse response is T or longer, then the system can select any one of the temporal modes.

Polarization Modes

There is no analogy in the theory of lumped electrical circuits to the concept of polarization, although it is important in transmission line theory. A given space-time mode associated with an optical source can have an arbitrary combination of two polarization modes. An arbitrarily polarized stochastic wave can be expressed as a sum of two uncorrelated polarization modes, generally of different strengths. For a fully polarized wave, one of the two modes carries no power. For a totally unpolarized wave, the power is equally divided between the two modes.

FAN-IN LIMITATIONS WITH LINEAR LUMPED ELECTRICAL CIRCUITS

The losses associated with fan in will first be discussed in the context of lumped linear electrical circuits. Later the analogy with fan-in loss in optical circuits will be explored.

Circuits with only Resistive Elements

Attention is first restricted to purely resistive electrical circuits. Such circuits have impulse responses that have infinitesimal duration, and therefore are unable to select particular frequency modes present at the input. Consider first a simple electrical circuit consisting of a voltage source of voltage $v(t)$, having internal resistance R_S ohms, driving a purely resistive load of size R_L ohms, as illustrated in Fig. 1 (a).

The power delivered by the source to the load resistor is

$$P = \frac{R_L}{(R_S + R_L)^2} \langle v^2 \rangle, \tag{6}$$

where $\langle v^2 \rangle$ is the mean-squared voltage of the source. The delivered power achieves its maximum value $\langle v^2 \rangle / 4R_S$ when the load is matched to the impedance of the source, $R_L = R_S$.

Now suppose that two identical but uncorrelated voltage sources, $v_1(t)$ and $v_2(t)$, each with internal resistance R_S, drive a common load resistor, again of resistance R_L. The circuit is shown in Fig. 1 (b). The power delivered to the load by one of the two sources is

$$P_1 = \frac{R_L \langle v_1^2 \rangle}{(2R_L + R_S)^2}, \tag{7}$$

which has maximum value $\langle v_1^2 \rangle / 8R_S$, half the result obtained in the previous case. Clearly a portion of the power delivered by this source has been

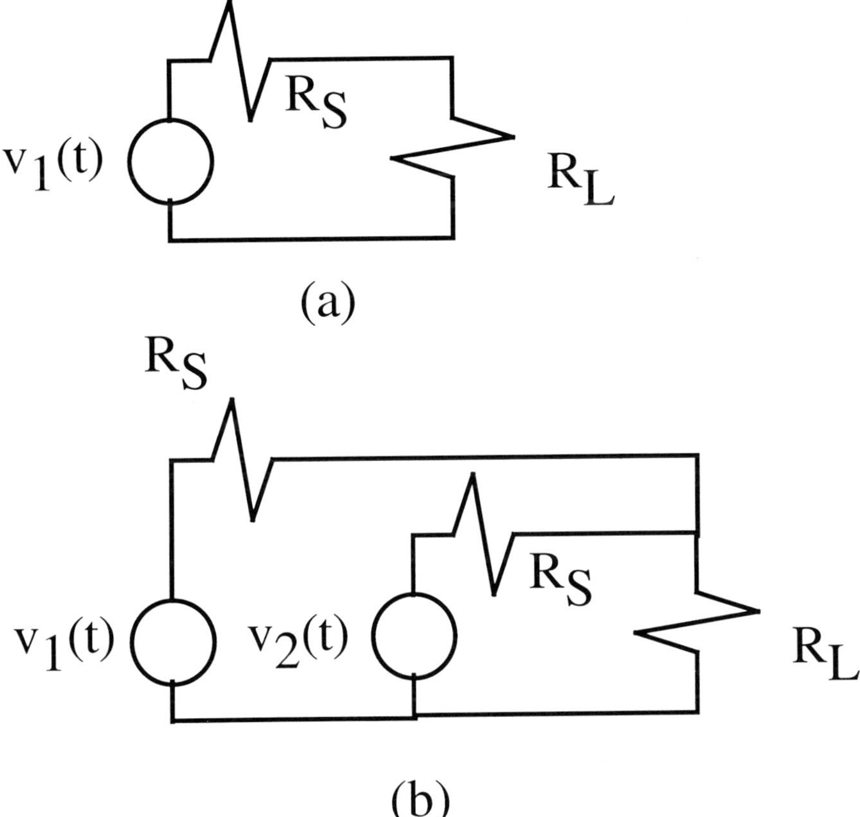

Fig. 1. Resistive circuits.

dissipated in the source impedance of the second source, rather than being delivered to the load. A fan-in power loss of a factor of two occurs for each of the sources.[2]

More generally, when Q different uncorrelated voltage sources, each with internal resistance R_S, are connected in parallel to the same load resistance R_L, an optimum choice of R_L results in only $1/Q$th of the power from any one of the sources being delivered to the load resistor. The total power delivered to the load is independent of how many voltage sources are applied in parallel, provided the load resistance is always chosen to maximize the power it dissipates.

---

[2] A similar conclusion is reached if the two voltage sources are arranged in series rather than in parallel, although the detailed expression for the power delivered differs slightly.

The above arguments rest fundamentally on the assumption that the sources are uncorrelated.

Relation to the Second Law of Thermodynamics

One form of the second law of thermodynamics (due to Clausius) is the following statement:

It is impossible for a self-acting machine to convey heat continuously from one body to another at a higher temperature.

Consider the circuits of Figs 1(a) and (b) again, but this time set the two external voltage sources to zero, i.e. replace them by short circuits, while retaining their internal impedances in the circuit. We assume that the circuits are in thermal equilibrium at temperature T^* degrees. Each of the resistors will generate a thermal voltage by virtue of its finite temperature. Each resistance R is replaced by a series combination of a noiseless resistance R and a series noise voltage having a white spectrum with (two-sided) power spectral density

$$p(f) = 2kT^*R, \tag{8}$$

(dimensions watts Hz^{-1}), where k is Boltzmann's constant. If we consider a finite bandwidth from $-b$ to b Hz, then the mean squared noise voltage associated with resistance R within that bandwidth is

$$\langle v^2 \rangle = 4kT^*Rb. \tag{9}$$

Figure 2 illustrates the circuit for the particular case of two sources driving a single load, i.e. the case illustrated in Fig. 1(b). The thermal noise generated by

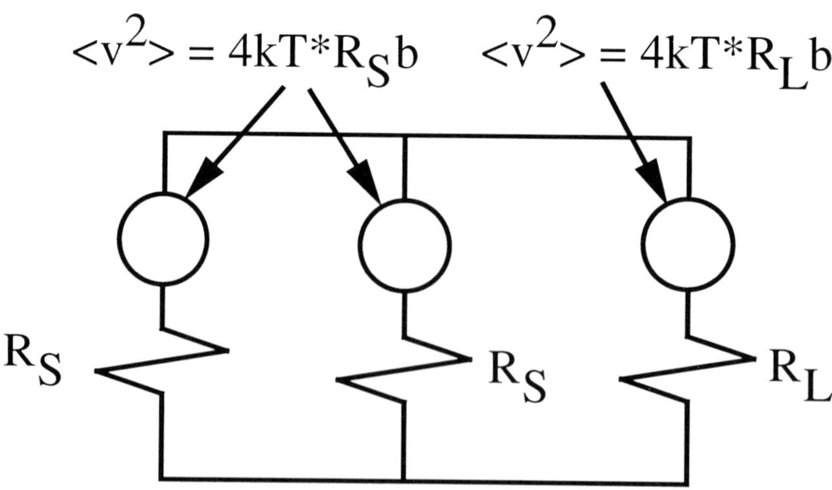

Fig. 2. Thermal noise sources.

a given noise source is dissipated in the resistances of the circuit, including its own self-resistance.

We can now show that the fan-in loss is an inevitable consequence of the second law of thermodynamics. Suppose that the circuit of Fig. 1(b) *were* able to deliver more than half the power from each source to the resistive load. Then it is a simple matter to show that, in the absence of any applied voltages, the thermal noise sources associated with the two source resistances would be supplying more thermal noise power to the load resistance than the load resistance is supplying to the two source impedances. Under such a condition, the temperature of the load resistance would have to rise indefinitely with respect to the temperature of the source resistances, violating the second law.

Furthermore, the above conclusion is valid even if in each branch of the circuit we insert a lossless narrow-band filter that selects only one *common* temporal mode for all three sources. More power will again be dissipated in the load resistor than the load resistor supplies to the source impedances. We conclude that *on a mode-by-mode basis*, the second law requires that no combination of Q uncorrelated sources can deliver more than $1/Q$ of the power from a single source to the common load. Stated another way, it is impossible for any passive linear network to increase the number of watts per Hz delivered by several uncorrelated sources to a single load beyond what can be delivered by one of those sources when it is the sole source.

Circuits with Resistive and Reactive Lumped Elements

Circuits containing both resistive and reactive lumped elements have impulse responses with finite duration, and as such may be able to selectively pass certain temporal modes of the input signal. This fact can change the amount of power that can be delivered to the load.

Consider the circuit shown in Fig. 3. In this case the terminals of one voltage source are connected to the resistive load through a capacitor, and the terminals of the other through an inductor. The sources are assumed to generate sinusoidal signals, each with the same mean-squared voltage $\langle v^2 \rangle$, and again they are assumed to have identical source resistances R_S. Let source #1 have a sufficiently high frequency f_1 and source #2 have a sufficiently low frequency f_2 that the following inequalities hold for the particular element values of the circuit:

$$\frac{1}{2\pi f_1 C} \ll R_L \qquad \frac{1}{2\pi f_1 C} \ll R_S \qquad \frac{1}{2\pi f_2 C} \gg R_L \qquad (10)$$

$$2\pi f_2 L \ll R_L \qquad 2\pi f_2 L \ll R_S \qquad 2\pi f_1 L \gg R_L.$$

Under these conditions, the inductor prevents current from source #1 from flowing into the internal impedance of source #2, and the capacitor prevents current from source #2 from flowing into the internal impedance of source #1.

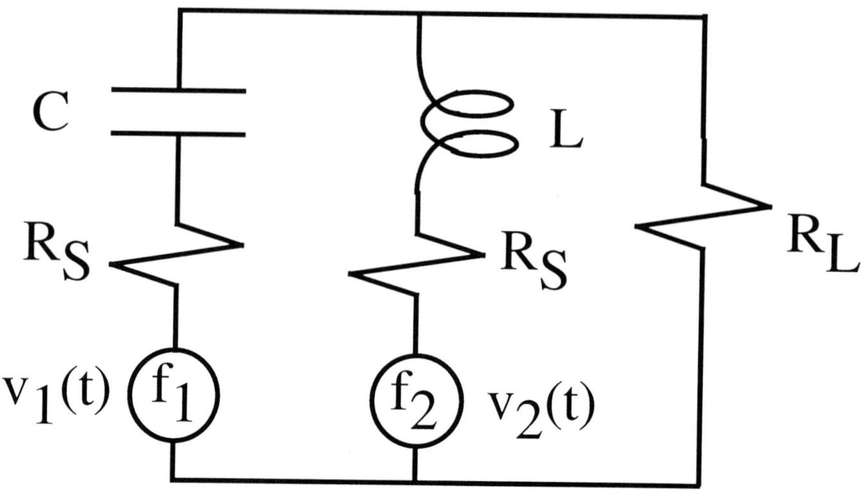

Fig. 3. Two sources with sufficiently different frequencies avoid fan-in loss.

Thus each source sees only the load resistor, and is not affected by the presence of the other source. The optimum load resistor is of value R_S, and the power dissipated in the load by any one source is $\langle v^2 \rangle / 4R_L$, as it was when only one source was attached to the load. The total power dissipated in the load is twice what it was for the purely resistive circuit with two sources. No fan-in loss has occurred!

The explanation for this behaviour lies in the ability of the circuit to select each of the two different temporal modes individually, one from each source. The addition of the capacitor and the inductor, together with the conditions we have imposed on their values, has created an impulse response with a duration satisfying $\tau > 1/|f_2 - f_1|$.

If the two sources in this case had been thermal noise sources, whose outputs consisted of the filtered noise resulting from the presence of the capacitor and the resistor, then those two filtered noise sources could be delivered to the load with little additional loss, at least to the extent that their spectra do not significantly overlap. Such delivery of power would not violate the second law, since there would be no significant increase in the power per temporal mode.

FAN-IN LIMITATIONS WITH LINEAR OPTICAL SYSTEMS

The laws governing the fan in of optical signals are quite analogous to those discussed above for electrical circuits, with the added complication that we must now consider not only temporal modes but also spatial and polarization modes.

The Role of the Second Law of Thermodynamics

In considering the implications of the second law of thermodynamics, the arguments are now based not on thermal sources associated with electrical resistors, but on black bodies in thermal equilibrium at temperature T^* emitting thermal light into the optical system and absorbing light from that system.

Consider the simple case, shown in Fig. 4, of two black bodies emitting light into two identical space-time-polarization filters that select only a single common space-time-polarization mode, and return all other modes to their originating source. (These filters are not simple devices, but they are possible in principle, and this is only a thought experiment.) We assume that the black bodies are perfect emitters and absorbers, and that they all reside in a medium of common refractive index. The selected light is combined by a passive linear optical system (as yet unspecified in detail), passes through a space-time-polarization filter identical to those already mentioned, and is incident on a third black body where light is absorbed. The entire apparatus is at a common temperature T^*. Note that the black body on the right not only absorbs radiation from those on the left, but also emits radiation, which then is filtered by the space-time-polarization filter (assumed bidirectional). The power transmitted by the mode filter is absorbed by the black bodies on the left, and the power rejected by the mode filter is returned to the source from which it came. So there is an interchange of energy between the two black bodies on the left and the black body on the right. Note that the combining system can deliver at most 50% of the light in the selected space-time-polarization mode from the source on the right to each of the sources on the left, by conservation of energy.

Suppose that the combining system were more than 50% efficient in delivering light in one space-time-polarization mode from both black bodies on the left to the black body on the right. Then the logical consequence would be that the black body on the right would absorb more energy than it is able to deliver to the two black bodies on the left, and as a consequence its temperature would rise, thus violating the second law. Now we see that the second law

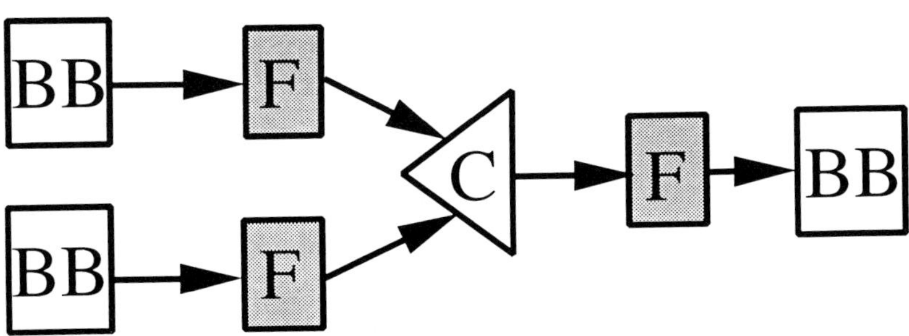

Fig. 4. Blackbody sources: BB = blackbody, F = filter, C = combiner.

implies that it is impossible for any passive linear optical system to increase the *brightness*, i.e., the number of watts Hz^{-1} steradian^{-1} per polarization mode in a single mode of an optical beam by combining two or more mutually incoherent (i.e. uncorrelated) optical beams.

Note that these conclusions are closely related to Liouville's theorem, which, however, is usually stated as a principle of geometrical optics (Marcuse 1972; Welford 1978). We see that the limitations are really a property of wave optics and do not rest on any geometrical optics approximation.

Specific examples of the impact of this *constant brightness* theorem above are plentiful. We illustrate below with some examples of cases in which fan in can and cannot be achieved without loss.

Fan In of Polarization Modes

A single spatial and temporal mode in general consists of a combination of two polarization modes. Polarization can be used as a parameter for achieving fan in without loss. Figure 5 shows a system, using a polarizing beam splitter as a combiner, that will take two different polarization modes having identical collections of spatial and temporal modes and combine them into a single beam with the same spatial and temporal mode occupancy as the constituent beams, without fan-in loss. (For the polarization combiner to work efficiently, it is necessary that the angular spread of the spatial modes not be too great.) The second law of thermodynamics is not violated because two modes are involved in the combining process, not one.

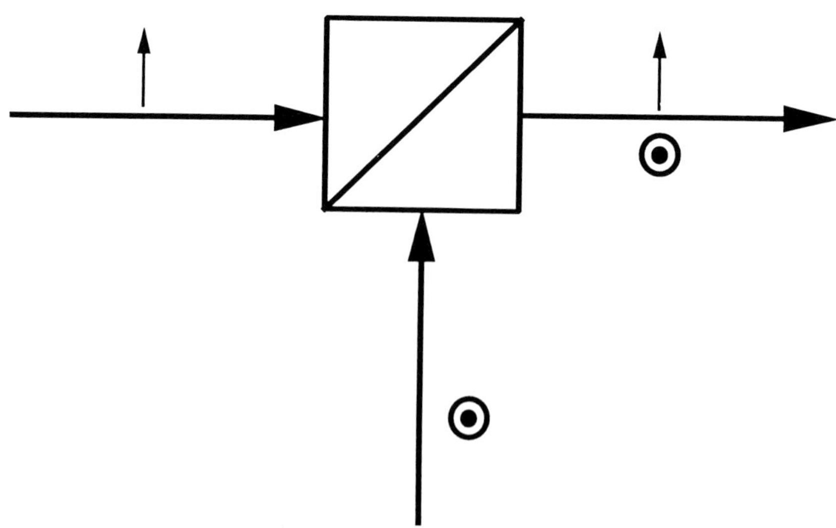

Fig. 5. Fan in without loss through polarization modes.

Fan In of Spatial Modes

As a first example illustrating the fan in of spatial modes, consider the lossless single-mode waveguide combining system shown in Fig. 6. Two single-mode waveguides come together in a 'Y', fanning in to one identical single-mode guide. If two uncorrelated optical waves are launched into these guides from the left, it can be shown (Rediker & Leonberger 1982) that half the power entering each guide is lost into radiation modes at the Y-junction, and half is transmitted on in the output waveguide. On the other hand, if the waveguide on the right has either a larger cross-sectional area or a larger numerical aperture than the individual waveguides on the left, implying that the output guide is multimode, then the 'modal volume' of the final waveguide may be large enough to accommodate two different spatial modes without superimposing them, in which case all the power delivered to the Y-junction can be delivered to the end of the output waveguide.

If the two waveguides are driven by mutually coherent optical signals, i.e. signals derived from the same optical source and highly correlated with one another, then with proper phasing of the two inputs it is possible to couple all the incoming power into a single outgoing spatial mode. The brightness of the output signal has been increased by this combining operation, but this is not forbidden by the second law, since our thermodynamic arguments used power superposition, whereas amplitude superposition holds for mutually coherent beams. It is also possible to choose the phasing of the inputs so that all the incoming power is coupled into radiation modes, and none is transferred to the output waveguide (Rediker & Leonberger 1982). A second example is illustrated by the beam-combining thick phase hologram shown in Fig. 7.

Two holograms, each made with monochromatic light of the same frequency, are superimposed in a thick photographic medium. The film is bleached to produce an approximately lossless combiner. The two light beams incident on the developed hologram are assumed to be narrowband and of the same centre frequency, but mutually incoherent. Each hologram individually would

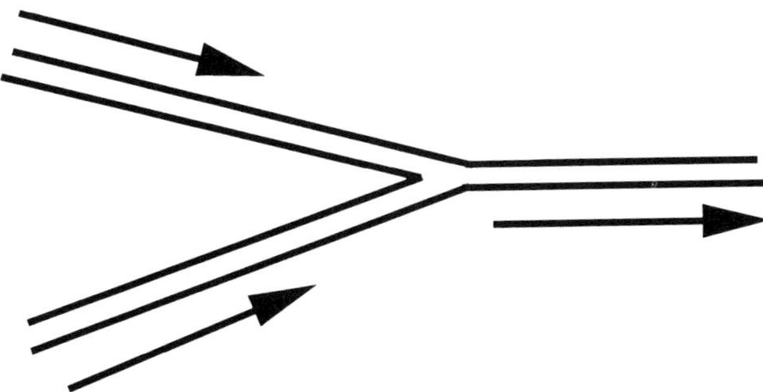

Fig. 6. Waveguide combiner.

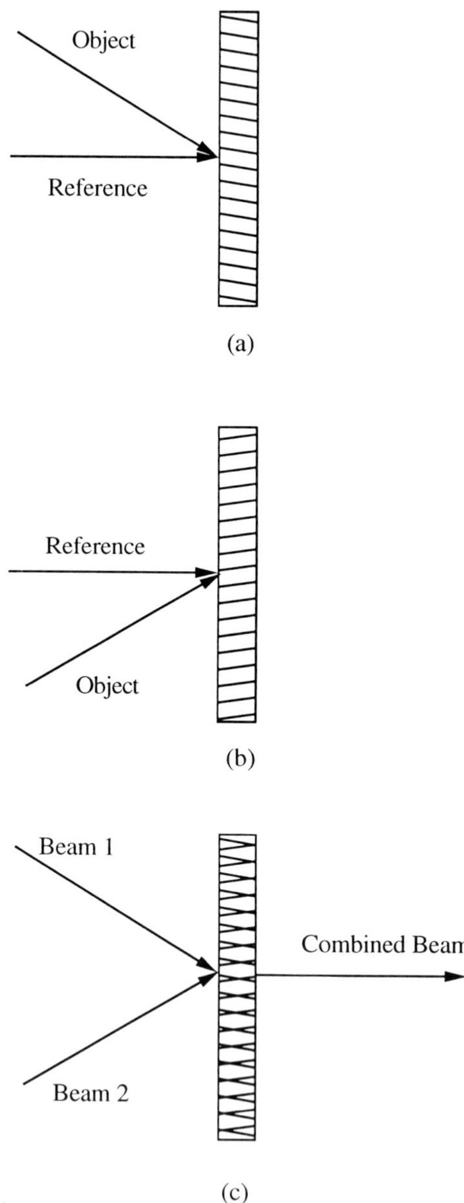

Fig. 7. Holographic beam combiner. (a) Record one hologram. (b) Superimpose a second hologram. (c) Use the doubly exposed hologram as a beam combiner.

(theoretically) be capable of 100% diffraction efficiency, but the combined hologram is capable of diffracting at most 50% of the light in any one input beam into the combined beam. The reason rests on the shared reference beam and the resultant coupling of diffraction orders that results. Light from the upper incident beam is coupled into the desired order of its own grating, but that order also couples into the first order of the second hologram that was made for the lower incident beam. The result is that at least half the light is lost in undesired diffraction orders. Further discussion can be found in Goodman (1985). Again, if the two incoming beams are mutually coherent, with proper phasing it *is* possible to increase the brightness of the outgoing beam.

Fan In of Temporal Modes

In the next example a thick phase holographic grating is again used, but this time the dispersive properties of the gratings are taken advantage of to selectively pass different temporal modes. The recording steps for the hologram are identical to those shown in Fig. 7, with one exception. The wavelengths used in recording the two gratings are significantly different from one another. The result is two holograms, each of which is Bragg matched to a different (non-overlapping) band of wavelengths. Now the developed hologram is used as a wavelength multiplexer. That is, if we illuminate the hologram with two narrow-band optical beams, each one with a different centre frequency and matched to only one of the recorded holograms, both of the incident beams can be transferred into a single outgoing beam with 100% diffraction efficiency, in principle. The power per mode has not been increased by this device, since each of the arriving beams corresponds to a different temporal mode or group of temporal modes.

Thus, just as in the case of electrical circuits, we have found that fan-in loss can be overcome by proper use of frequency as a parameter.

CONCLUDING REMARKS

We have seen that linear electrical circuits and optical systems share a common set of laws regarding fan in of signals. However, in the optical case several parameters are available that are not accessible in the case of lumped electrical circuits. Optical systems can support two polarization modes and many spatial modes. Methods can be devised to combine mutually incoherent optical beams without loss, if and only if they occupy different modal volumes in the temporal-spatial-polarization mode space.

Misleading results can be predicted if only one type of mode is considered (i.e. space, time or polarization). For example, if we consider only spatial modes, we are actually considering a projection of the full mode volume on to the spatial dimensions alone, and either polarization or temporal frequency can be used as a parameter to achieve what might otherwise appear to be a violation of the constant brightness theorem.

In both the electrical and the optical cases, fan-in loss can be avoided if non-linear elements are included in the circuits. Such techniques are easily applicable in the electrical case, but more difficult to apply in the optical case, due to the fact that non-linear behaviour is more difficult to realize in a simple way.

REFERENCES

J.W. Goodman, Fan-in and fan-out with optical interconnections, *Optica Acta*, **32**, 1489–1496 (1985).

D. Marcuse, *Light Transmission Optics* Van Nostrand-Reinhold, Princeton, New Jersey (1972).

W.T. Welford and R. Winston, *The Optics of Nonimaging Concentrators* Academic Press, Inc., New York, NY (1978) (see Appendix A).

R.H. Rediker and F.J. Leonberger, Analysis of Integrated-Optics Near 3dB Coupler and Mach-Zehnder Interferometric Modulator Using Four-Part Scattering Matrix. *IEEE J. Quantum Electronics* **18**, 1813–1816 (1982).

13 Signal processing and storage using hybrid electro-optical procedures

Joseph Shamir
Department of Electrical Engineering,
Technion – Israel Institute of Technology,
Haifa 32000, Israel

INTRODUCTION

After the invention of the laser, at the beginning of the 1960s, it was realized that the laser is a solution to many questions that still had to be asked. Among these questions were the applications of laser light for communication, signal processing and data storage. In principle, some of the high expectations materialized almost immediately: a former scientific curiosity, holographic recording became a technological reality and the holographic matched filter [*1, 2*] laid the ground for a flourishing research area that had as its objective the applications of optical methods in signal processing and information storage. While holography made large strides forward in the fields of display and non-destructive testing, fuelled by the demand of the wide public and the needs of modern technology, optical signal processing was left behind. The two main reasons for this staggering of optical signal processing were some technological stumbling blocks and, in particular, the rapid progress in digital computing which did not leave room for any other technology. Only the laser disk and the bar-code reader, that do not exploit the full capabilities of optics, managed to enter through a side door as a consumer product because they are compatible with simple digital technology. In view of these developments, many research establishments turned to the idea of digital-optical computing which intended to convert the concepts of the *electronic* computer into a *photonic* computer. This research direction is still in its infancy and has only a small probability of reaching maturity.

Throughout this chapter the term *photon* is applied loosely since a unique and rigorous definition is still lacking. Also, we shall not distinguish between the term computing and the term *signal processing* since the border line between the two is quite ambiguous.

Considering this short history of so-called *optical computing* one should ask the question: Does anything remain for optics to do with computing? The optical disk, the bar-code reader, the optical card reader and other devices are living proofs that optics can supplement digital electronics, at least in the field of information storage technology. Moreover, the rapid development of optical communication and the layout of the fibre-optic information 'superhighway' demonstrate the capabilities of light in information transfer. However, all these technologies are supplements to the digital electronic world. Is there anything else? Can optics be employed to perform actual computing tasks?

The objective of this chapter is to indicate that there are positive answers to the above questions and that the best strategy is to exploit the attributes of both electrons and photons in a hybrid system using an approach which is somewhat similar to the attitude of the maturing technologies mentioned above. One should look for a computing architecture where, in addition to information storage and communication, photonic processing complements (rather than competes with) electronic computing.

We start, in the next section, with some general considerations regarding the physics of electrons and photons. Examples of architectures for the combination of electronic and optical processors are then described implementing analogue as well as numeric processes. In the rest of the chapter it is demonstrated, with the help of examples, that these hybrid systems are especially suitable to perform iterative processes and, in particular, to solve optimization problems.

ELECTRONS AND PHOTONS

Considerations of the high propagation velocity, the high temporal frequency and the short wavelength of visible electromagnetic waves led to a flourishing research activity to exploit the laser. In particular, there were high expectations regarding the information handling capabilities of coherent light. It was soon realized, however, that the high propagation velocity of light did not contribute much toward applications in signal processing since the propagation speed of energetic electrons in space is not substantially less. Moreover, the real limiting bottlenecks were the input–output devices and the detection processes. The high temporal frequency, on the other hand, became the basis for wide-band communications which took off in the early 1970s by a lucky hit on to the optical fibre which matched the characteristics of existing laser sources that could be modulated at high frequencies. Considerations of the short wavelength led to efficient information storage technologies and, of course, contributed to other areas like high-accuracy measuring methods based on interferometry and holography. None of these breakthroughs, however, resulted in an all-optical computer.

The main reason for the unbelievable progress in electronic computing technology is the strict control one has on electrons *by* electrons due to the strong interaction among the electrons themselves. The strong electron–electron interaction is exploited in conventional digital computing to implement switching circuits that are the basis for digital operations. Such strong interaction does not exist among photons, making them unsuitable for digital operations. However, the same lack of interaction provides a property which is ideal for information transfer where interchannel interference must be minimal.

Are there any unique attributes of light that are better suited for the task of signal processing than electrons? In the following we give two views of one characteristic of light propagation which may turn out to be crucial for its applications in signal processing. It should be noted that, in principle, the physics of electrons satisfy similar rules but it will be more difficult to utilize them due to the differences already discussed.

Consider a light source at some point A in space (Fig. 1) and a detector at another point, B, separated by a distance *l* from A. By Fermat's principle, a photon emitted by the source and detected by the detector will take the path which has the shortest possible duration. In free space this time will be $t = l/c$ where c is the velocity of light. If there is matter between A and B, or relativistic effects become noticeable, *l* denotes a generalized *optical path* between A and B rather than the simple Euclidean distance. How did the photon 'know' which is the shortest path? Why not path a or path b in the figure? An acceptable answer to this question is that the photon 'measured' all

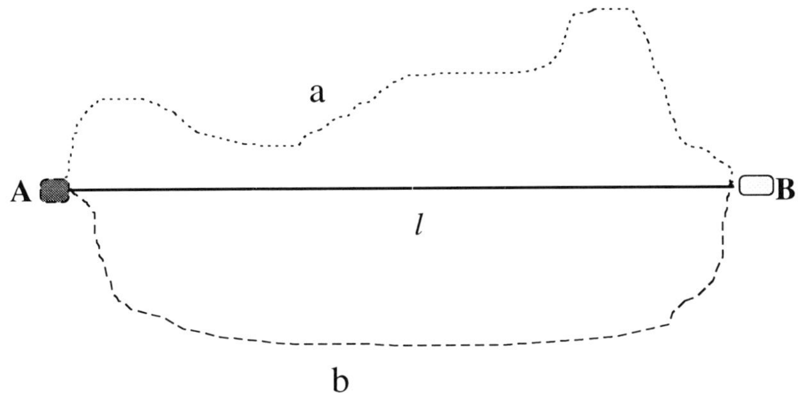

Fig. 1. Fermat's principle: Light will take the quickest path between A and B.

possible routes and chose the quickest one. All these measurements were performed within a time period t. If, according to quantum theory, these were performed by virtual measurements the theory of relativity is not violated. Even so, this picture is a little naive and oversimplified but, nevertheless, there must be some truth in it. The classical picture can be stated in simpler terms: the light emitted by the source solves the wave equation. This wave equation, which may have an infinite set of boundary conditions, is solved in a finite time duration t.

Which is the proper way to look at this problem is a matter of philosophy and a question of taste (many scientists will not agree to either one of the above). The important thing, however, is the fact that a photodetection event in the detector is the result of light propagating through space while collecting information from a huge number (probably infinite) of physical parameters. If one is able to present a problem in the form of a set of boundary conditions, the photon detection event represents the solution. This philosophy does not violate any fundamental law of physics or information theory, since, in this simple case, the result is a single bit of information. Although a large number of calculations were necessary to arrive at this solution, all the intermediary computation results are destroyed at the moment of the detection, leaving the single bit of information which is either one or zero. The energy dissipated in the whole process is $h\nu$ which occurs as the photon is annihilated by a photon detection event [3]. The same calculation performed by a digital computer (if at all possible) would require the dissipating of energy and the expense of time for all the intermediary calculations, even though the individual results of those calculations are of no interest.

The detection of a single bit may be an important piece of information about the state of a complicated system. The crucial attribute of light is the fact that a large number of calculations can be performed in parallel with no cross-talk among the individual processes. This is in contrast to the motion of electrons that can, in principle, also solve a wave equation with an infinite set of boundary conditions but with a substantial interaction among the electrons that propagate in parallel. To avoid any misunderstanding, one must emphasize that the seemingly infinite parallelism of optical channels is only with respect to photon–photon interaction but the actual number of parallel channels that can be propagated with negligible cross-talk is limited by the number of degrees of freedom [4] and other physical limitations [5]. The discussion of these limitations is outside the scope of this chapter but they must be considered for any system design.

Since photons practically do not interact with each other the solution of the wave equation is probably the only thing they can do without some interaction with a mediator. Thus there is little hope for the realization of a photonic-only computer to replace the electronic computer. However, a relatively strong interaction does exist between photons and electrons (or rather, electronic states). Such interaction was already practised in the early photographic process where photons interact with electronic states in the photosensitive material and, after development (chemical amplification), the photographic material modulates the light illuminating it. Although a different physical process, the situation is similar in photorefractive materials that are based on the diffusion of charge carriers excited by light, or photopolymers where the diffusion of molecules is responsible for the information storage. Thus, in principle, the marriage of optics and electronics dates back to the invention of the photographic process. Electronic processes based on chemical reactions or free diffusion are relatively slow. To increase the speed one needs the intervention of some accelerating force. Such forces exist when the excited electrons are subject to electric fields as is the case in electronic photodetection. In fact, photodetection dates back to the first living eye but this is only a one-way interaction. To make the system into a processor one needs a detector (photodiode, charge coupled device (CCD), etc.) that has a controllable response and can be coupled to the outside world. The generated information can now be reintroduced into the system or transferred to the next stage of the process. For our purpose we shall apply the generic term, 'spatial light modulator' (SLM) to a large class of devices that include actual modulators and also active light emitters. The combined technology of photodetection and spatial light modulation constitutes the heart of a hybrid system which is able to exploit the attributes of both electronics and photonics. If the detection and the modulation are included in the same device, it is called optically addressed SLM. Of course, one should not confuse such a device, even if it contains light amplification based on stimulated emission, with a hypothetical direct interaction of light with light, as sometimes is claimed.

The rest of this chapter is devoted to ideas and examples that provide some insight into this relatively new field of hybrid systems for signal processing.

SELECTED ARCHITECTURES OF ELECTRO-OPTICAL PROCESSORS

An important contribution for speeding up computational processes is the utilization of a hardware environment which helps in the calculation of the main mathematical operations with appropriate accuracy. An electro-optical architecture with its intrinsic parallelism provides a favourable hardware environment for massive computations at high speed. Earlier examples of electro-optical processors fitting into this category are vector-matrix or matrix-matrix multipliers. Although these processors are essentially digital, they exploit the massive parallelism provided by optics. We start the discussion with the class of analogue processors that are more natural to optics.

Analogue Optical Processors

The best-known optical processors are imaging systems that date back to the 'invention' of the living eye. Imaging systems are all-optical but they are only

one-way and *hard-wired* processors in the sense that no other useful manipulations of the information are possible. Nevertheless, in addition to regular
imaging functions such an optical system is the first step in many computationally intensive processes such as robotic vision and control systems. In these
applications optics performs the imaging operation while the digital processor
complements this operation to extract the needed information.

Considered from a signal processing point of view, the Fourier transformation (FT) performed by an ideal lens illuminated with a coherent light
distribution constitutes a fundamental optical process and can be the building
block for more elaborate signal manipulations. The $4f$ optical correlator (Fig.
2) is a combination of the two processes, FT and imaging. These systems
exploit the global parallelism of the photonic processor and, therefore, are a
good starting point for our discussions.

In the $4f$ optical correlator each lens performs a FT between its two focal
planes. Thus, a two-dimensional complex function imprinted on a coherent
wave front at the input plane is Fourier transformed to the FT plane. A second
FT, performed by the second lens, results in an inverted image on the output
plane. This image can be modified by a spatial filter (a transparency) placed at
the FT plane. The filter function multiplies the complex amplitude distribution
over the FT plane to obtain, over the 'correlation plane', a correlation between
the input object and the FT of the filter function. One of the best-known filter
functions is the classical matched filter used for pattern recognition. However,
while optimal for the detection of signals in additive white Gaussian noise, the
matched filter is not optimal for optical systems. The reason is that optical
correlators are mainly used to discriminate among different objects and if noise
is present it is not white Gaussian and usually is even signal dependent. After
this fact was realized it became common practice to use synthetic filters [6]

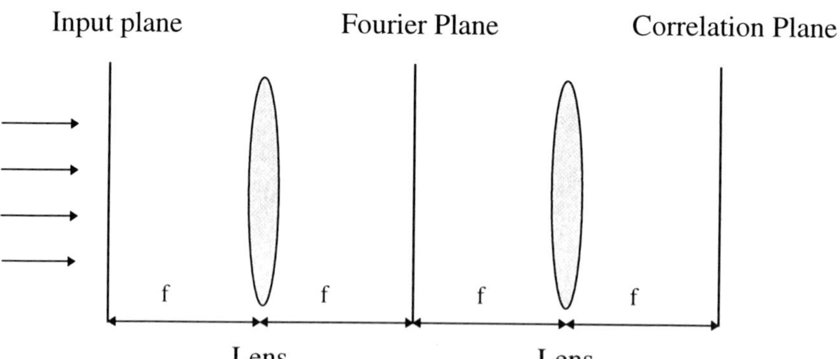

Fig. 2. The $4f$ optical correlator is composed of two cascaded FT systems each
composed of a lens and two regions of free space of length f, the focal length of the
lens.

designed as computer generated holograms (CGH) that required sophisticated numerical processing. We see that already at this level, a strong link is established between the optical and the electronic system.

The next two steps for hybridization are obvious. First, insert SLMs into the input and filter planes. Next, follow the detector at the output plane by an electronic computer and, eventually, close the loop from the computer to the SLMs (Fig. 3). Many signal processing tasks can be performed employing this

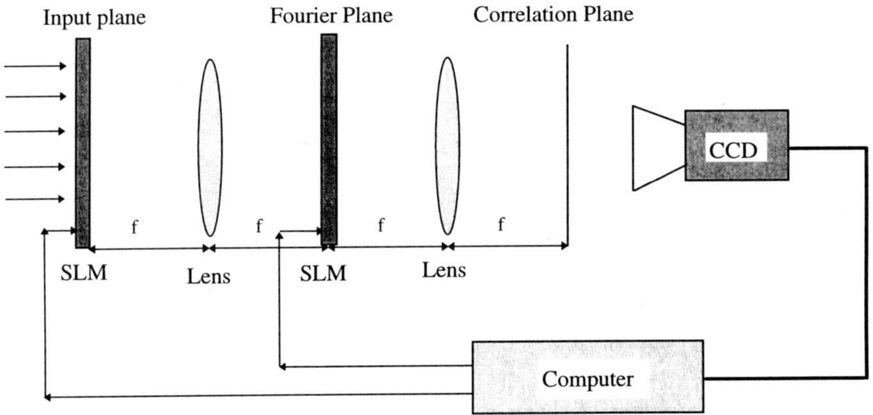

Fig. 3. The hybrid electro-optical 4*f* correlator is similar to Fig. 2 but the interfaces (input and filter) are controllable by electronic blocks (computer).

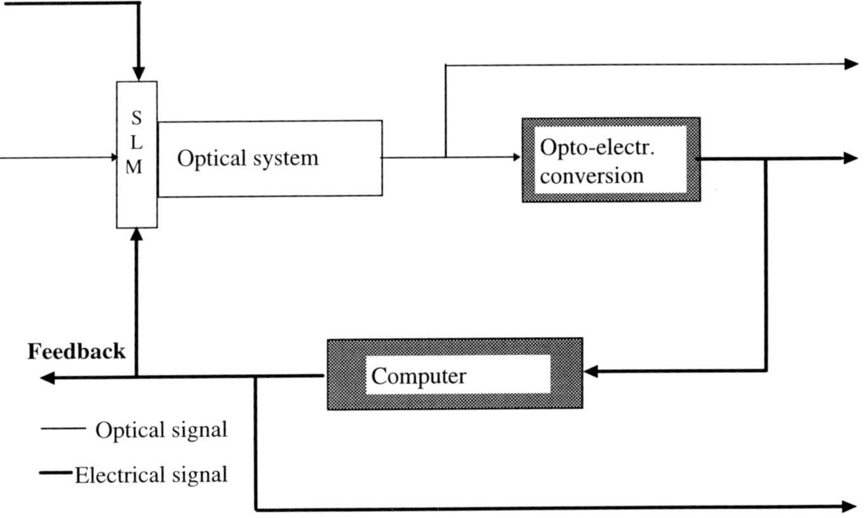

Fig. 4. A generic hybrid signal processing stage (see text for details).

system which can be considered to be a special case of an extended and more general architecture as indicated by the scheme of a generic stage shown in Fig. 4. This stage consists of an optical block (SLM and optical system), an electronic block (computer) and an opto-electronic converter (detector array, TV camera, etc.). Input signals (from the left) and output signals or feedforward signals (on the right) can be electronic and/or optical. Several stages like this can be cascaded to form a complete signal processing system and there is also a feedback option to previous stages.

A simple example of the generic architecture is a modified form of correlator, the joint transform correlator (JTC) of Fig. 5. In this architecture a spatial filter is not required; rather, a real reference pattern is placed over the input plane, side by side with the input pattern. Like in the $4f$ correlator, the

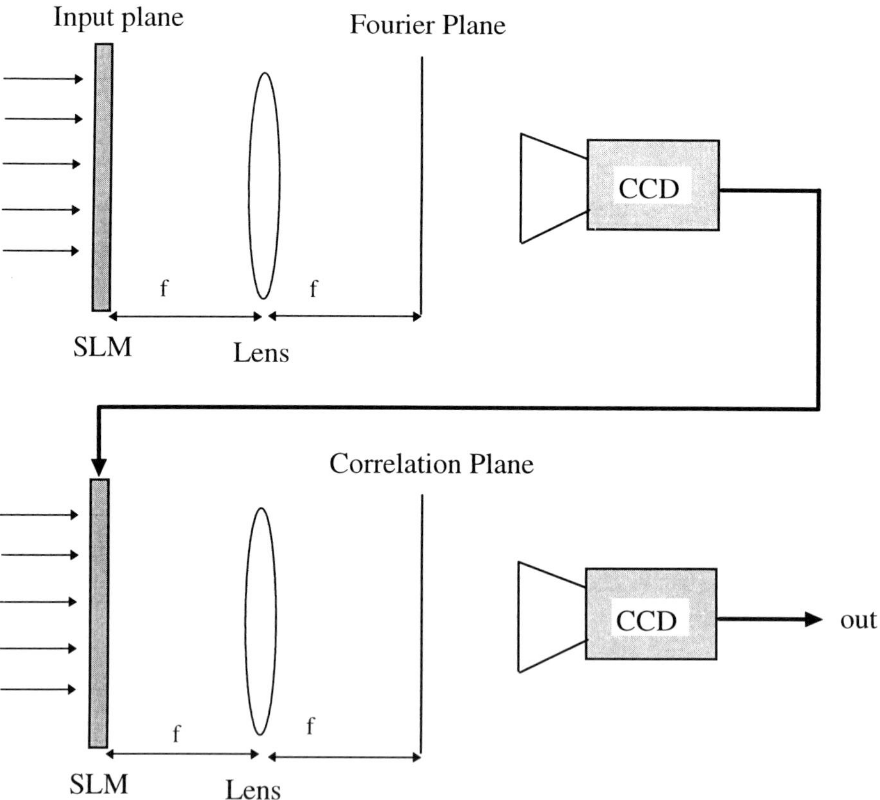

Fig. 5. The JTC architecture. Two lenses perform FTs and there are opto-electronic converters (the CCD cameras). An optional computer can be inserted in the line between the first camera and the second SLM. The input and reference functions are displayed over the first SLM side by side.

correlation is finally obtained over the output plane. The main attribute of the JTC as compared to the $4f$ correlator is the replacement of the tedious process involved in the implementation of a Fourier plane spatial filter by a simpler (usually real) reference pattern. Historically, the JTC was the first optical correlator architecture that combined the optical FT process with opto-electronic and electro-optic conversions. This was also the first optical correlator which could be operated in practically real time. Considering the JTC also as a special case of the generic hybrid architecture (Fig. 4), it has FT operators as the two optical blocks. These two optical blocks are separated by the electronic block containing an electronic photodetector sometimes combined with a digital computer for intermediary processing and a SLM which is the input to the second optical block. Another photodetector constitutes the output electronic block. The intermediary electronic block may be replaced by a single optically addressed SLM. However, since such a SLM contains also photon–electron and electron–photon conversion stages we shall not dwell on them separately. Adding a computer to the output block may provide postprocessing and eventually a feedback loop which can be utilized to implement iterative algorithms.

Matrix Processors

The global parallelism of optical propagation can also be exploited for processes that are basically digital in nature. The vector-matrix multiplier [7] is a good example of this class of processors. This, however, is a degenerate case of a more general matrix processor which uses optics for a massive intercon-nection net connecting two or more electronic processors. This architecture fits into our generic picture (Fig. 4) where the space between two electronic blocks is filled with the optical interconnection network. Probably the most complete network of this sort is what we call the N^4 interconnection net that uses free space propagation in the most efficient way.

One possible architecture for an N^4 interconnection network is shown in Fig. 6. The system consists of a hologram array, H, containing $N_{hx} \times N_{hy}$ holo-graphic optical elements, a SLM with $N_{sx} \times N_{sy}$ pixels sandwiched between two lenses with respective focal lengths f_1 and f_2 and a detector array, D, with $N_{dx} \times N_{dy}$ detector elements. The ijth hologram ($i = 1, 2, ..., N_{hx}$; $j = 1, 2, ..., N_{hy}$) in the array is imaged by the double lens configuration on to the ijth element of the detector array (for this task we assume that $N_{hx} = N_{dx}$; $N_{hy} = N_{dy}$). This hologram diffracts light from a reconstruction beam with an efficiency t_{ijkl} toward the klth pixel ($k = 1, 2, ..., N_{sx}$; $l = 1, 2, ..., N_{sy}$) in the SLM. The same pixel receives a weighted fraction of the light diffracted also from all other holograms in the array but, assuming a linear interaction in the SLM, these are separated again on arrival at the detector array. Thus, ideally, each detector receives the sum of all the weighted beams just from a single hologram element. Mathematically, if the power transmittance of the klth pixel

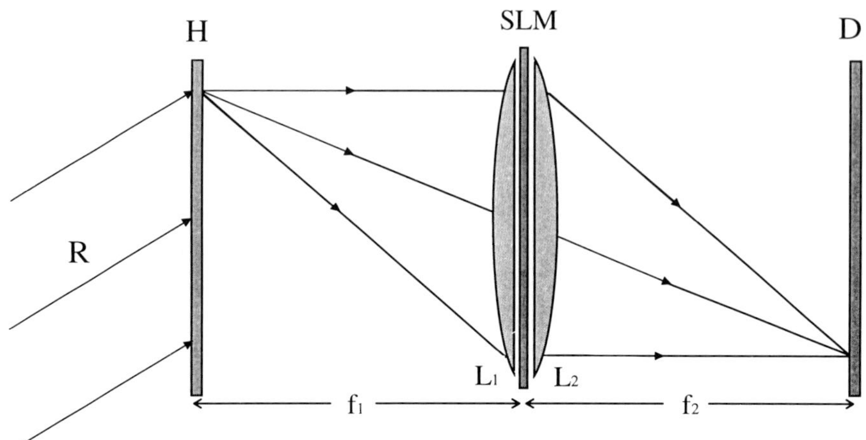

Fig. 6. Architecture for an N^4 interconnection network. H is a hologram array illuminated by reconstruction beam, R; SLM is a spatial light modulator between two lenses L_1 and L_2 with their respective focal lengths, f_1 and f_2. D is a detector array or an array of non-linear optical devices.

in the SLM is a_{kl}, the total power received by the ijth detector will be

$$b_{ij} = \sum_{kl} t_{ijkl} a_{kl}. \tag{1}$$

This system, in its ideal form, may be viewed either as a matrix–matrix multiplier of a four-dimensional matrix by a two-dimensional matrix or as a vector–matrix multiplier with vectors of $N_{sx} \times N_{sy}$ dimensions, $\| T \| \, A = B$. The elements of the input vector (or matrix) are introduced by the transmittance of the SLM pixels with the hologram providing the fixed matrix, $\| T \|$. The output vector is read out from the detector array.

An important application of an architecture like that of Fig. 6 is the optical implementation of one layer in a neural network [8]. The input vector is written on the SLM while the interconnection weight matrix is contained in the hologram array.

In a different application we may view this architecture as a communication interconnection network with $N_{sx} \times N_{sy}$ input channels that are connected by $N_{sx} \times N_{sy} \times N_{hx} \times N_{hy}$ interconnections that are 'hard wired' for a given hologram array. In this context, each SLM pixel is an input channel. Illuminating it with several holograms, say M, spreads the light transmitted by this pixel into the M corresponding detector elements. Thus the M non-zero matrix elements generate an M-fold fan-out (or broadcasting) of this communication channel. Alternatively, if a single hologram element illuminates M channel pixels, we obtain a fan-in of M channels into the corresponding detection channel.

The signal processing part of the architecture described above is all-optical. However, for the process results to be of any meaning, they have to be read by the detector array which must be supplemented by a digital computing system which can arrive at proper decisions. Iterative processes can be implemented if the computer decisions are fed back into the input SLM or by dynamically addressing proper sections in a large hologram array. Eventually, the hard wiring can also be made more flexible by displaying the hologram array also on SLMs or dynamic recording media that can be updated and controlled.

Optimization and adaptation

An important family of signal processing tasks is optimization, usually implemented as an iterative process. Iterative methods are preferential for solving non-linear and 'noisy' problems and signal reconstruction tasks from incomplete information. There are also several additional incentives to motivate the applications of iterative processes that can, in principle, accommodate to various environmental situations. For example, animal eyes appear to work, at least partly, in an iterative fashion. Seeing involves the optical equipment (cornea, lens, retina, etc.) which is not ideal and the neural equipment (optic nerve, brain, etc.) which operate in conjunction with the optical one. The neural network contains numerous readily adjustable variables (connections) which evolve as the individual learns to see. This arrangement is very desirable because it can readily accommodate rather gross defects in the optical system. Good performance does not necessarily require good optics, a situation which may be quite advantageous for many applications.

In any mass production of items there are two ways to assure satisfactory performance. First, we can build, assemble, and maintain all components and the relationships between them to within close, predetermined tolerances. Most, but not all, technologically mass produced equipment is so designed. Second, we can assume that some errors will inevitably creep in and, therefore, adaptive 'fixes' should be built in to allow use of imperfect hardware. Many integrated circuits and computers are designed for such error tolerance. Naturally this increases yield and reliability. Optical correlators made of inferior components imperfectly assembled might be much less expensive than 'perfect' correlators. If they could work almost as well, this would be a highly desirable situation.

Finally, one should mention the rapidly evolving field of adaptive optics that is being developed, primarily, for astronomical telescopes and atmospheric optical energy transmission. The objective in this field is the adaptive correction of the distortions induced on the propagating light by atmospheric turbulence. Adaptive optics is based on dynamic optical elements that can be controlled to compensate the atmospheric distortions using optimization and control based on the combination of optics and electronics.

In the next section we describe some experimental realizations applicable to pattern recognition; applications to information storage are then given.

APPLICATIONS FOR PATTERN RECOGNITION

As noted in the previous section, interest in FT based optical correlators received a substantial boost from Van der Lugt's first use of a holographic filter. Significant progress in the field was marked by the introduction of spatial filters based on synthetic filters that were implemented as CGHs. The difficulty in the earlier implementations of optical pattern recognition methods stems from the interface problem. Input patterns and filters had to be recorded on photographic film and only after tedious and time-consuming chemical processing could the task be completed. Recent advances in SLM technology changed this state of affairs and viable solutions for many pattern recognition tasks became feasible. Nevertheless, limited dynamic range, limited resolution, various distortions and noise still hinder the progress toward practical applications. To overcome these difficulties more sophisticated approaches are needed. Much of the current work is devoted to advanced optimization procedures for designing filters with improved performance.

In this section we review one possible procedure that proved to be very efficient in the generation of spatial filters for pattern classification using hybrid electro-optical systems.

Outline of some theoretical principles

The outlined theoretical considerations are quite general but, for the sake of clarity, we shall refer to a specific system, the $4f$ optical correlator of Fig. 2. A spatial filter carrying the FT of a filter function $h(x, y)$ is inserted in the FT plane. Placing a function $f(x, y)$ over the input plane leads to a complex amplitude distribution

$$c(x_0, y_0) = \int_{-\infty}^{\infty} \int_{-\infty}^{\infty} f(x,y) h^*(x + x_0, y + y_0) \, \mathrm{d}x \, \mathrm{d}y \tag{2}$$

over the output plane.

Assuming a set of possible input patterns $\{f_n(x, y)\}$ we define our goal as the detection of the presence of patterns out of a subset $\{f_n^D(x, y)\}$ while rejecting all other patterns denoted by the subset $\{f_n^R(x, y)\}$. To execute a detection one must define a criterion for detection. A convenient criterion is the appearance of a strong and narrow peak for a match between the input and the filter function as contrasted with a uniform distribution for a pattern to be rejected. One important attribute of this criterion, as compared to some other possible ones, is that it takes into account the distribution over the whole output plane.

With pixellated physical devices in mind (SLMs and detector arrays) we use the discrete form of the amplitude distribution over the output plane $c(x_0, y_0)$ in the form,

$$c(m, n) = \sum_{i=1}^{N} \sum_{j=1}^{N} f(i,j) h^*(i + m, j + n); \quad m, n = 1, 2, \dots, (2N - 1) \tag{3}$$

where $f(i, j)$ and $h(i, j)$ are the sampled representations of $f(x, y)$ and $h(x, y)$ respectively.

The complex amplitude distribution over the output plane is usually detected electronically, thus the phase is lost in the process. In a more general sense one may apply a non-linear operator $N\mathcal{L}$ to generate a new, non-negative function on (m, n),

$$\varphi(m, n) = N\mathcal{L}[c(m\,n)]; \quad \varphi(m, n) \geqslant 0 \quad \forall (m, n). \tag{4}$$

The normalized form of φ

$$\Phi(m, n) = \frac{\varphi(m, n)}{\displaystyle\sum_{j=1}^{2N-1} \sum_{l=1}^{2N-1} \varphi(j, l)} \tag{5}$$

has all the properties of a probability density, i.e.

$$0 \leqslant \Phi(m, n) \leqslant 1 \quad \forall (m, n) \quad \text{and} \quad \sum_{m, n} \Phi(m, n) = 1. \tag{6}$$

If one operates the system at very low light levels or with a very fast photon counting detection system the quantity $\Phi(m, n)\, \Delta_m \Delta_n$ is the probability that a given photon detection event will occur in an area $\Delta_m \Delta_n$ centred at (m, n).

Define the quantity

$$S = \sum_{m=1}^{N} \sum_{n=1}^{N} \Psi[\Phi(m, n)] \tag{7}$$

where Ψ is a non-linear functional of $\Phi(m, n)$ such that $S_{\min}$ is obtained for

$$\Phi(m, n) = \text{Const} \quad \forall (m, n) \tag{8}$$

while $S_{\max}$ is obtained for

$$\Phi(m, n) = \begin{cases} 1 & m = m_0; \; n = n_0; \; (m_0, n_0) \in (\text{Domain of } \Phi) \\ 0 & \text{otherwise.} \end{cases} \tag{9}$$

In addition, the function Ψ satisfies:

$$|\Psi[\Phi(m, n)]| < \infty \quad \forall \Phi(m, n) \in [0, 1] \quad \text{and} \quad \Psi \in \mathscr{C}^1 \tag{10}$$

where $\mathscr{C}^1$ is the space of continuous functions having at least first derivatives. It turns out that strictly convex functions Ψ satisfy the above requirements.

Eq. (7) is sometimes called a generalized entropy and is usually defined with a minus sign.

Denote by S_k^D and S_k^R the following quantities:

$$S_k^D = \sum_{m=1}^{2N-1} \sum_{n=1}^{2N-1} \Psi[\Phi_k^D(m,n)] \tag{11}$$

$$S_k^R = \sum_{m=1}^{2N-1} \sum_{n=1}^{2N-1} \Psi[\Phi_k^R(m,n)] \tag{12}$$

where $\Phi_k^D(m,n)$ corresponds to the distribution over the output plane for the kth pattern of the detected set D, and $\Phi_k^R(m,n)$ is derived from the output distribution for the kth pattern of the rejected set R. Since we regard Φ as a probability density function, uniform distribution gives minimum height on the average. This corresponds to the maximum value of the generalized entropy defined by Eq. (11), while a single peak, with the value of unity, corresponds to a minimum generalized entropy given by Eq. (12). All this can be combined in a generalized cost function of the form

$$M_{\mathrm{h}} = \sum_{\{k \in R\}} S_k^R - \sum_{\{k \in D\}} S_k^D. \tag{13}$$

We achieve our goal if this cost function is minimized. The subscript h indicates that these cost functions depend on the filter function, h. Given filter $h(i,j)$, we calculate S_k for each member of the training set, $f_n(i,j)$ and substitute into Eq. (13). The resultant generalized cost function is minimized by varying the components of $h(i,j)$.

An ideal filter function, h, would generate a steep peak for $f^D(i,j)$ represented by a distribution of the form given in Eq. (9) and a uniform distribution for $f^R(i,j)$ as given by Eq. (8). A minimization procedure performed on the generalized cost function tends to this ideal filter which satisfies

$$M_{h(\mathrm{ideal})} = N^R S_{\min}{}^R - N^D S_{\max}{}^D \tag{14}$$

where $S_{\min}$ and $S_{\max}$ are determined by the distributions given in Eqs. (8) and (9), respectively, while N^R and N^D are the respective numbers of members in each class.

Optimization using the genetic algorithm

In principle, any optimization procedure should lead to the proper solution. However, while computer simulations operate quite well for problems of small dimensions and deterministic algorithms, they are inadequate to treat large vectors and problems with insufficient information about system parameters. Thus, this is a typical case where the help of optics can be invoked. In one set of experiments [9] the optimization procedure was implemented in the hybrid $4f$ correlator using the genetic algorithm.

In biological evolution, it is not the individual but a population (species) that evolves. The success of the individual (phenotype) gives it an improved chance of breeding. In breeding, the genetic structure (genotype) of the offspring is made up of genotypic contributions from both parents. In addition, errors (mutations) occasionally occur. The offspring then competes for the right to reproduce in the next generation. Thus a gene pool evolves which not only governs future generations but also bears within it a memory of where it has been.

Optimization procedures based on the above are called genetic algorithms (GA) [*10*]. In GA a genome or vector is specified as a way to describe the system. It contains all (usually) of the information needed to describe the system. A figure of merit is then evaluated for each member of a pool of genomes. Winners are selected for genetic exchange (usually called 'crossover') and mutation. Losers are usually dropped from the pool to keep the pool size constant.

To use a GA one should have the following features:

A) A chromosomal representation of solutions to the problem, usually binary.
B) An evaluation function that gives the fitness of the population. This is, in principle, a distance function or cost function such as given in Eq. (13).
C) Combination rules (genetic operators) to produce new structures from old ones – reproduction, crossover and mutation.

There are several variants of GA. The algorithm used for generating spatial filters for pattern recognition in a hybrid electro-optical system [*9*] is summarized as follows:

1) Start:
Select at random a population of m members (binary functions) $\{h_1, h_2, ..., h_m\}$ and evaluate the values of the corresponding cost functions, $M_i\{i = 1, 2, ..., m\}$. Compute the average value of the cost function

$$\theta = \frac{1}{m} \sum_{i=1}^{m} M_i. \tag{15}$$

Set a discrete time parameter, t, to zero. Define a probability P for a mutation to occur and set it to some $P_{\max}$.

2) Crossover/mutate:
Select the function h_l which corresponds to the minimal cost function, M_l. Pick from the population a function h_j at random. The two functions, h_l and h_j are the parents to be used for generating an offspring function. Select a random integer k between 0 and n, where n is the dimension of the vectors h representing the filter functions. Create the offspring function, h_c, by taking the first k elements from one of the parents, randomly, and the remaining $n - k$ elements from the other parent. Induce a mutation (inverting the sign of the elements '1' to '0' or '0' to '1') with probability P on each element of the offspring vector h_c. Evaluate the offspring cost function M_c.

3) Reproduce:
Pick at random a function h_d from the population subject to the constraint: $M_d \geq \theta$. Replace h_d in the population with the new offspring h_c and update the average value of the cost function,

$$\theta \rightarrow \theta + \frac{1}{m}\,(M_c - M_d). \tag{16}$$

4) Setting parameters:
Set the new parameters, $t \rightarrow t + 1$ and $P \rightarrow P_{max}(1/t)^r$. Terminate the procedure when adequate discrimination is achieved (a predetermined value of M). Else, if $P > P_{min}$ go to 2, otherwise go to 1. Selection of the parameters r, P_{min}, P_{max} depends on the particular problem at hand.

Synthetic filter design on hybrid systems

An example of the results obtained is shown in Fig. 7: a hybrid $4f$ correlator was used to 'instruct' the filter for discrimination between the two faces displayed as grey level pictures on 100×100 pixels. This problem was too computing intensive to be handled by a conventional digital computer. While a smaller problem like this took about a week of computation time on a medium size computer, the learning process on the hybrid system was completed in about 20 minutes. The single correlation peak in Fig. 7(b) when both faces were presented to the system indicates the high discrimination capability. The shift invariance of the filter is demonstrated in Fig. 7(c) where the object to be detected was presented at three different positions. Similar procedures were successfully implemented also on the JTC architecture where the filter function is replaced by a reference function. In these experiments the main speed-limiting factors were the write and read times on the SLM and TV camera, respectively.

Adaptive pattern recognition

A well-designed spatial filter in an optical correlator is extremely sensitive to any changes in the input object. While a favourable attribute for pattern discrimination, this is a destructive feature if the orientation or size of the objects are not uniquely defined. One major approach to handle this severe intolerance to distortions of the input function is to induce some controlled deterioration of the discriminating capability of the filter. For example, rotation invariance can be achieved by decomposing the input function into its circular harmonics and designing the spatial filter for a single circular harmonic component. The result is a filter with reduced information content, less discriminating in general, and completely insensitive to the orientation of the object. The same procedure can be followed for scale changes by decomposing

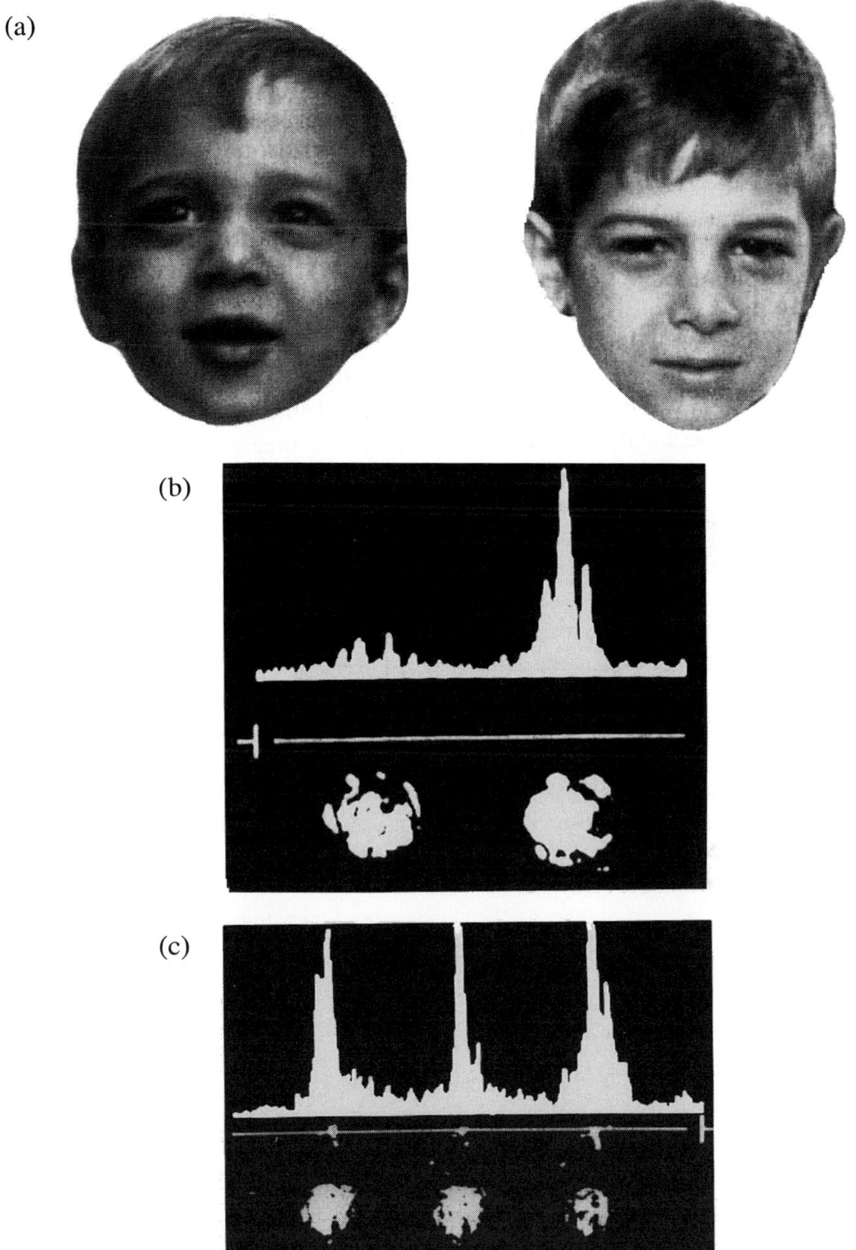

Fig. 7. Using several pictures like those in (a), the GA was employed to instruct a hybrid correlator to distinguish among the pictures. The intensity distribution across the correlation peaks, masked by the solid line in (b) indicates good discrimination. Shift invariance of the input is demonstrated in (c). The blobs below the scan line are the zero-order distributions.

the function into logarithmic radial harmonics. Unfortunately, since scale and rotation are in a sense orthogonal to each other, a filter which is totally rotation and scale invariant becomes also 'object invariant'. That is, it loses discrimination. Nevertheless, filters can be made partially rotation invariant and partially scale invariant while still maintaining a fair capability of discrimination. One efficient way to design such partially distortion-invariant filters is to use an iterative process as described above. It should be noted that all this discussion is relevant to the regular correlator architecture which is position invariant. If one is willing to give up this position invariance, more tolerance can be obtained for distortions.

Using hybrid electro-optical systems a higher degree of invariance is possible by employing adaptive processes. Mutually orthogonal pattern distortions can be handled by a system which can measure one or more of the distortion parameters and adapt the system parameters to the recognition of the object with the remaining distortions. The system of Fig. 8 is a double channel architecture where one channel is used to measure the scale of the object while the second channel implements shift invariance and rotation invariance after the filter is adjusted to the measured scale. As a result, pattern recognition with rotation, scale and shift invariance is implemented in almost real time.

It is important to note that any measurement has finite accuracy and this problem will always be much more severe if the measurement is performed on an object before it is recognized. This means that the recognition process must be robust to the measured parameter within a significant error range. This is not the case for conventional filters and, therefore, special filters having limited scale robustness and rotation invariance must be designed for this system to operate properly. As for other special types of filters, here too, iterative optimization methods are useful.

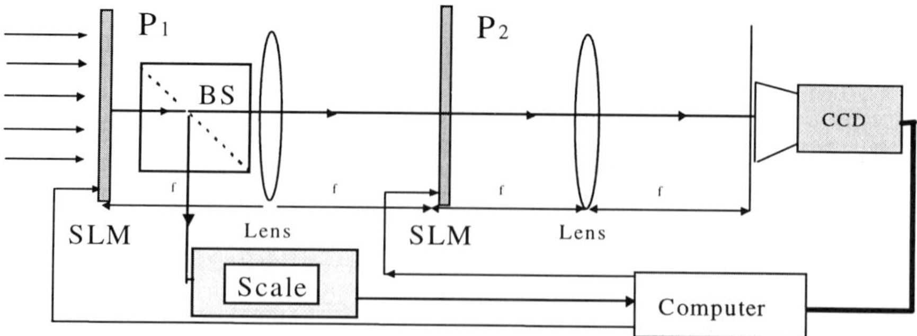

Fig. 8. A schematic description of the double channel recognition system. The lower channel is the scale estimation channel while the upper channel is an adaptable (controlled filter) 4 *f* correlator.

APPLICATIONS FOR INFORMATION STORAGE

The synthetic filters described in the previous section contain adequate information to discriminate among various objects. As such, one may view them also as a kind of memory device. Moreover, the procedures described lead to the storage of information with an extremely highly efficient use of the storage resources. In one of the examples presented the information stored on an array of 100×100 binary pixels was adequate to discriminate among complete grey level pictures. In this section we carry this idea further and describe several applications where the procedures are used for the direct recording of information in an efficient way. The efficiently recorded information is equivalent to a highly compressed version of the data and it may be stored in a digital computer memory to be extracted with the help of SLMs or recorded as a CGH. Without loss of generality we shall use the generic term CGH to represent all these records.

Practically all available processes for generating a CGH rely on binary recording of the information. Therefore, various binary encoding procedures were developed to contain the information of the amplitude as well as the phase. These procedures, however, do not use the available information capacity in an efficient way, making them unsuitable for recording on state of the art SLMs – a requirement for dynamic processors. The situation can be improved by using iterative optimization procedures, preferably in a hybrid electro-optic architecture. The resulting binary code no longer consists of a deterministic scheme and it is implemented as a stochastic process with high recording efficiency. Using this method for holographic recording leads to a new kind of hologram which may be termed a random carrier hologram.

Holographic recording of two-dimensional information

One experimental procedure demonstrates the generation of a FT hologram employing the system of Fig. 9. The transmittance of the SLM may be represented by a vector H with elements $H_{kl} \in \{0, 1\}$. The lens performs an optical FT, $\mathscr{F}$, which yields the two-dimensional vector $h = \mathscr{F}H$, with elements h_{kl}. The squared absolute values of these vector elements, $I_{kl} = |h_{kl}|^2$, are observed by the CCD camera.

Storing a reference pattern, f, we intend to generate a vector, H, which will reconstruct, under coherent illumination, an output intensity distribution, I_{kl}, close to $|f_{kl}|^2$ within a given region of the output plane. The difference between I_{kl} and $|f_{kl}|^2$ can be characterized by an error function (cost function) which should be minimized:

$$e_n = \frac{1}{N^2} \sum_{k,l=1}^{N} [|f_{kl}|^2 - \gamma_n I_{kl}^{(n)}]^2 \tag{17}$$

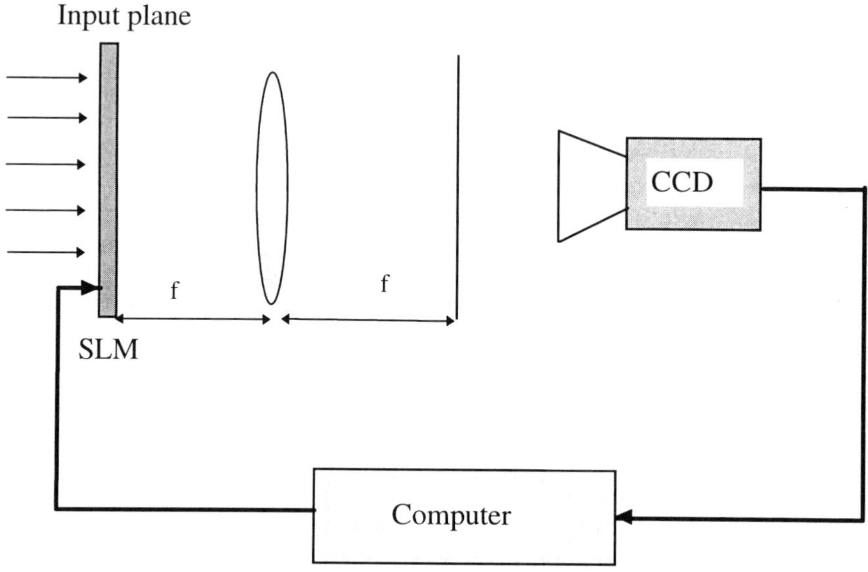

Fig. 9. Laboratory setup for recording a FT hologram on the SLM. The hologram is Fourier transformed by the lens and observed by the CCD camera. The reconstructed image is compared to a reference stored in the computer for evaluating the value of the cost function.

where the inspected region of the output plane contains N^2 pixels and γ_n is a parameter which is adapted to match the difference between the optically reconstructed intensity and the digitally stored reference pattern.

In the next iteration, a small perturbation is induced on the vector H to yield a new value for the error function, e_{n+1}. This iteration is accepted or rejected following suitable statistical criteria.

The above procedure is adequate to generate a CGH which reconstructs the desired intensity distribution. If one requires a reconstruction which contains also the phase information an additional measurement is necessary. The phase measurement may be performed with the help of a reference wave or some other sophisticated procedure. For example, it was shown that inspection of the correlation between the CGH and the reference function together with the intensity comparison provides adequate information to determine the phase as well as the amplitude deviation from the desired distribution. Thus, one may extend the optical system into a $4f$ correlator and measure also the correlation function. A series of iteration results in the process to record a FT hologram of an airplane is shown in Fig. 10.

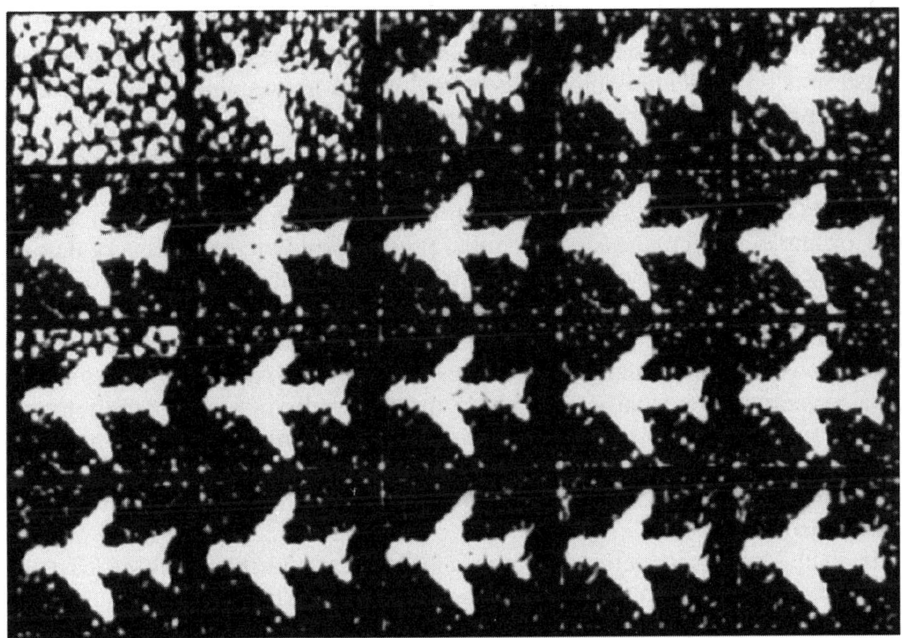

Fig. 10. A series of reconstructions after various iterations in the generation of a FT CGH for an airplane. First iteration is in the upper left corner and the last one is at bottom right.

Holographic recording of three-dimensional information

Among the various applications of holography the characteristic which mostly excites the imagination of the general public is the capability to record three-dimensional information using an optical holographic recording setup. In many laser and optics applications it is necessary to govern the wavefront propagation through long distances. These include display of information, spatial distribution of energy, optoelectronic interconnections, precision measurements and alignment. For these applications a real object does not exist and the CGH must be artificially synthesized.

The problem, in its most general form, can be stated as follows: *given a known wavefront incident on a diffractive element, design this element to obtain a desired intensity distribution within a given three-dimensional domain.* It is obvious that physics does not allow a solution for all arbitrary distributions but it does in many cases of practical interest. Moreover, if a solution does not exist, it still may be valuable to derive a solution which is closest to the required constraints. This is clearly an optimization problem.

We consider a region of space behind the diffractive element where we want to control the beam propagation, and in this region we impose the necessary constraints on sufficiently close transverse planes, assuming that the field does not change significantly between these planes. An optimization algorithm especially suitable for solving problems like this belongs to the family of projection on to constraint sets (POCS) algorithms [*11*].

In a POCS algorithm a function is transferred from one domain to another (for example, from plane to plane in the present problem) and in every domain it is projected on to one or several constraint sets. To implement the projections one must define a distance function that determines the deviation of the solution h^t obtained at the tth iteration with respect to the sets of constraints. These constraints represent the conditions that must be fulfilled by the solution. The constraints are usually determined by the results of some measurements, the physical characteristics of the experimental system and some demands on the required solution. The procedure is repeated in a cyclic way until the solution converges to a function h that satisfies all the constraints simultaneously.

If all the constraint sets are closed and convex, and they have at least one common domain, then the process converges weakly. In our context it is difficult, and may be impossible, to fulfil these conditions, so the convergence is not always guaranteed. Nevertheless, if the algorithm is based on a proper generalized distance function it will not increase from one iteration to the next. Using a modified POCS algorithm [*12*], the distance function, $d(h')$, is defined as some weighted, generalized distance that contains the distance between the actually measured intensity distribution and the required distribution, as well as the distance to the predefined constraints on the CGH function.

The first experiment to demonstrate the power of the method in storing 3D information was the generation of a light intensity distribution that presents a strong and narrow peak that persists for a relatively long distance without substantial expansion. This solution resembles the analytic solution for a 'Bessel beam' but it can be implemented on a low-resolution device which is inadequate for generating a real Bessel beam. The iteratively generated beam also had much better performance than the analytic one (longer propagation distance without expansion and more constant beam intensity along the direction of propagation). Moreover, the POCS algorithm also made it possible to generate *arrays* of such beams with a single element [*13*]. An example of such an array is shown in Fig. 11 where an array of five beams propagates a distance of 20 cm with each beam having a cross-section of about 100 μm and 18% maximum amplitude deviations. This example was a computer simulation and the diffractive element contained amplitude as well as phase information. Similar procedures were used in the laboratory, with additional constraints, to make amplitude-only or phase-only elements that are more practical for actual implementation. The experimental results obtained with the latter procedure were quite similar to the computer simulations.

(a)

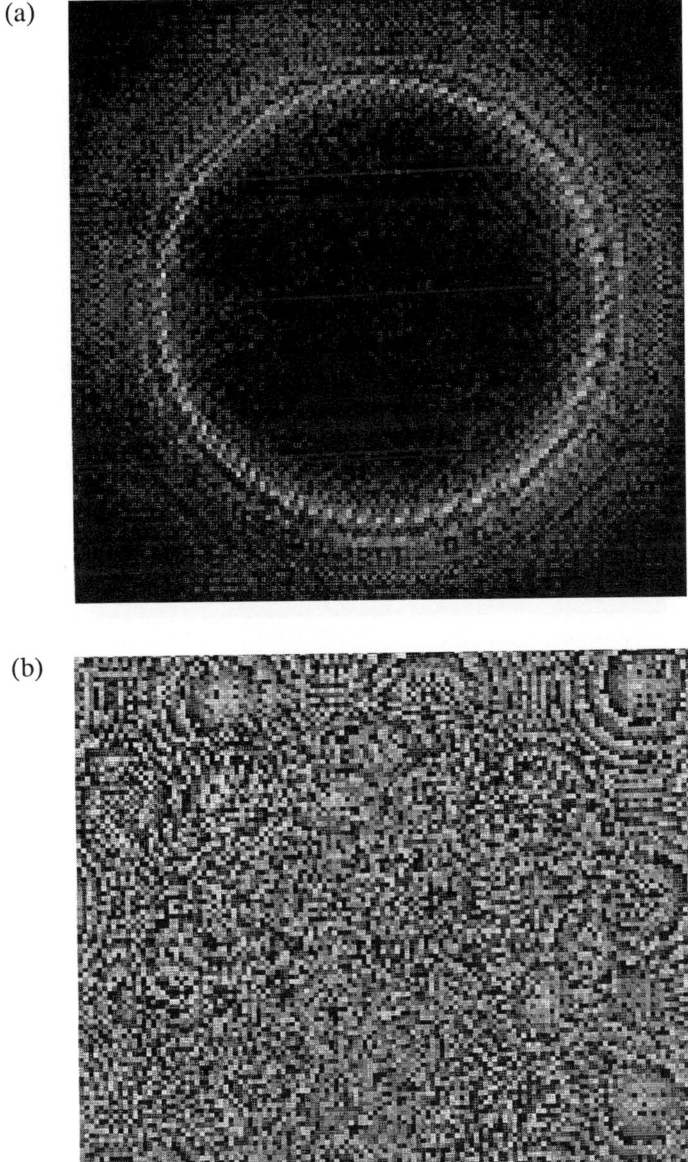

(b)

Fig. 11. (*continued*).

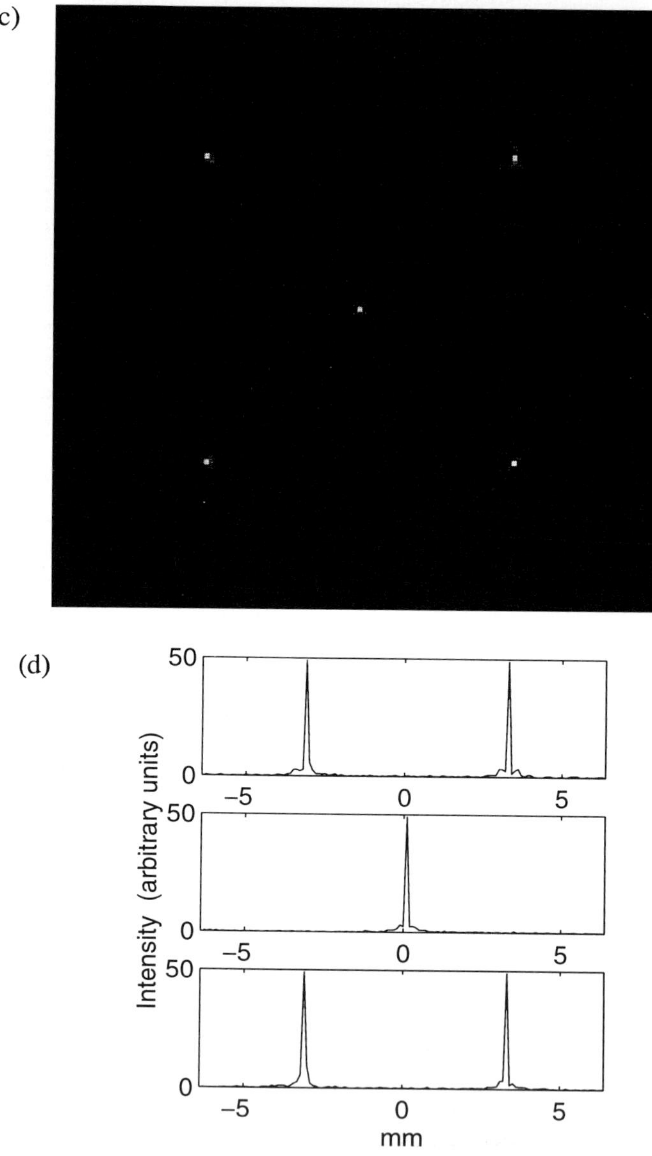

Fig. 11. Generating an array of five non-expanding beams. (a) Amplitude distribution of the diffractive element and (b) phase distribution. (c) The five beams and their respective intensity profiles (d).

CONCLUSIONS

Due to the lack of interaction among photons, they are not suited for digital operations. Such operations can only be implemented with the help of a mediator, usually electrons or electronic states. In a general sense any process requires at least two photon–electron interactions. The photon–electron interaction in the first stage of the process transfers energy from the photon field to a modulation of the electronic states. In the second stage, the modulation of the electronic states in the medium is transferred to the photon field incident on it – the electron–photon interaction. If an electronic device (photodetector) is used for the first stage, the second stage will also be a light modulation (SLM) or the activation of a more complicated device like a surface-emitting laser array or laser amplifier.

In this chapter we discussed several architectures for exploiting such two-stage processes for computational purposes. We demonstrated many advantages of such systems but one should keep in mind that several fundamental and practical limitations exist that should be considered. A detailed discussion of these limitations is outside the scope of this chapter but they can be found in the literature [5].

ACKNOWLEDGMENT

It is a pleasure to thank my past and present students who performed much of the work that served as a basis for this chapter. In particular I would like to note J. Rosen, U. Mahlab (who also donated the photographs of his children for Fig. 7), T. Kotzer and R. Piestun. I am also indebted to H. John Caulfield for the collaboration that stimulated some of the ideas discussed here.

REFERENCES

1. A.B. Van der Lugt, Signal detection by complex spatial filtering, *IEEE Trans. Inf. Theory* **IT-10**, 139–145 (1964).
2. J.W. Goodman, *Introduction to Fourier Optics*, McGraw-Hill Book Co., San Francisco (1968).
3. H.J. Caulfield and J. Shamir, Wave-particle duality processors – characteristics, requirements and applications. *J. Opt. Soc. Am. A* **7**, 1314–1323 (1990).
4. D. Gabor, Light and Information, in: *Progress in Optics* **1**, ed. E. Wolf (North-Holland, Amsterdam), 111–152 (1961).
5. J. Shamir, H.J. Caulfield and R. B. Johnson, Massive holographic interconnections and their limitations, *Appl. Opt.* **28**, 311–324, (1989); reprinted in the SPIE Milestone series *Optical Computing* SPIE Vol. **1142**, 546–559 (1989).
6. B.V.K.V. Kumar, Tutorial survey of composite filter designs for optical correlators, *Appl. Opt.* **31**, 4773–4801 (1992).

7. J.W. Goodman, A.R. Dias and L.M. Woody, Fully parallel, high-speed incoherent optical method for performing discrete transforms, *Opt. Lett.* **2**, 1–3 (1978).
8. T. Kohonen, An introduction to neural computing, *Neural Networks* **1**, 3–16 (1988).
9. U. Mahlab and J. Shamir, Comparison of iterative optimization algorithms for filter generation in optical correlators, *Appl. Opt.* **31**, 1117–1125 (1992).
10. D. Lawrence, *Genetic Algorithm and Simulated Annealing*, Morgan Kaufmann, Los Altos, CA (1987).
11. D.C. Youla and H. Webb, Image restoration by the method of convex projections: Part 1 – Theory, *IEEE Trans. on Medical Imaging* **TMI-1**, 81–94 (1982).
12. T. Kotzer, J. Rosen and J. Shamir, Application of serial and parallel projection methods to correlation filter design, *Appl. Opt.* **34**, 3883–3895 (1995).
13. R. Piestun and J. Shamir, Control of wave front propagation with diffractive elements, *Opt. Lett.* **19**, 771–773 (1994).

14 Young's experiment in signal synthesis

Jorge Ojeda-Castañeda
School of Science, University of Las Americas,
Exhda. Santa Catarina Martir, 72820 Cholula, Puebla, México

Adolf W. Lohmann
Physikalisches Institut der Universität Erlangen-Nürnberg,
D-91058 Erlangen, Germany

INTRODUCTION

Young's double-slit experiment, in Fig. 1, is a cornerstone in the development of physical optics, as well as in quantum physics. The basic idea of generating interference patterns of wavefields, by using a rather simple setup, has been usefully exploited in several applications.

Among the many applications of using Young's experiment as a conceptual tool, one can mention: Rayleigh's two-beam interferometer [1], Michelson's stellar interferometer [2], the van Cittert–Zernike theorem [3], Duffy's speckle encoding [4], and the experiments of interference between two independent sources [5, 6]. Here, we explore the possibility of using slits, as in Young's experiment, for synthesizing 1D optical signals.

We show that a narrow slit in a Fourier transformer can be applied to generated 1D aberrated wave fronts with continuously variable aberration coefficient. We describe how to modify the previous result for achieving a parallel display of several members of a family of special functions. Next we discuss the use of two narrow slits in an optical processor, for generating complex amplitudes. Then we consider the possibility of employing wide slits with variable borders and discuss briefly our previous results in the context of

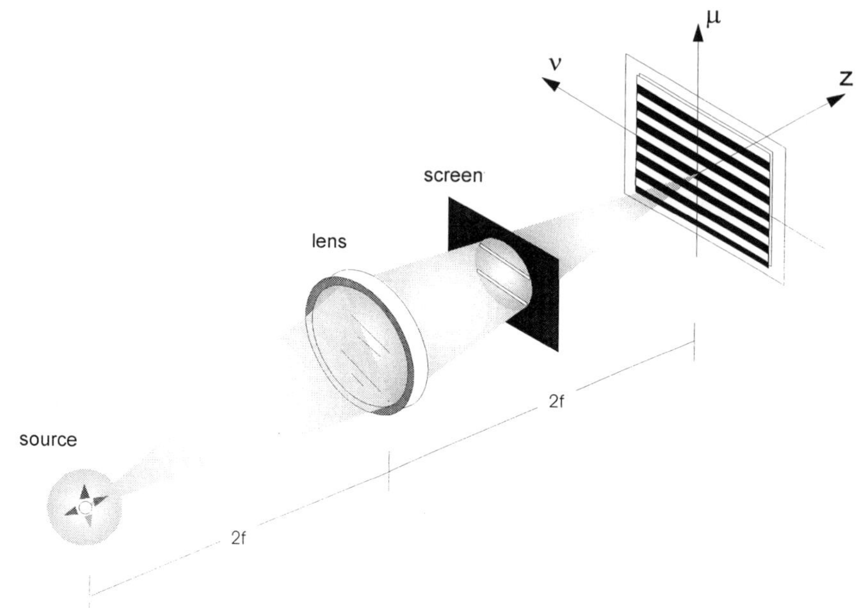

Fig. 1. Classical setup of Young's double-slit experiment.

optical processors. Finally, we extend the previous results to polar coordinates with the purpose of synthesizing signals along the optical axis.

All the above results have been presented recently, in a scattered number of publications [7–15]. Here, we attempt to present them in a unified form emphasizing the basic underlining idea.

SYNTHESIS OF 1D ABERRATIONS

Hill and Felstead [16] appear to be the first scientists to realize that a narrow slit that follows the curve $y = F(x)$ can be applied to encoded phase variations as follows.

In the Fourier transformer, shown in Fig. 2, we place in contact with the lens a screen whose binary amplitude transmittance is

$$T(x, y) = \delta[y - F(x)], \tag{1}$$

where $F(x)$ is a planar curve, and $\delta(\bullet)$ denotes the physical implementation of Dirac's delta. Then, except for an irrelevant phase factor, the complex amplitude transmittance, $\tilde{T}(v, \mu)$, at the image of the point source is propor-

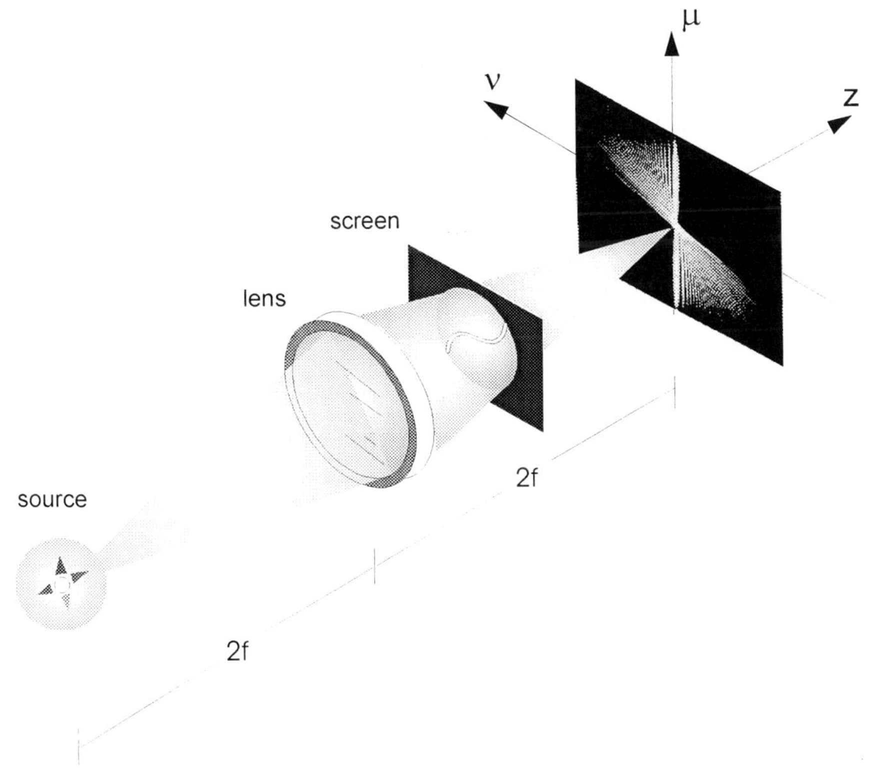

Fig. 2. Fraunhofer diffraction pattern of a generalized narrow slit that follows the curve $y = Ax^3$, for generating the Airy function with variable parameter.

tional to the Fourier transform of Eq. (1), that is

$$\tilde{T}(\nu,\mu) = \iint_{-\infty}^{\infty} T(x,y)\exp[-i2\pi(\nu x + \mu y)]\, dx\, dy$$

$$= \int_{-\infty}^{\infty} \exp[-i2\pi(\nu x + \mu F(x))]\, dx, \tag{2}$$

where $\nu = x'/2\lambda f$, $\mu = y'/2\lambda f$ are spatial frequency coordinates.

It is apparent from Eq. (2) that the spatial variation described by $F(x)$, along the y-axis, is converted into a phase variation in a similar fashion to the pioneer procedure for computer-generated holography [17, 18].

It is straightforward to consider that the phase variation generated inside the integrand in Eq. (2) can represent a 1D wave aberration. In this latter case the complex amplitude in Eq. (2) can be thought of as a 1D Point Spread Function (PSF) in the variable ν, with variable aberration coefficient μ of a wave aberration function $F(x)$, where x is the pupil coordinate. Let us consider an example. If we set

$$F(x) = Ax^3, \tag{3}$$

where A is a constant, then Eq. (2) becomes

$$\tilde{T}(\nu,\mu) = \int_{-\infty}^{\infty} \exp[-i2\pi(\nu x + \mu A x^3)]\, dx, \tag{4}$$

which represents the PSF of a 1D pupil that suffers from coma. The coma coefficient, μA, changes continuously along the μ-axis. We will see that this feature of continuously changing the optical path difference of the generated phase signal, is of great value.

PARALLEL DISPLAY OF SPECIAL FUNCTIONS

The above result can also be related to the optical synthesis of some special functions, for example the so-called Airy function, as we discuss next.

Airy function

The Airy function, according to reference [19], is defined as

$$H[(3c)^{-1/3}t] = \frac{(3c)^{1/3}}{2\pi} \int_{-\infty}^{\infty} \exp[-i(c\nu^3 + t\nu)]\, d\nu. \tag{5}$$

By comparison between Eqs. (4) and (5) we can express the complex amplitude in the Fraunhofer diffraction pattern, in Fig. 2, as

$$\tilde{T}(\nu,\mu) = 2\pi(6\pi A\mu)^{-1/3}\, H[(6\pi A\mu)^{-1/3}\, 2\pi\nu]. \tag{6}$$

Consequently, by displacing a narrow scanning slit, along the ν-axis, at $\mu = \Omega$ one is able to produce optically the Airy function, along the ν-axis with variable parameter μ. In other words, the Fraunhofer diffraction pattern displays in parallel Airy functions, with continuous variation of its parameter c [8]. This is the first application of continuously changing the optical path of the phase signal $F(x)$. Next we discuss other possibilities.

Bessel functions

In this example we consider that, as is shown in Fig. 3, the binary screen is equal to unity along a sinusoidal curve; then the trajectory $F(x)$ is

$$F(x) = A\sin(2\pi x/d), \tag{7}$$

where A is the amplitude of the sinusoidal curve with period d.

By substituting Eq. (7) into Eq. (2) we find that the complex amplitude in the

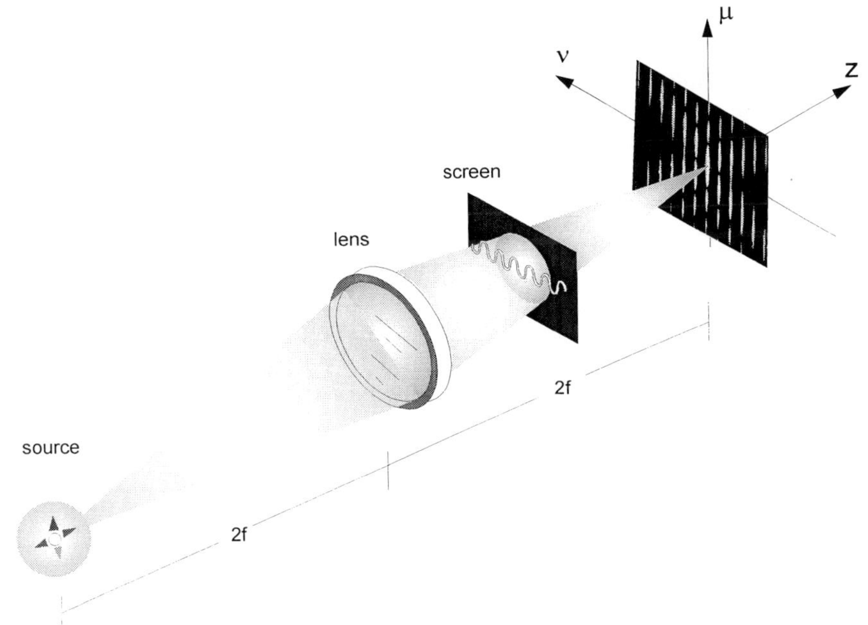

Fig. 3. Fraunhofer diffraction pattern of a narrow slit that follows a sinusoidal curve, for generating the Bessel function with variable order.

Fraunhofer pattern is

$$\tilde{T}(v,\mu) = \int_{-\infty}^{\infty} \exp[-i2\pi(vx + A\mu \sin(2\pi x/d))]\, dx, \tag{8}$$

and now by using Jacobi's identity,

$$\exp(iz \sin \theta) = \sum_{m=-\infty}^{\infty} J_m(z)\exp(im\theta), \tag{9}$$

where $J_m(z)$ denotes the Bessel function of the first kind and order m, we obtain

$$\tilde{T}(v,\mu) = \sum_{m=-\infty}^{\infty} J_m(2\pi A\mu)\delta(v + m/d). \tag{10}$$

It is apparent from Eq. (10) that the Fraunhofer diffraction pattern displays in parallel, along the μ-axis, the Bessel functions $J_m(2\pi A\mu)$, with variable order $m = vd$ which can be changed along the v-axis. Of course, one can select one line to obtain a fixed order of Bessel function, for performing certain signal processing operations, as is discussed in [9]. In this application the variation of μ is employed in the argument of the Bessel function.

Derivatives of the Laguerre polynomials

Our third example shows that the same procedure applies for optically implementing the generation function of derivatives of certain special functions. For this application the trajectory is

$$F(x) = A \cot(2\pi x/d), \tag{11}$$

as is depicted in Fig. 4. The complex amplitude in the Fraunhofer diffraction pattern is obtained by substituting Eq. (11) in Eq. (2) to give

$$\tilde{T}(\nu,\mu) = \int_{-\infty}^{\infty} \exp[-i2\pi(\nu x + A\mu \cot(2\pi x/d))]\, dx. \tag{12}$$

It is shown in reference [5] that a sort of Jacobi identity holds for the cotangent function:

$$\exp(ix \cot \theta) = \sum_{m=0}^{\infty} \exp(-z)L_m(2z)[\exp(i2m\theta) - \exp(i2(m+1)\theta)], \tag{13}$$

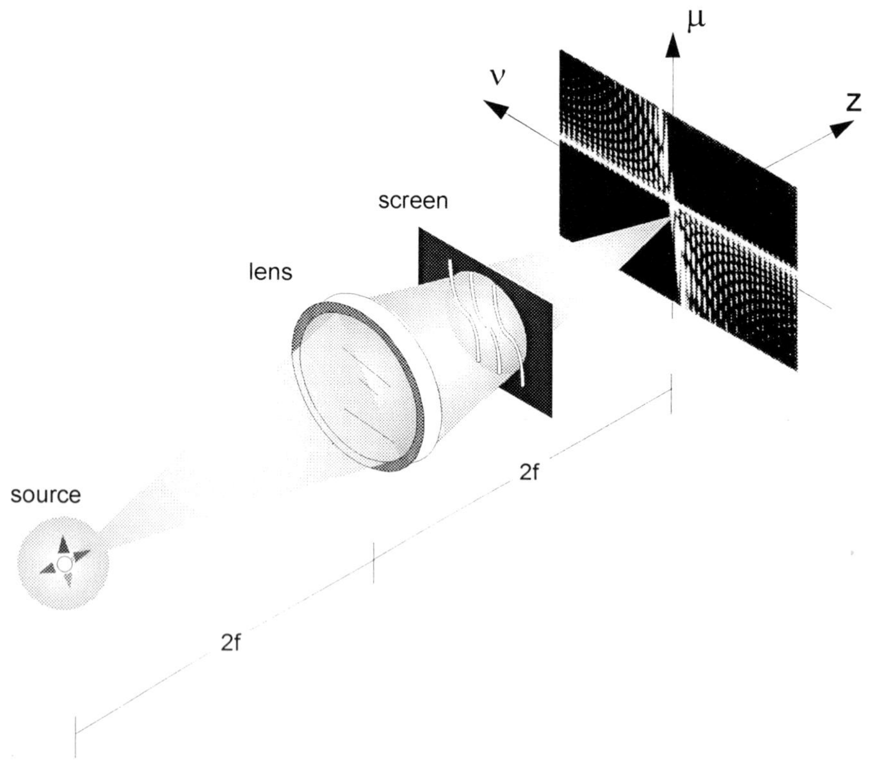

Fig. 4. Same as Fig. 3 but for a cotangent curve, for generating the first derivative of the Laguerre polynomial with variable order.

where $L_m(z)$ denotes the Laguerre polynomial of order m. Using the result in Eq. (13), it is straightforward to show that Eq. (12) becomes

$$\tilde{T}(v,\mu) = \sum_{m=0}^{\infty} \exp(-2\pi A\mu)L_m(4\pi A\mu)[\delta(v + 2m/d) - \delta(v + 2(m+1)/d)]. \quad (14)$$

Hence, the Fraunhofer diffraction pattern shows, along the lines $v = 2M/d$, the special function

$$\begin{aligned}
\tilde{T}(2M/d, \mu) &= \exp(-2\pi A\mu)\,[L_M(4\pi A\mu) - L_{M-1}(4\pi A\mu)] \\
&= \exp(-2\pi A\mu)(4\pi A\mu/M)L'_M(4\pi A\mu),
\end{aligned} \quad (15)$$

where L'_M denotes the first derivative of the Laguerre polynomials of order M. The last result in Eq. (15) is based in the recurrence relationship

$$(z/M)L'_M(z) = L_M(z) - L_{M-1}(z), \quad (16)$$

as is shown in reference [19].

In the three examples above, we are able to synthesize in parallel several members of a special function, whose generating function is a pure phase signal. In other words, some 1D wave aberrations that are generated with a binary screen can produce in parallel certain special functions. Next, we discuss how to implement optically complex amplitude signals.

GENERALIZED TWO-SLIT EXPERIMENT

In Fig. 5 we show schematically that the binary screen is composed by two narrow slits, which follow respectively the planar trajectories $y = F(x)$, and $y = G(z)$. Hence, the amplitude transmittance of the screen is

$$T(x, y) = \delta[y - F(x)] + \delta[y - G(x)]. \quad (17)$$

Since the two slits are no longer parallel, then one can consider this screen as a generalized version of Young's experiment. In this case the complex amplitude in the Fraunhofer plane is

$$\tilde{T}(v,\mu) = \int_{-\infty}^{\infty} \{\exp[-i2\pi\mu F(x)] + \exp[-i2\pi\mu G(x)]\}\exp(-i2\pi vx)\,dx. \quad (18)$$

It is convenient now to recognize that the integrand in Eq. (18) can be rewritten as follows

$$\begin{aligned}
&\exp[-i2\pi\mu F(x)] + \exp[-i2\pi\mu G(x)] \\
&= 2\exp\left\{-i2\pi\mu\left[\frac{F(x) + G(x)}{2}\right]\right\} \times \cos\{\pi\mu[F(x) - G(x)]\} \\
&= 2\exp[-i2\pi\mu\Phi(x)]\cos[\pi\mu W(x)],
\end{aligned} \quad (19)$$

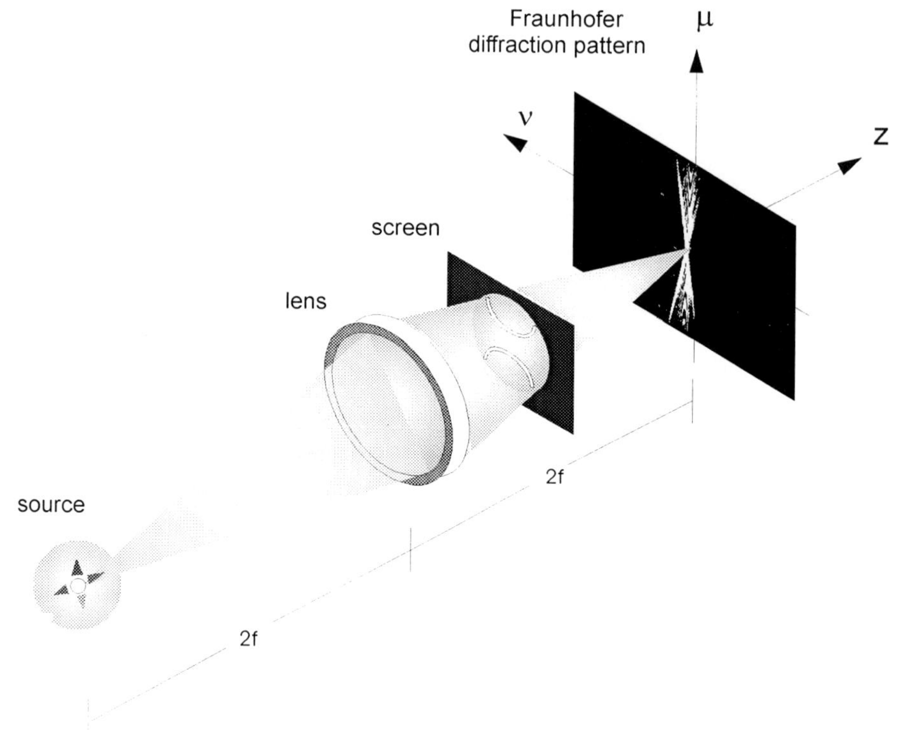

Fig. 5. Generalized setup of Young's double-slit experiment.

where we recognize the average trajectory between the two slits as the function

$$\Phi(x) = [F(x) + G(x)]/2, \tag{20}$$

and the separation, or width, between the two slits as the function

$$W(x) = F(x) - G(x); \tag{21}$$

see Fig. 6. Thus, the complex amplitude in the Fraunhofer diffraction pattern, in Eq. (18), can be expressed as

$$\tilde{T}(\nu,\mu) = \int_{-\infty}^{\infty} \cos[\pi\mu W(x)]\exp[-i2\pi(\nu x + \mu\Phi(x))]\, dx. \tag{22}$$

The average trajectory synthesizes a phase variation, $\Phi(x)$, while the separation between the slits synthesizes an amplitude variation, $\cos[\pi\mu W(x)]$. And hence, we are now able to generate, with binary screens, 1D wave aberrations, $\Phi(x)$, as well as 1D apodizing pupils, $\cos[\pi\mu W(x)]$, by suitably selecting the trajectories of the two narrow slits.

From the point of view of synthesizing special functions, we have shown the capability of using binary screens for optically producing generating functions

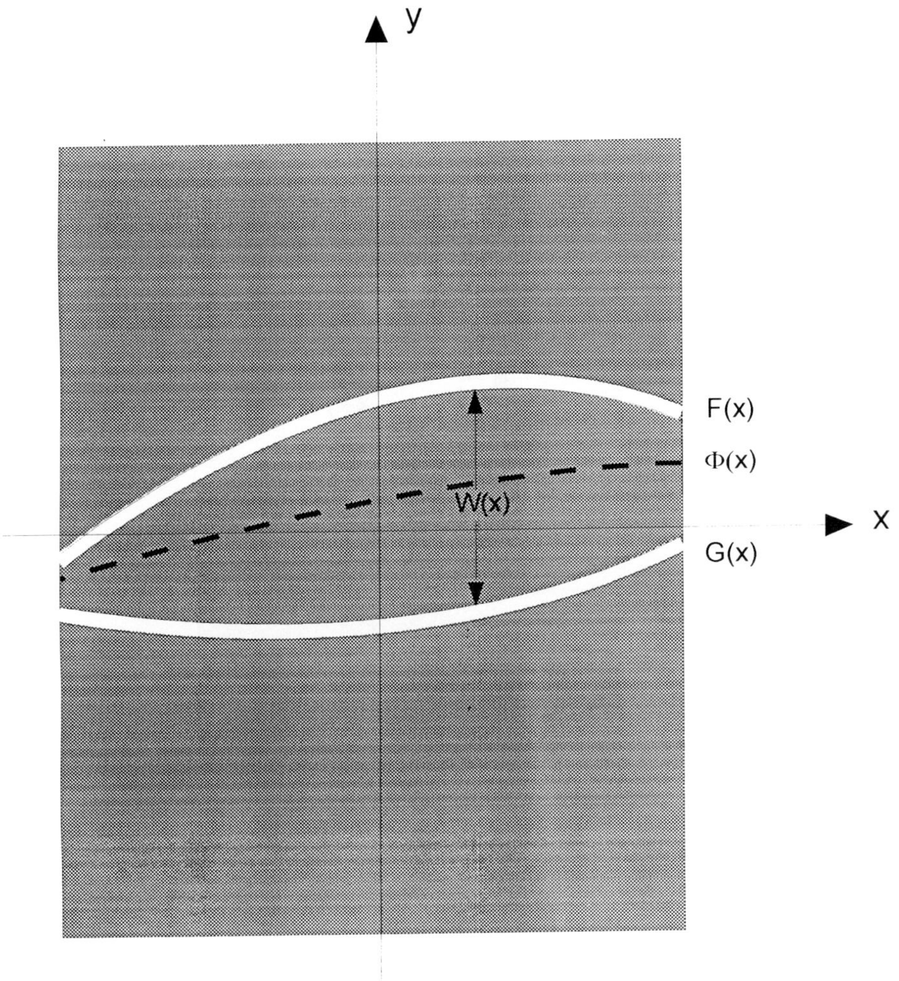

Fig. 6. Average trajectory $\Phi(x) = [F(x) + G(x)]/2$, and separation function $W(x) = F(x) - G(x)$ in the generalized setup of Young's double-slit experiment.

that have amplitude variations, in addition to the cases treated above, which are generating functions having phase-only variations. So far, this capability has not been exploited in specific examples.

GENERALIZED WIDE SINGLE SLIT

From our previous result one can infer the following possibility. What happens if the two narrow slits are substituted by the borders of a wide slit? Here we consider this situation. In Fig. 7 we show a wide slit whose borders follow the trajectories $F(x)$ and $G(x)$, respectively. Moreover, we note that the centre of the slit follows the curve $\Phi(x)$ in Eq. (20), and that the width of the slit is given by Eq. (21).

Hence, the amplitude transmittance of the screen in Fig. 7 can be expressed as

$$T(x, y) = \text{rect}\left[\frac{y - \Phi(x)}{W(x)}\right], \tag{23}$$

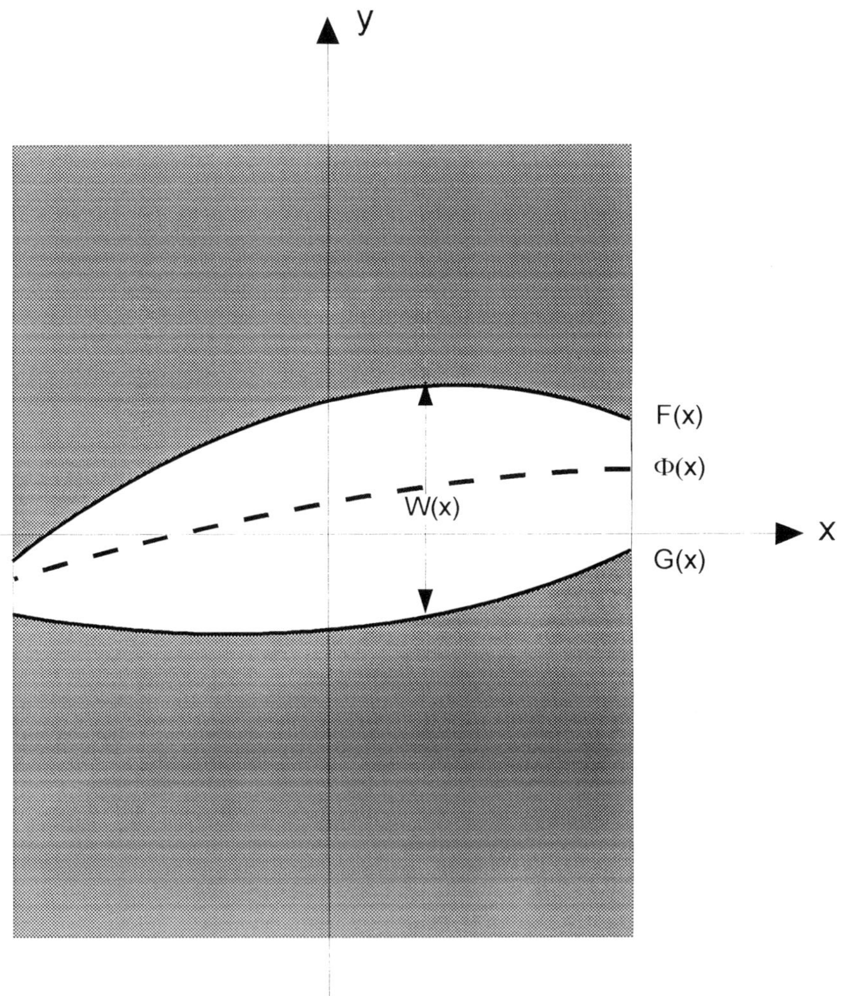

Fig. 7. Same as Fig. 6 but for a wide-slit experiment.

where the rect function is $\mathrm{rect}(t) = 1$ only if $|\,t\,| \leqslant \frac{1}{2}$, and zero otherwise. By substituting Eq. (23) in Eq. (2) we find that the complex amplitude in the Fraunhofer plane is

$$\tilde{T}(v,\mu) = \int_{-\infty}^{\infty} \int_{\Phi(x)-W(x)/2}^{\Phi(x)+W(x)/2} \exp(-i2\pi\mu y)\, dy \, \exp(-i2\pi v x) \, dx.$$

$$= \int_{-\infty}^{\infty} [\sin(\pi\mu W(x))/\pi\mu] \exp[-i2\pi(v x + \mu\Phi(x))] \, dx. \qquad (24)$$

It is apparent from Eq. (24) that again, by using a suitable binary screen, we are able to generate 1D phase variations, $\Phi(x)$, as well as 1D amplitude variations, $\sin[\pi\mu W(x)]/\pi\mu$. In particular it is interesting to note that by setting $\mu = 0$ in Eq. (24), we have that along the v-axis the complex amplitude in the Fraunhofer plane is

$$\tilde{T}(v,\mu = 0) = \int_{-\infty}^{\infty} W(x)\exp(-i2\pi v x) \, dx. \qquad (25)$$

The result in Eq. (25) indicates that we are able to optically generate 1D impulse responses from analogue spatial filters, $W(x)$, which are obtained from a binary screen. According to Jacquinot and Roizen-Dossier [20] this particular case of Eq. (24) was first noticed and used by Couder and Jacquinot [21]. Below we consider a polar coordinate variation of this result.

OPTICAL PROCESSORS

In our previous discussion we have emphasized that simple binary screens are able to generate unconventional Fraunhofer diffraction patterns for signal synthesis. An interesting feature of our results is that the μ-axis is employed for changing the optical path difference in the generated phase signal, and this fact was exploited for changing the parameter of the Airy function, or the argument of the Bessel functions, as well as the argument of the derivative of the Laguerre polynomial. Furthermore, the μ-axis can also be used to weight the argument of amplitude variations, as was pointed out in Eqs. (22) and (24).

Due to the above feature our previous results can be discussed in the context of trading one dimension in an optical processor; see reference [15]. Here, we comment briefly that our previous results can also be described as an imaging process in an optical processor. This extension follows from the fact that the μ-axis can conveniently be scanned to select optical path differences, or to set weighting factors for introducing amplitude variations.

In this section we consider that the μ-axis is scanned with a narrow slit, along the v-axis, located at $\mu = \Omega$. In other words, we consider that at the Fraunhofer diffraction pattern we have the spatial filter represented by

$$\tilde{p}(v,\mu) = \delta(\mu - \Omega). \qquad (26)$$

The Fourier transform of the pupil function in Eq. (26) is the impulse response

$$p(x, y) = \delta(x) \exp(i2\pi\Omega y). \tag{27}$$

Hence, the complex amplitude in the image of the binary screen with amplitude transmittance $T(x, y)$ is

$$u(x, y) = \iint_{-\infty}^{\infty} T(x', y')p(x - x', y - y')\, dx'\, dy'$$

$$= \exp(i2\pi\Omega y) \int_{-\infty}^{\infty} T(x, y')\exp(-i2\pi\Omega y')\, dy'. \tag{28}$$

The latter formula allows us to emphasize the following image formation properties.

1. For a single narrow slit, $T(x, y) = \delta[y - F(x)]$, the complex amplitude in the image is a phase-only signal

$$u(x, y) = \exp[i2\pi\Omega(y - F(x))], \tag{29}$$

    whose phase delay, $2\pi\Omega$, can be changed continuously by displacing the spatial filter along the μ-axis [7].
2. For two narrow slits, $T(x, y) = \delta[y - F(x)] + \delta[y - G(x)]$, the complex amplitude in the image is the complex signal

$$u(x, y) = 2\exp[i2\pi\Omega(y - \Phi(x))] \cos[\pi\Omega W(x)], \tag{30}$$

    where $\Phi(x)$ and $W(x)$ are defined in Eqs. (20) and (21) respectively. Again the phase delay and the argument of the amplitude variations can easily be changed by displacing the spatial filter to another value of Ω, as discussed in reference [10].
3. Note that if the above two slits are symmetrically located along the x-axis, $F(x) = -G(x)$, then Eq. (30) becomes

$$u(x, y) = 2\exp(i2\pi\Omega y) \cos[2\pi\Omega F(x)], \tag{31}$$

    which indicates that except for a phase tilt the synthesized signal is an amplitude-only variation. An example of this result is shown in Fig. 8 for $F(x) = -G(x) = Ax^2$, which generates a 1D zone plate whose spatial frequency, ΩA, can be easily tuned by displacing the spatial filter along the μ-axis. Obviously for $W(x) = 0$ we obtain Eq. (29).
4. If the binary input is a wide slit with variable borders, $T(x, y) = \text{rect}[(y - \Phi(x))/W(x)]$, then the complex amplitude in the output is

$$u(x, y) = \exp[i2\pi\Omega(y - \Phi(x))] \sin[\pi\Omega W(x)]/\pi\Omega. \tag{32}$$

It is apparent from Eq. (32) that the output's amplitude distribution is a complex signal. Here again, as in Eq. (31), for $W(x) = 0$ we obtain a phase-only signal, and for $\Phi(x) = 0$ we obtain an amplitude-only signal.

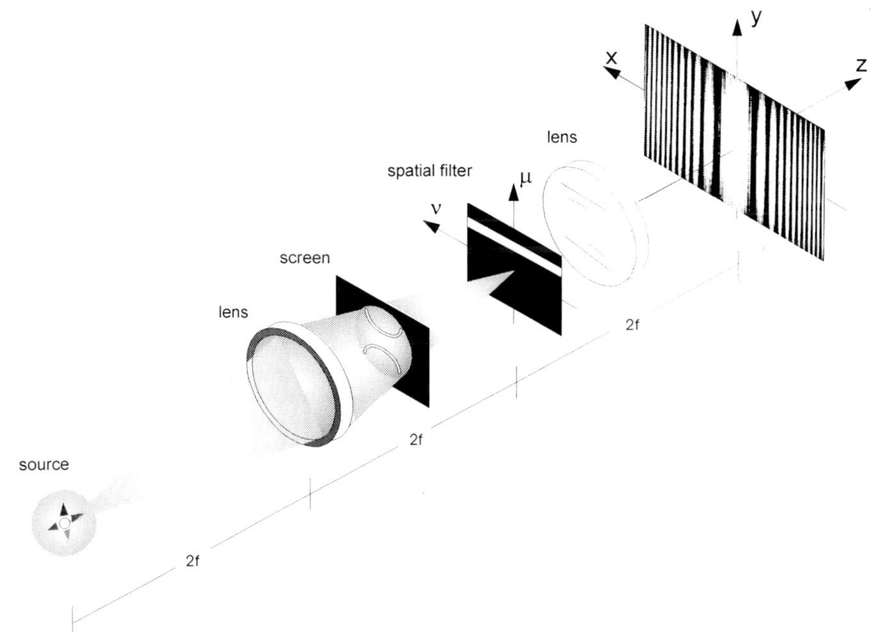

Fig. 8. Fraunhofer diffraction pattern of double-slit experiment with $F(x) = Ax^2$ and $G(x) = -Ax^2$.

But in addition, it is worth noting that if the spatial filter is located at $\mu = 0$, then the output's amplitude distribution is

$$u(x, y) = W(x). \tag{33}$$

In Fig. 9 we show the generation of a 1D zone plate by exploiting this latter result [11].

As a final comment, before our last application, note that the spatially filtered images can also be obtained with cylindrical lenses, with the purpose of increasing light gathering power. However, the use of the spatial filter gives the freedom of selecting the phase delay and/or the carrier frequency.

AXIAL SIGNAL SYNTHESIS

According to reference [22], the amplitude distribution along the optical axis of an optical system, within the paraxial approximation, is

$$u(z) = -\frac{i}{\lambda s(s+z)} \exp(ikz) \int_0^{2\pi} \int_0^a T(r, \theta) \exp\left[-i2\pi \frac{z}{2\lambda s(s+z)} r^2\right] r \, dr \, d\theta, \tag{34}$$

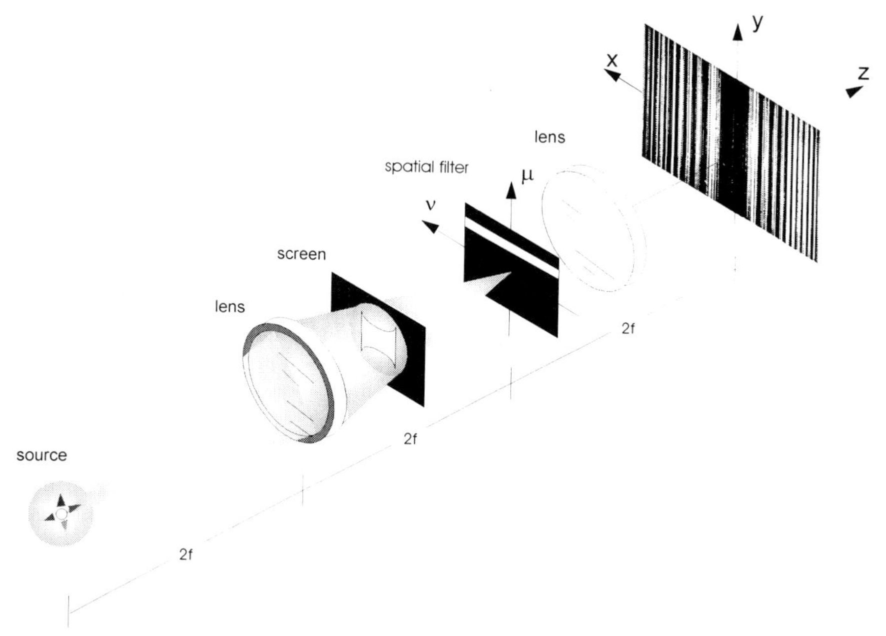

Fig. 9. Generation of a variable spatial frequency, $\mu = \Omega$, 1D zone plate using the screen in Fig. 8.

where z represents the axial coordinate in the image domain, λ is the wavelength of the incident light, $k = 2\pi/\lambda$, s stands for the radius of the converging wavefront as shown in Fig. 10, and $T(r, \theta)$ is the amplitude transmittance of the pupil function.

Some interesting features of Eq. (34) are rendered apparent by using the following suitable change of variables [12, 13]. For the pupil plane we employ the geometrical mapping

$$\zeta = \left(\frac{r}{a}\right)^2 - 0.5, \qquad \tilde{Q}(\zeta, \theta) = T(r, \theta). \tag{35}$$

Furthermore, in our formula we use the Fresnel number

$$N = \frac{a^2}{\lambda s}, \tag{36}$$

which indicates the number of Fresnel zones that are covered by the pupil as viewed from the geometrical image, $z = 0$. And, finally, we denote the defocus

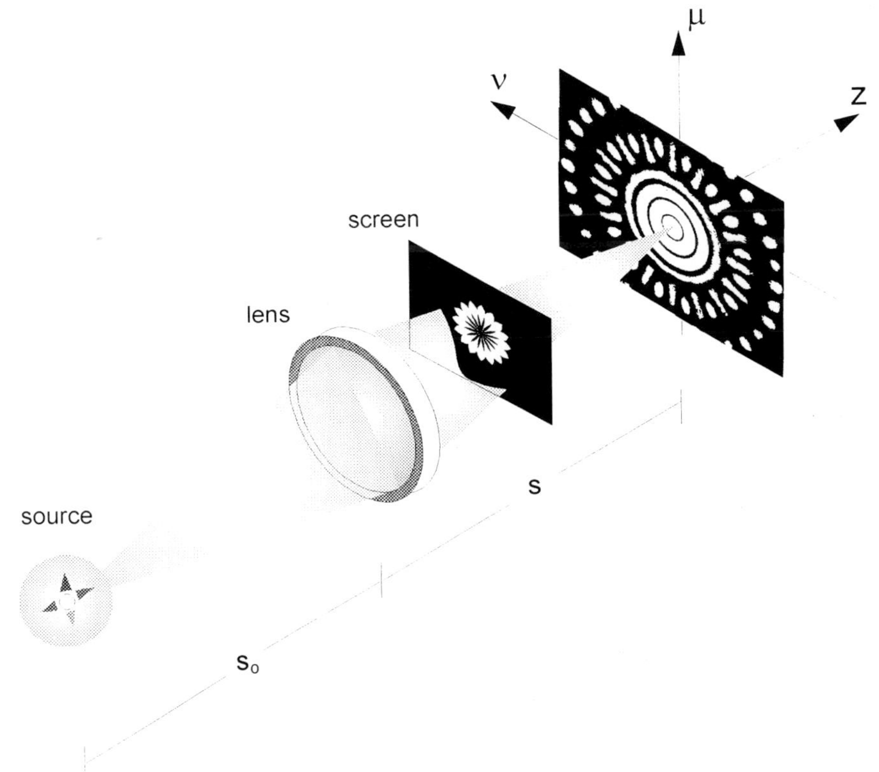

Fig. 10. Axial synthesis of 1D optical signals using 2D screens.

coefficient, measured in units of wavelength, as

$$W_{20} = \frac{Nz}{2(s+z)}.$$ (37)

By substituting Eqs. (35), (36) and (37) in Eq. (34) we obtain that, except for an irrelevant phase factor,

$$u(z) = Q(W_{20}) = -\frac{i\pi}{s}(N - 2W_{20})\int_{-0.5}^{0.5}\frac{1}{2\pi}\int_0^{2\pi}\tilde{Q}(\zeta, \theta)\,d\theta\,\exp(-i2\pi W_{20}\zeta)\,d\zeta.$$

(38)

Now, from Eq. (38) the following features become apparent. The axial amplitude response is proportional to the 1D Fourier transform of the azimuthal average of the pupil function [23]. The factor $(N - 2W_{20})$, with W_{20} as the functional variable, is responsible for the loss of symmetry of the axial irradiance distribution when using optical systems with low Fresnel numbers.

In this case the axial point of maximum irradiance does not coincide with the geometrical image, but it is located closer to the aperture [24]. Finally, we remark on the following angular average operation. The zero-order circular harmonic of the pupil function is of course the angular azimuthal average

$$\tilde{Q}_0(\zeta) = \frac{1}{2\pi} \int_0^{2\pi} \tilde{Q}(\zeta, \theta) \, \mathrm{d}\theta. \tag{39}$$

$Q(W_{20})$, in Eq. (38), and $\tilde{Q}_0(\zeta)$ are uniquely related by a 1D Fourier transform. But since $\tilde{Q}_0(\xi)$ is an angular average of $\tilde{Q}_0(\xi, \theta)$, then there can be several pupil functions with different $T(r, \theta)$ but equal $\tilde{Q}_0(\xi)$, and consequently, with equal axial response $Q(W_{20})$.

In what follows we exploit the azimuthal average in Eq. (39) to generate, from binary screens, analogue apodizers to shape the axial response.

For that purpose we consider first binary screens of the form

$$\tilde{Q}(\zeta, \theta) = \mathrm{rect}\left[\frac{\theta - \alpha}{\beta W(\zeta)}\right], \tag{40}$$

where α and β are two arbitrary angles, and $W(\xi)$ is a radial function such that $0 < W(\xi) \leqslant 1$. This type of screen has unity transmittance inside the angular sector $-\beta W(\xi)/2 \leqslant \theta - \alpha \leqslant \beta W(\xi)/2$. Hence, the azimuthal average in Eq. (39) becomes

$$\tilde{Q}_0(\xi) = (\beta/2\pi)W(\xi), \tag{41}$$

which clearly indicates that from the viewpoint of the axial impulse response we have an analogue apodizer, $W(\xi)$, that was generated from the borders of a binary angular sector.

Our previous proposal generates highly asymmetrical diffraction patterns outside the optical axis. If for some applications this feature is undesirable, then one can reduce the asymmetric behaviour substantially by using several (say, $2M$) sectors in the form of a pie or a daisy [25, 26]. Each sector contains the binary pattern in Eq. (40) but now centred at $\alpha = (2m - 1)\pi/2M$, for $m = 1, \ldots, M$, and with $\beta = (2\pi/2M)$. Hence, Eq. (40) becomes

$$\tilde{Q}(\zeta, \theta) = \sum_{m=1}^{2M} \mathrm{rect}\left[\frac{\theta - (2m - 1)\pi/2M}{(\pi/M)W(\zeta)}\right], \tag{42}$$

and consequently Eq. (38) becomes

$$Q(W_{20}) = -\frac{\mathrm{i}\pi}{s} (N - 2W_{20}) \int_{-0.5}^{0.5} W(\zeta)\exp(-\mathrm{i}2\pi W_{20}\zeta) \, \mathrm{d}\zeta. \tag{43}$$

It is apparent from Eq. (43) that the amplitude distribution along the optical axis is proportional to the Fourier transform of the analogue spatial filter $W(\xi)$, which was implemented with a binary screen, as is shown schematically in Fig. 10. This result finds several applications: for reducing the focal shift [*12*], for analysing non-centrally obscured pupils [*13*], for extending the range of non-zero axial distributions [*14*], for generating a null test for alignment [*25*], and for designing novel families of zone plates [*26, 27*].

We have discussed several examples of using binary patterns, like Young's experiment for synthesizing 1D optical signals.

ACKNOWLEDGMENTS

We thank our colleagues Pedro Andrés, Victor Arrizón, Juan G. Ibarra, Manuel Martínez-Corral, Gustavo Ramírez and Alfonso Serrano-Heredia for their fruitful collaboration. J.O.C. is grateful to the University of Las Américas, Puebla, for partially supporting several projects, here reported.

REFERENCES

1. Lord Rayleigh (J.W. Strutt), On some physical properties of argon and helium, *Proc. Roy. Soc. A* **59**, 198–208 (1886).
2. A.A. Michelson, On the application of interference methods to astronomical measurements, *Phil. Mag.* **30**, 1–21 (1890).
3. F. Zernike, The concept of degree of coherence and its applications to optical problems, *Physica* **5**, 785–795 (1938).
4. D.E. Duffy, Moire gauging of in-plane displacement using double aperture imaging, *Applied Optics* **11**, 1778–1781 (1972).
5. G. Magyar and L. Mandel, Interference fringes produced by superposition of two independent laser light beams, *Nature* **198**, 255–256 (1963).
6. F. Louradour, F. Reynaud, B. Colombeau and C. Froehly, Interference fringes between two separate lasers, *American Journal of Physics* **61**, 242–245 (1993).
7. J. Ojeda-Castañeda and V. Arrizón, Synthesis of 1D phase profiles with variable optical path, *Microwave and Optical Technology Letters* **5**, 429–432 (1992).
8. A.W. Lohmann, J. Ojeda-Castañeda and J.G. Ibarra, Airy function and Laguerre polynomials: optical display and processing, *Optics Communications* **109**, 361–367 (1994).
9. A.W. Lohmann, J. Ojeda-Castañeda and A. Serrano-Heredia, Bessel functions: parallel display and processing, *Optics Letters* **19**, 361–367 (1994).
10. A.W. Lohmann, J. Ojeda-Castañeda and A. Serrano-Heredia, Synthesis of 1D complex amplitudes using Young's experiment, *Optics Letters* **19**, 55–57 (1994).
11. A.W. Lohmann, J. Ojeda-Castañeda, Computer generated holography: novel procedure, *Optics Communications* **103**, 181–184 (1993).
12. M. Martínez-Corral, P. Andrés and J. Ojeda-Castañeda, On-axis diffractional behaviour of two dimensional pupils, *Applied Optics* **33**, 2223–2229 (1994).
13. J. Ojeda-Castañeda, M. Martínez-Corral, P. Andrés and A. Pons, Strehl ratio

versus defocus for noncentrally obscured pupils, *Applied Optics* **33**, 7611–7616 (1994).

14. A.W. Lohmann, J. Ojeda-Castañeda and A. Serrano-Heredia, Synthesis of analog apodizers with binary angular sectors, *Applied Optics* **34**, 317–322 (1995).

15. A.W. Lohmann, J. Ojeda-Castañeda and A. Serrano-Heredia, Trading dimensionality in signal processing, *Optics and Laser Technology* **28**, 101–107 (1996).

16. K.O. Hill and E.B. Felstad, Optical simulation of one dimensional complex-valued functions, *Applied Optics* **10**, 1693–1694 (1971).

17. B.R. Brown and A.W. Lohmann, Complex spatial filtering with binary mask, *Applied Optics* **5**, 967–969 (1966).

18. A.W. Lohmann and D.P. Paris, Binary Fraunhofer holograms generated by computer, *Applied Optics* **6**, 1739–1748 (1967).

19. M. Abramowitz and I.A. Stegun, *Handbook of Mathematical Functions*, Dover, New York, (1970) p. 447.

20. P. Jacquinot and B. Roizen-Dossier, Apodisation, Chapter II, *Progress in Optics* Ed. E. Wolf, Vol. III (North Holland, Amsterdam, 1964).

21. A. Couder and P. Jacquinot, *C. R. Acad. Sci. Paris* **208**, 1639 (1939).

22. Y. Li and E. Wolf, Focal shift in focused truncated Gaussian beams, *Optics Communications* **68**, 317–323 (1988).

23. C.W. McCutchen, Generalized aperture and the three dimensional diffraction image, *J. Opt. Soc. Am.* **54** 240–244 (1964).

24. Y. Li and E. Wolf, Focal shifts in diffracted converging spherical waves, *Optics Communications* **39**, 211–215 (1981).

25. J. Ojeda-Castañeda, P. Andrés and M. Martínez-Corral, Zero axial irradiance by annular screens with angular variation, *Applied Optics* **31**, 4600–4602 (1992).

26. J. Ojeda-Castañeda and G. Ramírez, Zone plates for zero axial irradiance, *Optics Letters* **18**, 87–89 (1993).

27. A.W. Lohmann, J. Ojeda-Castañeda, and G. Ramírez, Zone plates encoding Limaçon variations, *Optics Communications* **114**, 30–36 (1995).

15 Resolution enhancement by data inversion techniques

Christine De Mol
Département de Mathématique, Université Libre de Bruxelles,
Campus Plaine C.P. 217, Bd du Triomphe, B—1050 Brussels,
Belgium

INTRODUCTION

The development of micro-informatics and of computer-assisted devices has deeply modified the classical concept of resolving power of an optical instrument. We want to justify this assertion in the light of some work done in the field of inverse problems. Indeed, in modern instruments numerical algorithms can be implemented to process, i.e. invert, the recorded data in order to get estimates of the probed object with enhanced resolutions. The resolution limits then arise as practical limitations due to the noise amplification inherent in all inversion procedures. We will show how such limits can be assessed and under which circumstances a significant amount of super-resolution is achievable. In this framework the resolving power no longer appears as a purely intrinsic characteristic of the instrument itself, but rather as a combined property of the hardware (the optical components) and of the implemented inversion software. As will be shown, it depends then on the available signal-to-noise ratio and, to a lesser extent, on the choice of the data inversion algorithm. We start by recalling how classical resolution limits are introduced and defined.

In the framework of geometrical optics, the image of a point source through an aberration-free optical system is a perfectly sharp point. Accordingly, the resolving power of the system, which is a measure of its ability to separate the images of two neighbouring points, is unlimited. In practice, however, diffraction effects always occur and, consequently, the image is never just a point, but instead a small light patch called the *diffraction pattern*. The intensity distribution in the diffraction pattern corresponding to a point source is called the (intensity) *point response* or *point spread function*. When two point sources get closer and closer, their diffraction patterns progressively overlap, until it is no longer possible to discriminate between one or two point sources. The limit distance between the two sources down to which the discrimination is still possible characterizes the resolving power of the apparatus. The ability to resolve point sources may depend in practice on many factors which can hardly be quantified (as e.g. the sensitivity of the human eye). Nevertheless, it is useful to have some simple and objective criteria to compare the performances of optical systems. The most famous one is the Rayleigh criterion (Rayleigh 1879, Born and Wolf 1980). According to its original formulation, the diffraction patterns are considered as just resolved if the central maximum of the first coincides with the first minimum of the other. For the so-called Airy pattern, which is the Fraunhofer (far-field) intensity diffraction pattern at a circular aperture, the distance between the principal maximum and the first zero or dark ring is given by $R = 1.22 \, (\lambda/2\alpha)$, where λ is the wavelength of the light and α is equal to the radius of the diffracting aperture divided by the distance between the aperture and image planes. The distance R is known as the *Rayleigh resolution limit*.

Following the development of Fourier methods in optics (see e.g. Goodman 1968), it has become more usual to characterize the resolution of an optical system in terms of its *bandwidth*. Diffraction effects are indeed responsible for the existence of a *cut-off* for the spatial-frequency components that are transmitted by the system. In such a case, the optical system is said to be *band-limited* or *diffraction-limited*. In other words, the sharpest details in the object, which are smaller than the resolution limit, are smoothed off and no longer noticeable in the image, which appears as a blurred version of the object. Rayleigh's criterion can then be reinterpreted in the framework of communication and information theory (Toraldo di Francia 1969), the number of Rayleigh resolution elements being identified with the Shannon number, which is also called in optics the *number of degrees of freedom*.

The idea of trying to beat the Rayleigh limit is quite old and has been abundantly discussed in the literature (see e.g. Toraldo di Francia 1955, McCutchen 1967 and Rushforth and Harris 1968). The corresponding enhancement in resolution, when feasible, is called *super-resolution*. When dealing with a given imaging system using a fixed probing wavelength, the only way to enhance its resolving power is to try to process the image in order to reconstruct the object *beyond the Rayleigh limit*, i.e. to restore spatial frequencies which were not transmitted by the band-limited system. Such an *out-of-band* extrapolation seems to be feasible, at least in principle, under the assumption that the object has a finite spatial extent, i.e. vanishes outside some finite domain. Indeed, in such a case, the Fourier transform of the object (the object spectrum) is an entire analytic function and hence could be uniquely recovered from that piece of spectrum transmitted by the optical system. This argument, however, does not take into account the instability of analytic continuation in the presence of noise, which prevents in practice an easy recovery of the object spectrum beyond the cut-off frequency. The appropriate framework to address this question and to assess quantitatively the achievable resolution improvement appears to be the regularization theory for inverse problems. In such a framework it has been shown that, because of the ill-posedness of the extrapolation problem in the presence of noise, only very little super-resolution could be achieved in most practical situations (Bertero *et al.* 1979, 1980). Such a conclusion, however, no longer holds for problems characterized by a very small *Shannon number* or equivalently *space-bandwidth product*. This condition is fulfilled for example in scanning microscopy, as noticed by Bertero and Pike (1982) who suggested the possibility of using inversion techniques based on singular-system expansions to enhance the resolution of such microscopes. In the last section we will show how to assess the amount of achievable super-resolution in the case where the object is uniformly illuminated over some finite slit. The imaging problem corresponding to more general object illumination profiles was investigated later by Bertero *et al.* (1984) and the results were applied to the kind of illumination realized in a confocal scanning microscope. The above situations refer to object restoration from far-field data. We will show that the diffraction limit can also be beaten when restoration can be made from near-field data, using the information conveyed by evanescent waves.

Before analysing the conditions under which super-resolution can be achieved, we briefly sketch the required framework and show how to define a mathematical model for the imaging process and how to solve the associated inverse problem.

HOW TO DESCRIBE THE IMAGING PROCESS

Let us recall that in the framework of Fourier optics (Born and Wolf 1980, Goodman 1968, Papoulis 1968), an optical system is viewed as a linear system, i.e. as a black box fully characterized by its impulse response $S(x, y)$, which is called in optics the point spread function (PSF) or point response. This function, which is assumed to be known from a good theoretical model or from reliable measurements, represents the image at point x of a point source situated at point y. Hence the imaging process is described by the following equation

$$g(x) = \int \mathrm{d}y \, S(x, y) f(y) \tag{1}$$

relating the object (input) $f(y)$ to its image (output) $g(x)$. The quantities f and g represent scalar light amplitudes in coherent imaging, and light intensities in incoherent imaging. The variables x and y are in general two-dimensional and denote vector coordinates in the image and object planes, respectively (notice that as usual the magnification factor will be reduced to 1 by appropriate scaling). In some cases, these variables can be three-dimensional, including depth information along the optical axis. For simplicity, however, we will focus here on one-dimensional equations. Then, unless stated otherwise, integration in (1) and in similar expressions hereafter is made over the whole real axis. All the tools introduced in the following can be easily generalized to two- and three-dimensional situations although at the price of increased numerical complexity.

If the system is shift invariant, i.e. invariant for translations in space, the integral kernel in (1) depends only on the difference of the variables x and y: $S(x, y) = S(x - y)$. The image is then the convolution of the object by the PSF $S(x)$. By the convolution theorem, the corresponding imaging equation becomes a simple multiplication in conjugate space: the Fourier transform of the image

$$\hat{g}(\omega) = \int_{-\infty}^{+\infty} dx\, e^{-i\omega x} g(x) \tag{2}$$

is then just the product of the Fourier transform $\hat{f}(\omega)$ of the object by the Fourier transform $\hat{S}(\omega)$ of the PSF $S(x)$, i.e.,

$$\hat{g}(\omega) = \hat{S}(\omega)\hat{f}(\omega). \tag{3}$$

The function $\hat{S}(\omega)$ is called the *optical transfer function* of the system.

As a consequence of diffraction effects, the transfer function $\hat{S}(\omega)$ vanishes outside some finite region in frequency space, called the *band*, the *aperture* or else the *pupil* of the optical system. The corresponding PSF $S(x)$ is then said to be *band-limited* to that pupil. We will denote the cut-off frequency by Ω so that the (one-dimensional) pupil is the interval $[-\Omega, +\Omega]$. The spatial frequencies of the object lying outside the pupil are totally filtered by the system and, therefore, are no longer visible in the image, which is a blurred version of the object. For perfect coherent imaging, i.e., in the absence of all kinds of aberrations, the spatial frequencies up to the cut-off are transmitted by the optical system without any attenuation or phase distortion. Hence, the transfer function of such an *ideal diffraction-limited optical system* is just the characteristic function of the pupil $[-\Omega, +\Omega]$,

$$\hat{S}(\omega) = \begin{cases} 1 & \text{for } |\omega| \leq \Omega \\ 0 & \text{for } |\omega| > \Omega, \end{cases} \tag{4}$$

and the corresponding PSF is given by

$$S(x) = \frac{\sin(\Omega x)}{\pi x} \tag{5}$$

(in the case of a circular lens, the transfer function is equal to 1 in the circle of radius Ω and to 0 outside).

In incoherent imaging, optical systems are still linear if f and g represent light intensities (i.e., the squared moduli of the corresponding amplitudes) and the PSF to be used is given by

$$S_{\text{incoherent}}(x) = \mid S_{\text{coherent}}(x) \mid^2 \tag{6}$$

where $S_{\text{coherent}}(x)$ is the coherent PSF of the same optical system. For perfect incoherent 1D imaging, this corresponds to a triangular transfer function which is the convolution of Eq. (4) with itself and hence is different from zero on the interval $[-2\Omega, +2\Omega]$. In the presence of aberrations, distortion occurs and the transfer function (4) differs from 1 on the band $[-\Omega, +\Omega]$.

Since the PSF is a band-limited function, the image $g(x)$ is also band-limited and therefore, by the famous *sampling theorem* due to Whittaker (1915) and later popularized by Shannon (1949) in communication theory (see also Papoulis 1968 or Goodman 1968), it is entirely determined by its sampled values at a certain (sufficiently high) sampling rate. For a one-dimensional band-limited function g of bandwidth Ω, the theorem states

$$g(x) = \sum_{n=-\infty}^{+\infty} g\left(n\,\frac{\pi}{\Omega}\right) \frac{\sin(\Omega x - n\pi)}{\Omega x - n\pi}. \tag{7}$$

The corresponding sampling distance $R = \pi/\Omega$ is known as the Nyquist distance in signal theory and as the Rayleigh resolution distance in optics. According to this theorem, there is only a finite number of independent samples in any finite portion of a band-limited function and details on a finer scale than the sampling distance are not resolved.

In classical instruments, the resolution available in the band-limited data determines the ultimate resolution capabilities of the instrument itself. In computer-assisted instruments, however, one can try to recover (restore) the actual object from its blurred image by solving numerically the imaging equation (1). This requires the image to be recorded by means of some detection scheme and then processed digitally on an associated computer. Hopefully, one should thus get an enhanced resolving power for the instrument supplemented with an appropriate computer algorithm. Solving equation (1) is a typical *inverse problem* and one has to devise a method to perform the inversion, i.e. to reconstruct the object. A common mathematical feature of inversion methods is their sensitivity to noise in the data: if appropriate tools are not applied to control the noise amplification, the computed solution can be completely unstable and therefore meaningless. The inverse problem is then said to be *ill-posed* or *ill-conditioned.* Inversion theory, also referred to as *regularization theory*, precisely deals with such stability problems and allows us to define meaningful stable approximate solutions of first-kind integral equations like the imaging equation (1). Looking at equation (3), we now distinguish between two problems. The first one, which we call *deblurring*, consists in restoring the object inside the band to suppress distortion (aberrations) and noise. This leads to better imaging but not to a corresponding increase in resolution, since this procedure affects only the spatial frequencies which are also present in the band-limited image. On the contrary, the second problem, which we call *out-of-band extrapolation*, consists precisely in trying to go beyond the diffraction limit, i.e., to restore at least a part of the object spectrum that has been filtered by the instrument and is no longer present in the image. We will see in the last section that we need extra a priori information about the object to be able to perform this task, i.e. to achieve what is called *super-resolution*.

HOW TO SOLVE THE INVERSE IMAGING PROBLEM

We briefly recall in the present section some basic features of inversion theory, which are useful for assessing the resolution limits. By introducing an additive noise term $n(x)$ in the imaging equation (1), we take into account the fact that the recorded image is inevitably contaminated by measurement errors:

$$g(x) = \int dy \, S(x, y) f(y) + n(x). \tag{8}$$

Given some recorded data function $g(x)$, the inverse imaging problem consists in estimating the original object $f(y)$, i.e. in solving the imaging equation (8). This equation can be rewritten more concisely in operator form as

$$g = Lf + n \tag{9}$$

where L denotes the imaging operator mapping the object on its noise-free image. As is common for such problems, we will restrict ourselves to linear estimation and denote throughout by $\tilde{f}(y)$ the resulting estimate of the object. The most general linear estimate of the object is obtained as a linear superposition of the values of the recorded image, which we write in the following form:

$$\tilde{f}(y) = \int dy \, g(x) M(x, y). \tag{10}$$

This estimator is characterized by the function $M(x,y)$, which we call the *reconstruction kernel* or else the *restoration function*. Substituting in Eq. (10) the imaging equation (8), we get

$$\tilde{f}(y) = \int dx \, M(x, y) \int dy' \, S(x, y') f(y') + \int dx \, n(x) M(x, y). \tag{11}$$

Reversing the order of integration and putting

$$T(y, y') = \int dx \, M(x, y) S(x, y') \tag{12}$$

we get equivalently

$$\tilde{f}(y) = \int dy' \, T(y, y') f(y') + \int dx \, n(x) M(x, y). \tag{13}$$

In the absence of noise, perfect restoration, with unlimited resolution, of the object $f(y)$ would correspond to $T(y, y') = \delta(y - y')$, where $\delta(y)$ denotes the Dirac distribution. This cannot be achieved in practice, however, because of the diffraction limit and of the presence of noise. Indeed, we have to keep the second term – the noise term – sufficiently small and, as we will see later, this implies a certain widening of the function $T(y, y')$. In the absence of noise or with a negligible noise term, this function represents the restoration at point y of a point source situated at point y'. Typically, for fixed y', $T(y, y')$ looks like a point spread function, namely a central peak with a few side-lobes. Its width characterizes the achievable resolution in the restored object, taking into account both the resolving power of the instrument, given by the PSF S, and the improvement due to the use of a restoration algorithm, characterized by the reconstruction kernel M. Therefore, following Gori and Guattari (1985), we call T the *overall PSF* or *overall impulse response*.

There are many possible choices for the restoration kernel M or, in other words, there are many linear restoration algorithms on the market. In the following, we will review some of the most common ones. As a matter of fact, there is no such thing as 'the best inversion algorithm'. Nevertheless, one can try to devise the optimal one according to some specific and well-defined criteria. The selection of the relevant criteria is then rather a matter of taste and should be matched with the specificities of the particular inverse problem one has to deal with. As we will see, the most important parameter in the game appears to be the signal-to-noise ratio.

For the case where the imaging operator is a true convolution, namely that we have to solve the equation

$$g(x) = \int dy\, S(x - y)f(y) + n(x), \tag{14}$$

a well-known restoration kernel is provided by the Wiener filter. In Fourier space this equation becomes

$$\hat{g}(\omega) = \hat{S}(\omega)\hat{f}(\omega) + \hat{n}(\omega). \tag{15}$$

Because of the space-invariance of the system, let us restrict ourselves to space-invariant restoration, i.e.

$$\tilde{f}(y) = \int dx\, M(y - x)g(x) \tag{16}$$

or in Fourier space

$$\hat{\tilde{f}}(\omega) = \hat{M}(\omega)\hat{g}(\omega). \tag{17}$$

In the case where the noise term is zero, one sees that perfect deblurring on the band is achieved by means of the inverse filter $\hat{M}(\omega) = [\hat{S}(\omega)]^{-1}$. Of course, this filter can be applied only on the frequency band $[-\Omega, +\Omega]$ where $\hat{S}(\omega)$ is different from zero. Outside this band, there is no hope of recovering the spectrum of the object, which is totally lost in the data. In other words, even in the noise-free case, the solution of the imaging equation (14) is clearly non-unique: there are many *transparent* or *invisible objects* (with zero spectrum on the band and arbitrary spectrum outside) that produce a zero image and, accordingly, there are many different objects corresponding to the same data. Moreover, if we apply the inverse filter to the noisy data of Eq. (15), the noise term will be considerably amplified for those frequencies where $\hat{S}(\omega)$ is very small. This will result in uncontrolled instabilities in the restored object. Therefore, the formal solution given by the inverse filter cannot be applied straightforwardly; it has to be replaced by a so-called *regularized algorithm* for which stability with respect to noise is guaranteed. An example is the so-called Wiener filter, which provides the best linear estimate in the least-squares sense of the solution of equation (14) (see e.g. Tikhonov and Arsenin 1977). It is given by

$$\hat{M}_\alpha(\omega) = \hat{W}_\alpha(\omega)\, \frac{1}{\hat{S}(\omega)} \tag{18}$$

with

$$\hat{W}_\alpha(\omega) = \frac{|\hat{S}(\omega)|^2}{|\hat{S}(\omega)|^2 + \alpha} \tag{19}$$

where $\alpha = \varepsilon^2/E^2$ is the ratio of the variance of the noise to the variance of the object. The quantity α^{-1}, sometimes called the signal-to-noise ratio, is assumed to be known. To derive this result one needs to assume that both the noise and the object are white-noise stochastic processes (if this is not the case, the parameter α is no longer a constant and should be replaced by the ratio of the corresponding power spectra). The function $\hat{W}_\alpha(\omega)$ can be viewed as a spectral window which has to be used to apodize the inverse filter $[S(\omega)]^{-1}$ in order to prevent noise amplification. Because of the presence of the positive parameter α in the denominator, and contrary to the inverse filter, the Wiener filter can be used for all frequencies, even where $\hat{S}(\omega)$ vanishes. For those frequencies where $\hat{S}(\omega)$ is very small, the presence of α prevents the uncontrolled amplification of the data noise that would occur with the inverse filter.

In a purely deterministic setting, i.e., without reference to stochastic concepts, the Wiener filter estimate given by Eqs. (18–19) can also be viewed as the solution of the variational problem which consists in minimizing the following functional

$$\Phi_\alpha[f] = \int_{-\infty}^{+\infty} dx\, |(Lf)(x) - g(x)|^2 + \alpha \int_{-\infty}^{+\infty} dx\, |f(x)|^2. \tag{20}$$

In other words, it coincides with the regularized solution provided by another well-known regularization method, the Tikhonov method, with a *regulariz-ation parameter* α. The second term in this expression can be interpreted as a penalization functional aimed at stabilizing a pure least-squares solution obtained by minimizing the first term alone (Tikhonov and Arsenin 1977, Groetsch 1984). Hence Tikhonov's method can be viewed as a constrained least-squares method. Various methods for choosing the regularization parameter have been proposed in the literature (Groetsch 1984). A common feature is that, as for the Wiener filter, α should always be appropriately related to the error level in the data.

It is easily seen that the overall PSF corresponding to Eq. (18) is space invariant and is given by the Fourier transform of the apodizing function (19). Clearly, this overall transfer function cannot be equal to one in the presence of noise and vanishes where the transfer function $\hat{S}$ is zero. Hence perfect deblurring is not achievable in the presence of noise and moreover we cannot recover the object spectrum outside the band $[-\Omega, +\Omega]$. Super-resolution – in the sense of out-of-band extrapolation as defined above – cannot be achieved in the case of a pure convolution.

To be able to extrapolate outside the band, we need something more, namely some further constraint on the object expressing some a priori information about its properties. For example, if we know that the object has finite spatial extent, i.e., vanishes outside some finite known domain, then, by the Paley–Wiener theorem, its Fourier transform $\hat{f}(\omega)$ is an entire analytic function (Paley and Wiener 1934). Thanks to the uniqueness of analytic continuation, such a function is (at least in principle) entirely determined by its values on the finite band $[-\Omega, +\Omega]$. The a priori knowledge about the domain of the object can be expressed by rewriting the imaging equation (14) as follows

$$g(x) = \int \mathrm{d}y\, S(x - y)P(y)f(y) + n(x) \qquad (21)$$

where $P(y)$ is the characteristic function of the domain of the object, i.e., $P(y) = 0$ where the object vanishes and $P(y) = 1$ where the object is expected to be different from zero. Let us remark that the localization of the object can also be expressed through more general and smooth profile functions $P(y)$ such as, e.g., a Gaussian function (Bertero *et al.* 1984). Then the imaging equation (21) also describes the case where the object is illuminated non-uniformly by a laser spot with Gaussian profile. For such a profile, the solution of the imaging equation with $n = 0$ is still unique, but uniqueness no longer holds for certain profiles such as band-limited functions (see below).

When $P(y)$ is zero outside some finite domain or decreases sufficiently fast at infinity, the imaging operator acquires the mathematical property of being compact in an appropriate function space (such as, e.g., the space of square-integrable objects). Compact operators have spectral properties similar to those of finite-dimensional matrices, namely they have a discrete spectrum. Compact self-adjoint operators can be expressed in diagonal form by means of their eigensystem. Imaging operators, however, are not necessarily self-adjoint (e.g., because the image and object have different domains). Then, instead of its eigensystem, one uses the so-called *singular system* of the imaging operator. Let us recall that the singular system of L solves the following 'shifted eigenvalue problem' (L^* denotes the adjoint operator):

$$L v_k = \sigma_k u_k; \qquad L^* u_k = \sigma_k v_k. \tag{22}$$

The singular values σ_k are real and positive by definition, and ordered as follows: $\sigma_1 \geqslant \sigma_2 \geqslant \sigma_3 \geqslant \dots$. The singular functions or vectors u_k, eigenvectors of LL^*, constitute an orthonormal basis in the space of all possible images, whereas the singular functions or vectors v_k, eigenvectors of L^*L, form an orthonormal basis in the set of all visible objects. Hence, data and solutions can be expanded on the singular functions of L, and restoration kernels as well. A widely used kernel corresponds to sharp truncation of the singular-system expansion

$$M(x, y) = \sum_{k=1}^{N} \frac{1}{\sigma_k} u_k(x) v_k(y) \tag{23}$$

corresponding to the following overall PSF

$$T(y, y') = \sum_{k=1}^{N} v_k(y) v_k(y'). \tag{24}$$

The need for truncation arises from the fact that the singular values of a compact operator usually accumulate to zero, resulting in an unacceptable amplification of the noise term in Eq. (13), unless the singular values close to zero are ruled out. For example, one keeps only the terms corresponding to singular values larger than the square root of α (it can be shown that among all truncated solutions obtained through Eq. (23), this stopping criterion yields the minimum of the least-squares error – see Bertero *et al.* 1979). The resulting number N of terms in the expansion of the restored object can be considered as its number of degrees of freedom. Indeed, since the expansion is orthogonal, N represents the number of independent 'pieces of information' about the object which can be reliably (i.e., stably) retrieved from the noisy data. As we will see in the next section, the number of degrees of freedom, which depends on the operator L and on the value of α, can be viewed as a generalization of the Shannon number expressing the information content of a band-limited signal.

Let us mention that the method also applies to the very important case where the data are discrete. In practice, data are always recorded by a finite set of detectors and, instead of a continuous image, one measures a finite-dimensional data vector g. The imaging operator is then automatically compact (even without any assumption on the domain of the object) and the above formulas still hold provided that the integrals on the x-variable are replaced by discrete sums on the components of the data vector and of the singular vectors u_k. In such a case, it is in general rather easy to compute the singular system numerically. Since, moreover, for a given imaging operator L, it can be computed once for all and then stored, the restoration kernel, Eq. (23), provides fast numerical algorithms for recovering an estimate of the object.

Instead of sharp truncation of the singular-system expansion, one can filter the smallest singular values more smoothly, by introducing gently decreasing weighting factors $W_k(\alpha)$ in the expansions (23) and (24). For example, one can use the classical window shapes of Hanning or Hamming, a triangular filter, and also the following Tikhonov filter, analogous to Eq. (19) and providing the minimum of the functional (20) for a compact operator L:

$$W_k(\alpha) = \frac{\sigma_k^2}{\sigma_k^2 + \alpha}. \tag{25}$$

Again the ultimate resolution capabilities of the instrument can be assessed by looking at the overall PSF $T(y, y')$. When the imaging equation $Lf = g$ has a unique solution, then, in the absence of noise, N tends to $+\infty$ and the singular system provides a resolution of the identity, i.e., $T(y, y') = \delta(y - y')$, which yields a perfect restoration. In the presence of noise, however, the expansion has to be truncated to avoid instabilities, and accordingly, $T(y, y')$ acquires a certain width depending on the signal-to-noise ratio. Notice that whenever the solution of the imaging equation is not unique (and in particular in the case of discrete data), the overall PSF always has a finite width, even in the absence of noise, because of the existence of transparent objects that can never be retrieved from the data.

HOW TO GET SUPER-RESOLUTION

Object restoration beyond the Rayleigh limit

Consider the problem of restoring a uniformly illuminated object with finite extent $[-Y, +Y]$ from its image through an ideal diffraction-limited system of bandwidth Ω. This means that we have to solve the imaging equation (21) where S is given by Eq. (5) and where the profile $P(y)$ is the characteristic function of the interval $[-Y, +Y]$. We want to perform an out-of-band extrapolation, i.e., to recover at least part of the spatial frequencies beyond the cut-off frequency Ω. As already observed, the Fourier transform of a finite-extent object is an entire analytic function and hence the uniqueness of analytic continuation ensures that the solution of the noise-free imaging equation is also unique. The singular system of the compact operator L is constituted by the so-called prolate spheroidal wave functions and the singular values are the square roots of the prolate eigenvalues λ_k of Slepian and Pollack (1961). Thanks to the published numerical values of these eigenvalues (see e.g. Frieden 1971), we can easily estimate the number of degrees of freedom for a given signal-to-noise ratio, i.e., the number of values λ_k greater than α, the inverse of the signal-to-noise ratio. The values λ_k depend on Ω and Y only through the space-bandwidth product $c = \Omega Y$. When c is sufficiently larger than one, they present a characteristic stepwise behaviour: approximately equal to 1 for $k < S$, they fall off very rapidly to zero for $k > S$. The number

$$S = \frac{2c}{\pi} = \frac{2\Omega Y}{\pi}, \tag{26}$$

called the Shannon number by Toraldo di Francia (1969), represents the number of sampling points, spaced by the Rayleigh distance $R = \pi/\Omega$, interior to the domain of the object $[-Y, +Y]$. Hence, for large c and realistic signal-to-noise ratios, the number of degrees of freedom N is approximately equal to S. Consequently, in that case, no significant out-of-band extrapolation can been achieved and the Rayleigh limit cannot really be improved. This conclusion no longer holds, however, when c is small, i.e., when the size $2Y$ of the object is small compared to the Rayleigh distance π/Ω. Then the fall-off of the eigenvalues λ_k is much smoother and the resulting number of degrees of freedom can be significantly greater than S. If $N = N(E/\varepsilon, c)$ is the number of degrees of freedom, the new achievable resolution distance is given by $R' = 2Y/N(E/\varepsilon, c)$ and the amount of super-resolution can be quantified by the ratio

$$\frac{R'}{R} = \frac{S}{N(E/\varepsilon, c)}. \tag{27}$$

In Table 1, we report some numerical figures for this ratio.

For example, we see from this table that for $c = 1$ and $E/\varepsilon = 10^2$, we get a

Table 1. Resolution ratio R'/R for the problem of out-of-band extrapolation, as a function of the signal-to-noise ratio $(E/\varepsilon)^2$ and for different values of the space-bandwidth product c.

$c = 1, S = 0.64$		$c = 2, S = 1.27$		$c = 5, S = 3.18$		$c = 10, S = 6.36$	
E/ε	R'/R	E/ε	R'/R	E/ε	R'/R	E/ε	R'/R
10	0.32	10	0.42	10	0.64	10	0.71
10^2	0.21	10^2	0.32	10^2	0.45	10^2	0.64
10^3	0.16	10^3	0.25	10^3	0.40	10^3	0.53
10^4	0.13	10^4	0.21	10^4	0.35	10^4	0.49
10^5	0.13	10^5	0.18	10^5	0.32	10^5	0.42
10^6	0.11	10^6	0.16	10^6	0.29	10^6	0.40

resolution of one fifth of the Rayleigh distance. Equivalently, it means that $\hat{f}(\omega)$ is extrapolated from $[-\Omega, +\Omega]$ into $[-\Omega', \Omega']$ with $\Omega' = 5\Omega$. Notice that the amount of achievable super-resolution depends in a critical way on the space-bandwidth product $c = \Omega Y$. It also depends on the signal-to-noise ratio, even if this dependence is rather weak. Indeed, it is possible to show that the number of degrees of freedom behaves as a linear function of the logarithm of the signal-to-noise ratio. Similar features also hold true for incoherent and 2D imaging.

The confocal scanning light microscope

According to the conclusions of the above analysis, significant super-resolution can be achieved only when the space-bandwidth product is rather low. This certainly does not happen in ordinary optical microscopes, for which the resolving power is essentially determined by the Rayleigh criterion. In a scanning microscope, however, for each scanning position, a very small portion of the object is illuminated, resulting in a low space-bandwidth product. In a confocal scanning microscope (see e.g. Wilson and Sheppard 1984), the specimen is put under a sharp laser spot focused by a first lens – the illumination lens – and then the image is formed by means of a second lens – the *imaging* or *collector lens* – and recorded by a detector (e.g. a photo-multiplier). A small pinhole placed before the detector allows us to achieve almost pointwise detection. A scan through the object is then performed by translating the specimen in the object plane, which is the common focal plane of the lenses. Confocal microscopes are operated either in transmission or reflection mode, with coherent image formation, or else in fluorescence mode, with incoherent image formation.

For a fixed scanning position, the imaging process can be modelled by an equation similar to Eq. (21), namely:

$$g(x) = \int \mathrm{d}y \, S_2(x - y)S_1(y)f(y) \tag{28}$$

where the object is first multiplied by the illumination profile, which is the PSF S_1 of the first illuminating lens, and then convolved with the PSF S_2 of the second lens. In many situations, and surely in the reflection mode where the two lenses coincide, one can assume that these two PSF are equal. In 1D ideal coherent imaging, both S_1 and S_2 are then given by Eq. (5) and in this case, a nice feature happens, namely that the singular system of the compact imaging operator defined by Eq. (28) can be determined analytically (Gori and Guattari 1985).

The images formed at the different scanning positions are given by

$$g(x, t) = \int \mathrm{d}y \, S_2(x - y)S_1(y)f(y + t) \tag{29}$$

where t is the variable indicating the scanning position. In a conventional confocal scanning microscope, a single detector is used and, for each scanning position, the image is recorded only on the optical axis, i.e., at $x = 0$. No data processing is performed and the unknown object at a given point t is reconstructed just as $\tilde{f}(t) = g(0, t)$, i.e., as the central value of the image for the same scanning position. Putting $x = 0$ in Eq. (29), we see that

$$\tilde{f}(t) = \int dy \, S_2(-y)S_1(y)f(y + t) \tag{30}$$

or by a change of variables

$$\tilde{f}(t) = \int dt' \, H(t - t')f(t') \tag{31}$$

where $H(t) = S_2(t)S_1(-t)$. Hence we see that the recorded image $\tilde{f}(t) = g(0, t)$ is simply the convolution of the object by the PSF $H(t)$. Accordingly, the spatial-frequency cut-off of such a confocal system is determined by the cut-off of the convolution of the functions $\hat{S}_1(-\omega)$ and $\hat{S}_2(\omega)$. This leads to a resolution enhancement with respect to an ordinary microscope using uniform illumination of the object and only one imaging lens with PSF S_2. For example, in coherent imaging, if the two transfer functions $\hat{S}_1$ and $\hat{S}_2$ are both given by Eq. (4), $\hat{H}$ is a triangle with basis 2Ω, i.e., we get twice the cut-off frequency of an ordinary microscope with the same imaging lens. Unfortunately, this transfer function becomes very small near the edges of its domain so that the achievable resolution improvement is limited in practice by the noise, and the effective gain in resolving power is close to a factor 1.4 (Cox *et al.* 1982; Brakenhoff *et al.* 1979).

A way of further enhancing the resolving power of a confocal microscope has been suggested and investigated theoretically in a paper by Bertero *et al.* (1984) and in a series of subsequent papers (1987a, 1987b, 1989, 1991, 1992). The idea is to replace the single on-axis detector by an array of detectors in order to measure, at each scanning position, the complete diffraction image (and not only its value on the optical axis). In other words, one should record the image g as a function of both variables x and t and use the imaging equation (29) to reconstruct the object. The linear estimator we use for this is assumed to be shift invariant with respect to the scanning position

$$\tilde{f}(t') = \int dx \int dt \, M(x, t' - t)g(x, t) \tag{32}$$

but not with respect to the detector position since the integration on x may be limited to some finite domain D. In the absence of noise, one can devise different choices of the restoration function leading to the following overall PSF

$$T(y, y') = \frac{\sin[2\Omega(y - y')]}{\pi(y - y')} \tag{33}$$

i.e., to an overall transfer function which is the characteristic function of the band $[-2\Omega, +2\Omega]$. This corresponds to an enhancement of the resolving power by a true factor 2 with respect to an ordinary microscope with a PSF given by Eq. (5). A first possibility relies on the exact inversion formula for the imaging equation (28) derived by Bertero *et al.* (1987b):

$$M(x, t) = \frac{4\pi}{\Omega} \cos(\Omega x)\delta(t). \tag{34}$$

In this case, the object at point t is reconstructed only from the data at the same scanning position t and the multiplication by $(4\pi/\Omega)\cos(\Omega x)$ could also be implemented optically by means of a mask (Bertero *et al.* 1992). However, besides Eq. (34) and because of the redundancy of the data in Eq. (29), a whole family of reconstruction kernels can be constructed, all yielding Eq. (33) as overall PSF (Defrise and De Mol 1992), including the following one,

$$M(x, t) = \frac{\pi}{\Omega} \delta\left(t - \frac{x}{2}\right) \tag{35}$$

first proposed by Sheppard (1988). In principle, these reconstruction formulas hold true only for continuous noise-free data, but numerical simulations have been done in more realistic situations: noisy data and a finite (small) number of detectors. In such cases, one uses the truncated singular-system kernel Eq. (23) and the simulations show that in practice the resulting overall PSF Eq. (24) is quite close to the corresponding ideal case Eq. (33) (Bertero *et al.* 1987a). Numerical simulations have also been done for 2D imaging with circular pupils (Bertero *et al.* 1991) and for incoherent imaging (Bertero *et al.* 1989, 1991). All these results lead to the conclusion that a significant enhancement in resolving power can be obtained in confocal microscopy through the use of a multidetector data recording scheme and the implementation of adequate data inversion algorithms.

Near-field imaging

As a third example, let us briefly mention, another possibility for enhancing the resolution limits. It consists in recording – when possible – near-field data, instead of using far-field data as in the previous examples. Strictly speaking, this is not true super-resolution in the sense defined above of out-of-band extrapolation. Indeed, the near-field data contain information – conveyed by the so-called evanescent waves – about the spatial frequencies of the object that are no longer visible in the far-field region. Nevertheless, we think that this alternative way of beating the Rayleigh limit is worth mentioning to conclude this chapter, because of the growing importance of near-field imaging techniques, such as scanning near-field optical microscopy (SNOM – see e.g., Betzig and Trautman 1992, Pohl and Courjon 1993, and the references therein) and near-field acoustic holography (NAH – see Williams and Maynard 1980).

A good theoretical laboratory for investigating the resolution enhancement arising from the effect of evanescent waves is provided by the so-called inverse diffraction problem. It consists in back-propagating (towards the sources) a scalar field propagating in free space according to the Helmholtz equation. The simplest geometry is the case where the field propagates in the half-space $z \geq 0$ and where one has to recover the field on the boundary plane $z = 0$ from its values on the plane $z = a > 0$. The properties of the corresponding imaging operator are studied in a paper by Bertero *et al.* (1983). For the one-dimensional situation of a field amplitude invariant with respect to one of the lateral variables and contained in a slit of width $2Y$ in the boundary plane, the imaging equation is of the form Eq. (21) with $P(y)$ equal to the characteristic function of $[-Y, +Y]$ and $S(x)$ to the forward propagator

$$S(x) = \frac{1}{2\pi} \int_{-\infty}^{+\infty} d\omega \, e^{i\omega x} \, e^{ima} \tag{36}$$

where $m = \sqrt{k^2 - \omega^2}$ for $\omega^2 \leq k^2$ (homogeneous waves) and $m = i\sqrt{\omega^2 - k^2}$ for $\omega^2 > k^2$ (evanescent waves). The constant $k = 2\pi/\lambda$ is the wavenumber.

When the data plane is situated in the far-field region ($a \gg \lambda$), the inverse imaging problem is essentially equivalent to the object restoration problem considered above, with $c = kY$ and the Rayleigh distance given by $R = \pi/k = \lambda/2$. For near-field data, the number of degrees of freedom $N(E/\varepsilon, c, a)$ and the corresponding resolution ratio R'/R can be assessed by means of a numerical computation of the singular values of the imaging operator (Bertero *et al.* 1983). To quote an example, for $Y = \lambda/2$ and $E/\varepsilon = 10^2$, this ratio is equal to 0.4 as long as $a \geq \lambda/2$, to 0.25 for $a = \lambda/4$ and drops to 0.125 for $a = \lambda/10$. For a larger size of the object, the effect of evanescent waves can be felt further away from the boundary plane, but not further than a few wavelengths. As in the far-field, super-resolution is made possible by the a priori knowledge of the domain of the object and appears to be much easier for small values of $c = kY$, i.e., for objects of size comparable to the wavelength of the field.

ACKNOWLEDGMENTS

The author is *Maître de recherches* with the Belgian National Fund for Scientific Research.

REFERENCES

Bertero, M. and E.R. Pike, *Optica Acta* **29**, 727–46 and 1599–611 (1982).
Bertero, M., C. De Mol and G.A. Viano, *J. Math. Phys.* **20**, 509–21 (1979).
Bertero, M., C. De Mol and G.A. Viano, in: *Inverse Scattering Problems in Optics* ed. H.P. Baltes (Berlin: Springer-Verlag): Chapter V, pp. 161–214 (1980).

Bertero, M., C. De Mol, F. Gori and L. Ronchi, *Optica Acta* **31**, 1051–65 (1983).

Bertero, M., C. De Mol, E.R. Pike and J.G. Walker, *Optica Acta* **31**, 923–46 (1984).

Bertero, M., P. Brianzi and E.R. Pike, *Inverse Problems* **3**, 195–212 (1987a).

Bertero, M., C. De Mol and E.R. Pike, *J. Opt. Soc. Am. A* **4**, 1748–50 (1987b).

Bertero, M., P. Boccacci, M. Defrise, C. De Mol and E.R. Pike, *Inverse Problems* **5**, 441–61 (1989).

Bertero, M., P. Boccacci, R.E. Davies and E.R. Pike, *Inverse Problems* **7**, 655–74 (1991).

Bertero, M., P. Boccacci, R.E., Davies, F., Malfanti, E.R., Pike and J.G. Walker, *Inverse Problems* **8**, 1–23 (1992).

Betzig, E. and T.D. Trautman, *Science* **257** 189–95 (1992).

Born, M. and E. Wolf, *Principles of Optics* (Sixth Edition) (Oxford: Pergamon Press) (1980).

Brakenhoff, G.J., P. Blom and P. Barends, *J. Microsc.* **117**, 219–32 (1979).

Cox, I.J., C.J.R. Sheppard and T. Wilson, *Optik* **60**, 391–96 (1982).

Defrise, M. and C. De Mol, *Inverse Problems* **8**, 175–85 (1992).

Frieden, B.R., Evaluation, design and extrapolation for optical signals, based on the use of prolate functions, in: *Progress in Optics* vol IX, ed. E. Wolf (Amsterdam: North Holland): pp. 311–407 (1971).

Goodman, J.W., *Introduction to Fourier Optics* (New York: McGraw-Hill) (1968).

Gori, F. and G. Guattari, *Inverse Problems* **1**, 67–85 (1985).

Groetsch, C.W., *The Theory of Tikhonov Regularization for Fredholm Equations of the First Kind* (Boston: Pitman) (1984).

McCutchen, C.W., *J. Opt. Soc. Am.* **57**, 1190–92 (1967).

Paley, R.E.A.C. and N. Wiener, *Fourier Transforms in the Complex Domain* (Providence, RI: Amer. Math. Soc.) (1934).

Papoulis, A., *Systems and Transforms with Applications in Optics* (New York: McGraw-Hill) (1968).

Pohl, D.W. and D. Courjon, eds, *NATO Advanced Research Workshop on Near-Field Optics* (Dordrecht: Kluwer) (1993).

Rayleigh, J.W.S., *Phil. Mag.* **8**, 261 (1879).

Rushforth, C.K. and R.W. Harris, *J. Opt. Soc. Am.* **58**, 539–45 (1968).

Shannon, C.E., *Proc. IRE* **37**, 10 (1949).

Sheppard, C.J.R., *Optik* **80**, 53–4 (1988).

Slepian, D. and H.O. Pollack, *Bell Syst. Tech. J.* **40**, 43–64 (1961).

Tikhonov, A.N. and V.Y. Arsenin, *Solutions of Ill-posed Problems* (Washington: Winston/Wiley) (1977).

Toraldo di Francia, G., *J. Opt. Soc. Am.* **45**, 497 (1955).

Toraldo di Francia, G., *J. Opt. Soc. Am.* **59**, 799 (1969).

Whittaker, E.T., *Proc. Roy. Soc. Edinburgh, Sect. A* **35**, 181 (1915).

Williams, E.G. and J.D. Maynard, *Phys. Rev. Lett.* **45**, 554–7 (1980).

Wilson, T. and C. Sheppard, *Theory and Practice of Scanning Microscopy* (London: Academic Press) (1984).

16 Electronic speckle pattern interferometry: an aid in cultural heritage protection*

Giuseppe Schirripa Spagnolo
*Dipartimento di Energetica, Università di L'Aquila,
Località Monteluco di Roio, 67040 Roio Poggio, (AQ-Italy)*

INTRODUCTION

The proper time of a work of art is a never-ending present but its matter (e.g., marble, stone, wood and so on) is ruined by the passing of the years. Even if mankind has been restoring art for centuries, it was often done in the wrong way. Since Roman times restoring meant remaking and the most important evidence of this approach is the systematic sack of ancient art (Lanciani 1901). Without examining the problems of modern restoring theory, we recall some fundamental points. According to C. Brandi (1977), who founded the Italian Istituto Centrale per il Restauro in 1939, only the matter of a work of art has to be restored and the restoring itself has to aim at re-establishing the potential unity of the work of art, if possible, without realizing an artistic or historic fake and without cancelling the work of art marks due to its passage through time. This very difficult task needs a knowledge of the artefact as complete as possible. In the last 35 years advances in technology have created tools and diagnostic methods that are revolutionizing the art conservators' field. Scanning electron microscopy, X-ray fluorescence and radiography, mass spectrometry, ultraviolet and infrared imaging, computer-aided classification and image enhancement, colorimetry, tomography, neutron activation, and ultrasound are just some of the 'Space Age' tools being used by the art conservation community, as well as by museums, researchers, and historians.

---

* The colour plate section for this chapter appears between p. 304 and p. 305.

Optical techniques, such as holographic and speckle interferometry, provide means by which one can obtain full-field, quantitative measurements of deformations of objects under different loading conditions. The non-destructive, non-intrusive and non-contact nature of such techniques makes them suitable and particularly attractive as diagnostic tools.

We first observe that some researches in art conservation are of a rather general nature. They refer, for example, to the study of the behaviour of typical materials used in artworks or in restoration, say marble, wood and canvas, under the action of chemical or biological agents as well as mechanical or thermal stresses. Very often, the modifications induced in these materials by the applied agent finally lead to morphological changes of the specimen under test. In these cases, holographic and speckle interferometry can be used as a tool for detecting such changes.

Wooden panels and frescoes are artefacts particularly suitable for testing using optical techniques. In fact a painting on wood or on a wall can be considered as a layered structure with a support. The wooden or mural support is coated with some priming layers or plaster, which serve as a base for the painting. These layers, normally made of a mixture of gesso and glue, are thinner and more fragile than the support. Expansion and contraction of the support due to daily fluctuations of ambient parameters can produce large strains and eventually cracks in the priming layers, as they become less flexible with age. Furthermore, abrupt changes of temperature and humidity, traffic induced vibrations, and heat exposure may also cause unpredictable stress distributions in the heterogeneous materials of the support with consequent damage of the painted surface. All these mechanisms may lead to the formation of detachment and cracks which are very common in mural and wooden paintings. For conservation purposes the knowledge of incipient and invisible flaws and of the artefact deformation caused by ambient drifts is very important, as well as the understanding of how the presence of the support cracks or discontinuities alter the movements of the painted surface. A further necessity is to prove the effectiveness of the repair work after restoration.

Since 1974, holographic interferometry has been applied to non-destructive testing problems on a variety of works of art (Amadesi *et al.* 1974, Westlake *et al.* 1976, Amadesi *et al.* 1982, Bertani *et al.* 1982). Generally, the visual observation of the fringe pattern covering the three-dimensional holographic image of the specimen furnishes qualitative information about the overall deformation of the complex structure as well as the location of strain or stress concentration (defects). However, the artworkers have seldom applied these techniques for routine inspections. This can be attributed to a variety of factors, from the sceptical attitude towards new technologies to economical aspects. A major problem was the necessity of moving the object to optical laboratories. In fact, the practical application of holographic interferometry has some disadvantages, which have, in general, hindered its widespread use *in situ* as a non-destructive testing tool. The mechanical stability demanded by holographic measurements is not readily compatible with in-field conditions unless pulsed lasers are used; furthermore, the optical alignment of a holographic set-up, and the development and reconstruction of holograms, are time-consuming processes, requiring skilled operators. To have a testing technique able to visualize deformations with interferometric sensitivity, that provides real-time analysis of micro displacements, and offers the possibility of operating *in situ* during the several phases of the restoration process is very important to the people involved in conservation and restoration of works of art. It is quite obvious that the test technique should also be economic and easy to use.

The only candidate capable of fulfilling these requirements is the technique called electronic speckle pattern interferometry (ESPI) (Løkberg 1980). Measuring instruments based on ESPI have been proposed recently (Gülker *et al.* 1990, Paoletti and Schirripa Spagnolo 1993, Paoletti *et al.* 1993, Zanetta 1994) and might become very popular among artwork restorers, allowing a rapid and real-time diagnosis of several types of defects with the same sensitivity of holographic techniques also in non-laboratory conditions. The aim of this chapter is to give a survey of the several practical applications of an electronic speckle pattern interferometer in the artwork preservation field. By starting from a rapid analysis of a typical experimental apparatus, we report on the results of investigations carried out on some test models and subsequently on some ancient artefacts with a particular emphasis on the possibility of operating *in situ* during a restoration.

BASIC PRINCIPLES OF ESPI

The potentialities of electronic speckle pattern interferometry (ESPI) for diagnostics in the art conservation field are of interest not only for optics, but also for other branches of science; for this reason, and for completeness, a brief description of the basic notions of ESPI is given; for a more thorough explanation see, e.g., Jones and Wykes (1989).

Speckle patterns are random intensity distributions, produced by a coherent light source, such as a laser, as it illuminates a diffusely reflecting body (Goodman 1975). There are two types of fundamentally different speckle patterns; photographic speckle patterns which contain information only about the light wave amplitude, and holographic speckle patterns which contain both phase and amplitude information simultaneously. Photographic speckle patterns are obtained when a diffusely scattering surface of an object is illuminated by a single coherent beam and the resulting scattered wavefront is recorded on a photographic emulsion. Holographic speckle patterns are obtained when the scattered wavefront emanating from the diffuse surface of the object is recorded simultaneously with a reference wavefront on the photographic emulsion.

The principle of the standard electronic speckle pattern interferometry (ESPI) technique is based on the recording of a holographic speckle pattern sequence on the photosensor of a TV camera.

ESPI was developed in the early 1970s as a method of producing interferometric data without using traditional holographic techniques (Butters and Leendertz 1971, Macovski *et al.* 1971, Schwomma 1972, Wykes 1982).

It is possible to describe ESPI as image holography (an image hologram is recorded of a real image of the object instead of the object itself), with an in-line reference beam, where the TV target has replaced film as the recording medium. Obviously, the photosensor of the TV camera is not suitable for optical reconstruction of the hologram, therefore the reconstruction process is performed electronically and visualized on a monitor.

The image interferogram is converted into a corresponding video signal by the scanning action of a video camera. This video signal is electronically processed, through an intermediate recording medium (commonly a frame grabber), before being displayed on a TV screen, so that the variations in the texture of the speckles are converted into a variation of brightness. This image is entirely equivalent to a holographic reconstructed image and possesses the same interferometric sensitivity.

Any deformation of the object under test changes the distribution of the speckles; two consecutive holographic speckle patterns combined together will generate correlation fringes corresponding to the deformation field. These correlation fringes, in which each fringe represents a line of constant displacement, can be displayed in real time as the object is deformed.

We consider now an interferometer like the one presented in Fig. 1.

The complex amplitudes of the object and reference beams can be expressed,

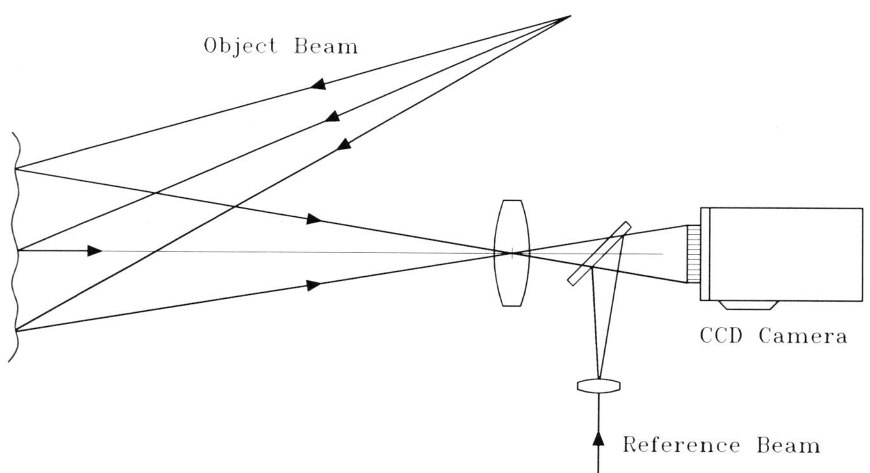

Fig. 1. A typical arrangement to realize standard ESPI interferograms.

at any point (x, y) on the image plane, as

$$U_O(x, y) = |U_O(x, y)| \exp[i\phi_O(x, y)] \tag{1}$$

$$U_R(x, y) = |U_R(x, y)| \exp[i\phi_R(x, y)] \tag{2}$$

respectively, where $|U_O(x, y)|$ and $|U_R(x, y)|$ are the real amplitudes, $\phi_O(x, y)$ is the phase of the light scattered from the object surface, randomly varying across the image plane, and $\phi_R(x, y)$ is the phase associated with the reference beam. The resulting holographic speckle pattern, obtained with the test surface in its equilibrium condition (e.g., in the condition that we assume as a reference state), presents an intensity I_{eq}. This intensity distribution, picked up by the camera target, is given by the expression

$$I_{eq}(x, y) = |U_O(x, y)|^2 + |U_R(x, y)|^2$$
$$+ 2|U_O(x, y)||U_R(x, y)|\cos[\phi_O(x, y) - \phi_R(x, y)]. \tag{3}$$

When the object surface is displaced or deformed, the object beam is represented by

$$U_O(x, y) = |U_O(x, y)| \exp\{i[\phi_O(x, y) + \theta(x, y)]\}. \tag{4}$$

Consequently, the intensity becomes I_{def}, where

$$I_{def}(x, y) = |U_O(x, y)|^2 + |U_R(x, y)|^2$$
$$+ 2|U_O(x, y)||U_R(x, y)|\cos[\phi_O(x, y) - \phi_R(x, y) - \theta(x, y)]. \tag{5}$$

If the phase change $\theta(x, y)$ is not significantly large, the scattered object beam amplitude ($|U_O(x, y)|$), the reference beam amplitude ($|U_R(x, y)|$), and the phase difference between object and reference beam $[\phi_O(x, y) - \phi_R(x, y)]$ are assumed to be the same in Eqs. (3) and (5).

It can be seen that where $\theta(x, y) = 2n\pi$, then I_{eq} and I_{def} are equal, but for other values of $\theta(x, y)$ the two values are different. Thus, if one image is subtracted from the other, those regions where $\theta(x, y) = 2n\pi$ will have zero intensity, while the intensity in other regions varies randomly (i.e., a speckle pattern is obtained) (Wykes 1987). To subtract one image from the other, the video signal from the TV camera is digitized and stored by a frame buffer and a computer. If the video signal detected by the TV camera is proportional to the light intensities of the speckle image, the output signal from the TV camera, $B_n(x, y)$, corresponding to the light exposure $I_n(x, y)$ of the nth video frame, may be expressed as:

$$B_n(x, y) \propto I_n(x, y). \tag{6}$$

These image patterns can be digitized and stored in a frame grabber, so that

$$B_{eq}(x, y) \propto I_{eq}(x, y) \quad \text{and} \quad B_{def}(x, y) \propto I_{def}(x, y) \tag{7}$$

can be subtracted to obtain speckle fringes. The magnitude of the difference between the two speckle patterns can be displayed on a video monitor. Since the video monitor responds only to positive signals, the subtraction signal is square-law detected before being displayed on the TV monitor. The brightness on the video monitor can be expressed by

$$
\begin{aligned}
B(x, y) &= [B_{\text{def}}(x, y) - B_{\text{eq}}(x, y)]^2 \\
&\propto [I_{\text{def}}(x, y) - I_{\text{eq}}(x, y)]^2 = N(x, y)S(x, y),
\end{aligned} \tag{8}
$$

where

$$
N(x, y) = 8\,|U_{\text{O}}(x, y)|^2\,|U_{\text{R}}(x, y)|^2\,\sin^2\left\{[\phi_{\text{O}}(x, y) - \phi_{\text{R}}(x, y)] + \frac{\theta(x, y)}{2}\right\} \tag{9}
$$

$$
S(x, y) = 1 - \cos[\theta(x, y)]. \tag{10}
$$

Here $N(x, y)$ represents the speckle noise, and $S(x, y)$ describes the fringe pattern.

Equation (8) represents a typical ESPI interferogram obtained by subtraction techniques. This interferogram is afflicted with speckle noise that appears as a multiple of tiny spots of varying intensity, superimposed on the true image (the fringe pattern).

However, assuming that the fringe spacing is large enough compared with the mean speckle size, this noise can be considered to be a high-frequency random noise term. Therefore, random speckle noise can be smoothed by using a low-pass filter and a convolution or median window. If we perform an average of the Eq. (8) over all speckle phases, and consider the statistical properties of the speckle pattern we obtain

$$
\begin{aligned}
\langle B_{\text{s}}(x, y)\rangle &\propto \langle [I_{\text{def}}(x, y) - I_{\text{eq}}(x, y) + \Delta E(x, y)]^2\rangle = \langle N(x, y)S(x, y)\rangle \\
&= \langle\,|U_{\text{O}}(x, y)|^2\rangle\langle\,|U_{\text{R}}(x, y)|^2\rangle\{1 - \cos[\theta(x, y)]\}.
\end{aligned} \tag{11}
$$

The angular brackets denote ensemble averaging.

Equation (11) is equivalent to the classical interferometry equation and describes perfect correlation fringes without speckles.

To obtain Eq (8) we have considered $(|U_{\text{O}}(x, y)|)$ to be the same in Eqs. (3) and (5). In practice, the speckles are decorrelated due to object deformation as the effective rays collected by the imaging lens are not the same for the first and the second frame. As a consequence, the object beam amplitude term $(|U_{\text{O}}(x, y)|)$ in Eqs (3) and (5) is not identical and does not cancel out in Eq. (8). Therefore, the visibility of the fringes is reduced. To take this problem into account, the function $\langle B_{\text{s}}(x, y)\rangle$ can be written as

$$
\langle B_{\text{s}}(x, y)\rangle = \langle\,|U_{\text{O}}(x, y)|^2\rangle\langle\,|U_{\text{R}}(x, y)|^2\rangle\{1 - \gamma\cos[\theta(x, y)]\} \tag{12}
$$

where γ is a visibility factor depending on the speckle size and the object deformation.

Equation (12) represents the smoothed ESPI interferogram. Since the phase change $\theta(x, y)$ is a function of the displacement of the surface, information about the relative displacement of different parts of the surface can be obtained.

Plate 16.1. Wooden panel Trionfo di Davide. (a) Inspected area; (b) ESPI interferogram before restoration; (c) ESPI interferogram after restoration.

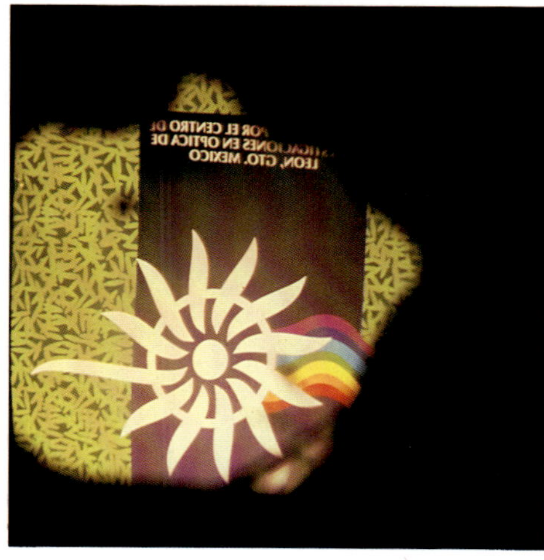

Plate 22.1. Image of printed letters and colour symbol reflected from an Olmec mirror. Photographed by the author.

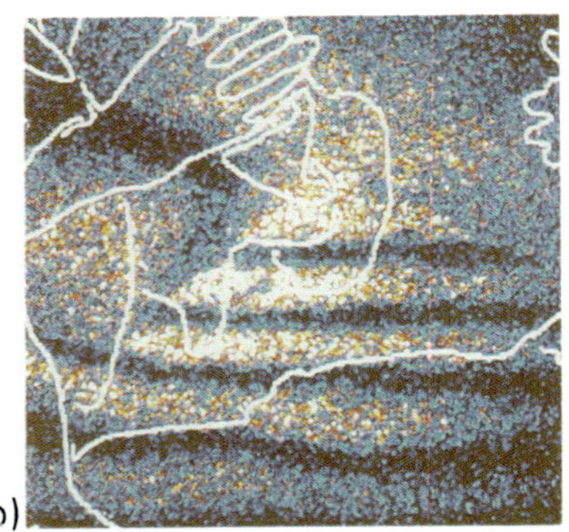

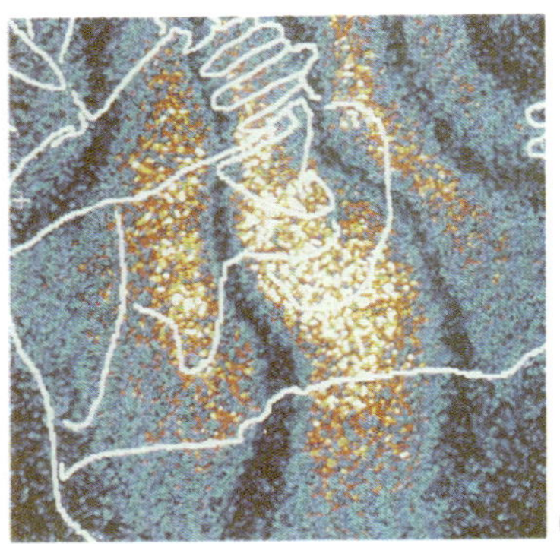

Plate 16.2. Fresco Adorazione dei Magi by Farrelli. (a) Inspected area; (b) ESPI interferogram before restoration; (c) ESPI interferogram after restoration.

Plate 29.1. The retrieved colour image from an encoded black-and-white film.

EXPERIMENTAL ARRANGEMENT FOR IN-FIELD OPERATION

The ESPI system, because of its speed and ease of use, may become a routine technique in diagnostics of the state of preservation of ancient artworks of several kinds. The use of optical fibres in digital speckle interferometry leads to a drastic reduction in the number of optical components (beam splitters, spatial filters, mirrors) and to a simpler experimental arrangement procedure in hostile conditions (Løkberg and Krakhella 1981, Davies *et al.* 1988, Paoletti and Schirripa Spagnolo 1993, Valera *et al.* 1993). Indeed, the realization of all-fibre optical components has allowed the implementation of compact and portable interferometer configurations, impractical using conventional optics, with resulting substantial improvements in system stability, ease of alignment and system cost, and hence not restricted to research laboratories. Besides, the possibility of using laser diodes, with a high level of temporal coherence (which can be obtained by controlling the temperature of the diode junction) as light sources frees us from the limitations on the path difference between reference and object beams so that the distance between the TV camera and the object can be changed over a wide range with a considerable increase in flexibility of in-field use (Virdee *et al.* 1990, Ford *et al.* 1993, Kato *et al.* 1993).

A basic fibre optic arrangement for a conventional ESPI system, with an object beam approximately collinear with the viewing axis combined with a conjugate reference beam, is shown in Fig. 2. The configuration is analogous to that of a Mach–Zehnder (Born and Wolf 1986) interferometer where one mirror is replaced by the object under study. This type of arrangement is commonly used to give fringes which are mainly related to out-of-plane displacement.

Coherent light coming from a laser diode is launched in a single-mode fibre, and with the help of a bidirectional coupler, the beam is amplitude divided to provide a reference and an object beam.

To avoid amplitude instabilities or frequency drifts of the optical radiation caused by reflections, the coupling arrangement includes an optical isolator based on a Faraday crystal. The output light from the optical fibre has a spherical wave front (no astigmatism).

One arm of the bidirectional coupler terminates at a point close to the imaging lens. From this point the light diverges to illuminate the object, so that we realize a quasi-normal incidence. The scattered light from the object is collected by the imaging lens and is focused on to the photosensor of a solid state camera. Light from the other arm of the interferometer is sent by a beam splitter cube to the camera image plane, where the object and reference beams interfere. The coupler splitting ratio is generally unequal, typically 90 : 10, with more power in the object beam to allow for the low light levels returned from the object. The monomode fibre acts as an excellent spatial filter and hence the emerging beams require no further spatial filtering.

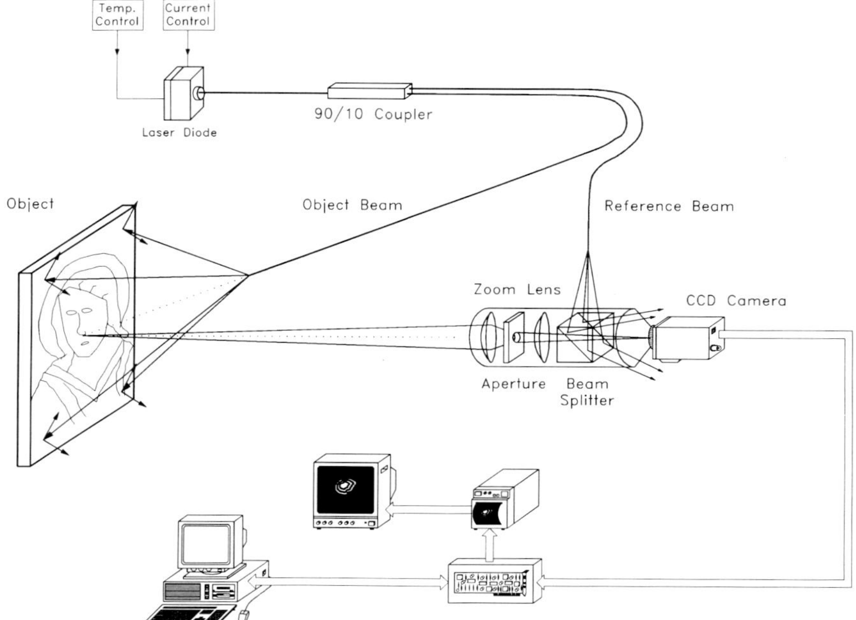

Fig. 2. Experimental set-up: practical arrangement.

With PC-based image processing, the video signal with the object in an initial position can be digitized, processed and stored in a reference frame. After deformation, the new stored pattern can be subtracted from the initial pattern, and the squared difference can be displayed on a video monitor as real-time fringes. The fringes represent the contour lines of change in the path difference of the interferometer caused by out-of-plane deformation of the object with a sensitivity of $\lambda/2$. The stored initial image can be refreshed by an outer control signal so that a large amount of deformation can be continuously tracked.

Furthermore, to obtain a good-quality ESPI interference pattern, the ESPI system as well as the surrounding environment must remain stable, during the exposure time, to at least $\lambda/8$ (Tyrer 1986).

Any relative motion of the order of the laser wavelength between the optical system and the object under investigation introduces unwanted fringes which, inevitably, alter the results. In particular, during in-field operation, the presence of mechanical vibrations is the major source of uncontrollable micro-movements, which might have destructive effects on the measurements. When vibrations influence the interferometric system, the deformation fringes become barely visible as they are no longer stable. Consequently, any attempt to record the fringe pattern leads to very poor results. In optical laboratories these problems can be solved by use of massive tables, whose design is based on drastic reduction of the structure resonance peaks. It is evident that heavy and bulky tables do not comply with the portability of an optical system and its in-field operation. A simple and effective solution could consist of a rigid mechanical coupling between the optical system and the inspected object. However, the required contact between the measuring head and the object can cause the transmission of mechanical interactions through the contact points, which will influence the results. Moreover, the mechanical coupling is likely to change from one object to another, according to its shape and location, causing a reduction of flexibility.

An alternative solution can be chosen for *in situ* tests. It is possible to assemble the video camera and the fibre ends in an optical head mounted on a tripod. This arrangement seems to provide the best flexibility. The tripod system, on the other hand, has to be as insensitive as possible to external vibrations. In a church, a museum or a conservation laboratory, floor resonance (5 to 50 Hz) and street traffic (5 to 100 Hz) are the main source of micro-vibrations. In general, in non-industrial environments, the common sources of mechanical instability have frequencies below 150 Hz. Therefore, if the measuring system has a resonance frequency greater than this value, vibration effects will be greatly reduced. In practice, this condition can be obtained with relative ease by using a high stiffness tripod, and an optical head with a mass as low as possible (Thomson 1972).

However, the effect of residual relative rigid displacements between the object and optical head will be to reduce the fringe contrast, which can be recovered by image enhancement techniques (Pratt 1978, Hall 1979).

DIGITAL PROCESSING OF ESPI IMAGES

The ESPI technique produces results in the form of images (fringe patterns), which, if properly interpreted, offer a whole-field description of the investigated physical properties.

A digital image can be considered as a numerical matrix, whose elements (pixels) are associated with the brightness. The resolution of this digital image is proportional to the number of pixels per unit area and to the number of grey levels. Several types of digitizing devices are commercially available; however ESPI interferograms are mostly digitized in a standard video system, so that the ESPI fringe pattern has a pixel resolution of $512 \times 512 \times 8$ deep (i.e., 256 grey levels). Image digitization is made by sampling the signal from the TV camera with a frame grabber hosted in a computer. A frame grabber card is commonly equipped with memories (frame buffers) where the digitized images can be temporarily stored. Dedicated hardware circuits such us look up tables (LUTs) and an arithmetic logic unit (ALU) are sometimes available in order to control the display features and perform real-time operations between digital images.

Fundamental digital operations are often performed on ESPI images in order to improve the display or to facilitate the use of fringe analysis algorithms.

Contrast enhancement

Contrast generally refers to a difference in luminance or grey level values in one image or in some particular regions of it. One of the most common defects of ESPI images is poor contrast. A reduction of fringe quality in fact may occur for many reasons, which are often related to the experimental conditions (e.g., the presence of micro-vibrations, ambient light, speckle decorrelation, non-uniform object reflectance, incorrect ratio between the intensities of the reference and object beams, etc.). Contrast enhancement techniques are therefore very useful and are commonly applied for preliminary processing of ESPI interferograms. Contrast manipulation is based on the principle of image transformation or mapping. In this respect, a transformation is defined as an operation which performs a redistribution of the image grey levels. In practice, the grey level of a pixel will be changed into a new value according to a defined transformation function. Image contrast can often be improved by amplitude rescaling of each pixel. This procedure is particularly effective when the pixel values are grouped in a small interval of grey levels. The scaling transformation distributes the pixel values over all the range of available levels with positive effects on the image contrast. For ESPI images the rescaling procedures are very useful before the next elaboration.

In Fig. 3 the original fringes recorded during *in situ* inspection on a fresco and the result obtained after manipulation with scaling transformation techniques are both presented. Other types of grey level transformations, e.g.,

Fig. 3. Speckle fringes with enhanced contrast over a portion of the original picture.

piecewise-linear transformation, logarithmic transformation, exponential transformation, histogram equalization, histogram specification, etc. (Pratt 1978, Hall, 1979), can be performed on ESPI interferograms but in general they do not lead to significant improvements.

Moreover, the human visual system can simultaneously appreciate and distinguish only 16–32 different shades of grey from black and white (Sheppard *et al.* 1969, Tukey 1971). Colours can be very helpful in highlighting details which are scarcely visible in grey-scaled images. For this reason pseudo-colour techniques are commonly used for the final display of ESPI fringes (pseudo-colour processing consists in assigning a correspondence between grey levels and colours).

Filtering

Speckles carry the displacement information in ESPI, but, after subtraction and square-law operation, they represent pure noise (see Eq. 8). This speckle noise is the main source of the random errors in ESPI (Nakadate and Saito 1985) and some care should be taken to suppress it to obtain Eq. (11) (or Eq. 12). Several methods have been proposed to reduce speckles, the main difficulty being to suppress speckles while preserving image information such as edges or textures. The filtering operations can be performed either in the spatial or in the frequency domain (Pratt 1978). Spatial filtering consists in the discrete convolution of the image with a small matrix (kernel), which performs a weighted average in the neighbourhood of each image point. Obviously, direct application of an averaging spatial filter is inadequate, because it smooths not only the speckles but also the contours. For this reason local statistical smoothing methods have been proposed (Lim and Nawab 1981, Wu and Maitre 1992, Kato *et al.* 1993). However, these methods are either time consuming or unsatisfactory. An alternative method is to perform the convolution operation indirectly by Fourier filtering techniques (Pratt 1978). This is an efficient procedure if the fringe spacing is considerably larger than the mean speckle size so that the overlap between the frequency distributions of the fringes and the speckle pattern is small; usually this is the case for ESPI fringe patterns. The method consists in the multiplication of the image Fourier spectrum with a filtering function which suppresses some frequency components of noise. The cleaned image is recovered by performing the inverse transform operation.

In Fig. 4 is shown the image of Fig. 3 processed by the Fourier filtering techniques using a Butterworth function as filter (Kerr *et al.* 1989).

Image composition

Because of the subtraction nature of ESPI measurements, the interferograms contain no visible details of the object. As the restorers need to know the exact

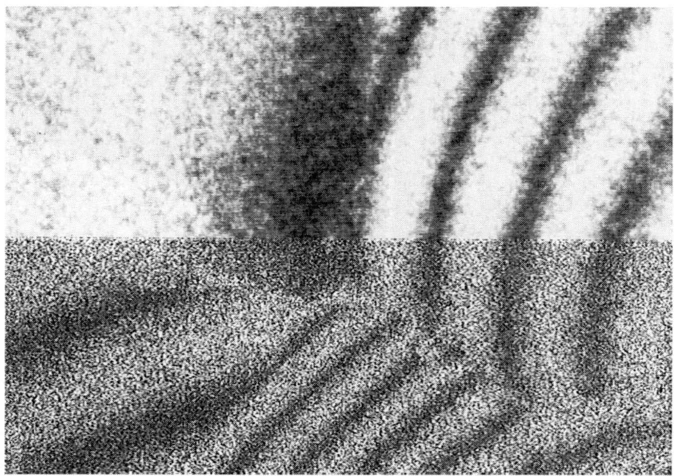

Fig. 4. Filtered fringes of Fig. 3 over a portion of non-filtered fringes.

location of the defects on the artwork, digital techniques can be used to solve this problem.

It is possible, before realizing the ESPI interferogram, to capture the details of the object by illuminating the object itself with incoherent light. This capture must be realized by the TV camera of the ESPI system, located in the same position that will be used to record the speckle pattern. Subsequently, by edge detection processing, it is possible to obtain the edge map of the object under investigation. Figure 5(a) shows a detail of the fresco Madonna con Bambino of the XIV century located in the Piazza del Popolo of Fontecchio (AQ-Italy). Figure 5(b) shows the edge map, relative to the fresco of Fig. 5(a), obtained using a Sobel operator (Pratt 1978, Hall 1979) and thresholded edge image with a threshold level that produces a binary image. In Fig. 5(c) is shown an ESPI interferogram relative to the fresco of Fig. 5(a); the closed fringes suggest the presence of a detachment. By digital addition of Fig. 5(b) to Fig. 5(c), we obtained the exact location of the defect with respect to the fresco (see Fig. 5(d)).

PHASE MEASUREMENT TECHNIQUES

For most applications only a qualitative interpretation of the correlation fringes is required. However, for particular studies of the stress methods or of calibration experiments effected on models simulating panel paintings and frescoes with the most common defects, a quantitative interpretation of the fringe pattern can be useful.

(a)

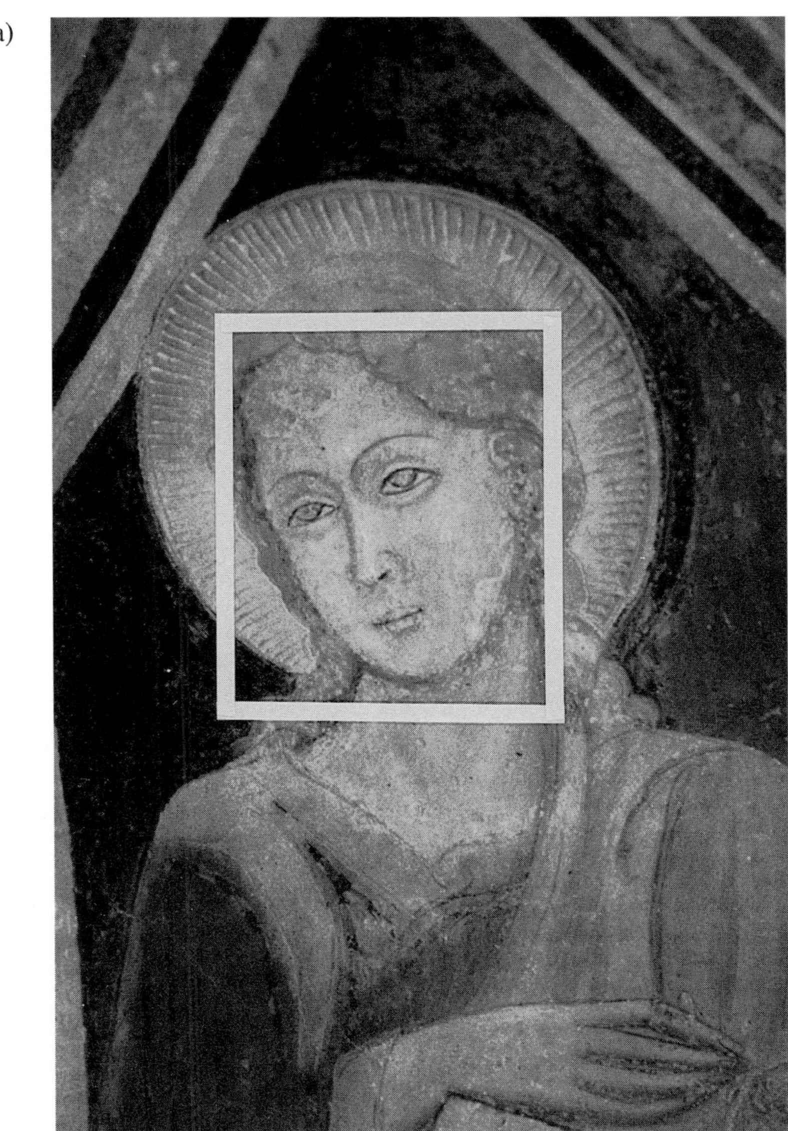

Fig. 5. (a) Detail of a fresco (Madonna con Bambino) of the XIV century situated in the Piazza del Popolo – Fontecchio (AQ-Italy). The marked zone is relative to the detected area. (b) Edge map. (c) ESPI interferogram. (d) Composition of image (b) and image (c).

(b)

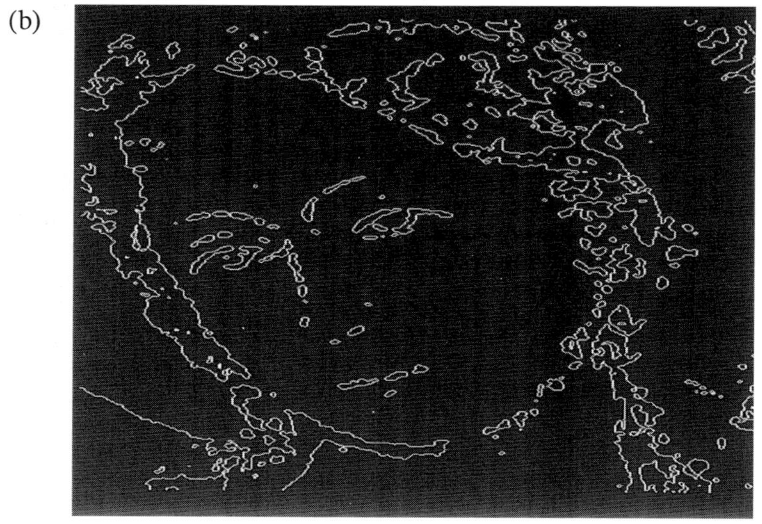

(c)

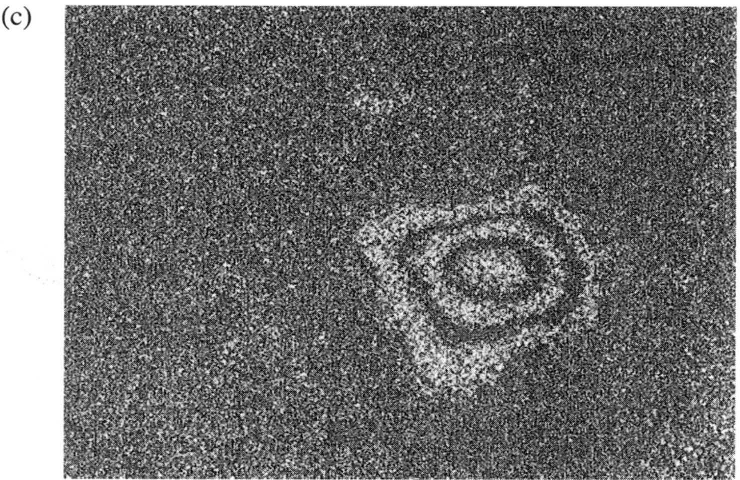

Fig. 5. *(continued)*.

(d)

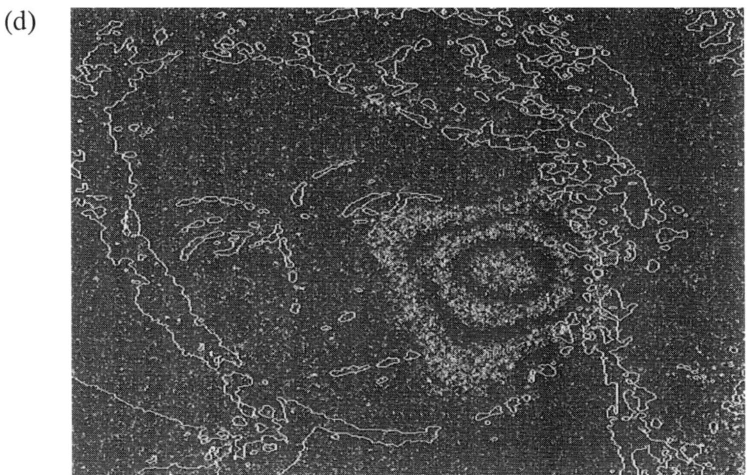

Fig. 5. (*continued*).

Fourier analysis with additional carrier fringes is suitable for *in situ* measurements. In fact, this method is particularly attractive for a quantitative analysis of the fringes when the measurement is made in difficult environmental conditions (for example with transient phenomena or in the presence of external disturbances) (Kreis 1986, Kujawinska and Robinson 1988).

The Fourier transform method (Takeda *et al.* 1982, Macy 1983, Bracewell 1986) is a global operation based on retrieving the phase function in the spatial frequency plane. A phase variation, between complex amplitudes of the current image and of the reference one, is introduced by translation of the object beam. To realize the translation, the fibre, which illuminates the object, is mounted upon an (x, z) translation stage, adjustable by a piezoelectric device. This shifting causes a carrier spatial frequency modulation to appear, added to the phase difference due to the deformation.

The resulting intensity distribution, $B_s(x, y)$ of a two-dimensional fringe pattern, can be rewritten as follows:

$$\begin{aligned}
B_s(x, y) &= N(x, y)E(x, y) = N(x, y)\{1 - \gamma \cos[2\pi(f_x x + f_y y) + \theta(x, y)]\} \\
&= N(x, y)\{1 - \gamma \tfrac{1}{2} \exp[i\theta(x, y)] \exp[i2\pi(f_x x + f_y y] \\
&\quad - \gamma \tfrac{1}{2} \exp[-i\theta(x, y)] \exp[-i2\pi(f_x x + f_y y)]\}
\end{aligned} \tag{13}$$

where f_x and f_y are the linear components of the imposed translation and $\theta(x, y)$ is the phase change related to the object surface deformation.

The Fourier transform $B_s(\eta, \xi)$ of the function $B(x, y)$ is the convolution of the signal and noise spectra, i.e.,

$$B_s(\eta, \xi) = E(\eta, \xi) \otimes N(\eta, \xi). \tag{14}$$

If $f_0 = \sqrt{f_x^2 + f_y^2}$ is made larger than twice the largest spatial frequency component of the phase variation, then the Fourier transform of $E(x, y)$ will have three distinct regions of non-zero values centred at $\pm f_0$ and at 0 (i.e., a DC term),

$$E(\eta, \xi) = A_E(\eta, \xi) - C(\eta - f_x, \xi - f_y) - C^*(-\eta - f_x, -\xi - f_y). \tag{15}$$

$A_E(\eta, \xi)$ and $C(\eta, \xi)$ denote the Fourier transform of 1 and $C(x, y)$, respectively. Roughly speaking, $A_E(\eta, \xi)$ can be thought as an impulse so that it will be considered to be a Dirac function $\delta(\eta, \xi)$ in the following.

For a typical speckle pattern, $N(x, y)$ will be a function with rapid local variations but with a slowly varying local mean value. The Fourier spectrum of $N(x, y)$ is

$$N(\eta, \xi) = A_N(\eta, \xi) + S(\eta, \xi). \tag{16}$$

In this case too, the sharply peaked function $A_N(\eta, \xi)$, corresponding to the slowly varying part of $N(x, y)$ can be assimilated to a Dirac function $\delta(\eta, \xi)$. $S(\eta, \xi)$ denotes the Fourier transform of the fluctuating part of $N(x, y)$.

Hence the spectrum of $B_s(x, y)$ can be written in the form:

$$B_s(\eta, \xi) = A(\eta, \xi) + C(\eta - f_x, \xi - f_y) + C^*(-\eta - f_x, -\xi - f_y) + Q(\eta, \xi). \tag{17}$$

$A(\eta, \xi)$ is again an impulse-like function arising from the convolution between $A_E(\eta, \xi)$ and $A_N(\eta, \xi)$. $Q(\eta, \xi)$ is the noise floor (the convolution of $S(\eta, \xi)$ with $C(\eta, \xi)$ and $C^*(\eta, \xi)$). If the carrier frequency is small, the speckle noise and the fringe information can be separated in the Fourier plane; in this case it can be seen that, for values of η and ξ where either the function $C(\eta, \xi)$ or the function $C^*(\eta, \xi)$ is appreciable, the contribution from $A(\eta, \xi)$ and $Q(\eta, \xi)$ can be neglected.

The components centred at f_0 can be isolated with a band-pass window. If these components are now shifted by a spatial frequency of $-f_0$, the complex function

$$C(x, y) = \gamma_2^{\frac{1}{2}} \exp[i\theta(x, y)] \tag{18}$$

is obtained. From this equation the phase $\theta(x, y)$ can be calculated pointwise by the relation:

$$\theta(x, y) = \tan^{-1} \frac{\Im[C(x, y)]}{\Re[C(x, y)]} \tag{19}$$

where $\Im[C(x, y)]$ and $\Re[C(x, y)]$ represent the imaginary and real part of $C(x, y)$. The desired phase measurements are thus obtained without sign ambiguity.

The phase calculated by Eq. (19) gives principal values ranging from $-\pi$ to π; the phase distribution is wrapped into this range and 2π jumps occur for variations of more than 2π. These discontinuities can be corrected by adding or subtracting 2π according to the phase jump ramping $-\pi$ to π or vice versa (Robinson 1993).

EXPERIMENTAL RESULTS

As is known, conventional ESPI techniques can be used for non-destructive testing whenever the parameter of interest (crack, void, diebond, detachment) can be made manifest as a surface deformation discontinuity, revealed as local changes of the interferometric fringe pattern. For opaque objects, ESPI provides information about the displacement of the object's surface only. In the case of debonded regions or weak areas in the object's interior, we must find excitation mechanisms which transform interior inhomogeneity into interpretable surface movements. Let us assume the object suffers some kind of deformation. Regardless of the detailed features of the process, it is to be expected that the detached regions behave differently from the non-detached ones. As a consequence, the region of fault can be identified.

Figure 6(a) shows, schematically, a flaw inside a typical layered structure on a wooden panel of the type used in paintings since the XII–XIII century. An external thermal load, applied to the object, causes a perturbation in the surface-strain field (Fig. 6(b)). By recording one holographic speckle pattern relative to the surface without stress and subsequently another with the application of the stress, it is possible to produce, by subtracting one from the

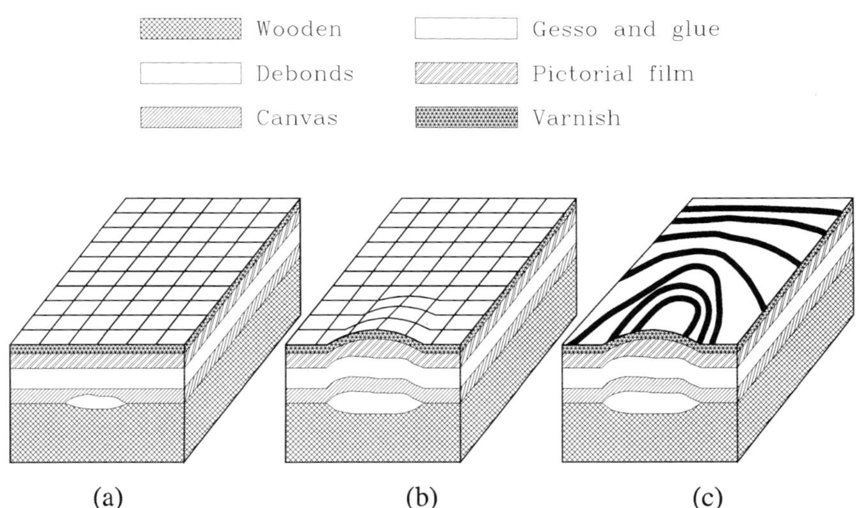

Fig. 6. Typical layered structure on a wooden panel.

other, an ESPI interferogram corresponding to the relative object deformation (Figure 6(c)).

The interpretation of ESPI correlation fringes can be made on the basis of experiments effected on models simulating panel paintings and frescoes with the most common defects. As a rule, ancient paintings were made on poplar boards consisting of several separate pieces glued together, which is a typical layered structure. In the majority of cases delamination occurs, due to poor cohesion between the priming and wooden panel, and the priming and paintings, under ambient parameter variations (temperature and humidity), while irregular wood grain or support structural anomalies can cause the formation of subsurface cracks or fissures. The frescoes in ancient churches or in open environments present the same problems as wooden paintings. The presence of internal forces in the fresco materials (lime plaster, paintings and their base) can cause detachment between the layers or cracks in the painted surface.

The existence of ambient drift suggests that ESPI techniques can be used without further stress application. However, the sensitivity of this procedure is rather poor. In fact, as room parameter variations act on both the wood support and the layered structure, the method must be thought of as differential. The sensitivity can be increased by producing some kind of stress that acts directly on the surface layers only, for example surface heating through the application of infrared radiation. Due to the lack of adhesion between the priming layers and the wood support, the detached regions disperse heat at a lower rate than the surrounding regions. Therefore, both the local temperature rise and the thermal expansion effects are higher for the detached regions. An ESPI interferogram can be made during the heating process or during the subsequent cooling process.

Usually, the stressing method (a temperature gradient obtained by heating the surface) is chosen empirically with guidance provided by an analysis of previous results obtained from models. Finding the best stressing method is generally a function of the object nature and of the expected kind of fault. By comparing the induced defects of the test samples with the results, the potentiality of the technique is established for known conditions.

Preliminary experiments were carried out with a model simulating a wooden panel painting. This model was prepared by the same technique and the same material, as was used in paintings since the XII–XIII century (Herringham 1899, Cennino Cennini 1982). Inside the layered structure, a number of detached regions of known form and extension were provoked by the insertion of suitable plastic sheets. From these studies, it is reasonable to assume that where there is a high spatial density or islands of fringes, strain concentrations or defects are present, while discontinuities along the trend of some fringes indicate the presence of cracks.

Figure 7 shows an ESPI interferogram of a model with simulated defects situated between the pictorial film and the gesso and glue layer. A subsurface detachment is revealed by closed fringes. Figure 8 shows a new interferogram realized in the same experimental condition with added carrier fringes, while a 3D plot of the defect is shown in Fig. 9.

Figure 10 shows ESPI fringes relative to a model with simulated defects situated between a wooden support and canvas while Fig. 11 shows correlation fringes relative to a typical crack on a wooden panel.

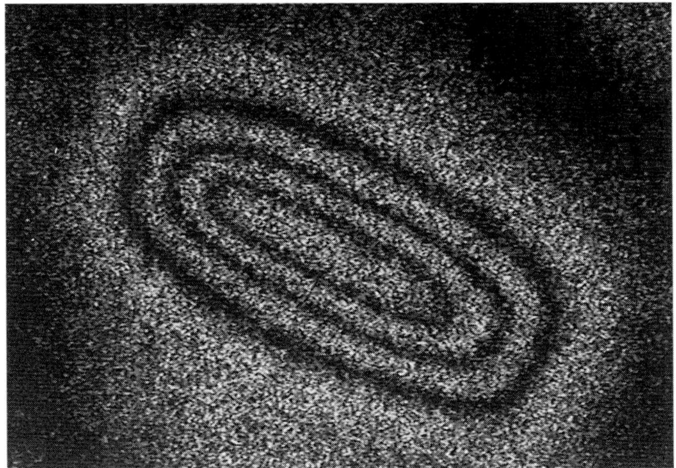

Fig. 7. ESPI interferogram of a wooden model with subsurface detachment between the pictorial film and the gesso and glue layer.

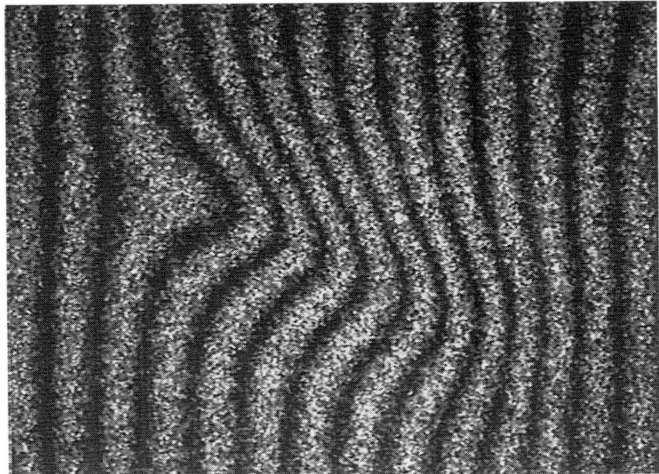

Fig. 8. Speckle interferogram relative to the same specimen of Fig. 7 with carrier fringes.

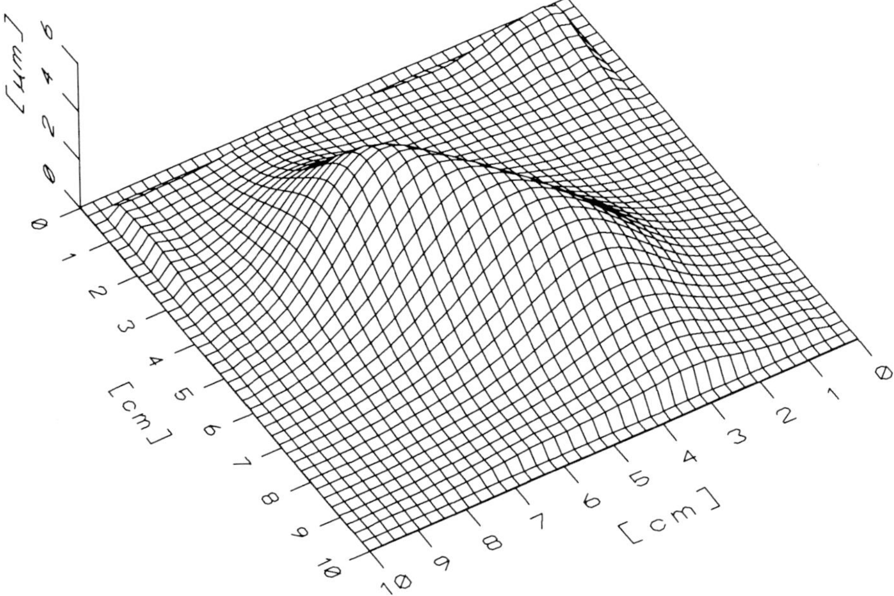

Fig. 9. 3D-plot of the defect of Fig. 7.

Fig. 10. ESPI interferogram of a wooden model with a subsurface detachment between wooden support and canvas.

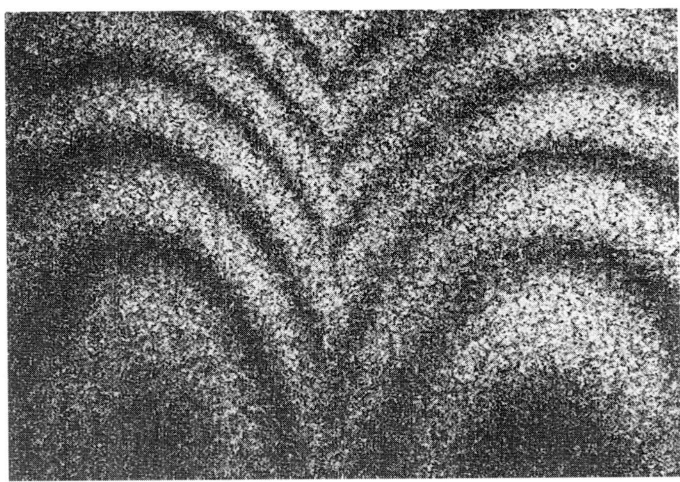

Fig. 11. ESPI fringes showing the presence of a crack on a model.

The investigation of the possibility of using the portable ESPI instrumentation *in situ* was effected in examining some ancient paintings on wood and some frescoes.

The portable ESPI instrumentation has been used to examine the condition of an ancient wooden panel painting (Fig. 12(a)) (Madonna in trono con Bambino by Giotto – wooden painting without frame (190 × 97 cm) XIII century – Church of St. George in Florence, under restoration at Opificio delle Pietre Dure and Laboratorio di Restauro of Florence).

Figure 12(b) presents an interferogram relative to an area in the middle of the panel where a nail was driven in from the back and bent down to the front surface. The anomalous density and shape of the fringes at the centre of the image clearly indicate the extent of the detachment, as compared with the model with simulated defects.

Figure 13(a) shows the ESPI system operating on the fresco Madonna con Bambino of the XIV century situated in the Piazza del Popolo of Fontecchio (AQ-Italy), while Fig. 13(b) shows a typical crack detected on this fresco.

An example of the versatility of the technique during the restoration process consists in determining whether or not repair work has been effective. Investigations were effected on a wood panel (42 × 125 cm) Trionfo di Davide by Imitatore del Pesellino (XV century), from the Museo Horne (Florence-Italy) and on a fresco in the church of St. Maria della Croce of Roio-L'Aquila Italy (Adorazione dei Magi by Farrelli 1667). The measurements were performed before and after repair work on areas of the paintings containing defects.

Plate 16.1(a) shows a detail of the wood panel Trionfo di Davide. From the interferogram of Plate 16.1(b), a large detachment was readily detected. The

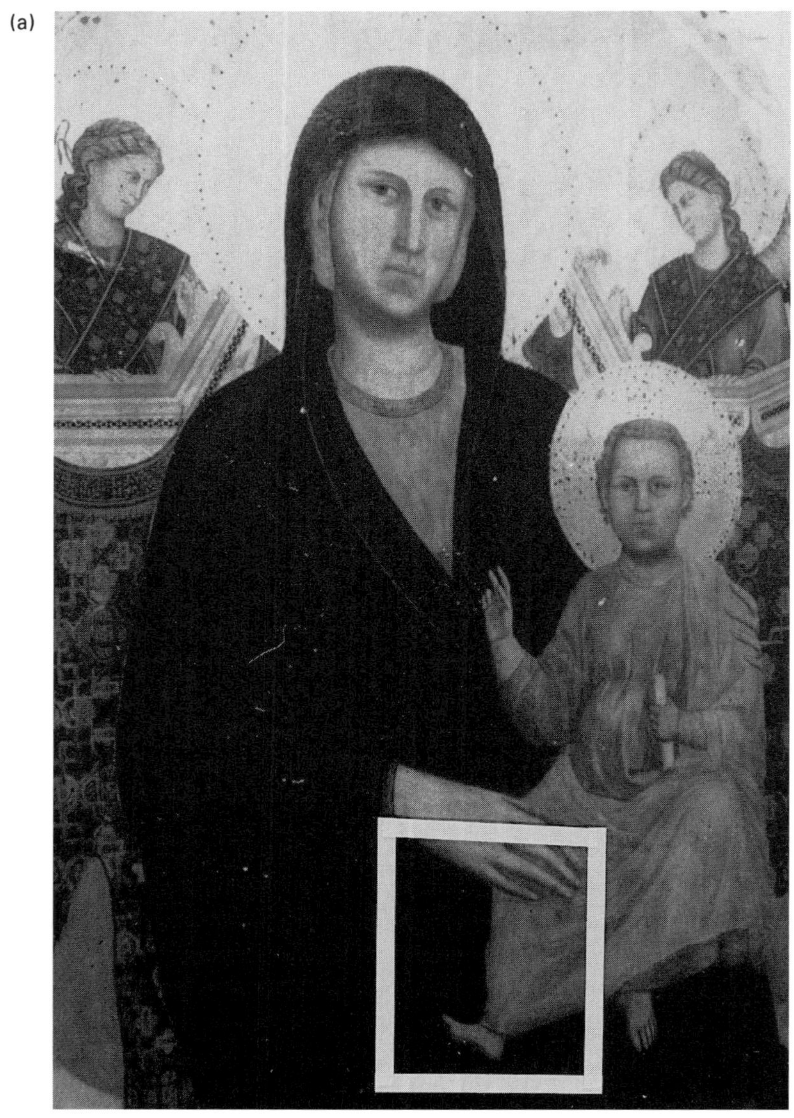

Fig. 12. (a) Panel painting Madonna in trono con Bambino by Giotto under restoration at Opificio delle Pietre Dure and Laboratorio di Restauro of Florence. (b) ESPI interferogram.

(b)

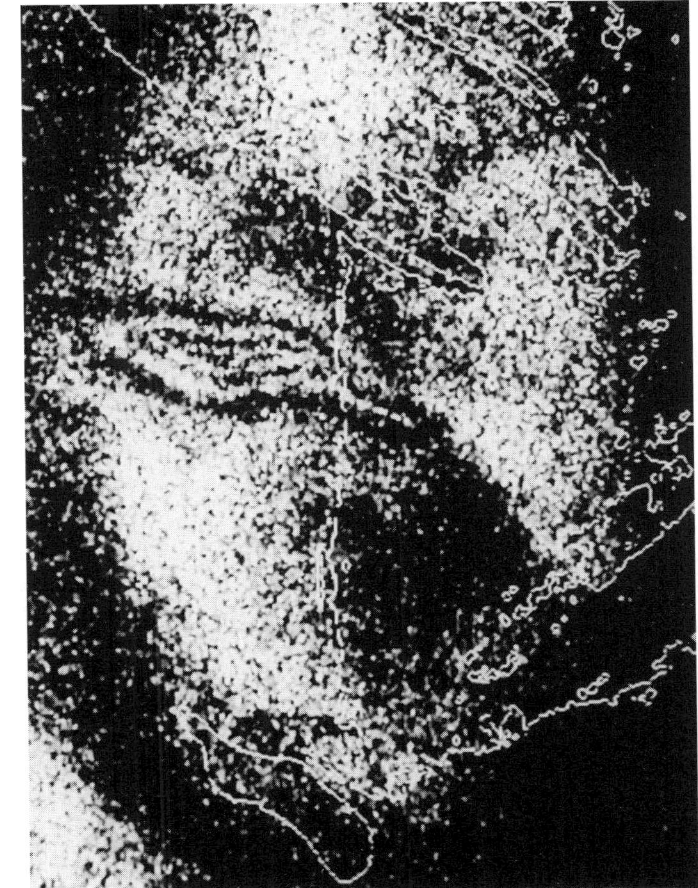

Fig. 12. *(continued)*.

same test was performed one day after the consolidation of the damaged area. The closed-fringes, typical of detachment, were absent and the fringe pattern showed no anomalies (see Plate 16.1(c)).

Plate 16.2(a) shows a detail of Farrelli's fresco. The interferogram of Plate 16.2(b) reveals the presence of a detachment surrounding a cracking area. Plate 16.2(c) shows how the crack and the surrounding area are repaired and a proper bond between the primer ground and the painting was reinstated for the detached region. The two interferograms were realized in the same experimental conditions.

The monitoring of on-going deterioration in time is also possible. Recording at time intervals of one or more months would show the deterioration rate of a fresco and the development of defects under environmental stress.

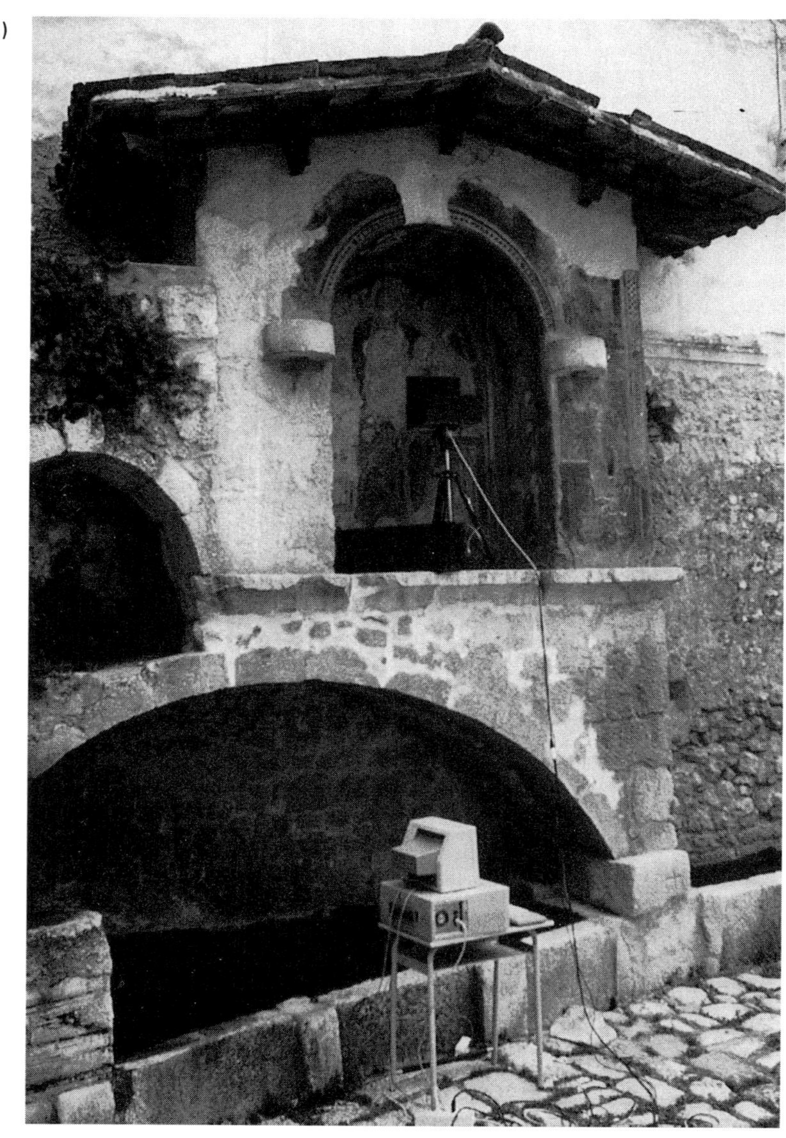

Fig. 13. (a) ESPI system operating *in situ*. (b) ESPI interferogram showing a typical crack detected on the fresco.

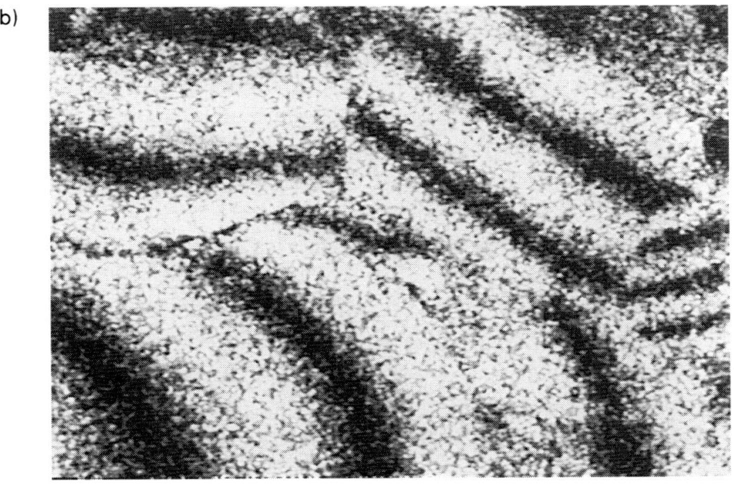

Fig. 13. (*continued*).

PROBLEMS AND FUTURE IMPROVEMENTS

The possibility of using a portable electronic speckle pattern interferometer (ESPI) in artwork diagnostics has been demonstrated by the detection of some typical defects in ancient wooden panel paintings and frescoes. Moreover, the results obtained before and after consolidation of detachment confirmed the importance of interferometric techniques for controlling the effectiveness of repair work. An interesting application could also be the evaluation of the dynamic interactions between the work of art and the surrounding environment, even if artificial. Consider, for example, the deformation effects associated with the thermal gradients induced by the illumination lamps of an exhibition in a museum.

The results presented above give an idea of the potential applications of ESPI technology, mainly developed for industrial diagnostics, to art restoration and conservation. The system is low cost but it has some limitations.

One typical limitation in ESPI is that the individual speckles must be resolved by the TV camera. If the separation of the maxima and minima of $\cos[\phi_0(x, y) - \phi_R(x, y)]$ in Eq. (5) is less than the spatial resolution of the TV camera, the last term in that equation will be averaged to zero and no variation in correlation will be observed. This requires matching the numerical aperture of the imaging system to a particular sensor. The mean speckle size must be equal to the size of a single pixel on the camera. This will optimize the performance as well as the fringe visibility. With a typical pixel size of $\approx 10~\mu\text{m}$, an F-number of ~16 is necessary. The use of relatively large F-numbers, giving large speckles, affects the performance of the system in several ways. Firstly, it reduces the intensity of light in the image with the consequence of limiting the possibility of analysing larger or less reflective objects, unless a higher-power laser and/or a higher-sensitivity camera are used. A second consequence is that the speckles are clearly seen in the monitor image and the clarity of the fringes is therefore poor. For this reason, it is necessary to use a despeckle function to analyse the ESPI pattern. Unfortunately, the actual algorithms are usually unsatisfactory, therefore further work needs to be done to improve the speckle-reducing procedures in ESPI interferograms.

A further limitation in the ESPI system is the spatial resolution of the camera. The number of points of an interferogram limits the number of fringes which can be seen and analysed. If we consider that a minimum of about 20 speckles is required for a resolvable fringe, with a typical detector array having 512×512 resolution elements, about 25 equally spaced fringes would be resolvable. Future developments of portable ESPI systems should consider a larger area of inspection, which for paintings and frescoes is now approximately 15×15 cm. In order to increase the dimensions of the inspected area, the parameters to be varied are the optical power and the sensitivity of the video camera. Compact frequency-doubled $(Nd^{3+}:YAG)$ lasers are now available with 400 mW continuous power, a wavelength of 532 nm, and a coherence length of approximately 150 m. Innovative performances are also expected from the single-mode laser diode. The CCD market is constantly growing, offering a variety of sensors with improved characteristics at reasonable price. Modern camera technology offers devices with spatial resolutions that are much larger than the standard television resolution (a CCD sensor of the last generation can have up to 5000×5000 elements (Janesick *et al.* 1990, Spooren 1994)). However these devices can be applied only with non-standard signal readout and transmission, as they demand an image processing system with a suitable resolution. The commercial introduction of high-definition television, being near at hand, will offer a new generation of standard video systems with larger spatial resolution compared to the current television standard. With the introduction of this new technology, it is likely that the surface of paintings or frescoes up to 80×80 cm will soon be investigated.

Finally, the major limitation of ESPI deals with decorrelation of the speckle pattern. In fact, the effect of decorrelation will be a decrease in fringe visibility until the limit of total decorrelation, when no fringes will be formed. This determines the maximum displacement and/or deformation that can be observed (Owner-Petersen 1991).

ACKNOWLEDGMENTS

The author wish to thank Prof. Franco Gori for his helpful suggestion. He is also grateful to Paolo Zanetta (Joint Research Centre, Institute for System Engineering and Informatics, Ispra, Italy) for use of his material.

REFERENCES

Amadesi S., A. D'Altorio and D. Paoletti, *Appl. Opt.* **21**, 1889–1992 (1982).

Amadesi S., F. Gori, G. Guattari and R. Grella, *Appl. Opt.* **13**, 2009–2013 (1974).

Bertani D., M. Cetica and G. Molesini, *Stud. Conserv.* **27**, 61–65 (1982).

Born M. and E. Wolf, *Principles of Optics* 6th (cor.) ed., Pergamon Press, Oxford (1986).

Bracewell, R.N., *The Fourier Transform and Its Applications* McGraw-Hill, New York (1986).

Brandi C., *Teoria del Restauro* Einaudi, Torino (1977).

Butters, J.N. and J.A. Leendertz, *Meas. Control* **4**, 349–354 (1971).

Cennino Cennini, *Il Libro Dell'Arte*, Neri Pozza Editore, Vicenza (1982).

Davies J.C., C.H. Buckberry, J.D.C. Jones and C.N. Pannell, *Proc. SPIE* **863**, 194–207 (1988).

Ford H.D., H. Atcha and R.P. Tatam, *Meas. Sci. Technol.* **4**, 601–607 (1993).

Goodman J.W., Statistical properties of laser speckle patterns, in *Laser Speckle and Related Phenomena* (ed. J.C. Dainty), pp. 9–75, Springer-Verlag, Berlin (1975).

Gülker G., K. Hinsch, C. Holscher, A. Kramer and A. Neunaber, *Opt. Eng.* **29**, 816–820 (1990).

Hall E., *Computer Image Processing and Recognition*. Academic, New York (1979).

Herringham C., *The Book of the Art of Cennino Cennini, a Contemporary Practical Treatise on Quattrocento Painting*, Macmillan London (1899).

Janesick J., T. Elliott, A. Dinigizian, R. Bredthauer, C. Chandler, J. Westphal and J. Gunn, *Proc. SPIE* **1242**, 223–237 (1990).

Jones R. and C. Wykes, *Holographic and Speckle Interferometry* 2nd ed., Cambridge University Press, Cambridge (1989).

Kato J., I. Yamaguchi and Q. Ping, *Appl. Opt.* **32**, 77–83 (1993).

Kerr D., F. Mendoza Santoyo and J.R. Tyrer, *J. Mod. Optics* **36**, 195–203 (1989).

Kreis T.M., *J. Opt. Soc. Am. A* **3**, 847–855 (1986).

Kujawinska M. and D.W. Robinson, *Proc. SPIE* **1026**, 93–103 (1988).

Lanciani R., *The Destruction of Ancient Rome*. Macmillan, London (1901).

Lim J.S. and H. Nawab, *Opt. Eng.* **20**, 472–480 (1981).

Løkberg, J.O., *Physics in Technology* **11**, 16–22 (1980).

Løkberg, O.J. and K. Krakhella, *Opt. Commun.* **38**, 155–158 (1981).

Macovski A., S.D. Ramsey and L.F. Schaefer, *Appl. Opt.* **10**, 2722–2727 (1971).

Macy, Jr. W.W., *Appl. Opt.* **22**, 3898–3901 (1983).

Nakadate, S. and H. Saito, *Appl. Opt.* **24**, 2172–2180 (1985).

Owner-Petersen, M., *J. Opt. Soc. Am. A* **8**, 1082–1089 (1991).

Paoletti D. and G. Schirripa Spagnolo, *Meas. Sci. Technol.* **4**, 614–618 (1993).

Paoletti, D., G. Schirripa Spagnolo, M., Facchini and P. Zanetta, P., *Appl. Opt.* **32**, 6236–6241 (1993).

Pratt, W.K., *Digital Image Processing*. John Wiley & Sons, New York (1978).

Robinson, D.W., Phase unwrapping methods, in *Interferogram Analysis: Digital*

Fringe Pattern Measurement Techniques (eds D.W. Robinson and G.T. Reid), IOP Publishing, Bristol (1993).

Schwomma, O., Austrian Patent 298830 (1972).

Sheppard, Jr. J.J., R.H. Stratton and C. Gazley, Jr., *Am. J. Optom.* **46**, 735–754 (1969).

Spooren, R., *Opt. Eng.* **33**, 889–896 (1994).

Takeda, M., H. Ina and S. Kobayashi, *J. Opt. Soc. Am.* **72**, 156–160 (1982).

Thomson, W.T., *Theory of Vibration with Applications.* Prentice-Hall, New York (1972).

Tukey, J.W., *Exploratory Data Analysis.* Addison-Wesley, Reading, Massachusetts (1971).

Tyrer, J.R., Critical review of recent developments in Electronic Speckle Pattern Interferometry, *Proc. SPIE* **604**, 95–111 (1986).

Valera, J.D., A.F. Doval and J.D.C. Jones, *Meas. Sci. Technol.* **4**, 578–582 (1993).

Virdee, M.S., D.C. Williams, J.E. Banyard and N.S. Nassar, *Opt. Laser Tech.* **22**, 311–316 (1990).

Westlake, D., R.F. Wuerker and J.F. Asmus, *SMPTE Journal* **85**, 84–89 (1976).

Wykes, C., *Opt. Eng.* **21**, 400–406 (1982).

Wykes, C., *J. Mod. Opt.* **34**, 539–554 (1987).

Wu, Y. and H. Maitre, *Opt. Eng.* **31**, 1785–1792 (1992).

Zanetta, P., Ph.D. Thesis, Loughborough University of Technology (UK) Technical Note ISEI/IE/2777/94, Commission of the European Communities, Joint Research Centre, Ispra Site (1994).

17 Numerical simulation of irradiance fluctuations for optical waves through atmospheric turbulence

Stanley M. Flatté

*Physics Department and Institute of Tectonics,
University of California at Santa Cruz, Santa Cruz, California 95064*

INTRODUCTION

Optical waves traversing turbulence develop dramatic spatial patterns of irradiance fluctuations. These patterns have been observed with optical propagation in the atmosphere [1], in the laboratory [2], and in numerical simulations [3]. The spatial patterns on an observing screen can be understood only partially from knowledge of their wavenumber spectra; a description of the complete pattern requires knowledge of all significant higher moments of irradiance, which is very difficult to obtain either theoretically or experimentally.

A measure of the strength of medium fluctuations in this type of wave propagation is β_0^2, the ln-irradiance variance calculated from weak-fluctuation theory for Kolmogorov turbulence with zero inner scale [4].

Weak fluctuation theory applies for $\beta_0 \leqslant 0.5$; we will define the saturated region as $\beta_0 \geqslant 3$. The intermediate region, particularly β_0 between 1 and 2, is known as the strong focusing regime. The combination of the strong focusing regime and the saturated regime is called the strong fluctuation regime. It is well known that the observed irradiance variance for optical propagation through the atmosphere rises above unity in the strong focusing regime, and then slowly decreases toward unity as β_0 increases in the strong fluctuation regime. The ln-irradiance variance behaves in a similar fashion, except that the variance approached as β_0 gets very large is $\pi^2/6$ instead of unity.

Another important parameter of medium fluctuations is the inner scale, which affects irradiance fluctuations sufficiently that it must be measured reasonably well if irradiance fluctuations are to be predicted [5]. Early parametrizations of the spectrum of medium fluctuations near the inner scale used a simple empirical Gaussian form; however it has been shown that a proper fluid-dynamical treatment of viscosity reveals a more complicated form that rises above the pure power law before being cut off [6]. Both spectra of refractive index can be expressed by

$$\Psi(K) = 0.033 C_n^2 K^{-11/3} F(Kl_0) \tag{1}$$

where K is the magnitude of the spatial wavenumber, C_n^2 is a standard measure of turbulence strength, and where the function $F(Kl_0)$ is unity at very small Kl_0 and goes to zero when Kl_0 is much larger than unity. The length scale l_0 is called the inner scale. The traditional-spectrum expression for $F(x)$ is $\exp[-(x/5.9)^2]$, while the 'atmospheric' spectrum has a more complicated form. It has been clearly established that the atmospheric spectrum is compatible with the behaviour of the actual atmosphere, while the traditional spectrum is not [7].

Because l_0 appears only in the combination Kl_0 it can be shown that the only parameter is actually l_0/R_f where R_f is the Fresnel length relevant to the optical propagation.

Prior to the application of numerical simulation to this field, a number of theoretical approaches were attempted in order to calculate the irradiance variance and also higher moments of irradiance. It would be desirable here to compare those calculations with experimental observations to see how well they did. It is very difficult to do that for several reasons: first, the calculations are most easily done with zero inner scale, but experimental observations cannot be done there; second, if we need to know the value of the inner scale, we must also know which spectrum we are using as explained above, and the experiment must have an independent means of measuring that inner scale. The situation generated a great deal of argument until finally two important contributions came together to provide a resolution. First an extensive and careful experiment was done that measured both the strength and inner scale of turbulence simultaneously with measurements of the irradiance variance [8]. Second numerical simulations were carried out with high spatial resolution and a large number of realizations [5].

The excellent agreement between the new experiment and the simulations improves our understanding in two areas: first, we can systematically discuss the effects of inner scale and spectrum shape on irradiance variance (next section); second, we can judge other theoretical calculations against simulations that have been done with the same assumptions about spectrum shape, inner scale, and initial conditions. Several following sections take up different theoretical approaches separately.

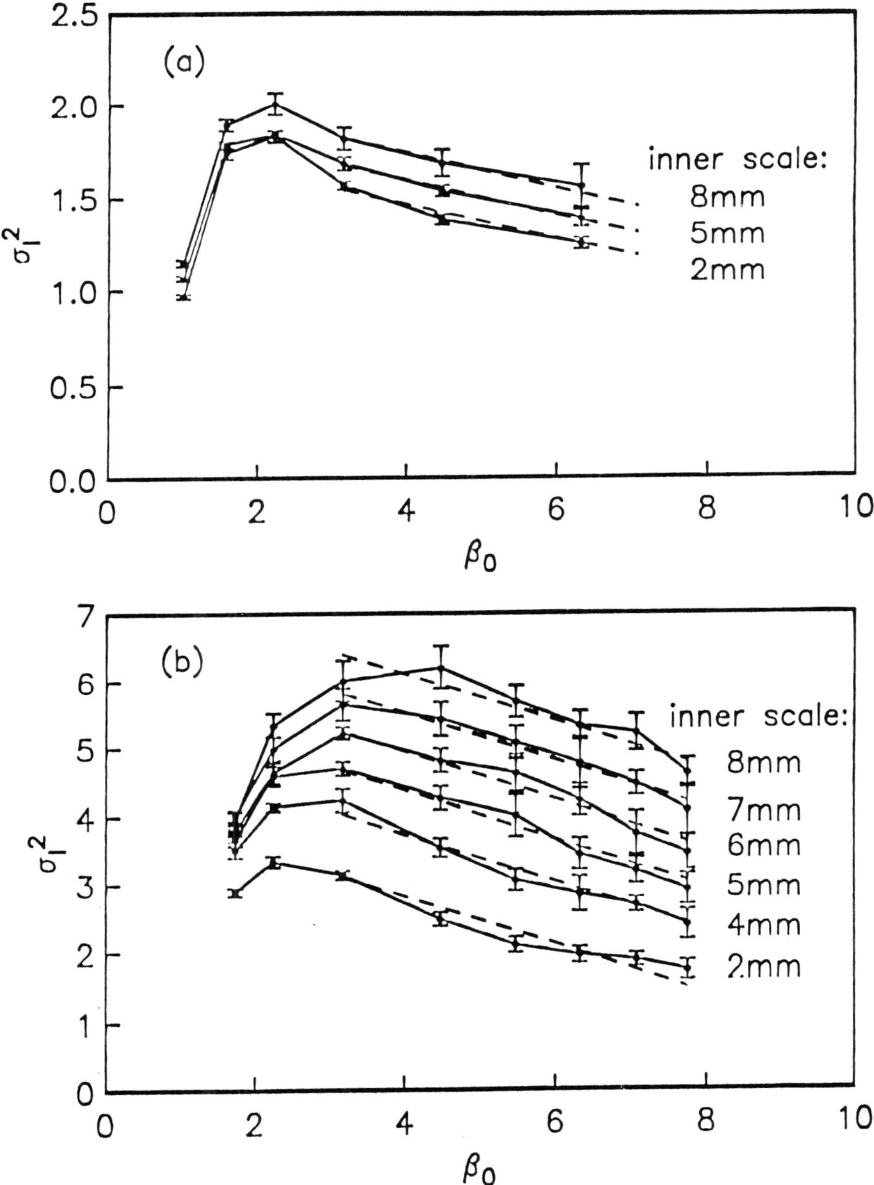

Fig. 1. Irradiance variances from numerical simulation with the atmospheric spectrum at different inner scales and as a function of turbulence strength. The solid lines connect the points to guide the eye; the dashed lines are best straight-line fits to the data with $3 < \beta_0 < 8$, constrained to a common slope. (a) Plane wave; the Fresnel scale $R_f = 13$ mm, the same as the experiment of Gurvich *et al.* [2]; (b) point source; the Fresnel scale is 10 mm, the same as the experiment of Consortini *et al.* [8].

SYSTEMATICS OF TURBULENCE STRENGTH AND INNER SCALE

Figures 1 and 2 show the irradiance-variance results from a large number of numerical simulations for plane wave and point source initial conditions and different values of turbulence strength (β_0) and inner scale (l_0) [5]. These results can be summarized in the strong-fluctuation regime (near $\beta_0 = 5$ and $l_0/R_f = 0.5$) by a simple empirical equation:

$$\sigma_I^2 = s_0 - s_b\beta_0 + s_1(l_0/R_f) \tag{2}$$

where the constants s_0, s_b, and s_1, which are different for the plane wave and the point source, are given in Table 1.

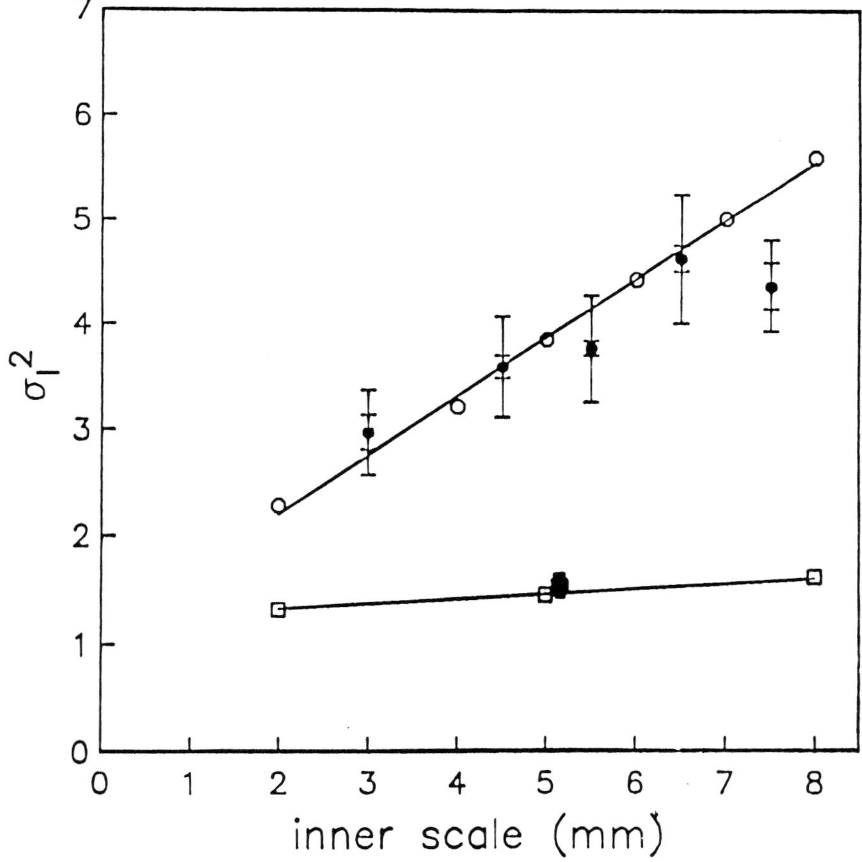

Fig. 2. Irradiance variance as a function of inner scale. Values obtained from the intercept of the straight-line fits in Fig. 1 with $\beta_0 = 5.48$. Open circles: point source; open squares: plane wave. Solid straight lines are from the best-fit empirical formulas. Closed circles: data from Consortini *et al.* [8]; closed square: data from Gurvich *et al.* [2].

Table 1. Empirical constants in the formula for irradiance variance.

	Plane wave	Point source
s_0	1.74	3.02
s_b	0.092	0.35
s_l	0.60	5.56

It will be useful in discussing systematics to recast the above formulas by picking a specific value of β_0 in the middle of the treated region. For zero inner scale, the plane wave and point source variances intersect at $\beta_0 = 4.96$. We therefore select $\beta_0 = 5$ as our reference value, and recast the formula in terms of the deviation of the irradiance variance from unity:

$$d_1 = \sigma_1^2 - 1,0 = d_0 - s_b(\beta_0 - 5.0) + s_1(l_0/R_f) \tag{3}$$

where the constant d_0 has nearly the same value for the plane wave (0.28) as for the point source (0.27).

We now see that the point source and the plane wave, for zero inner scale, have nearly identical deviations of about 0.28 at $\beta_0 = 5$. As one moves away from $\beta_0 = 5$, the point-source deviation changes much more rapidly than the plane-wave deviation. The β_0-slopes of 0.35 and 0.092 differ by a factor of 3.8. As one increases the inner scale, the point-source deviation also changes (increases) much more rapidly than the plane-wave deviation. The l_0-slopes of 5.56 and 0.60 differ by a factor of 9.3. These systematics of the strong-focusing regime just above the peak in irradiance variance are now established by the excellent agreement between the simulations and the experimental data [5].

However, we can now see that the empirical straight-line behaviour, at least as a function of β_0 cannot continue to very high β_0 because saturation tells us the deviation will not go below zero. Since the straight-line formula for zero inner scale has the deviation going to zero at β_0 values of 8.0 and 5.8 for the plane-wave and point-source cases, we see that our empirical formula is really not valid for inner scales very close to zero. In fact, the simulations, the experiment, and hence the formulas can only be applied for $l_0 > 2.0$ mm when β_0 is above 5.0.

Spectrum Shape

The agreement between the simulations and the experiments was obtained using the atmospheric spectrum. Many previous calculations have been made using the traditional spectrum. It therefore becomes important to know how much difference it would have made if the traditional spectrum were to have been used.

In order to answer this question we first must be careful in our definition of the inner scale. In the Consortini *et al.* [8] experiment, the inner scale is measured from the ratio of two variances found from beams propagating through about 150 metres of turbulence, one averaged over a large aperture and the other having a small aperture. It has been shown that as long as l_0 is above 5 mm, the numerical values of l_0 resulting from that measurement are not affected by which spectrum is used.

The simulations of Flatté *et al.* [5] who used the atmospheric spectrum, can be compared with those of Martin and Flatté [3,9], who used the traditional spectrum.

For an incident plane wave with $l_0 = 0.8\ R_f$ and $\beta_0 = 2$, the atmospheric spectrum gives $\sigma_I^2 = 2.0$ while the traditional spectrum gives 2.36. Therefore, assuming that the atmospheric spectrum is the correct one, the traditional spectrum gives a deviation that is 136% of the correct deviation.

For an incident spherical wave (point source) with $l_0 = 0.6\ R_f$ and $\beta_0 = 3$, the atmospheric spectrum gives $\sigma_I^2 = 5.2$ while the traditional spectrum gives 3.5. Therefore, assuming that the atmospheric spectrum is the correct one, the traditional spectrum gives a deviation that is 60% of the correct deviation.

Thus, the traditional spectrum gives errors for the deviation of the irradiance variance from unity that approach a factor of two: either 50% or 100% depending on your point of view. Since we are now in the realm of agreement between the experiment and the simulations of less than 10% discrepancy, the traditional spectrum needs to be abandoned, especially in view of the above numbers showing that even the sign of the effect is in doubt!

ASYMPTOTIC EXPANSIONS

First-order asymptotic theory for a plane wave in a pure power-law medium has been calculated long ago, and the results are well known. A new asymptotic expansion has been suggested by Dashen and Wang [10], and that suggestion has been compared successfully with the numerical simulations in Dashen *et al.* [11]. We refer to the latter paper for all the results for this type of expansion.

At $\beta_0 = 5$, the standard first-order result is about 57% of the 'correct' result, where 'correct' here refers to the numerical simulations. Second- and third-order standard-expansion results, presented in Dashen *et al.* [11], increase the agreement to 75% and 85%, respectively.

The new asymptotic scheme of Dashen and Wang [10] at first order yields 85% and at third order yields 92% of the simulation results.

Thus, asymptotic expansions for the irradiance variance must be carried out to third order to achieve accuracies better than 10%; first-order calculations yield errors in the 50–100% region.

NUMERICAL MOMENT EQUATION SOLUTION

A numerical solution of the fourth-moment equation gave a result for the variance deviation for an incident plane wave and zero inner scale [12]. At $\beta_0 = 5$ their result yielded 65% of the first-order asymptotic result; therefore their result was about 37% of the numerical simulation result.

TWO-SCALE EXPANSIONS

Whitman and Beran, in a series of papers [13–15], have done extensive calculations with the two-scale expansion technique. They have applied their technique to treating both plane wave and point source, and they have also accounted for the inner scale. They have carried their calculations to second order for the plane wave. They have provided the most extensive set of results of any treatment other than numerical simulation. So far they have carried out their numerical calculations for the traditional spectrum only. Therefore, in order to evaluate the accuracy of their approach, we must use the results of Martin and Flatté [3,9].

For the plane-wave case, their calculated results at $\beta_0 = 4$ were 50% and 60% of the numerical simulation results of Martin and Flatté [3] for inner scales of zero and 2.4 R_f, respectively. The second-order correction they calculated later [15] increased their numbers to about 60% and 75%, respectively.

For the point-source case, their calculated results at $\beta_0 = 5$ were 35% and 49% of the numerical simulation results of Martin and Flatté [9] for inner scales of zero and 1.5 R_f, respectively.

Presumably the second-order correction will not improve the point-source two-scale results by any more than it did in the plane-wave case. Finally one must remark that it is not clear how the use of the atmospheric spectrum will affect the two-scale expansion results. However, the conclusion at this juncture must be that the two-scale expansion is not yielding significantly better results than the standard asymptotic treatment.

PROBABILITY DENSITY FUNCTIONS

The probability density function (PDF) of irradiance has been evaluated by means of numerical simulation for an incident plane wave, many values of β_0 and three values of l_0/R_f [16]. In the same reference, two simulations were carried out with a point-source initial condition, in order to compare with an available experimental result [17]. The general conclusions of the Flatté *et al.* article [16] will be summarized here.

The PDF changes from the weak-fluctuation result of log-normal to the strong-fluctuation result of exponential as β_0 is increased. One of the main unknowns has been the route taken by the PDF on its way from one asymptotic result to the other. The answer is that the route is complicated, and depends in a strange way on whether the initial condition is a point source or a plane wave. The comparison between the experimental result and the numerical simulation was successful to within errors. However, the sensitivity of the result to slight changes in the spectral parameters, and the spectrum itself, indicates that more accurate simulations are needed to pin down the exact behaviour.

One question of interest to the PDF comparisons is the use of lower-lying moments of irradiance or log-irradiance. The Flatté *et al.* [*16*] article presents a scaled version of the PDF of log-irradiance as a comprehensive way of looking at the information. It is clear from their results that the log-irradiance moments are dominated by the behaviour of the irradiance near zero. Since there is expected to be no singular behaviour near zero irradiance, that low-moment behaviour is simply an analytic consequence of a smooth two-dimensional probability distribution of the complex phasor of the optical field near the origin.

Therefore, of more interest is the behaviour of the moments of irradiance itself. However, it has been argued that even low-index moments of irradiance are difficult to observe experimentally. It is probably better to concentrate on the PDF itself, and look directly at the high-irradiance tail of the PDF to evaluate the usefulness of various analytic approaches.

CONCLUSIONS

Numerical simulations have been carried out for both plane-wave and point-source initial conditions, with a correct spectrum of atmospheric turbulence near the inner-scale high-wavenumber cut-off. These simulations have been used to calculate the irradiance variance as a function of the two other parameters in the system: the turbulence strength β_0 and the inner scale l_0. A simple empirical equation for irradiance variance in the region centred on $\beta_0 = 5$ and $l_0/R_f = 0.5$ has been presented.

Two experiments, one for a near-plane-wave initial condition [*2*], and the other with a point-source initial condition [*8*], provide irradiance variances with no free parameters, since in both cases the values of β_0 and l_0 were independently measured.

In the strong-fluctuation regime, the zeroth-order approximation for the irradiance variance is unity. Therefore in evaluating various techniques for estimating the irradiance variance we have considered the deviation of the irradiance from unity as the quantity of interest.

The irradiance deviations measured in the experiments and calculated from the numerical simulations agree to well within 10%. This establishes the numerical simulations as the benchmark against which all other theories can be tested.

A different spectrum of turbulence can be regarded as another 'theory'. In this light, we have shown that the use of a traditional inner-scale cut-off represented by a simple Gaussian in wavenumber space gives deviations that are of order 50% different from those obtained with the atmospheric spectrum.

Other theories that have been suggested include first-order asymptotic expansions, numerical evaluation of moment equations, and two-scale expansions. All of these theories mis-estimate the irradiance deviation by amounts that are of order 50%.

When asymptotic expansions are carried out to third order, they agree with the numerical simulations within about 10%.

A general evaluation at this point would say that the numerical simulation technique has provided us with an order-of-magnitude improvement in our ability to estimate the irradiance deviation for optical propagation through atmospheric turbulence. Previous methods had errors of order 50%; now the errors are about 5%.

A further important conclusion must be that use of the proper atmospheric spectrum, and correct knowledge of the value of the inner scale, are both necessary for accurate estimates of irradiance deviation. The numerical simulations now provide us with the means to estimate the necessary accuracy of measurement of both the turbulence strength and inner scale in order to achieve a desired accuracy.

Finally, the beam shape has a very significant effect on the irradiance deviation for given turbulence parameters. Knowledge of the beam shape is most important near conditions similar to a point source; plane-wave results are not as sensitive to any of the parameters.

Conclusions about the probability density function of irradiance are harder to come by at this point. One can say that we understand the formation of the lower-index moments of log-irradiance very well, but that the lower-index moments of irradiance are difficult to evaluate, and it may be better to concentrate on the tail of the PDF itself.

ACKNOWLEDGMENTS

This research was supported in part by the U.S. Office of Naval Research under Contract N0001490J1022. We are grateful for a grant from the W.M. Keck Foundation. The simulations were carried out on the INTEL iPSC/860 parallel-processing computer at the San Diego Supercomputer Center and a MASPAR MP-2 parallel-processing computer at the University of California at Santa Cruz.

REFERENCES

1. P. Deitz and N. Wright, Saturation of scintillation magnitude in near-earth optical propagation, *J. Opt. Soc. Am.* **59**, 527–535 (1969).
2. A. Gurvich, M. Kallistratova and F. Martvel, An investigation of strong fluctuations of light intensity in a turbulent medium at a small wave parameter, *Radiophys. and Quan. Elec.* **20**, 705–714 (1977).
3. J. Martin and S. Flatté, Intensity images and statistics from numerical simulation of wave propagation in 3-D random media, *Applied Optics* **27**, 2111–2126 (1988).
4. V. Tatarskii, Depolarization of light by turbulent atmospheric inhomogeneities, *Radiophys. and Quan. Elec.* **10**, 987–988 (1967).
5. S.M. Flatté, G.Y. Wang and J. Martin, Irradiance variance of optical waves through atmospheric turbulence by numerical simulation and comparison with experiment, *J. Opt. Soc. Am. A* **10**, 2363–2370 (1993).
6. R. Hill and S. Clifford, Modified spectrum of atmospheric temperature fluctuations and its application to optical propagation. *J. Opt. Soc. Am.* **68**, 892–899 (1978).
7. R.J. Hill, Review of optical scintillation methods of measuring the refractive-index spectrum, inner scale and surface fluxes, *Waves in Random Media* **2**, 179–201 (1992).
8. A. Consortini, F. Cochetti, J. H. Churnside, and R. J. Hill, Inner-scale effect on intensity variance measured for weak to strong atmospheric scintillation, *J. Opt. Soc. Am. A* **10**, 2354–2362 (1993).
9. J. Martin and S. Flatté, Simulation of point-source scintillation through three-dimensional random media, *J. Opt. Soc. Am. A* **7**, 838–847 (1990).
10. R. Dashen and G. Wang, Intensity fluctuations for waves behind a phase screen: a new asymptotic scheme, *J. Opt. Soc. Am. A* **10**, 1219–1225 (1993).
11. R. Dashen, G. Wang, S. M. Flatté and C. Bracher, Moments of intensity and log-intensity: new asymptotic results for waves in power-law media, *J. Opt. Soc. Am. A* **10**, 1233–1242 (1993).
12. A. Gurvich, B. Elepov, V. Pokasov, K. Sabelfeld and V. Tatarskii, *Radiophysics and Quantum Electronics* **22**, 135–142 (1979).
13. A. Whitman and M. Beran, Two-scale solution for atmospheric scintillation, *J. Opt. Soc. Am A* **2**, 2133–2143 (1985).
14. A. Whitman and M. Beran, Two-scale solution for atmospheric scintillation from a point source, *J. Opt. Soc. Am. A* **5**, 735–737 (1988).
15. A. Whitman and M. Beran, First-order correction for the scintillation index and correlation of intensity function, *J. Opt. Soc. Am. A* **9**, 974–977 (1992).
16. S.M. Flatté, C. Bracher and G. Wang, Probability density functions of irradiance for waves in atmospheric turbulence calculated by numerical simulation, *J. Opt. Soc. Am. A* **11**, 2080–2092 (1994).
17. J. Churnside and R. Hill, Probability density of irradiance scintillations for strong path-integrated refractive turbulence, *J. Opt. Soc. Am. A* **4**, 727–733 (1987).

18 Optical scintillation methods of measuring atmospheric surface fluxes of heat and momentum

Reginald J. Hill
Environmental Technology Laboratory,
Environmental Research Laboratories,
National Oceanic and Atmospheric Administration,
325 Broadway, Boulder, CO 80303, USA

INTRODUCTION

Optical scintillation has been used to remotely sense many environmental parameters such as the wind component across the propagation path, rain rates and drop-size distributions, and the refractive-index structure parameter. Recent progress has been made on optical scintillation methods of measuring the inner scale of turbulence, l_0. This progress is reflected in new methods of using optical scintillation to obtain the surface fluxes of heat and momentum between the ground and the atmosphere. The refractive-index structure parameter is important for quantifying optical scintillation and deriving the fluxes. The refractive-index structure parameter is defined by

$$C_n^2 = \langle (n_1 - n_2)^2 \rangle / r_{12}^{2/3},\tag{1}$$

where n_1 and n_2 are refractive indices measured at points separated by the distance r_{12}; the angle brackets in Eq. (1) denote an average. The distance r_{12} in Eq. (1) must lie within the inertial range, which is typically from several centimetres to one-half the height above ground. Other structure parameters such as those of a velocity component or the temperature are defined similarly.

The essential quantity that relates optical scintillation to clear-air turbulence is the refractive-index spatial power spectrum, $\Phi_n(K)$. The inertial-convective range of the spatial spectrum refers to those spatial wavenumbers for which the turbulence energy spectrum has an inertial range and for which the dominant influence on the spectrum is convection (convection in the sense of being carried by fluid motion); that is, in the inertial-convective range, production of fluctuations and their diffusive dissipation are unimportant. In the inertial-convective range, the level of $\Phi_n(K)$ is proportional to C_n^2; l_0 is the spatial scale parametrizing the transition between the inertial-convective and dissipation ranges of $\Phi_n(K)$. Knowledge of the spatial power spectrum of refractive-index fluctuations is needed for predicting scintillation effects. In particular, scintillation methods of measuring l_0 require knowledge of the shape of $\Phi_n(K)$ in its dissipation range. Use of an erroneous spectrum results in an erroneous value of the inner scale, or no predicted value at all. Experiments and theoretical modelling have shown that in the atmosphere the dissipation range of $\Phi_n(K)$ is very different from the traditional assumption that the dissipation range is described by a Gaussian cut-off. In particular, a bump exists in the spectrum between the inertial-convective and dissipation ranges.

SCINTILLATION METHODS OF MEASURING INNER SCALE AND REFRACTIVE STRUCTURE PARAMETER

The inner scale, l_0, of refractive-index fluctuations in turbulence is defined as the spacing, r, at which the inertial range formula of the refractive-index structure function equals its formula in the dissipation range. Hill and Clifford (1978) give the relation between inner scale and the energy dissipation rate as:

$$l_0 = 7.4 \, (v^3/\varepsilon)^{1/4}, \tag{2}$$

where v is kinematic viscosity and ε is the rate of dissipation of turbulence kinetic energy per unit mass of fluid.

Methods of measuring the inner scale are based on the measurement of one quantity that is significantly influenced by the dissipation range of the refractive-index fluctuations and another quantity that is more influenced by the inertial range. That is, at least two spatial sizes of fluctuations are measured. There are many ways of doing this using optical scintillation; a review is given by Hill (1992). The spatial correlation of irradiances could be obtained at two different receiver spacings. The phase-difference structure function or mutual coherence function could be obtained at several receiver spacings. One could measure irradiance variances of waves having different Fresnel zone sizes; the Fresnel zone sizes ($\sqrt{\lambda L}$) could differ because either the path lengths (L) differ or the wavelengths (λ) differ. Temporal spectra of irradiance and phase difference contain many spatial scales if Taylor's frozen flow hypothesis is invoked. The angular spectrum obtained in the focal plane of a telescope is equivalent to the measurement of the mutual coherence function because they are related by a Fourier transform. Spatial filtering using several aperture sizes and aperture configurations could be used. Of these many scintillation methods for measurement of inner scale, only those that have been applied to measurement of atmospheric fluxes are discussed further.

The basic criteria for judging the relative merits of such methods are signal-to-noise ratio and systematic errors, sensitivity to inner-scale variation and insensitivity to other parameters, freedom from violations of Taylor's hypothesis, a close match between the weighting of propagation-path positions that contribute to the two measurements influenced by differing spatial sizes, and the expense and durability of the instrument.

One source of systematic error is the use of weak scintillation theory to determine inner scale from scintillation data. One could more precisely determine inner scale and also use longer propagation paths if the effects of strong scintillation are quantified. Hill and Frehlich (1996) have recently quantified strong scintillation effects on statistics for a diverged laser beam. The effect of strong scintillation on the large-aperture part of the instrument has been known for many years. These advances have been incorporated into advanced instrument designs.

Each scintillation method for measurement of inner scale involves two or more statistics that arise from different spatial sizes of refractive-index fluctuations. From the foregoing it is clear that if these statistics have nearly the same weighting over propagation-path positions, then the method will have an advantage in overcoming the effects of both turbulence intermittency and horizontal inhomogeneity. The first such method was devised by Hill and Ochs (1978). They used large-aperture phase-incoherent transmitters and large-aperture receivers of the same diameter. The signal was the aperture-averaged irradiance; the variance of its logarithm was obtained. The ratio of the variance from a smaller-aperture instrument to that from a larger-aperture instrument gives the inner scale. Hill and Ochs (1978) performed an experiment with three large-aperture instruments operating simultaneously; their diameters were 2.1, 5, and 15 cm. They obtained inner scales between 8 mm and 12 mm, obtained C_n^2, and observed the effect of the bump in the refractive-index spectrum.

Another method for which the path weighting of the two measurements is nearly identical was developed by Ochs and Hill (1985). They obtained the inner scale from the ratio of spherical-wave, point-receiver, log-irradiance variance to the variance of aperture-averaged irradiance from a large-aperture scintillometer. The spherical wave was emulated by a diverged laser and the point receiver by a 1 mm-diameter circular aperture. The large-aperture scintillometer had a phase-incoherent, uniformly illuminated, circular transmitter aperture having a 4.4 cm diameter, and a single circular receiving aperture of the same diameter. The variance of the logarithm of the aperture-averaged irradiance was obtained. Ochs and Hill (1985) measured l_0 values from 2.5 to 10 mm and compared them with *in situ* measurements of l_0. Using the same method, Hill *et al.* (1992a) obtained data on l_0 and C_n^2; much more extensive micrometeorological data were obtained in this new experiment. The ability of the method to obtain reliable values of l_0 and C_n^2 was confirmed even for very non-ideal turbulence conditions.

Figure 1 shows the combined data from the latter two experiments (respectively: 5 min averages, 260 m propagation path, and 1.5 m height; and

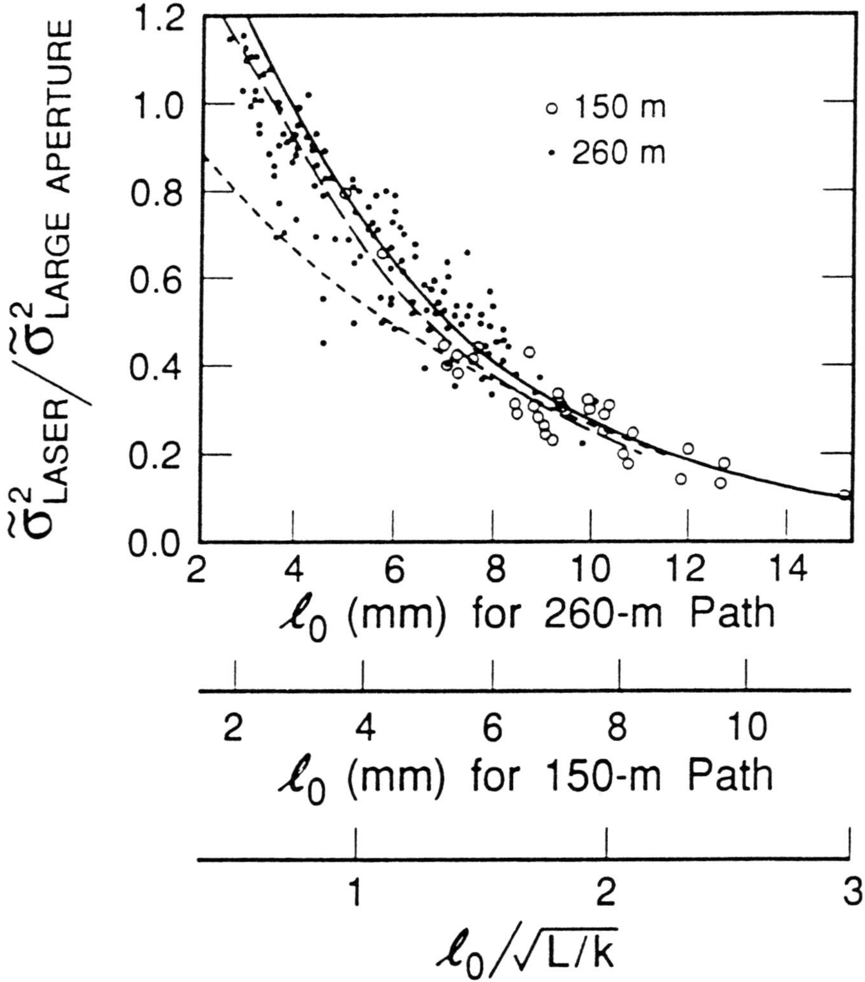

Fig. 1. Ratio of scaled variances as a function of $l_0/\sqrt{L/k}$, where l_0 is obtained from micrometeorological measurements. Values of l_0 are given on additional abscissae. Dots are from a 260 m path; open circles are from a 150 m path. Solid and long-dashed curves correspond to 150 m and 260 m paths, respectively, as calculated using the accurate refractive-index spectrum having the bump. The short dashed curve is calculated from the traditional spectrum for the 260 m path.

20 min averages, 150 m path, and 4 m height). The ratio of scaled variances from the laser radiation and large-aperture system is on the ordinate; the variances are scaled by their asymptotic formulas corresponding to $\Phi_n(K) \propto K^{-11/3}$, that is, for an inertial-convective range at all wavenumbers. The abscissa of Fig. 1 is $l_0/\sqrt{L/k}$, where l_0 is determined from micrometeorological data and $k = 2\pi/\lambda$. Separate abscissae give the values of l_0 in mm for the two paths; l_0 varied from 2.5 to 10 mm at the 1.5 m height, and from 3.5 to 12 mm at the 4 m height. The solid and long-dashed curves, corresponding to 150 m and 260 m paths, respectively, are calculated on the basis of weak scintillation and the accurate model of $\Phi_n(K)$. The short-dashed curve is calculated for the 260 m path using the traditional spectrum which has a Gaussian roll-off in the dissipation range. Clearly, the solid and long-dashed curves describe the data, but the short-dashed curve does not. To obtain l_0 from the scintillation measurement, one obtains the intercept of the ordinate of a datum with the appropriate theoretical curve and reads l_0 from the abscissa of this intercept. Erroneous values of l_0 would be obtained using the short-dashed curve for such a scintillation measurement of l_0. Graphing the ratio of scaled variances versus $l_0/\sqrt{L/k}$, as in Fig. 1, emphasizes that for shorter propagation paths, the larger wavenumbers within the dissipation range of $\Phi_n(K)$ contribute to the scintillation measurement. The data in Fig. 1 have contributions from the spectral bump on the left side of Fig. 1 to deep within the dissipation range on the right side. The decrease with increasing wavenumber of $K^{11/3}\Phi_n(K)$ within the dissipation range is reflected in Fig. 1 by the decrease of data and theoretical curves with increasing $l_0/\sqrt{L/k}$.

To summarize, three facts are shown in Fig. 1. First, the typical range of inner scales observed at a height of a few metres above the ground is from 3 to 12 mm. Second, the accurate refractive-index spectrum including the spectral bump is necessary to quantitatively predict the scintillation and hence to obtain the inner scale. Third, the dissipation range of the refractive-index spectrum, not the inertial-convective range, governs scintillation for heights of a few metres and propagation paths a few hundred metres in length.

If two waves of differing wavelengths propagate over the same path, then one can obtain the bichromatic correlation of irradiance fluctuations at the receivers as well as the two monochromatic variances. Azoulay *et al.* (1988) performed an experiment on both the bichromatic-correlation method and the ratio of monochromatic variances method using the wavelength pairs 0.6328 μm and 10.59 μm, as well as 0.6328 μm and 1.064 μm. They showed the humidity dispersion effect on the ratio of monochromatic variances for the wavelength pair 0.6328 μm and 10.594 μm. They also showed that the refractive-index spectrum obtained by Hill and Clifford (1978) is consistent with their data, but that the traditional spectrum having the Gaussian roll-off is not. A detailed comparison of the scintillation and micrometeorological data is given by Thiermann and Azoulay (1988). They observed inner-scale values in the range 3 to 7 mm.

MEASURING FLUXES USING SCINTILLATION

The atmospheric surface layer is within the lowest few tens of metres of the atmosphere. The basic dynamical interactions of the atmosphere with the solid or liquid surface are through the fluxes of heat, humidity, and momentum, all of which are driven by turbulence. For the horizontally homogeneous surface layer, these fluxes determine most turbulence statistics through empirical scaling relationships known as Monin–Obukhov similarity. Considering the vastness of the oceans and flat land areas, most of the world is covered by turbulent surface layers that are sufficiently homogeneous to allow useful applications of this similarity.

The three fluxes determine the stability of the surface layer. Stability is described by the Obukhov length. Both heat and humidity fluctuations determine the buoyancy fluctuations of the air, but humidity fluctuations often have a lesser or negligible role in affecting buoyancy; hence the stability is often determined from heat and momentum fluxes alone. These fluxes then determine the height profiles of many other surface-layer characteristics, such as the gradients of mean temperature and humidity, temperature variance and structure parameter (or those of humidity), vertical velocity variance and velocity structure parameters, and turbulence energy dissipation rate. Basically, all surface-layer turbulence statistics that obey Monin–Obukhov similarity are predicted from these three fluxes; the humidity flux is often unimportant for those quantities not involving humidity fluctuations.

Some terminology is needed. Unstable conditions occur when part of the turbulence energy is generated by upward convection of heat or humidity. Neutral conditions occur when the turbulence is generated by wind shear near the ground with convection providing no energy. Stable conditions occur when part of the turbulence energy is damped by stable stratification, but turbulence persists because of wind shear. The humidity flux is often multiplied by the latent heat of vaporization to produce the latent heat flux. The heat flux arising from temperature fluctuations is then referred to as the sensible heat flux. The Bowen ratio is the ratio of sensible heat flux to latent heat flux. These quantities are convenient for describing the surface energy balance.

Monin–Obukhov similarity relationships between the fluxes and other turbulence statistics allow estimation of the fluxes from measurements of some other turbulence statistics. For instance, the combined measurement of the near-surface gradients of temperature, humidity, and velocity can give the fluxes of heat, humidity, and momentum from the similarity relationships. Such flux estimates are called indirect, as opposed to direct measurements (i.e., so-called eddy correlation) which require correlation of vertical velocity fluctuations with fluctuations of heat, humidity, and horizontal velocity. The indirect dissipation method of estimating fluxes involves measuring the turbulence kinetic-energy dissipation rate and the temperature-variance dissipation rate to estimate heat and momentum fluxes via the similarity relationships.

Most studies using the indirect dissipation method do, in fact, measure the temperature structure parameter, C_T^2, the structure parameter of a velocity component, C_V^2, and sometimes the humidity structure parameter C_Q^2. The corresponding dissipation rates can be obtained from these structure parameters. In fact, most studies of the empirical similarity relationships between dissipation rates and fluxes have used simultaneous measurements of fluxes and structure parameters, with the dissipation rates subsequently derived from the structure parameters. Hence, the more directly measured similarity relationships are usually those relating structure parameters to the fluxes. Therefore, the method of estimating fluxes from structure parameters might better be referred to as the structure parameter method, rather than the indirect dissipation method, its more traditional name. Optical scintillation can be used to measure C_n^2, which is often nearly proportional to C_T^2. Hence, the structure parameter method can, for instance, use scintillation data to estimate fluxes.

If the Bowen ratio is sufficiently large that humidity flux has a negligible influence on stability, then values of C_T^2 and C_V^2 alone suffice to determine heat and momentum fluxes. If, in addition, the surface layer is very unstable, then C_T^2 alone suffices to estimate the heat flux. Quantities that are related to the turbulent kinetic energy dissipation rate, ε, such as the inner scale, can substitute for C_V^2 in estimating fluxes.

There are several advantages to measuring fluxes using scintillation relative to the eddy-correlation method. Scintillation averages spatially over the propagation path, whereas the eddy-correlation method is a point measurement. Eddy-correlation methods are subject to flow distortion caused by the instruments and their mountings, whereas there is no flow distortion for the scintillation method. The scintillation method can give flux values after only several minutes of averaging, whereas the eddy-correlation method requires about a 40-minute average. Thus, scintillation methods can resolve rapid changes such as the passage of clouds, which are problematic for the eddy-correlation method.

Consider the following example of estimating fluxes from scintillation measurements. Denote the surface fluxes of momentum and heat by τ and H, respectively. Denote a scaling velocity (the friction velocity) by

$$u_* = \sqrt{-\tau/\rho} \qquad (3)$$

where ρ is the mass density of air. Denote a scaling temperature by

$$T_* = -Q/U_*, \qquad (4)$$

where Q is the so-called temperature flux that is related to the heat flux by

$$H = \rho C_p Q, \qquad (5)$$

where C_p is the constant-pressure heat capacity of air. Q is often measured by correlating temperature fluctuations, T', with vertical velocity fluctuations, w', that is

$$Q = \langle T'w' \rangle, \tag{6}$$

where the angle brackets denote averaging. This is the so-called eddy correlation method of measuring Q, in which case H is obtained from (5). Likewise, u_* can be obtained using eddy correlation,

$$u_* = \sqrt{-\langle u'w' \rangle}, \tag{7}$$

where u' is the fluctuation of the horizontal wind component in the direction of the average wind vector. The Obukhov length is given by

$$L_{\mathrm{MO}} = \frac{u_*^2 T}{0.4 g T_*}, \tag{8}$$

where T is absolute temperature, and g is acceleration due to gravity. The effects of humidity fluctuations are neglected for simplicity. According to the Monin–Obukhov similarity,

$$\varepsilon = \frac{u_*^3}{z} h(z/L_{\mathrm{MO}}), \tag{9}$$

$$C_T^2 = \frac{T_*^2}{z^{2/3}} g(z/L_{\mathrm{MO}}), \tag{10}$$

where z is height. The dimensionless functions $h(z/L_{\mathrm{MO}})$ and $g(z/L_{\mathrm{MO}})$ have been determined by experiment. Given an optical scintillation measurement of C_n^2 and l_0, we have ε from (2), and C_T^2 follows from C_n^2. Thus, optical scintillation measurements give the left-hand sides of (9) and (10). Then, (8)–(10) can be solved by iteration to obtain T_* and U_*, and hence obtain H and τ from (3–5). Of course, simple measurements of temperature, pressure, and height above ground are also needed. In the very unstable limit, (10) reduces to

$$C_T^2 \propto Q^{4/3}, \tag{11}$$

so that heat flux is determined from C_T^2 alone.

Figure 2 compares the temperature flux, Q, obtained by using (8)–(10) from scintillation measurements of l_0 and C_n^2, with its value from eddy correlation [from Hill *et al.* (1992a)]. The solid dots indicate those very non-stationary cases for which the scintillation variances changed by more than a factor of eight during the 20 min data run. Good agreement is seen in Fig. 2, even for very non-stationary cases. The slight overestimate by the scintillation method is caused by an overestimate of C_n^2 by the scintillometer. For the same experiment that gives Fig. 2, Fig. 3 compares friction velocity obtained using (8)–(10)

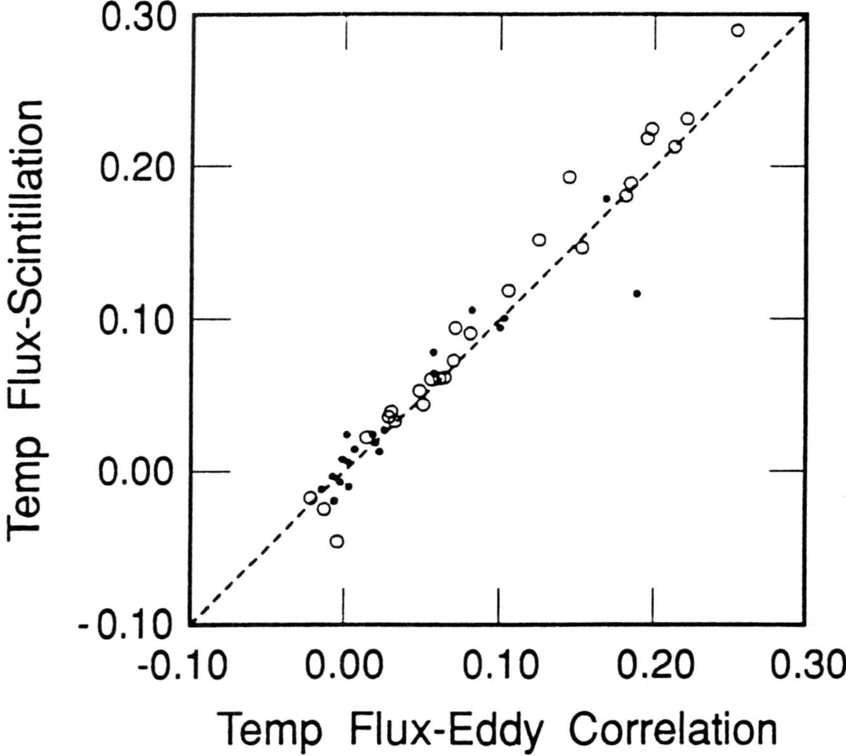

Fig. 2. Temperature flux obtained from scintillation measurements of l_0 and C_n^2 compared with values from eddy correlations. Units are °C m s^{-1}. Solid dots are very non-stationary data runs.

with its value from both eddy correlations as well as with the friction velocity inferred from wind speed, heat flux, and surface roughness. Good agreement is shown in Fig. 3.

EXPERIMENTS ON FLUXES FROM SCINTILLATION

Here, the experiments that used scintillation to measure fluxes are reviewed. The discussion is not exhaustive because there is an exhaustive review of this subject by Hill (1992), and a reprint volume containing all the important papers up to 1990 was prepared by Andreas (1990). In many cases, the fluxes were not measured using only scintillation; additional micrometeorological measurements were often used in combination with scintillation and often some simplifying assumption was used.

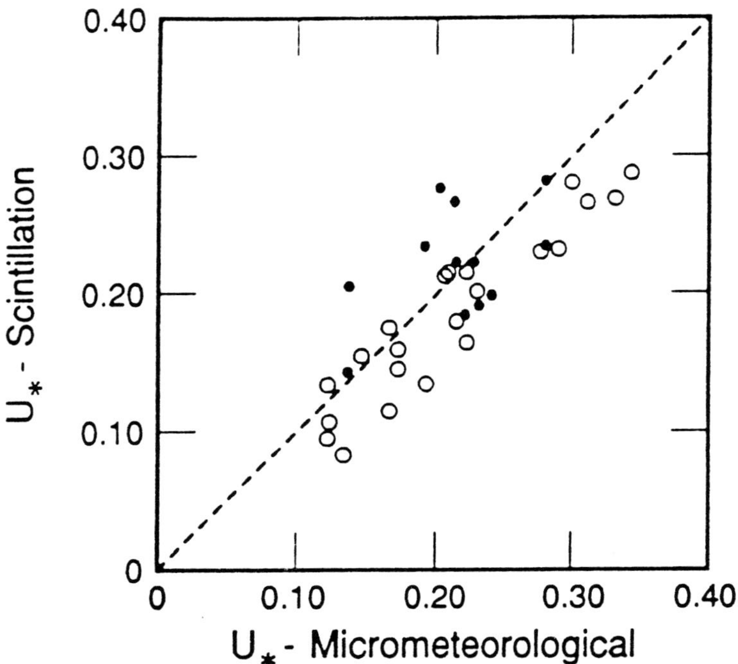

Fig. 3. Friction velocity obtained from scintillation measurements of I_0 and C_n^2 compared with values from eddy correlation (solid dots) and from wind speed, heat flux, and surface roughness (open circles). Units are m s^{-1}.

Wesely and Derzko (1975) used image blurring to determine the value of the refractive-index structure parameter, C_n^2. With the aid of supporting micrometeorological measurements, the value of C_n^2 can help determine other meteorological variables. Wesely (1976) used both image blurring and mirages (i.e., optical beam bending that yields the refractive-index gradient) with other supporting meteorological measurements to determine sensible and latent heat fluxes.

Wyngaard and Clifford (1978) gave the theoretical relationship between fluxes and C_T^2, C_Q^2, and C_V^2. They showed that measuring these structure parameters gives the fluxes of heat, humidity, and momentum for any value of stability. They suggested that radio-wave scintillation might be useful in determining C_Q^2, but one must also know the temperature–humidity cross-structure parameter, C_{TQ}. In a subsequent experiment Wyngaard and colleagues measured laser-scintillation variance during weak scintillation conditions to determine C_n^2 and hence C_T^2, from which they estimated heat flux using (11). Their scintillation method showed good agreement with the heat flux from eddy correlation, but their technique is limited to very unstable conditions.

Kohsiek and Herben (1983) used the fact that measurements of scintillation at three wavelengths can give values of C_T^2, C_Q^2, and C_{TQ}, and that if one assumes a relationship of C_{TQ} to C_T^2 and C_Q^2, then measurements using only two wavelengths are needed. They measured radio-wave scintillation and estimated humidity flux by assuming a value for C_{TQ} relative to $\sqrt{C_T^2 C_Q^2}$ and also assuming a value for C_T^2/C_Q^2. The latter is equivalent to assuming a value for the Bowen ratio.

Hill *et al.* (1988) obtained heat and humidity fluxes by three methods: two-wavelength scintillation, C_T^2 and C_Q^2 obtained from *in situ* resistance-wire thermometers and Lyman-α hygrometers, and eddy correlation. The two wavelengths were near-visible (0.94 μm) and millimetre-wave (173 GHz); for the first method, C_{TQ} was taken to be $\pm\sqrt{C_T^2 C_Q^2}$, which was verified from the *in situ* data. They obtained very good agreement of the fluxes measured by the three methods.

Thiermann and Azoulay (1988) described a two-wavelength method that, unlike other proposals, included measuring the bichromatic correlation coefficient of the irradiances as well as their variances. They showed that momentum, heat, and humidity fluxes can be obtained. To do so, they assumed some relationship between C_{TQ} and $\pm\sqrt{C_T^2 C_Q^2}$.

The fluxes of heat and momentum can be obtained from C_n^2 values measured at two heights; this is called the C_n^2-profile method, which was tested by Hill *et al.* (1992b). This method uses the fact that (10) evaluated at two different heights constitutes two equations from which the two unknown fluxes can be obtained; no measurement of inner scale is needed for this method. Andreas (1988) showed that the sensitivity of this C_n^2-profile method is such that stability can be obtained with an accuracy of about a factor of 2. Heat and momentum fluxes can be obtained with better accuracy.

Thiermann and Grassl (1992) used the optical bichromatic correlation method of measuring l_0 and C_n^2; they then used these values to obtain the heat and momentum fluxes from Monin–Obukhov similarity. The wind speed, wind profile, and heat flux were obtained using micrometeorological instruments. The results showed that reliable flux estimates were obtained. The experiment by Hill *et al.* (1992a) has already been discussed with regard to Figs 2 and 3. Both Hill *et al.* (1992a) and Thiermann and Grassl (1992) used essentially the same method to estimate the fluxes from l_0 and C_n^2; their optical scintillation techniques for measuring l_0 and C_n^2 differ.

Green *et al.* (1994) used laser-diode scintillation to obtain the refractive-index structure parameter and obtained wind speed from an anemometer. These two pieces of information with knowledge of the surface roughness were used to obtain the fluxes of heat and momentum. Good agreement was obtained with standard micrometeorological measurements. This experiment extended the range of conditions over which the scintillation methods have been verified.

CONCLUSIONS

Optical scintillation methods of measuring the surface fluxes of heat, momentum, and humidity are reviewed here. Future progress is possible, and directions of future developments are mentioned. Recent progress in measuring inner scale has led to unprecedented accuracy and has resulted in optical-scintillation measurements of heat and momentum fluxes, which show great promise for future micrometeorological applications. At the present time there are measurement programmes underway in New Zealand, the Netherlands, and the United States. The availability of a commercial scintillometer that measures inner scale and refractive-index structure parameter has encouraged these applications.

REFERENCES

Andreas, E.L., Atmospheric stability from scintillation measurements. *Appl. Opt.* **27**, 2241–2246 (1988).

Andreas, E.L., *Selected Papers on Turbulence in a Refractive Medium, SPIE Milestone Series* Vol. 25, Society of Photo-Optical Instrumentation Engineers, Bellingham WA (1990).

Azoulay, E., V. Thiermann, A. Jetter, A. Kohnle, and Z. Azar, Optical measurement of the inner scale of turbulence. *J. Phys.* **D21**, 541–544 (1988).

Green, A.E., K.J. McAneney, and M.S. Astill, Surface-layer scintillation measurements of daytime sensible heat and momentum fluxes. *Bound Layer Meteorol.* **68**, 357–373 (1994).

Hill, R.J., Review of optical scintillation methods of measuring the refractive-index spectrum, inner scale and surface fluxes. *Waves in Random Media* **2**, 179–201 (1992).

Hill, R.J., R.A. Bohlander, S.F. Clifford, R.W. McMillan, J.T. Priestley, and W.P. Schoenfeld, Turbulence-induced millimeter-wave scintillation compared with micrometeorological measurements, *IEEE Trans. Geosci. Remote Sens.* **26**, 330–342 (1988).

Hill, R.J. and S.F. Clifford, Modified spectrum of atmospheric temperature fluctuations and its application to optical propagation. *J. Opt. Soc. Am.* **68**, 892–899 (1978).

Hill, R.J. and G.R. Ochs, Fine calibration of large-aperture optical scintillometers and an optical estimate of inner scale of turbulence. *Appl. Opt.* **17**, 3608–3612 (1978).

Hill, R.J., G.R. Ochs and J.J. Wilson, Measuring surface layer fluxes of heat and momentum using optical scintillation. *Bound Layer Meteorol.* **58**, 391–408 (1992a).

Hill, R.J., G.R. Ochs and J.J. Wilson, Surface layer fluxes measured using the C_T^2-profile method. *J. Atmos. Oceanic Technol.* **9**, 526–537 (1992b).

Hill, R.J. and R.G. Frehlich, Onset of strong scintillation with application to remote sensing of turbulence inner scale. *Appl. Opt.* **35**, 986–997 (1996).

Kohsiek, W. and M.H.A.J. Herben, Evaporation derived from optical and radio-wave scintillation. *Appl. Opt.* **22**, 2566–2570 (1983).

Ochs, G.R. and R.J. Hill, Optical-scintillation method of measuring turbulence inner scale. *Appl. Opt.* **24**, 2430–2432 (1985).

Thiermann, V. and E. Azoulay, A two wavelength laser scintillometer for monitoring surface layer fluxes under near neutral conditions. *Proc. of the International Laser Radar Conf.* Innichen-San Candido, Italy, 20–24 June 1988, p. 70–73 (1988).

Thiermann, V. and H. Grassl, The measurement of turbulent surface layer fluxes by use of bichromatic scintillation. *Bound Layer Meteorol.* **58**, 367–389 (1992).

Wesely, M.L. A comparison of two optical methods for measuring line averages of thermal exchanges above warm water surfaces. *J. Appl. Meteorol.* **15**, 1177–1188 (1976).

Wesely, M.L. and Z.I. Derzko, Atmospheric turbulence parameters from visual resolution. *Appl. Opt.* **14**, 847–853 (1975).

Wyngaard, J.C. and S.F. Clifford, Estimating momentum, heat, and moisture fluxes from structure parameters. *J. Atmos. Sci.* **35**, 1204–1211 (1978).

19 Coherent Doppler lidar measurements of winds

Rod Frehlich
Cooperative Institute for Research in the Environmental Sciences (CIRES), University of Colorado, Boulder, CO 80309, USA

INTRODUCTION

Coherent Doppler lidar is an attractive method for remote measurements of winds. A short laser pulse is transmitted and the photons scattered by aerosol particles are collected by a telescope, focused on to a detector and mixed with a frequency-stable polarization-matched laser reference beam called the Local Oscillator (LO). Coherent or heterodyne detection is a sensitive method for detection of weak signals and is immune to the effects of background light [1]. The radial velocity v of the scatterers produces a Doppler frequency shift f given by

$$v = \lambda f / 2 \qquad (1)$$

where λ is the wavelength of the laser. For a 2 μm laser pulse, 1 m sec^{-1} of radial velocity corresponds to a 1 MHz Doppler frequency shift, which can be estimated as the pulse travels through the atmosphere.

The first coherent Doppler lidars were based on the CO_2 laser [2–5]. More recently, solid state Doppler lidars have been successfully operated [6–8]. The main use of these instruments is wind measurements in the atmospheric surface layer [9]. Coherent Doppler lidar is under consideration for measurements of the global wind field from space [4, 10, 11].

Incoherent Doppler lidar can also measure the Doppler frequency shift f using changes in the intensity of the received signal as a function of frequency. For coherent Doppler lidar, the Doppler frequency is estimated from an oscillating signal. A comparison of these two methods is given in [12].

We will review the design and operation of coherent Doppler lidar, the performance evaluation, the effects of refractive turbulence, the velocity estimation, examples of wind measurements, and the basic characteristics of the measurements.

COHERENT DOPPLER LIDAR DESIGN

Many different designs for coherent Doppler lidar have been produced. The basic schematic [*13*] is shown in Fig. 1. The laser pulse is transmitted through a transmitter aperture $W_T(u, 0)$ and propagates to the aerosol scatterers at location (p, R). The backscattered field is collected by a receiver aperture $W_R(v, 0)$ and focused on to a detector. The LO laser field is also focused on the detector using a beam splitter to produce the coherent or heterodyne Doppler lidar signal. Other optical elements are required to isolate the large transmitted pulse from the weak backscattered signal [*8*]. Typically, a monostatic lidar is used, i.e., the transmitter and the receiver aperture are the same. Bistatic lidar with separate transmitter and receiver telescopes have also been successfully operated [*14*].

The total backscattered field at the receiver is the superposition of the backscattered fields from many randomly located aerosol particles and therefore satisfies the central limit theorem to produce a Gaussian random process described by speckle statistics [*15, 16*]. Performance analysis requires an ensemble average over the random positions of the aerosol particles. An important measure of performance is the signal-to-noise ratio (SNR), which is the ratio of the signal power of the Doppler signal to the noise power. If the shot noise from the LO dominates the other detector noise sources, then [*13*]

$$\text{SNR} = \frac{P_D \eta_H}{h v B} \tag{2}$$

where P_D is the average direct detection power, η_H is the heterodyne efficiency, h is Planck's constant, v is the laser frequency, and B is the detector noise bandwidth. For a short pulse and a small range gate at range R [*13*]

$$P_D = \frac{\eta_Q \beta(R) K(R)^2 c U_T A_R}{2R^2} \tag{3}$$

where η_Q is the detector quantum efficiency, $\beta(R)$ is the aerosol backscatter coefficient, $K(R)$ is the one-way irradiance extinction, c is the speed of light, U_T is the transmitted laser pulse energy, and A_R is the receiver area. The receiver area and the pulse energy are the only lidar design parameters that influence the average direct detection power.

The heterodyne efficiency has a maximum value of unity when the spatial field of the LO on the detector is proportional to the spatial field of the

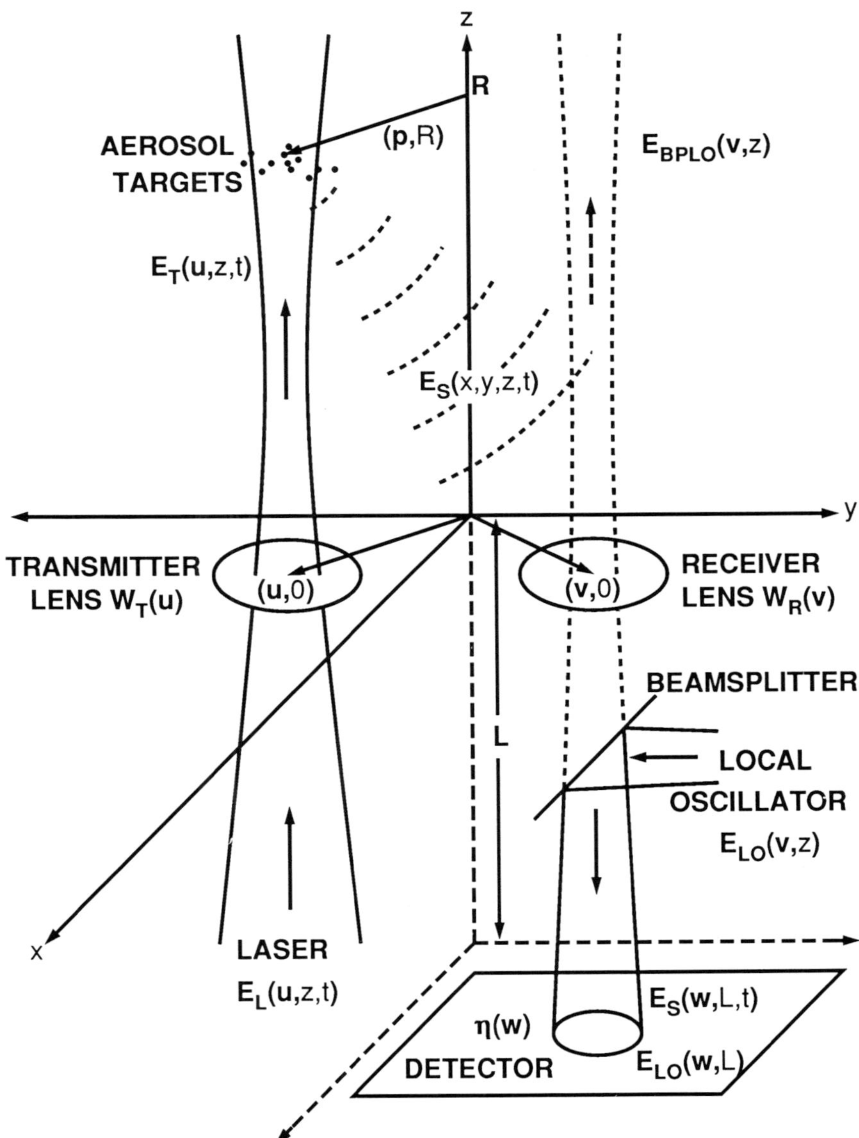

Fig. 1. Geometry for coherent detection Doppler lidar [13]. An actual system would have overlap of the transmitted and BPLO beams at the target.

backscattered radiation. Any mismatch of the amplitude and phase of the two fields will result in a loss in heterodyne efficiency and SNR.

For deterministic fields, calculations of SNR and heterodyne efficiency are determined from the fields on the detector surface. For coherent Doppler lidar, the calculations can be performed in any transverse plane [13, 17–21]. The transmitter and LO fields are usually defined in the receiver plane, i.e., on the target side of the two apertures $W_T(u, 0)$ and $W_R(v, 0)$. Calculations are particularly simple if the detector has uniform quantum efficiency η_Q and collects all the backscattered and LO radiation. The detector is then approximated as an infinite detector and calculations of SNR and heterodyne efficiency can be determined from the irradiance of the transmitted field on the target and the irradiance of the Back Propagated Local Oscillator (BPLO), i.e., the irradiance produced by propagating the LO field from the detector surface through the receiver optics to the target (see Fig. 1). This concept was proposed by Siegman [17] using far-field analysis, and extended to the Fresnel approximation by Rye [18]. Inclusion of the effects of general detector geometry with refractive turbulence were added later [13].

Collimated transmitter and BPLO beams [13] are typical for surface layer wind measurements because the SNR is constant with range R in the near-field region $(R\lambda < A_R)$ and heterodyne efficiency $\eta_H \propto R^2$. In the far-field or Fraunhoffer region $(R\lambda > A_R)$ SNR $\propto R^{-2}$ and η_H is a constant. Because the low SNR region is most critical for performance evaluation, lidar design optimization is determined for the far-field condition, or equivalently, the focused beam condition. Wang proposed an optimal design for a monostatic Doppler lidar with a circular aperture of diameter D, Gaussian transmitter beam, and Gaussian LO beam. Rye proposed an optimal design for the same configuration that has no truncation of the BPLO by the receiver aperture. These two optimal designs are compared in [22].

For a given monostatic lidar with a given transmitter field and a given aperture, the optimal LO field is the solution of an integral equation [23]. The solution for the optimal LO field distribution under the far-field or focused condition for a circular aperture and various Gaussian and super-Gaussian transmitter fields has been determined and a feasible optical design proposed [23]. The resulting performance is essentially the same as the design proposed by Rye. Calculations of performance for typical designs are usually numerical [20, 21]. However, analytic approximations have been produced [13, 24, 25].

The most crucial design constraint for coherent Doppler lidar is the alignment of the backscattered field and the LO field on the detector surface. This is equivalent to aligning the transmitter axis with the BPLO axis (see Fig. 1), i.e., the transmitter beam pattern in the target plane must match the BPLO beam pattern. In Fig. 2, the heterodyne efficiency for an optimal monostatic lidar with a circular aperture of diameter D and a target in the far field is plotted as a function of the parameter $\pi D \Delta\theta / \lambda$ where $\Delta\theta$ is the misalignment angle. The maximum $\eta_H = 0.46096$, which is independent of

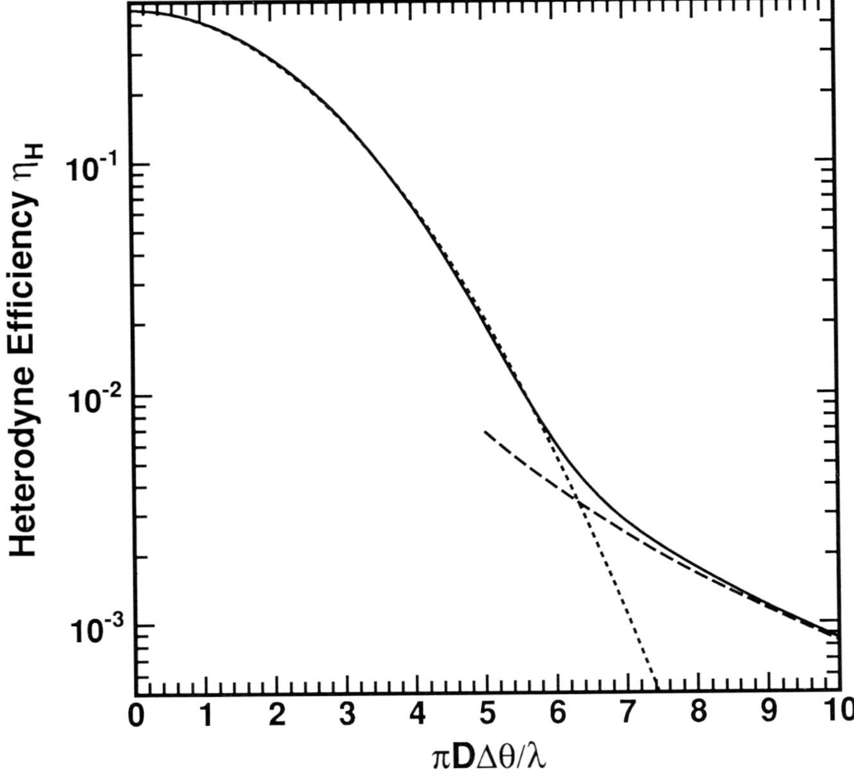

Fig. 2. Heterodyne efficiency η_H as a function of the normalized misalignment parameter $\pi D\Delta\theta/\lambda$ for the far-field (—). The best-fit Gaussian function $0.46096 \exp[-(\pi D \Delta\theta/2.8363\lambda)^2]$ is (…) and the best fit power-law model $0.9157(\pi D \Delta\theta/\lambda)^{-3.0415}$ is (---). Reprinted from Ref. [24] by permission of Taylor & Francis Ltd.

range R in the far-field. The $1/e$ loss in η_H and SNR from angle misalignment is given by $\Delta\theta = 0.9028\lambda/D$. For a 2 µm lidar and $D = 20$ cm aperture, $\Delta\theta = 9.028$ µrad. The beam misalignment tolerance is inversely proportional to λ. This is one of the main reasons 10 µm coherent Doppler lidars were the first operational lidars. For space-based operation, a coherent Doppler lidar operates in the far-field because the range to the aerosol targets near the earth's surface satisfy $R\lambda > A_R$. For metre-sized optics, the alignment criterion at 2 µm is $\Delta\theta < 2$ µrad.

Estimates of SNR, coherent power, direct detection power, and η_H are random because the backscattered field is a Gaussian random process. The Probability Density Function (PDF) for the coherent power from a diffuse or aerosol target is an exponential distribution [15, 24, 26, 27]. The PDF of the direct detection power is well represented as a gamma distribution [15, 24, 27].

Simulations of the normalized direct detection power and coherent detection power is shown in Fig. 3 for 10 000 shots. Note the correlation between the two signals. A realization of the random fields on the detector surface for a typical return is shown in Fig. 4. The heterodyne efficiency for this case is almost equal to the average heterodyne efficiency. Many statistically independent lidar signals must be averaged to produce a reliable estimate of the SNR or heterodyne efficiency [24]. An estimate for the heterodyne efficiency is the preferred measure of system performance because it is an absolute measure of beam alignment and independent of the transmitted laser power, the atmospheric attenuation, the target backscatter coefficient, the detector gain, and the detector quantum efficiency. However, an observable direct detection signal is required.

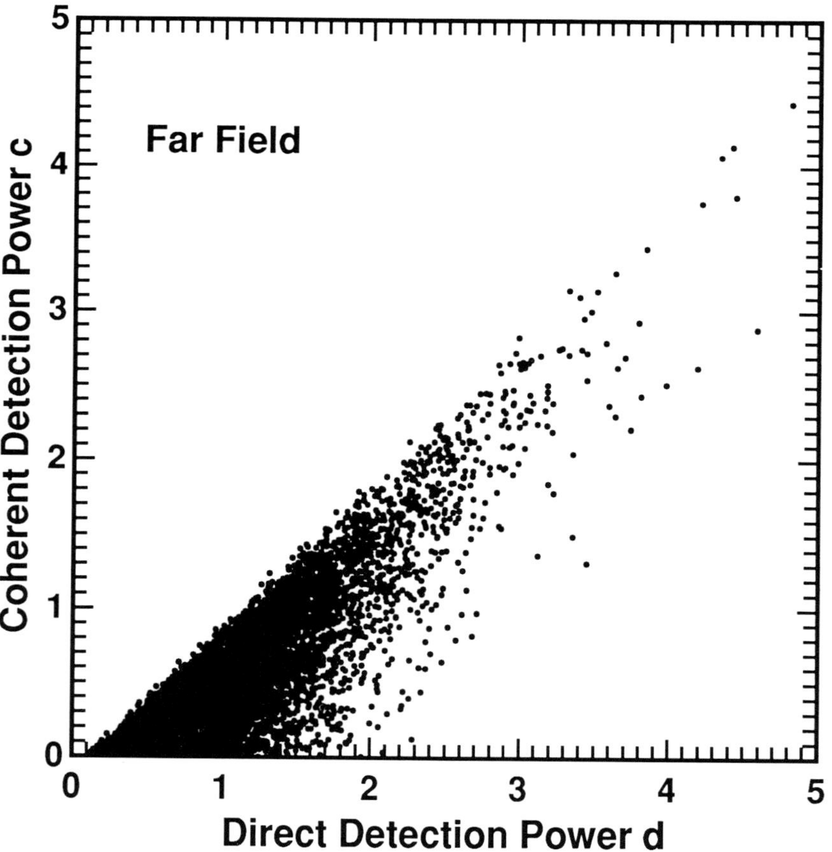

Fig. 3. Normalized coherent detection power *c* versus the normalized direct detection power *d* for a perfectly aligned coherent Doppler lidar and 10 000 shots with the target in the far field [24]. Reprinted from Ref. [24] by permission of Taylor & Francis Ltd.

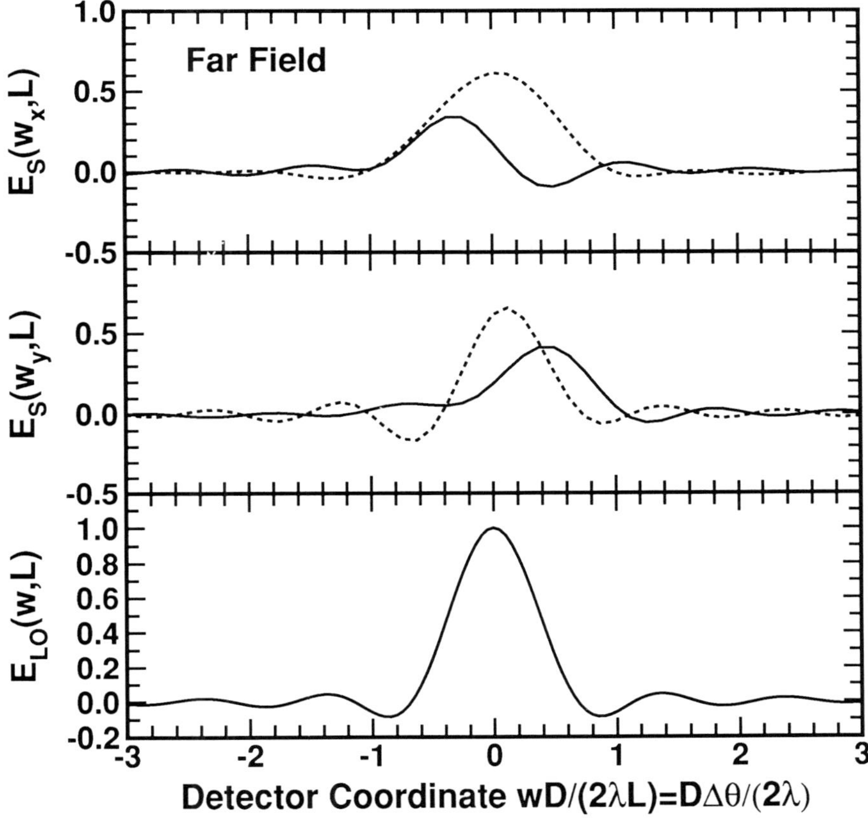

Fig. 4. Simulation [24] of the backscattered field on the detector surface in the x direction $E_S(w_x, L)$ and the y direction $E_S(w_y, L)$ for far field operation ($\eta_H = 0.460$). The real part of the complex field is (—) and the imaginary part is (...). The optimal LO field $E_{LO}(w, L)$ is also shown. Reprinted from Ref. [24] by permission of Taylor & Francis Ltd.

EFFECTS OF REFRACTIVE TURBULENCE

Atmospheric refractive turbulence reduces the phase coherence of propagating laser beams. This has a deleterious effect on coherent Doppler lidar because of the loss in heterodyne efficiency. However, the situation is more complex for the monostatic configuration. The propagation forward and backward through the same turbulence produces an enhanced backscattering mechanism [13, 28, 29]. In some cases, this enhancement can produce larger SNR than in the case of no refractive turbulence [29]. The enhanced backscatter mechanism is very sensitive to beam misalignment and in strong scattering situations can produce a variation in SNR and heterodyne efficiency η_H as a function of misalignment angle $\Delta\theta$ that is more sensitive than the case with no refractive turbulence [29]. These effects are more pronounced for the shorter wavelength lidars. For typical target ranges near ground level, the effects of refractive turbulence will be observable during sunny, clear, daytime conditions. This complicates calibration of coherent Doppler lidar.

VELOCITY ESTIMATION

The radial velocity v is estimated from the Doppler frequency of the lidar signal. For any time t, the lidar detector signal is a Gaussian random process because it is proportional to the integral of the backscattered field over the detector surface. For ideal conditions, the atmospheric parameters (β, K, v) and SNR are constant over the range gate of interest. Then the Doppler lidar signal $i_S(t)$ is a stationary Gaussian random process which is completely determined by its covariance function [30]

$$R(\tau) = \langle i_S(t) i_S(t + \tau) \rangle \tag{4}$$

A simulation [31] of a coherent Doppler lidar signal for ideal conditions is shown in Fig. 5 as well as the range weighting function for SNR of a Gaussian lidar pulse with temporal power profile

$$P_L(t) = \exp(-t^2/\sigma^2) \tag{5}$$

where σ is the pulse $1/e$ width. As the pulse propagates through the atmosphere, the Doppler lidar data is digitized and converted to complex data samples z_k with a sampling interval T_S. The range gate is determined by the total observation time per estimate MT_s where M is the number of complex data samples per estimate. The distance Δp that the pulse moves during the observation time is

$$\Delta p = MT_s c/2 \tag{6}$$

where c is the speed of light in the homogeneous atmosphere. The full width at half maximum of the range weighting function for SNR of the Gaussian pulse is given by

$$\Delta r = \sqrt{\ln 2} \, c\sigma \tag{7}$$

For ideal conditions, the covariance function R_k for the stationary complex data is determined by the SNR and the transmitted pulse [30]. For the Gaussian pulse (5)

$$R_k = \langle z_{k+i} z_i^* \rangle = R(kT_S) = \text{SNR} \, \exp[-k^2 T_S^2/(4\sigma^2) + 2\pi i T_S fk] + \delta_k \tag{8}$$

where δ_k is the Kronecker delta symbol and represents the uncorrelated shot noise of the detector. For stationary signals, the spectrum $S(f)$ is defined as the

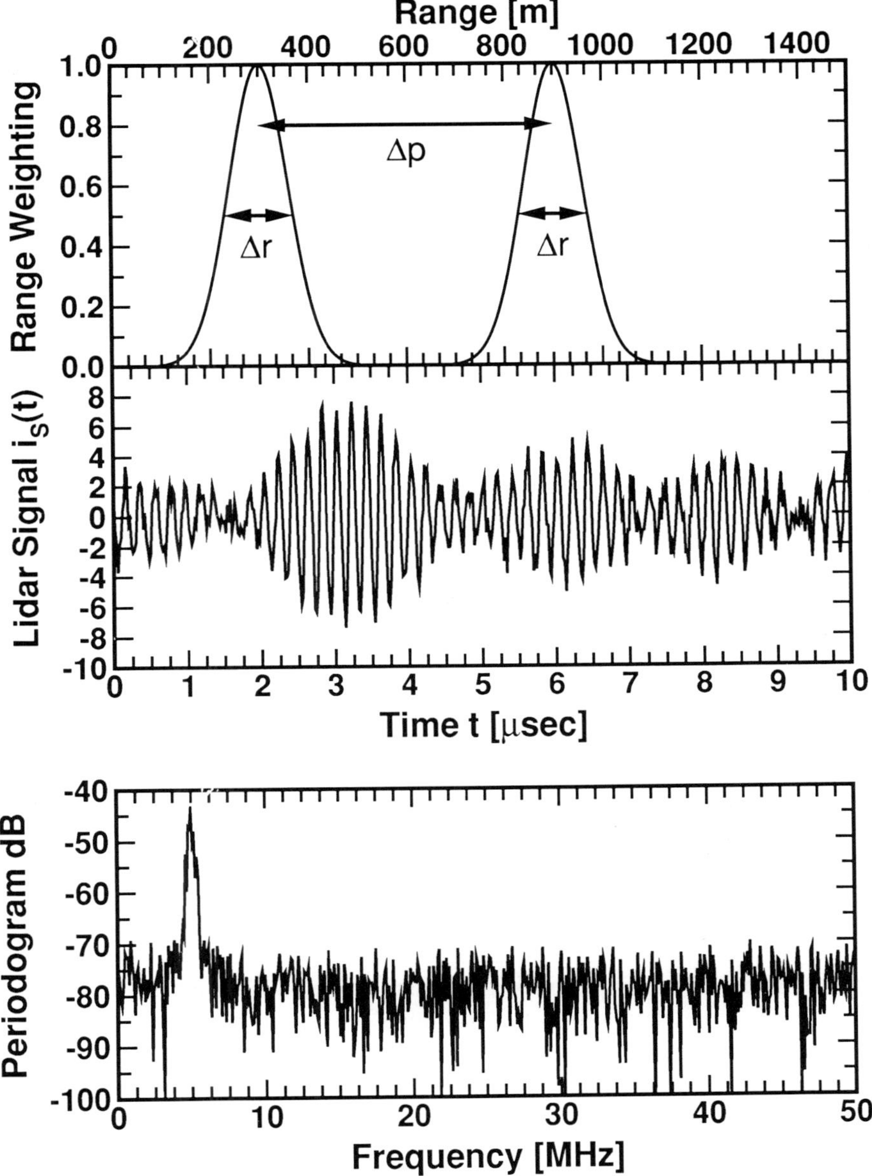

Fig. 5. Simulation of Doppler lidar data [*31*] for a 2 μm lidar [*8*] with a Gaussian pulse Eq. (16) and SNR = 10, f = 5 MHz, w = 0.2 MHz, T_S = 0.02 μsec. The range weighting function $P_L(t - 2R/c)$ is shown for the signal at 2 μsec and 6 μsec. The full width at half maximum Δr Eq. (12) of the pulse sensing volume, the distance Δp, Eq. (14), that the pulse moves in 4 μsec, and the periodogram estimate of the signal spectrum is also shown. Reprinted from Ref. [*31*] by permission of the American Meteorological Society.

Fourier transform of the covariance function, i.e.,

$$S(f) = \int_{-\infty}^{\infty} R(\tau)\exp(-2\pi i f \tau)\, d\tau = \frac{\text{SNR}}{\sqrt{2\pi}w}\exp[-f^2/(2w^2)] + 1/T_S \qquad (9)$$

where the spectral width w of the stationary Doppler signal is given by

$$w = 1/(\sqrt{8}\,\pi\sigma) \qquad (10)$$

In practice, it is difficult to estimate w with data from one range gate because the periodogram estimate of the spectrum (see Fig. 5) is a poor estimate due to small sample size effects.

For a given coherent Doppler lidar operating at wavelength λ, there are six parameters $(v, \lambda, \text{SNR}, w, M, T_S)$. By convenient normalization, these six parameters reduce to three basic parameters Φ, Ω, and M which are related to the important measurement parameters.

The parameter Φ is a measure of the signal energy per estimate. When the shot noise of the local oscillator dominates the detector noise [31]

$$\Phi = \eta_H N_e = \text{SNR}\ M \qquad (11)$$

where N_e is the average number of photo-electrons per estimate. Φ is the average number of coherent photo-electrons per estimate.

The parameter Ω can be written as

$$\Omega = wMT_S = \frac{[\ln(2)/2]^{1/2}}{\pi}\ \frac{\Delta p}{\Delta r} = 0.1874\ \frac{\Delta p}{\Delta r}. \qquad (12)$$

The parameter Ω is a measure of the number of independent samples of the Doppler lidar signal per range gate or the 'speckle count'. If $\Delta p = \Delta r$, then $\Omega = 0.1874 = \Omega_1$, which is a useful reference for high-resolution wind measurement. For $\Omega > \Omega_1$, the velocity estimates have excessive averaging of the wind field over the range gate by the moving pulse. In some cases, like measurements of tropospheric winds from space, this is not a disadvantage because the final data product is adequate and higher spatial resolution is possible in the boundary layer where the signal levels are higher. For $\Omega < \Omega_1$, adjacent velocity estimates are correlated because the spatial extent of the pulse Δr is larger than the distance Δp that the pulse moves per estimate. This correlation must be included in analysis of spatial wind statistics. The effective range resolution is defined as $\Delta R = \Delta p + \Delta r$. Fixed Ω implies fixed range resolution and fixed transmitted pulse length, the typical comparison case for Doppler lidar performance evaluation.

The choice of Δp and Ω determines MT_S and w which defines the pulse width σ. The sampling interval T_S is determined by the maximum search velocity v_{search} which is defined by

$$v_{\text{search}} = \lambda f_N/2 = \lambda/(2T_S) \qquad (13)$$

where $f_N = 1/T_S$ is the maximum observable frequency or Nyquist frequency for complex data.

Doppler lidar performance is completely described by the PDF of the velocity estimates as a function of the basic parameters Φ, Ω, and M. Many different algorithms have been considered for estimation of the Doppler frequency [31–34]. An example of a PDF of velocity estimates using the maximum likelihood [31] estimator is shown in Fig. 6 for a weak signal regime where the random fluctuations of the signal power produce realizations where the signal peak cannot be identified among the noise peaks, and a random noise peak is chosen as the Doppler frequency. These cases produce a uniform distribution of outliers. For the better velocity estimators, the PDF is characterized by a clump of good estimates with standard deviation g centred around the true mean velocity and a fraction b of uniformly distributed outliers.

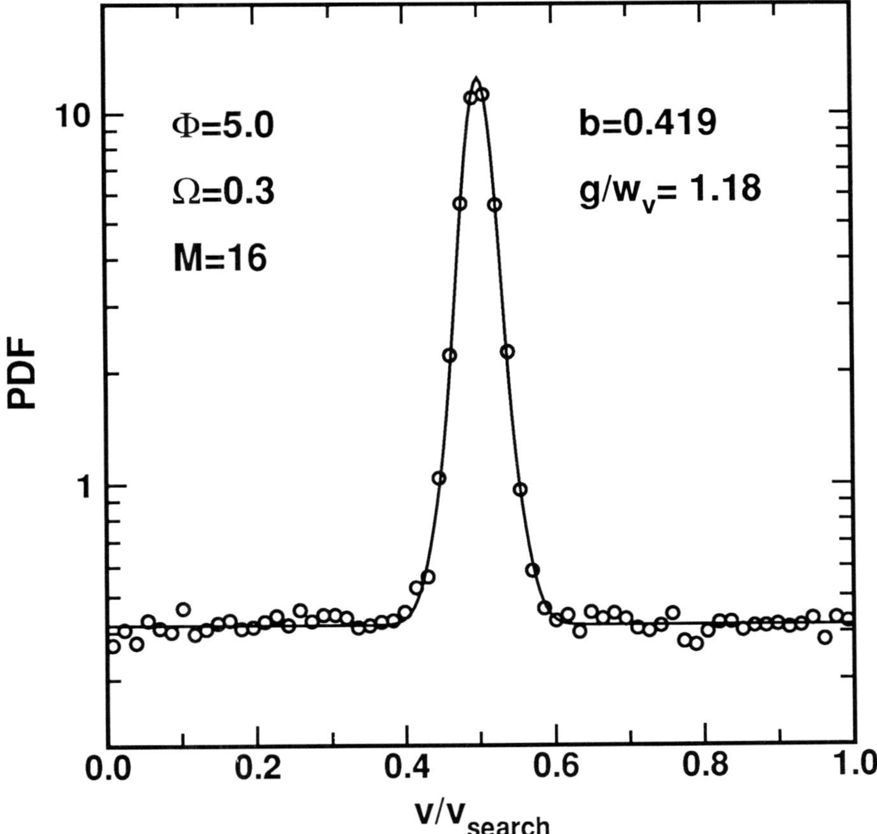

Fig. 6. Histogram (o) of 50 000 maximum likelihood estimates for velocity with $M = 16$, $\Omega = 0.3$, and $\Phi = 5.0$. The best fit PDF is indicated by a solid line for a two-component Gaussian model with a uniform distribution of outliers.

Approximately universal curves of performance of velocity estimates are produced [31] by plotting the standard deviation g normalized by w_v, the signal spectral width in velocity space, i.e.,

$$w_v = \lambda w/2 \tag{14}$$

versus the parameter Φ for fixed Ω and M, i.e., fixed range resolution and fixed velocity search space (see Fig. 7). For large Φ there are no outliers and the estimators are unbiased. In this regime, the performance approaches the ideal theoretical limit of the Cramer Rao bound [35] and performance is independent of M [31], i.e., velocity estimation error is independent of the velocity search space. In addition, if there are no wind fluctuations over the range gate, the

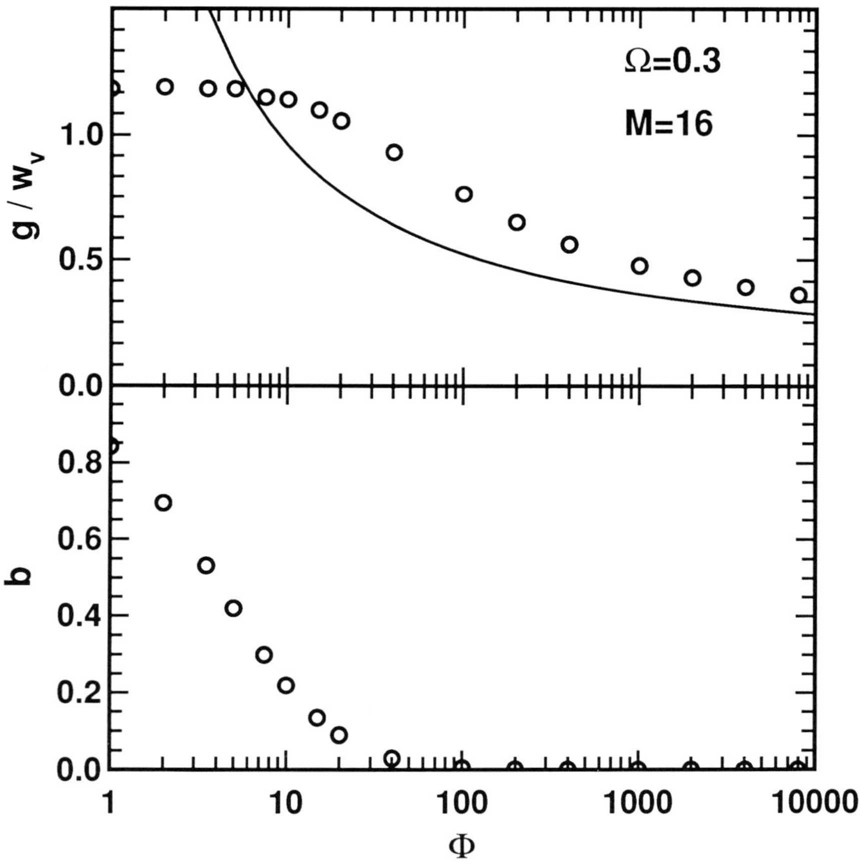

Fig. 7. The standard deviation g of the good estimates normalized by the velocity standard deviation of the pulse spectrum w_v and the fraction b of bad estimates (o) versus Φ. The theoretical minimum standard deviation for any unbiased estimate (the Cramer Rao Bound [35]) is (——).

velocity estimation error $g \propto \lambda$ and better performance is obtained with a smaller wavelength λ.

MEASUREMENTS OF WINDS

There have been many published measurements of winds using coherent Doppler lidar [9]. The early work was dominated by 10 µm lidars. The more recent work emphasizes the improved performance of the 2 µm lidar [36]. The 2 µm Doppler lidars measure the transmitted pulse for each lidar shot with an

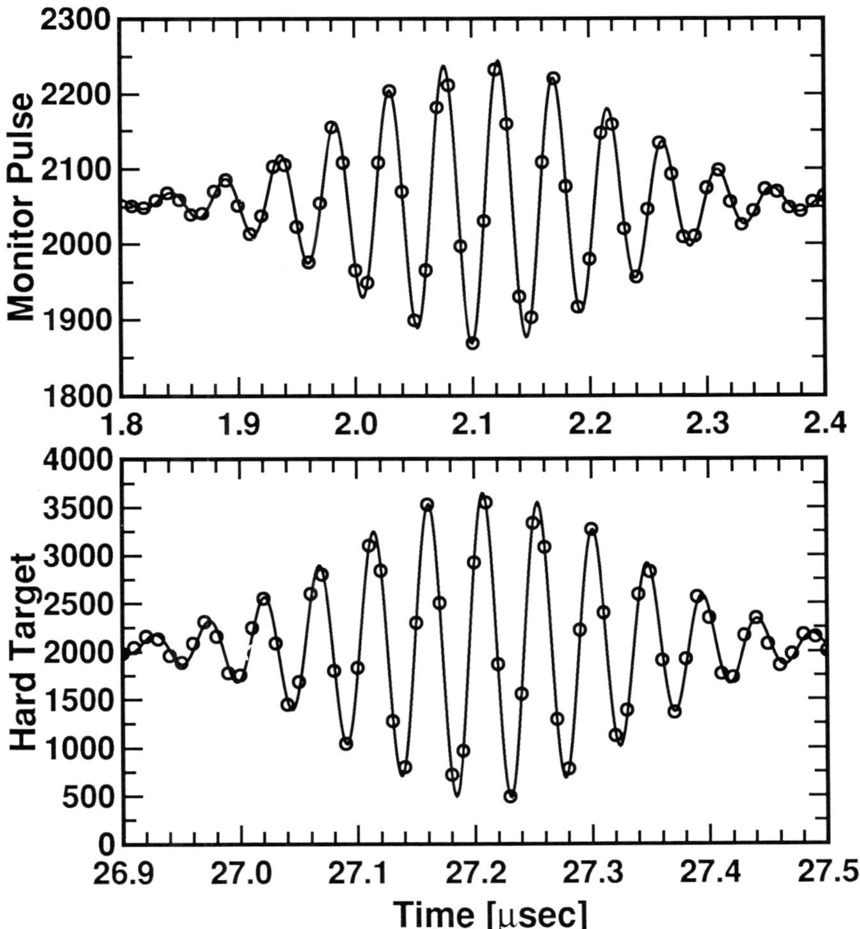

Fig. 8. Doppler lidar data [36] for the monitor pulse and hard target return (o). The maximum likelihood estimate of the model Eq. (15) is (——). Reprinted from Ref. [36] by permission of the American Meteorological Society.

additional heterodyne detector for the monitor pulse [8, 36]. This permits a correction for the frequency drift between the LO laser and transmit laser. The error in the Doppler frequency due to laser frequency drift can be estimated from the Doppler lidar signal of a non-moving hard target. Figure 8 displays an example of a monitor pulse and a hard target return as well as the best fit to the function

$$X(t) = A\exp[-t^2/(2\sigma^2)]\cos(2\pi f t + \pi\phi t^2 + \delta) \tag{15}$$

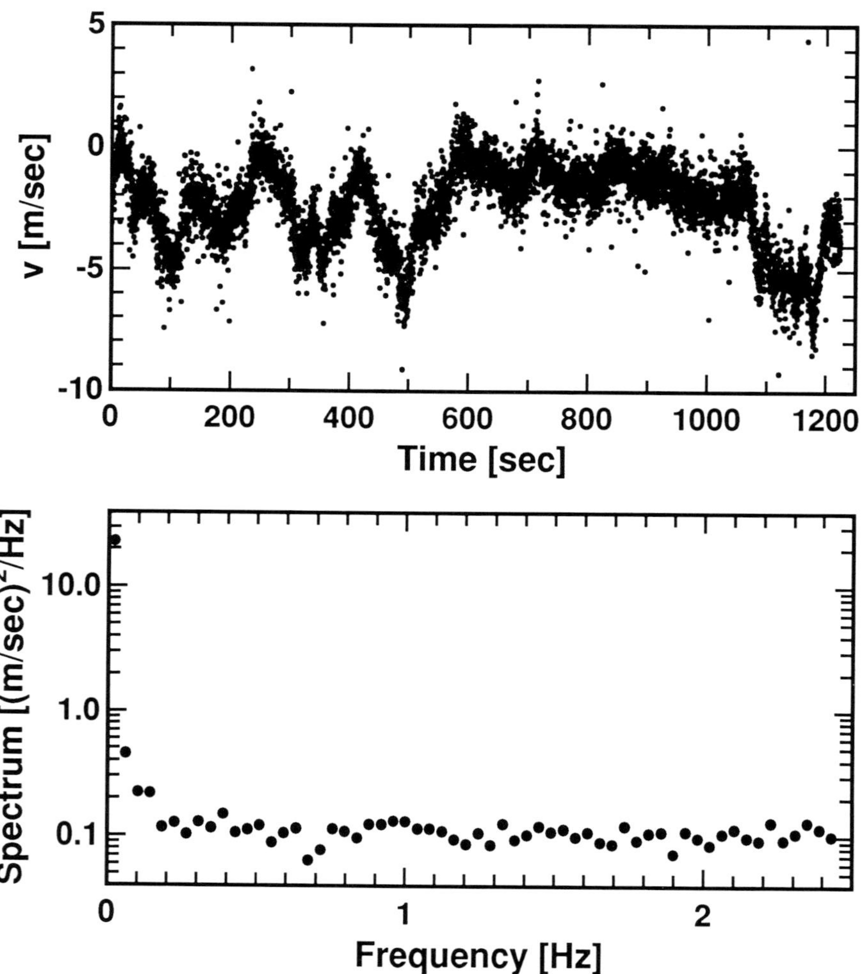

Fig. 9. The maximum likelihood estimates of the radial velocity v from 6000 shots of Doppler lidar data as a function of time for a range bin centred at 1 km [36]. The periodogram estimate of the velocity spectrum is also shown. Reprinted from Ref. [36] by permission of the American Meteorological Society.

where A is the pulse amplitude, ϕ is the linear frequency chirp, and δ is a random phase. The measurements in [36] attribute $3.6\ \mathrm{cm\,sec^{-1}}$ random error for the laser frequency drift for a single pulse.

2 μm Doppler lidar measurements of winds at range $R = 1$ km are shown in Fig. 9 with $M = 16$, $T_S = 0.02$ μsec, $\Delta p = 48$ m, $\Delta r = 31$ m, SNR $= 22.45$, $\Phi = 359$, $\Omega = 0.294$, and $w_v = 0.959\ \mathrm{m\,sec^{-1}}$. The velocity spectrum shows a constant level at high frequencies which can be used to produce reliable estimates of the estimation error g [36]. Velocity estimates for a range $R = 5$ km (SNR $= 0.36$, $\Phi = 5.8$) are shown in Fig. 10, using 1 pulse per estimate and 10 accumulated pulses per estimate. The accumulated pulses remove the outliers at the expense of temporal resolution [31, 33, 34, 36].

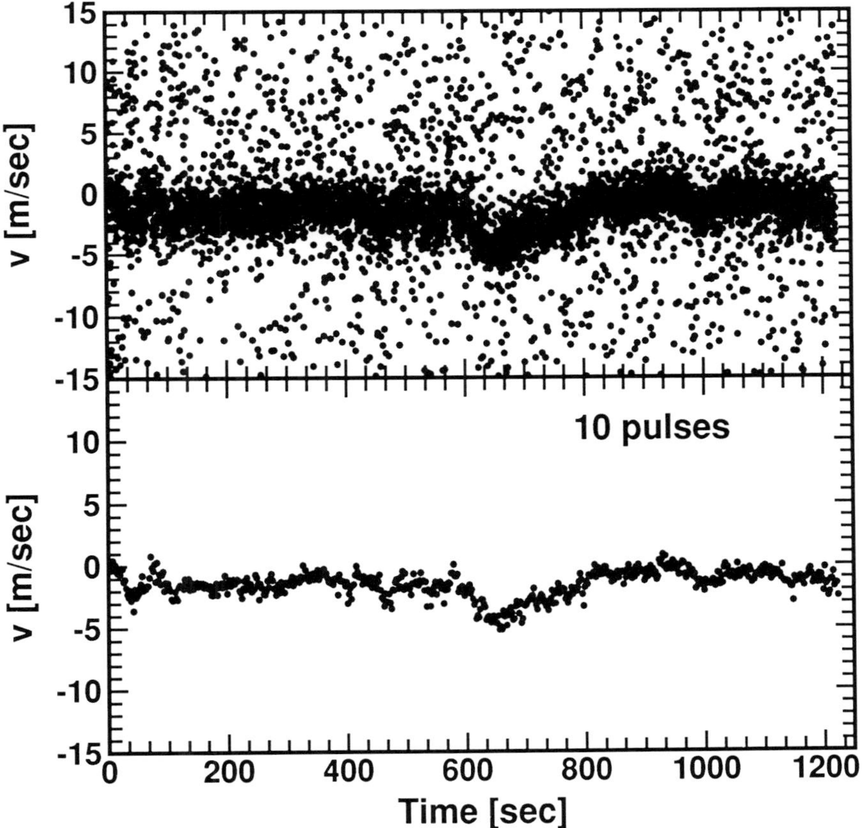

Fig. 10. The maximum likelihood estimates of the radial velocity v from Doppler lidar data as a function of time for a range bin centred at 5 km using 10 accumulated pulses per estimate and 1 pulse per estimate (top panel) [36]. Reprinted from Ref. [36] by permission of the American Meteorological Society.

The performance of the velocity estimates from 2 μm data is compared with the results of ideal computer simulations in Fig. 11. The larger estimation error g/w_v of the data is due to wind turbulence over the range gate and systematic errors. The deviation of the fraction b of outliers is due to variations in the noise spectrum over the velocity search space.

For single pulse operation, the velocity accuracy of the 2 μm Doppler lidar is typically better than 1 m sec^{-1} and using 10 accumulated pulses the velocity accuracy is better than 0.3 m sec^{-1}. Useful measurements can be made with $\Phi \approx 50$, which corresponds to $N_e \approx 100$ photo-electrons per estimate in far-field

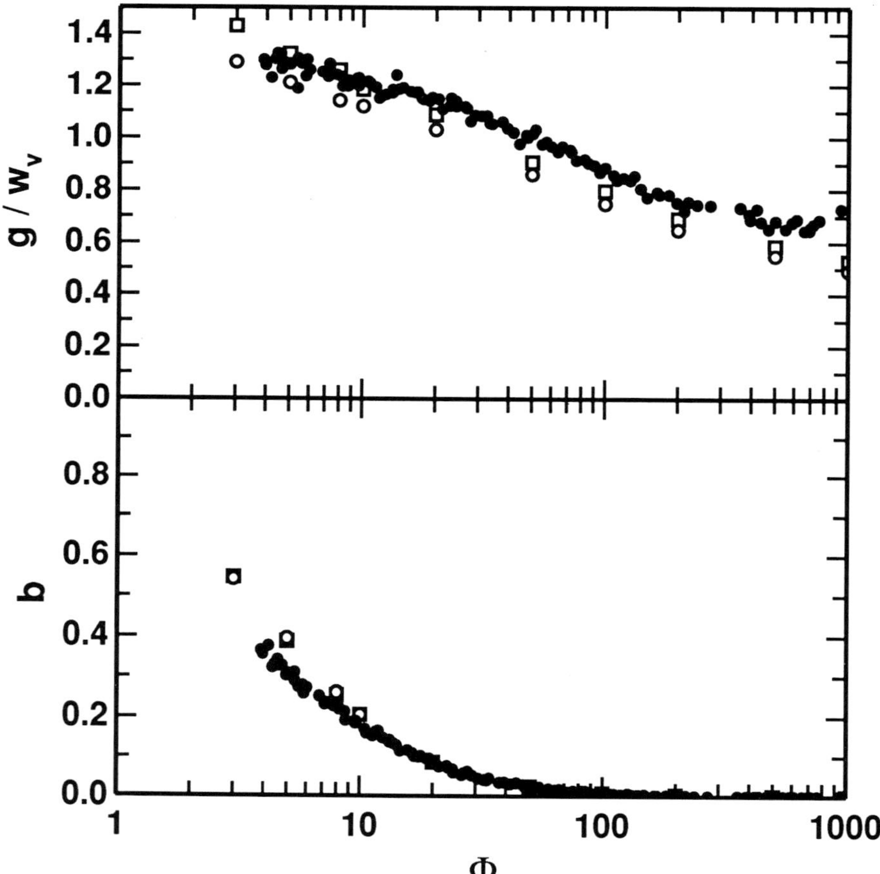

Fig. 11. Comparison of maximum likelihood estimator performance from simulations and actual data (•) from Fig. 8 of [36]. Simulations with no wind turbulence (o) and simulations (□) with Kolmogorov wind turbulence defined by $\varepsilon = 0.01$ m^2 sec^{-3} as estimated from the data. Reprinted from Ref. [36] by permission of the American Meteorological Society.

conditions where $\eta_{\mathrm{H}} \approx 0.5$. This implies useful measurements can be made with less than 200 photons per estimate since solid state detectors typically have quantum efficiencies greater than 0.5.

The performance of Doppler lidar estimates of the radial component of the wind is sufficient to produce estimates of the spatial statistics of the atmospheric wind field. An important statistical description of the velocity field is the structure function defined by

$$D_v(r) = \langle [v(r_0) - v(r_0 + r)]^2 \rangle = C_v \varepsilon^{2/3} r^{2/3} \tag{16}$$

for locally stationary turbulence described by a Kolmogorov spectrum where $C_v \approx 2$ is the Kolmogorov constant and ε is the energy dissipation rate. The

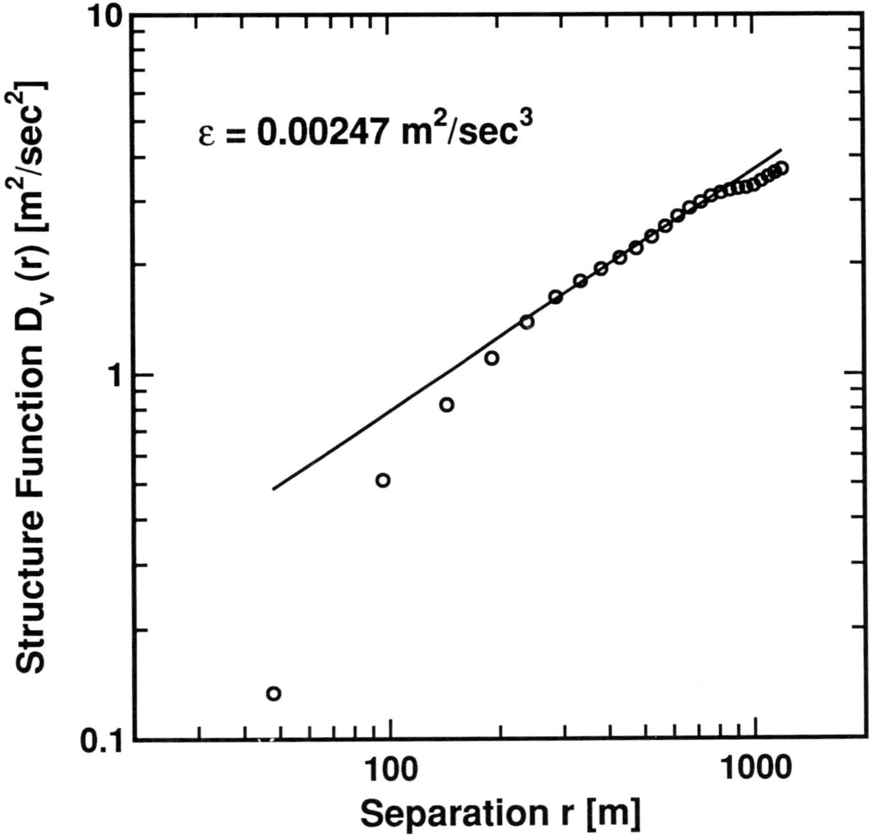

Fig. 12. Estimates of the vertical velocity structure function averaged over the height interval 1–2 km (o) using the maximum likelihood estimator with 10 accumulated pulses. The predictions of the Kolmogorov model, Eq. (16) using the structure function estimate at a separation of 384 m is indicated by (——) where $\varepsilon = 0.00247$ m^2 sec^{-3}. Reprinted from Ref. [36] by permission of the American Meteorological Society.

structure function of the vertical velocity fluctuations averaged over the region from 1–2 km in height above the surface is shown in Fig. 12 along with the theoretical prediction of Kolmogorov theory, Eq. (16). The effects of the spatial averaging of the pulse is clearly evident for small separations on the order of the pulse extent Δr. The effects of spatial averaging by the pulse can be predicted and in many cases removed. The coherent Doppler lidar provides photon efficient, accurate, reliable, high-resolution measurements of winds.

ACKNOWLEDGEMENTS

The author acknowledges useful discussions with M. J. Kavaya, B. J. Rye, W. Eberhard, M. Hardesty, and G. Spiers. Special thanks to S. Hannon and S. Henderson of Coherent Technologies, Inc. for a fruitful collaboration that produced comparisons of theoretical Doppler lidar performance with the results from actual data. This work was supported by the National Science Foundation, the Army Research Office, and the National Aeronautics and Space Administration, Marshall Space Flight Center (Michael J. Kavaya, Technical Officer).

REFERENCES

1. A.V. Jelalian, *Laser Radar Systems*, Artech House (1991).
2. T.R. Lawrence, D.J. Wilson, C.E. Craven, I.P. Jones, R.M. Huffaker, and J.A.L. Thomson, 'A laser velocimeter for remote wind sensing', *Rev. Sci. Instr.* **43** 512–518 (1972).
3. J.W. Bilbro, C. DiMarzio, D. Fitzjarrald, S. Johnson, and W. Jones, 'Airborne Doppler lidar measurements', *Appl. Opt.* **25**, 2952–2960 (1986).
4. J.C. Petheram, G. Frohbeiter, and A. Rosenberg, 'Carbon dioxide Doppler lidar wind sensor on a space station polar platform', *Appl. Opt.* **28**, 834–839 (1989).
5. M.J. Post and R.E. Cupp, 'Optimizing a pulsed Doppler lidar', *Appl. Opt.* **29**, 4145–4158 (1990).
6. M.J. Kavaya, S.W. Henderson, J.R. Magee, C.P. Hale, and R.M. Huffaker, 'Remote wind profiling with a solid-state Nd:YAG coherent lidar system', *Opt. Lett.* **14**, 776–778 (1992).
7. S.W. Henderson, C.P. Hale, J.R. Magee, M.J. Kavaya, and A.V. Huffaker, 'Eye-safe coherent laser radar system at 2.1 μm using Tm,Ho:YAG lasers', *Opt. Lett.* **16**, 773–775 (1991).
8. S.W. Henderson, P.J.M. Suni, C.P. Hale, S.M. Hannon, J.R. Magee, D.L. Bruns, and E.H. Yuen, 'Coherent laser radar at 2 μm using solid-state lasers,' *IEEE Trans. Geo. Remote Sensing* **31**, 4–15 (1993).
9. J.M. Wilzak, E. Gossard, W.D. Neff, and W.L. Eberhard, 'Remote sensing of the atmospheric boundary layer: 25 years of progress', *Boundary Layer Meteorology*, in press.

10. M.R. Huffaker, T.R. Lawrence, M.J. Post, J.T. Priestley, F.F. Hall, Jr., R.A. Richter, and R.J. Keeler, 'Feasibility studies for a global wind measuring satellite system (Windsat): analysis of simulated performance', *Appl. Opt.* **23**, 2523–2536 (1984).

11. W.E. Baker, G.D. Emmitt, P. Robertson, R.M. Atlas, J.E. Molinari, D.A. Bowdle, J. Paegle, R.M. Hardesty, R.T. Menzies, T.N. Krishnamurti, R.A. Brown, M.J. Post, J.R. Anderson, A.C. Lorenc, T.L. Miller, and J. McElroy, 'Lidar measured winds from space: an essential component for weather and climate prediction', *Bull. Amer. Met. Soc.* **76**, 869–888 (1995).

12. B.J. Rye, 'Comparative precision of distributed backscatter Doppler lidars', *Appl. Opt.*, **34**, 8341–8344 (1995).

13. R.G. Frehlich and M.J. Kavaya, 'Coherent laser radar performance for general atmospheric refractive turbulence', *Appl. Opt.* **30**, 5325–5352 (1991).

14. M.J. Kavaya and R.T. Menzies, 'Lidar aerosol backscatter measurements: systematic modeling, and calibration error considerations', *Appl. Opt.* **24**, 3444–3453 (1985).

15. J.W. Goodman, 'Some effects of target-induced scintillation on optical radar performance', *Proc. IEEE*, **53**, 1688–1700 (1965).

16. J.C. Dainty, *Laser Speckle and Related Phenomena*, Topics in Applied Physics **9**, Ed. Dainty, Springer-Verlag, 286 pp. (1975).

17. A.E. Siegman, 'The antenna properties of optical heterodyne receivers', *Proc. IEEE* **51**, 1356–1358 (1966) or *Appl. Opt.* **5**, 1588–1594 (1966).

18. B.J. Rye, 'Antenna parameters for incoherent backscatter heterodyne lidar', *Appl. Opt.* **18**, 1390–1398 (1979).

19. Shapiro, J.H., B.A. Capron, and R.C. Harney, 'Imaging and target detection with a heterodyne-reception optical radar', *Appl. Opt.* **20**, 3292–3313 (1981).

20. Y. Zhao, M.J. Post, and R.M. Hardesty, 'Receiving efficiency of pulsed coherent lidars. 1: Theory', *Appl. Opt.* **29**, 4111–4119 (1990).

21. Y. Zhao, M.J. Post, and R.M. Hardesty, 'Receiving efficiency of pulsed coherent lidars. 2: Applications,' *Appl. Opt.* **29**, 4120–4132 (1990).

22. B.J. Rye and R.G. Frehlich, 'Optimal truncation of Gaussian beams for coherent lidar using incoherent backscatter', *Appl. Opt.* **31**, 2891–2899 (1992).

23. R.G. Frehlich, 'Optimal design of monostatic coherent laser radar with a circular aperture', *Appl. Opt.* **32**, 4569–4577 (1993).

24. R.G. Frehlich, 'Heterodyne efficiency for a coherent laser radar with diffuse or aerosol targets', *J. Mod. Opt.* **41**, 1217–1230 (1994).

25. P. Salamitou, F. Darde, and P. Flamant, 'A semi-analytic approach for coherent laser radar system efficiency: the nearest Gaussian approximation', *J. Mod. Opt.* **41**, 2101–2113 (1994).

26. P.H. Flamant, R. T. Menzies, and M.J. Kavaya, 'Evidence for speckle effects on pulsed CO_2 lidar signal returns from remote targets', *Appl. Opt.* **23**, 1412–1417 (1984).

27. A. Dabas, P.H. Flamant, and P. Salamitou, 'Characterization of pulsed coherent Doppler LIDAR with the speckle effect', *Appl. Opt.* **33**, 6524–6532 (1994).

28. V.A. Banakh and V.L. Mironov, *Lidar in a Turbulent Atmosphere*, (Artech House, Boston and London, 1987).

29. R.G. Frehlich, 'Effects of refractive turbulence on coherent laser radar', *Appl. Opt.* **32**, 2122–2139 (1993).

30. R.G. Frehlich, 'Coherent Doppler lidar signal covariance including wind shear and wind turbulence', *Appl. Opt.* **33**, 6472–6481 (1994).
31. R.G. Frehlich and M.J. Yadlowsky, 'Performance of mean frequency estimators for Doppler radar and lidar', *J. Atm. Ocean. Tech.* **11**, 1217–1230 (1994).
32. D.S. Zrnic, 'Estimation of spectral moments of weather echoes', *IEEE Trans. Geosci. Electronics* **GE-17**, 113–128 (1979).
33. B.J. Rye and R.M. Hardesty, 'Discrete spectral peak estimation in incoherent backscatter heterodyne lidar. I. Spectral accumulation and the Cramer–Rao lower bound', *IEEE Trans. Geo. Sci. Remote Sensing* **31**, 16–27 (1993).
34. B.J. Rye and R.M. Hardesty, 'Discrete spectral peak estimation in incoherent backscatter heterodyne lidar. II. Correlogram accumulation,' *IEEE Trans. Geo. Sci. Remote Sensing* **31**, 28–35 (1993).
35. R.G. Frehlich, 'Cramer–Rao bound for Gaussian random processes and applications to radar processing of atmospheric signals', *IEEE Trans. Geo. Sci. Remote Sensing* **31**, 1123–1131 (1993).
36. R. Frehlich, S. Hannon, and S. Henderson, 'Performance of a 2 μm coherent Doppler lidar for wind measurements', *J. Atmos. Oceanic Technol.* **11**, 1517–1528 (1994).

20 Doing coherent optics with soft X-ray sources

Denis Joyeux
*Institut d'Optique Théorique et Appliquée and CNRS,
Campus d'Orsay, BP 147, 91403 Orsay, FRANCE*

Pierre Jaegle
*Laboratoire de Spectroscopie Atomique et Ionique,
Université Paris XI, 91405 Orsay, FRANCE*

Anne L'Huillier
*Department of Physics, Lund Institute of Technology,
P.O. Box 118, S 221 00 Lund, SWEDEN
and
CEA/DRECAM/SPAM, CEN Saclay,
91191 Gif sur Yvette Cedex, FRANCE*

INTRODUCTION

From the beginning of X-ray history, in 1896, and until synchrotron radiation sources became available in the 1970s, X-ray physics was based almost exclusively on two basic phenomena: energy exchange with atoms, and diffraction/diffusion by organized sets of atoms. These two fields relate respectively to both aspects of the well-known dual description of the interaction of light with matter, i.e., the photon and the electromagnetic (EM) wave.

Indeed, the wave description holds for X-rays. Since they are truly EM waves, X-rays can be used, at least in principle, to perform all those classical experiments of 'visible' optics that are better described by the wave formalism.

This is the case for experiments that directly involve wavefront phase determination, and particularly all sorts of interference devices. These require some degree of spatial and temporal coherence in the X-ray EM field. As the source and optics technology varies throughout the wide X-ray spectral range, we shall restrict ourselves from now on to the XUV to soft X-ray (SXUV) domain which ranges from $\lambda \approx 100$ nm ($h\nu \approx 10$ eV) to $\lambda \approx 1$ nm ($h\nu \approx 1$ keV), i.e., that spectral range in which optics can only be made with mirrors and artificial diffractive devices (not natural crystals).

Within that range, the main limitation to the early interferometric experiments was probably the small amount of *coherent flux* that was available from the classical bremsstrahlung X-ray sources. As will be reviewed in this chapter, this coherent flux depends only on the spectral luminance of the source and on the wavelength. Classical sources were dramatically insufficient in this respect, and therefore only the very basic principles were demonstrated, under very difficult experimental conditions [1, 2].

This situation began changing in the 1970s, with the advent of synchrotron radiation (SR) sources. The first-generation machines were basically particle accelerators, not designed for optimal photon production. Although designed to be highly collimated photon sources, second-generation machines, in particular insertion devices such as undulators, cannot be considered as coherent, at least in the soft X-ray range and for harder X-rays. But their high degree of collimation (linear milliradians, i.e., microsteradians in solid angle), compared to the 2π sr of classical sources, resulted in a very large gain in spectral luminance. This gain by itself made some coherent experiments possible, mainly in the field of imagery: diffraction limited scanning imagery (with Fresnel zone plates), and Gabor holography.

Since then, with progress in machine design and realization, orders of magnitude are still being gained in source luminance, with the most recent machines and the near-future third-generation of SR sources. This opens the way to a true SXUV interferometry research field. As a matter of fact, a few such experiments have been carried out already or are in progress.

The general goal of this chapter is to show that coherent interferometry can be a new tool for physical studies, even if all possible applications have not yet been identified. We begin with a short discussion to remind the reader of the source properties that are required to allow the practical performance of interferometric experiments in the SXUV range. Speaking of real sources, three main classes should dominate the field in the near future, namely *SR sources*, the *high-harmonic generation* sources and the *X-ray lasers*. As the last two are quite new, we have included a presentation of their physical principles and of their relevant properties, and a discussion of near-future progress. Finally, the design and realization of instruments are discussed, and two particular examples are presented.

WHAT SOURCE FOR SXUV COHERENT EXPERIMENTS?

Source Coherence and Coherent Flux

Performing coherent experiments generally implies that the incoming field has a fairly good spatial and temporal coherence. As temporal coherence is related to spectral width $\Delta\lambda$, it should be adjusted, if necessary, by the use of some monochromator, in order to ensure that the maximum path difference involved is smaller than $\lambda^2/\Delta\lambda$. In the field of soft X-rays, this may be quite a severe constraint for the design of interferometers, as will be shown later.

Concerning the spatial coherence, the usual question is: do I have a (spatially) coherent source? However, this is not the pertinent question: according to the partial coherence theory, it is always possible (using aperture and stops at the appropriate places) to get from any spatially incoherent source, at any couple of points in space, as much spatial coherence as desired. Therefore, spatially coherent experiments can be performed from incoherent sources, as is well known: Gabor recorded his first holograms without any laser. But the cost of this coherence 'synthesis' is a loss of flux. Therefore, the crucial point for setting up a coherent experiment is merely: how much coherent flux (i.e., fulfilling the coherence requirements) can be obtained from that source?

Answering this question implies dealing with beam optics. The relevant physical quantity is beam *étendue*,[1] a concept which is deeply connected with spatial coherence. For a one-dimensional analysis (necessary when only 1-D coherence is required), emittances should be used instead. As even a short treatment of this topic would extend this chapter to an unreasonable size, we shall only review the main results, assuming waves are propagating in the vacuum (refractive index 1).

(1) The étendue of a beam, and therefore the flux it carries, remains constant upon propagation through any optical system free of absorption, diffusion and diaphragmation. As a consequence, the *luminance*,[2] i.e., the flux radiated per unit étendue, is the relevant photometric quantity to assess the way energy

---

[1] The *étendue* of a uniform pencil of rays is the product $\varepsilon = \mathrm{d}S_n \times \mathrm{d}\Omega$ of its normal section by its solid angle of divergence. It is expressed in m^2 sr, or mm^2 μsr. As non-circular X-ray beams are often encountered, the one-dimensional equivalent of étendue is used, i.e., the emittance in one particular longitudinal section , with $E = \sigma\sigma'$, the product of the normal half width of the beam by its angular half width of divergence. Emittance is expressed in mm mrad. Two orthogonal sections are needed to characterize a non-circular beam. The exact relationship between the étendue and the two emittances depends on the exact beam geometry.

[2] Unfortunately, this basic quantity of photometry carries various names depending on the physicist's field or even country. To our knowledge, the official international term is presently luminance, although 'brilliance' (in the SR field), 'radiance' (infra-red), or even 'brightness' may be found in the literature.

propagates in optical systems from a source to a detector. When spectral width must be accounted for, the *spectral luminance* (per unit spectral bandwidth) must be used.

(2) Due to diffraction, a real pencil of rays cannot have an étendue smaller than a minimum value ε_m, equal to λ^2 times a geometrical factor depending on the source amplitude profile. When this minimum étendue is reached, the source and the beam are fully coherent. For narrow Gaussian beams characterized by their energy profile parameters at $1/\sqrt{\mathrm{e}} \approx 0.6$ (the usual case with modern SXUV sources), and using as a definition the equation: flux = (maximum luminance × étendue), the étendue is $\lambda^2/4$.

Conversely, the beam étendue can be greater than the minimum value both when the source is not coherent, and when the source is fully coherent but with an uneven phase repartition (e.g., 'diffuse'). Therefore, the simple observation of the étendue radiated by a source is not in general sufficient to assess its intrinsic spatial coherence, unless theoretical considerations allow us to settle the issue. This is the case of SR sources, which are basically incoherent, due to the way photons are generated at the microscopic level. In this latter case, the partial coherence theory [3] shows that the coherence factor γ of the radiated field is greater than some given value inside a so-called *coherent étendue* ε_c ('coherence cell'), which is again scaled by λ^2. For incoherent Gaussian beams, $\varepsilon_\mathrm{c} = \lambda^2/2$ for $\gamma \geqslant 1/\sqrt{\mathrm{e}} \approx 0.6$.

(3) From the above, the coherent energy a source can deliver always has the form (source luminance) × (coherent étendue), and therefore the total radiated flux is of no concern. In other words, the coherent flux does not depend on the source intrinsic coherence: from two sources having the same luminance but very different étendues (i.e., different coherence properties), one may get exactly the same flux for use in a given coherent experiment. The difference is that the source with the largest étendue radiates much more energy. But, there is no way to put that energy into a smaller étendue and its major part is therefore lost.

Finally, it must be kept in mind that an actual beam can be highly directive although nearly incoherent, at least for soft X-rays. For example, at 4 nm, the étendue of the coherent part of an incoherent beam is $\varepsilon_\mathrm{c} = 8 \times 10^{-6}\ \mathrm{mm}^2\,\mu\mathrm{sr}$, whereas the étendue of a source with Gaussian intensity profiles with parameters 50 μm linear, and 0.1 mrad in angle is $10^{-3}\ \mathrm{mm}^2\,\mu\mathrm{sr}$. Therefore, less than $1/100$ of the radiated energy is contained in one coherence cell.

What Source?

Following this discussion, it is clear that 'classical' X-ray sources (i.e., based on bremsstrahlung, and plasma sources) are extremely handicapped by their large angular radiation pattern, essentially close to 2π steradians. This creates a giant handicap which can be as large as 10^5–10^6 or more when comparing to naturally collimated sources (even if incoherent in the sense above).

Although the most powerful plasma sources can be used for image scanning (a coherent experiment when using Fresnel zone plates as focusing optics), it must be admitted that only three different kinds of SXUV sources can be considered for near future coherent experiments. Only SR sources are routinely used at the present time, because the other two, namely the X-ray laser and the high-order harmonic source (HHG), are still in their infancy. But, these have recently shown impressive progress, and, in our opinion, some of their characteristics are unique and could shortly make them much more attractive. As a matter of fact, some projects involving these sources in true interferometric experiments have been presented very recently (mid-1995).

These three sources differ essentially in the way their intrinsic coherence is obtained. Whereas the X-ray laser and the HHG are fundamentally (i.e., at the microscopic level) coherent processes, the SR source is fundamentally an incoherent process, because the generating particles (accelerated electrons or positrons) have no constant phase relationship to each other. The partial (possibly close to total) spatial coherence of the SR photon beams results from the high degree of collimation of the radiation (a pure relativistic effect), and from the properties of the particle beam optics.

The SR source will not be described here, as extensive literature exists on that topic [4]. A somewhat detailed overview of the other two is given below.

High-order Harmonic Generation

A typical high-order harmonic generation (HHG) experiment is sketched in Fig. 1. A high-energy pulsed (1 ps) Nd-glass laser ($\lambda = 1.053\ \mu$m) is focused across an atomic jet of xenon. As non-linear interaction occurs between the focused field and the atoms of the jet, the radiated field contains, as expected, harmonics of the pump laser, the intensity of which decreases rapidly for the first harmonics. However, unlike in classical up-conversion, the harmonic intensity of the next odd orders remains constant across a broad plateau which ends up by an abrupt cut-off, from about the seventh order to the 21st in the case shown in Fig. 1. This plateau behaviour is one of the main features of HHG sources. It was first reported in 1987, in two experiments, in Saclay (up to the 33rd harmonic of a YAG laser) [5] and Chicago (up to 17th harmonic of a KrF laser) [6]. HHG sources now produce radiation with energy from a few eV to more than 170 eV [7, 8]. The generated light has similar geometrical properties to the pump laser, with a considerable luminance even for quite a small pulse energy.

Optimizing the conversion process

Understanding the mechanism involved in the production of high-order harmonics has been the goal of most experiments to date, in view of the optimization of the radiated field with respect to the plateau extension, the conversion efficiency, and the coherence properties of the beam. As in any

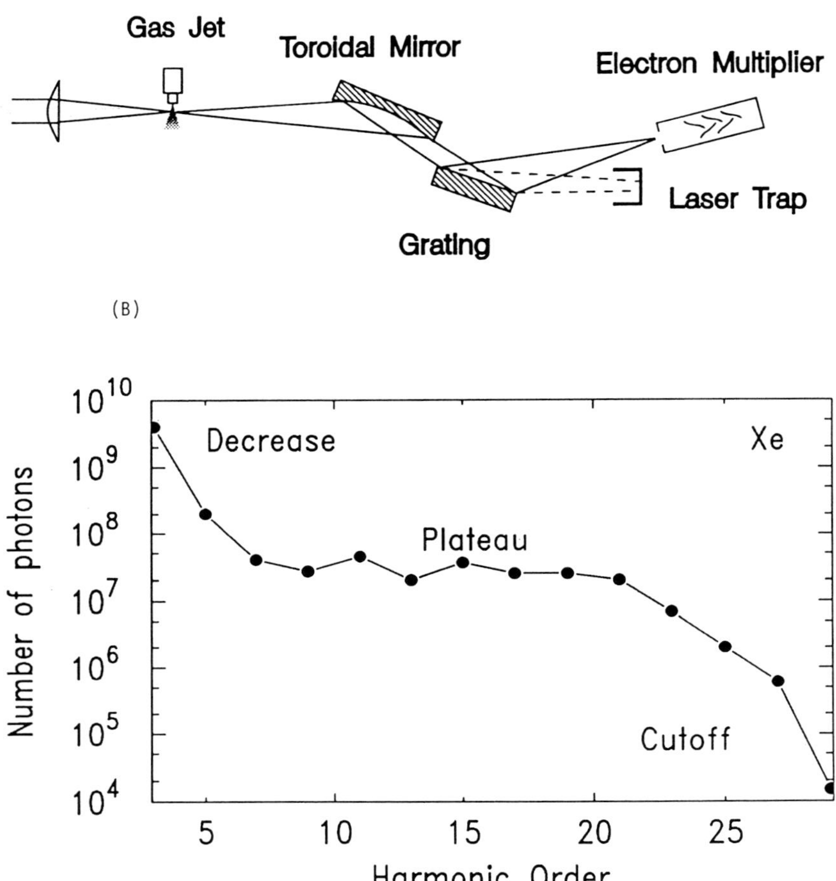

Fig. 1. A typical and historic harmonic spectrum experiment. (A) The set-up; (B) the spectrum obtained in xenon with a 1 ps Nd-glass laser. The plateau can actually extend up to orders as high as 143 and energies up to 180 eV.

non-linear optical process, HHG involves two aspects: (1) the response of a single atom to the incident radiation field, which depends on the laser wavelength and intensity, and on the atomic system; (2) the production of a macroscopic harmonic field by the coherent addition of the fields emitted by each atom ('phase matching'), which depends on the atomic density and the focusing geometry. Both aspects, however, are closely intertwined, in the sense that the characteristics of the single-atom emission strongly affects the macroscopic properties (efficiency, coherence) of the generated radiation.

A major breakthrough in the understanding of HHG came in 1992, with the 'cut-off law' [9] concerning the extension of the harmonic plateau. Theoretical calculations, performed for a single atom, showed that the width of the plateau varies as $I_p + 3U_p$, where I_p is the field-free ionization energy of the atom and U_p is the 'ponderomotive' energy (i.e., the mean kinetic energy of an electron in the laser field). $U_p^{(eV)}$ is $9.33 \times 10^{-14} I^{(W\,cm^{-2})} \lambda^{2(mm^2)}$, where I is the laser intensity and λ the laser wavelength. This cut-off law can be interpreted using classical arguments [10]. The electron may tunnel from its ground state through the potential barrier (the sum of the atomic potential and the potential energy due to the electromagnetic field). Beyond the barrier, the electron is accelerated by the laser field. After one half cycle, the field changes its sign and drives the electron back towards the ionic core. The electron may then recombine to the ground state, emitting a harmonic photon. The maximum kinetic energy acquired by the electron in the laser field is equal to the '$3U_p$' of the cut-off law.

The predictions of the cut-off law are well verified by experiment. In order to produce high-energy photons, one should use light rare gases (with a high ionization energy), low-frequency lasers and short pulses (such that neutral atomic species can survive to a higher intensity). The experimental plateau width is found to be equal to $I_p + aU_p$, with $a = 2$–2.5, smaller than that predicted by the cut-off law. This smaller cut-off energy does not mean that theory and experiment disagree, but that, obviously, a quantitative description of the process must include also the macroscopic response of the medium. Numerical calculations including both aspects of the problem reproduce well the experimental data [11]. The level of understanding achieved, both qualitatively and quantitatively, is now good enough for numerical simulations to play a significant role in the source optimization.

According to the cut-off law, conversion in ionized gases should produce extremely high-order harmonics (both the I_p and U_p terms are much higher than for neutral gases). From the calculation for a single He$^+$ ion [9], one expects the harmonic plateau to reach the water window (under $\lambda = 4.4$ nm). However, experiments have not succeeded so far in producing very high-order harmonics in ionic media, using low-frequency lasers. The reason is most probably the effect of the free electrons on the propagation, which induces wavefront distortion and deteriorates phase matching conditions.

The next parameter to optimize is the conversion efficiency of a given harmonic. At first sight, it should improve by increasing the pump field and the number of atoms interacting. However, limitations occur, mainly owing to ionization. The best conditions for the laser field are weak focusing (in order to optimize phase matching) and intensity just below the intensity at which the medium begins to ionize. The conversion efficiency is expected to vary as the square of the atomic density (typically of the order of 10^{17}–10^{18} atoms cm^{-3}), owing to the coherent character of the harmonic generation process. However, the spatial variations of the index of refraction throughout the medium (essentially due to free electrons), which increase with the density, alter this simple scaling, because the fundamental wavefront is distorted. The conversion efficiency can also be improved by using short-wavelength lasers and heavy rare gases, but to the detriment of the plateau width, as indicated by the cut-off law. Under fairly optimized conditions, for instance using a 825 nm, 140 fs LiSAF laser [*12*], the efficiency is estimated to be about 10^{-6} in the range 10–40 eV (10^9 up-converted photons per harmonic pulse with argon as generating gas) and 10^{-9} from 40 to 150 eV (~10^6 photons, with neon as generating gas). Energy yields of up to 60 nJ at 40–50 eV (21st to 27th harmonics of 526 nm laser light, corresponding to a 10^{-7} conversion efficiency) have been measured using an absolutely calibrated X-ray charged coupled device detector [*13*]. The estimation of the flux and luminance of the source requires a knowledge of the temporal and spatial characteristics of the radiation, which is only beginning to be addressed.

Temporal and spatial coherence

The pulse width is an important parameter, as time-resolved experiments are an important application. Most of the measurements have been done for quite low-order harmonics of long-pulse lasers. Recently, the pulse width of the 21st harmonic of a 150 fs titanium sapphire laser has been measured to be 80 fs, using a cross-correlation technique. In all the measurements and in numerical calculations, the harmonic pulse duration is found to be shorter than the incident laser pulse and larger than the perturbative limit (laser pulse width)/$\sqrt{q}$ (where q is the harmonic order), obtained by assuming, as in traditional up-conversion processes, that the laser-induced polarization varies as the qth power of the incident laser field. In the 'fairly optimized' conditions considered previously, the flux per harmonic, averaged on one pulse, can be estimated to be ~10^{22} photons s^{-1} (60 kW) at 40 eV and ~10^{19} photons s^{-1} (200 W) at 150 eV.

In accordance with Heisenberg's principle, these short pulse durations produce relatively large spectral widths. For example, the linewidth of the 27th harmonic of a 140 fs, LiSAF laser has been measured to be 0.1 nm [*12*], resulting in a $\Delta\lambda/\lambda = 1/300$. It may even be broader when the amount of ionization becomes more important (for large laser intensities). A smaller spectral width may be obtained by using longer-pulse lasers, in the picosecond range (10^{-4} at $\lambda = 53$ nm). In the absence of interferometric measurements yet to be done, these spectral profile measurements allow us to estimate the *temporal coherence* of the harmonics. In all cases, the coherence is good enough to allow one to perform interferometry experiments with an order of interference of up to 100.

As explained previously, the spatial coherence is related to geometrical properties of the radiated *complex* field. Thus, it must be kept in mind that the radiation angle by itself is not a test of the spatial coherence of the field: even a perfectly coherent field may radiate in a large angle, depending on the phase repartition in the source. Therefore, only interferometric measurements can provide a safe estimation of the beam coherence. The radiation aperture and profile may provide a useful indication only when the profile is even and the angular width close to the theoretical minimum. The spatial coherence of the HHG source has not been experimentally studied yet. The following discussion is based on emission profile measurements, combined with theoretical calculations, which, as explained previously, provide a quite accurate description of the HHG properties.

The harmonic radiation is emitted on the laser axis, with a small divergence, depending on the focusing conditions of the incident laser. Figure 2 displays measured normalized spatial intensity profiles of the 23rd, 25th, 27th and 29th harmonics (end of the plateau), generated in argon from a LiSAF laser ($\lambda = 825$ nm) [12]. The profiles are symmetric and show no substructure. The distributions are close to Gaussian functions for the highest harmonics and become more triangular as the order decreases. The divergence of the harmonics increases from about 6 mrad (full width at $1/e^2$) for the 29th harmonic (last harmonic visible in the cut-off) to 13 mrad for the 23rd (plateau region). The divergence of the (focused) incident laser is 55 mrad. Similar results have also been obtained in neon, for higher harmonic orders and photon energies.

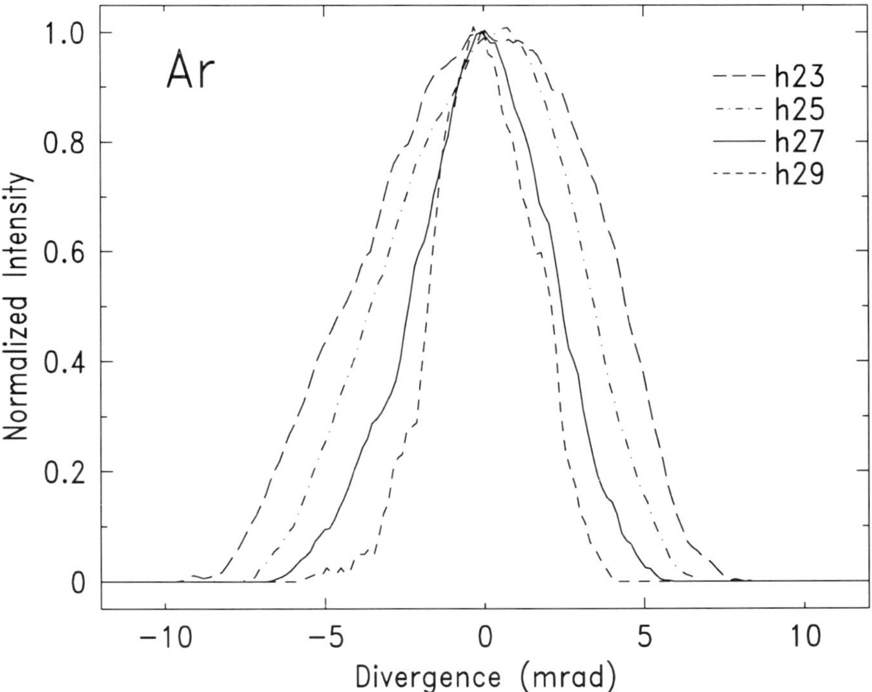

Fig. 2. Normalized spatial profiles of the 23rd to 29th harmonics generated in argon at an intensity of 1.8×10^{14} W cm^{-2}.

The understanding of the geometrical, as well as temporal, properties of the harmonics, which is very important in view of applications, has stimulated several theoretical studies. Unlike for low-order harmonic generation, the properties of the incident laser may not be transmitted to the generated fields, because, in the high-order case, the amplitude and phase of the dipole moment (responsible for the harmonic emission) vary in a complicated way with the laser electric field. The non-linear polarization induced in the medium is therefore a complicated function of space and time. All of the theoretical studies point out the influence of phase perturbation effects (phase of the dipole emission, index variations due to free electrons, etc.) on the angular emission. The spatial and spectral characteristics depend crucially on the focusing conditions, and more precisely on the position of the laser focus relative to the atomic medium [14]. When the laser is focused before the atomic medium, the emitted radiation might be close to perfect coherence, both in spectrum and in space. The beam divergence is rather large, with a 'virtual' focus spot calculated to be a few μm. When the laser is focused in or beyond the medium, the spatial profile as well as phase front of the harmonics is quite distorted, and the spectrum is considerably chirped and broadened. This implies a loss in spatial and temporal coherence. These calculations, supported by experimental results, predict that the coherence properties of the HHG source can be optimized and controlled by appropriately choosing the geometry of the interaction.

Conclusion and perspectives

In the present state of the art, high-order harmonic sources provide XUV radiation from a few eV to about 170 eV (from near UV to 7 nm in wavelength), with a laboratory size, pulsed source at a repetition rate of typically 10 Hz. In all cases, the spatial collimation of the harmonic beams is good, and related to the collimation of the pump laser. If the laser is properly focused in front of the atomic medium and if the density of free electrons in the medium is kept low, the radiation is expected to be highly coherent, Gaussian-like, both in space and spectrum.

Due to these properties, although the flux per pulse is generally small, the luminance can be fairly large, much larger than that of classical bremsstrahlung, or even plasma discharge sources. In the same 'fairly optimized' conditions as before, using the measured angular divergence and the calculated focal spot, the pulse-averaged luminance per harmonic is estimated to be 10^{25} ph $(\text{mm mrad})^{-2}\text{s}^{-1}$ at 40 eV and 10^{22} ph $(\text{mm mrad})^{-2}\text{s}^{-1}$ at 150 eV. The corresponding time average spectral luminance, for a 10 Hz repetition rate, is 10^{12} $(\text{mm mrad})^{-2}(0.1\%$ spectral width$)^{-1}\text{s}^{-1}$ at 40 eV (10^{9} at 150 eV), which is, of course, smaller than that obtained with synchrotron sources. The rapid technological progress in short-pulse lasers, concerning either the repetition rate (1 kHz) or the pulse duration (20 fs) and energy (1 J) will improve the performances of the HHG sources for the average or peak powers by at least two orders of magnitude.

Experiments involving the focusing of the radiated energy over very small spots (for high X-ray energy density), or the production and use of interference patterns, can therefore be performed. Besides, the harmonic pulses can be easily synchronized with a laser pulse, which offers the possibility of performing pump/probe-type studies with a high temporal resolution. The radiation can be tunable if the pump laser is tunable, or if wave-mixing techniques are used (an intense fixed-frequency laser and a less intense but tunable second colour) [15]. It is linearly polarized if the incident laser is linearly polarized. Finally, it is a laboratory sized source, which is much more manageable than large facilities like synchrotron radiation facilities. High-order harmonics of short-pulse lasers generated in gases have begun to be used as a source of radiation in some non-interferometric applications [12].

The extension above 200 eV (6 nm), expected by using low-frequency lasers and using ions as generating media, has not been demonstrated so far. Propagation effects in a plasma have to be better understood in order to make some progress in this direction. The influence of the different parameters (geometry, atomic density, generating gas, frequency, etc.) on the conversion efficiency has been investigated and is quite well understood. Concerning the understanding and optimization of the coherence, much work remains to be done. However, anticipating near-term progress of this very young XUV source, we believe that it will become a useful source, even for true coherent experiments, including interferometry.

The X-ray Lasers

Most of the recent X-ray laser development is based on the X-UV emission of hot plasmas produced in vacuum by pulsed infrared (or visible) lasers of large power [16]. Such plasmas can generate a gain coefficient from 1 to 5 cm^{-1} at wavelengths between 4 and 40 nm approximately. Intense saturated laser emissions have been obtained at various wavelengths between 15 and 25 nm. The laser emission duration, a little shorter than the plasma lifetime, ranges from a few tens up to a few hundreds of picoseconds.

Figure 3 shows the principle of the generation of the amplifying plasma column. The beam of the pumping laser ($\lambda = 1.06$ or 0.53 μm) is focused on the surface of a solid target by means of cylindrical lens systems. The plasma column length is $1-5$ cm and the transverse size is a few hundreds of microns. This basic scheme gives rise to various actual realizations according to the number of pumping beams, target material and target structure (slab, thin foil, strip, fibre, ...).

Two plasma pumping schemes – plasma recombination and ion collisional excitation by plasma free electrons – have been investigated extensively. Actually, only the so-called collisional pumping of neon-like ions can generate gain coefficients large enough for practical use. Figure 4 displays a simplified diagram of the levels involved in the pumping process [17]. The Ne-like 1s^2

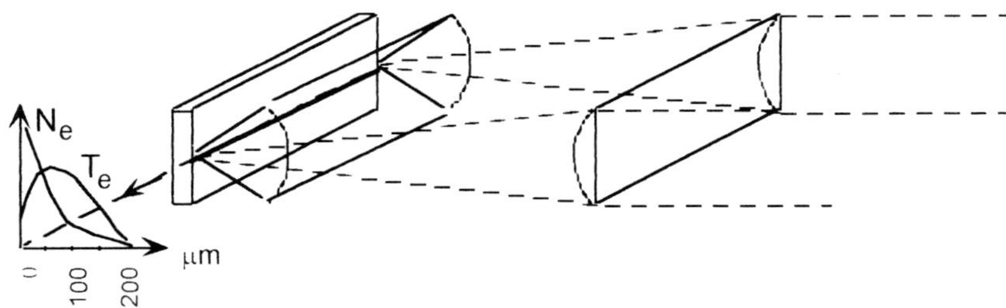

Fig. 3. Amplifying plasma column produced by cylindrical focusing of a laser beam (IR or visible) on to the surface of a solid target. The X-ray laser beam propagates along the plasma axis. On the left, a qualitative diagram of plasma density and temperature is shown.

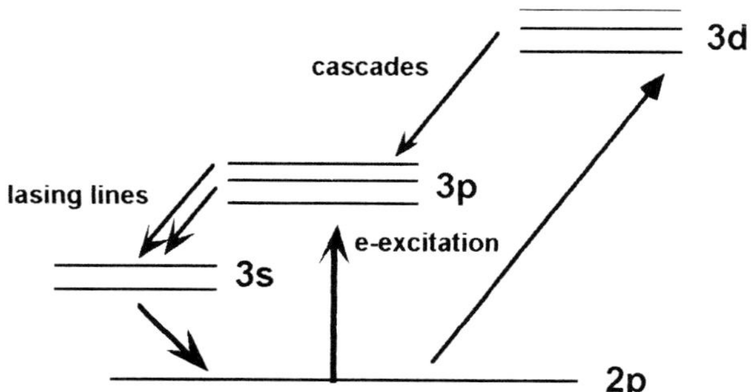

Fig. 4. Scheme of neon-like ion pumping by collisions with plasma free electrons. Population inversions appear between the 3p and the 3s levels.

$2s^2\, 2p^2$ ground state is excited to 3s, 3p and 3d levels by collisions with plasma free electrons. The 3p–2p transition is forbidden while the 3s–2p has a large radiative probability. Under suitable temperature and density conditions, the resulting population kinetics leads to 3s–3p population inversions which are additionally increased by population transfer due to cascades from the 3d levels. Thus plasma pumping does not require an energy contribution other than that necessary for heating the matter to the proper temperature in the proper density plasma zone. Population inversions occur at a density of 10^{20}–10^{21} electrons cm^{-3}. To be produced in the plasma, Ne-like ions require a plasma temperature ranging between 500 and 800 ev (1 ev $\approx 1.1 \times 10^4$ Celsius).

The short plasma lifetime severely limits the number of passes that the X-ray laser beam could achieve inside an optical cavity. However a double-passing configuration is possible by putting a single X-ray multilayer mirror perpendicularly to the X-ray laser beam. Figure 5 [*18*] shows a sketch of the double-pass system, designed to produce saturated emission at 21.2 nm from the Zn^{20+} plasma laser which has been used for the optical coherence study

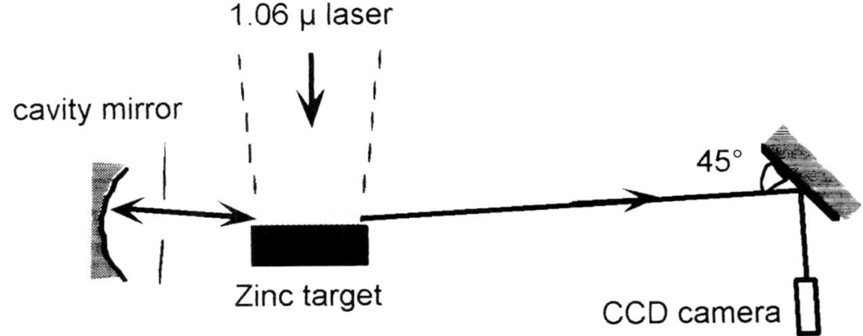

Fig. 5. Diagram of the double-pass system.

presented below. The Mo/Si cavity mirror is aligned to return the beam which is refracted by the plasma at ≈ 7 mrad from the axis to the amplifying region. The mirror reflectivity is about 30% but the cavity geometry limits the single-pass beam fraction which can be re-injected in the plasma so that the effective reflectivity is 5–10%. The gain factor that this 'half-cavity' can supply at 21.2 nm reaches approximately 17.

Table 1 summarizes the beam characteristics from two X-ray lasers taken as examples [18, 19]. The lasing-line FWHM is $0.5\text{–}2 \times 10^{-3}$ nm. This approximately corresponds to a 0.01% bandwidth. The source size is $2\text{–}10 \times 10^{-3}$ mm^2. Then, for comparison with other sources, one can deduce from Table 1 that the pulse-averaged luminance of X-ray lasers, in the 60–80 ev photon energy range, is $10^{25}\text{–}10^{26}$ ph (mm mrad)$^{-2}$s^{-1} for the 0.1% bandwidth generally used for such estimations. Let us mention that the beam can be linearly polarized. A polarization degree of 98% has been obtained on the germanium X-ray laser at the Rutherford Laboratory [20].

The repetition rate is fixed by the pumping laser. At LULI (Palaiseau, France), the interval between two shots is 20 minutes. It would be possible to

Table 1. Main characteristics of two typical X-ray laser emissions.

	Zinc (Zn^{20+})	Yttrium (Y^{29+})
Target material	Zinc (Zn^{20+})	Yttrium (Y^{29+})
Pump energy per pulse	0.4 kJ	20 kJ
Wavelength	21.2 nm	15.5 nm
Photons per pulse	4×10^{13}	5×10^{14}
Pulse duration	80 ps	200 ps
X-ray power (pulse average)	5 MW	32 MW
Solid angle (steradians)	2.5×10^{-5}	2×10^{-4}

considerably reduce this time by using new laser technology in order to shorten the solid amplifier cooling time. Future X-ray lasers could probably supply more that ten shots per minute.

The optical coherence of the beam is an important property for applications such as interferometry, holography and microscopy. The temporal coherence can be estimated from the lasing linewidth. One finds that the longitudinal coherence length is in the order of 1 mm.

As for the spatial coherence, it has been investigated in several laboratories [21–24]. It has been shown, for instance, that the X-ray laser makes interferometric diagnostics of plasma density quite achievable [25]. Here we report on the experimental observation of diffraction patterns obtained with the neon-like zinc laser. By putting a knife edge in the beam, Fresnel diffraction can be observed at various distances from the edge, by changing the radius of curvature of a multilayer mirror. Figure 6 displays a photometric profile of fringes obtained at 50 cm downstream from the diffracting edge, the latter cutting across the beam far from its axis. For the photography of Fig. 7, instead of a single edge, the diffracting object includes two crossed wires at a short distance from a mirror surface. The diffraction pattern is recorded near the beam axis, at 6 cm from the wires. One can see Fresnel linear fringes due to wire edges together with annular fringes due to small defects of the mirror surface.

The derivation of the transverse coherence width is generally performed

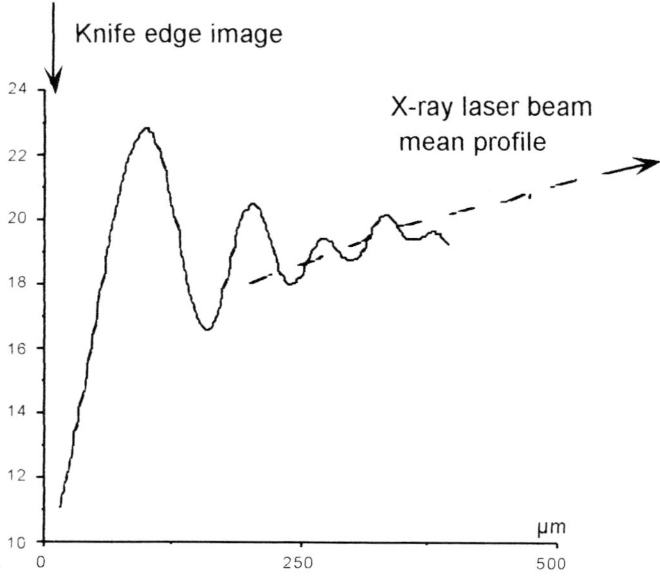

Fig. 6. Fresnel fringes observed in the beam footprint at 50 cm from a diffracting knife edge.

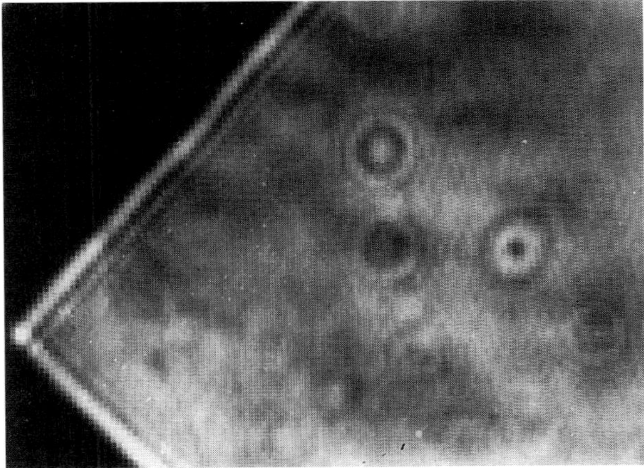

Fig. 7. Diffraction pattern at 6 cm in front of cross-wires set near the surface of a multilayer mirror. Annular diffraction fringes are due to small defects of the surface.

from an interference pattern obtained from a Young's double-slit experiment. In this case, the coherence width is defined as the distance between the slits for which the observed fringe contrast is better than a given value. B. Rus [24] has extended this determination to Fresnel diffraction patterns obtained from semi-infinite screens. This can be made by modelling the laser radiation by means of a remote uncorrelated source of fixed-intensity angular distribution. Intensity distribution fixing is carried out by fitting the calculated edge fringe system to an experimental pattern such as that shown in Fig. 6. Then, estimating the contrast loss of the fringe pattern when increasing the distance from the edge, the coherence width can be defined as the distance for which the loss is greater than a reasonable value, say 0.5.

Numerical modelling applied to the wire-edge fringes of Fig. 7 gives for the HWHM q_m of the Gaussian angular distribution 0.1 mrad, in the central region of the beam and at 50 cm from the end of the X-ray laser. The calculation gives a corresponding coherence width of about 45 µm. At the same distance but far from the centre of the beam, the plot of Fig. 6 gives [26] $q_m = 0.025$ mrad, i.e., about 180 µm for the coherence width. It is not surprising that the spatial coherence varies across the beam. However part of the divergence between the two observations could be due to casual factors having an effect on image recording in the above experiment.

There is no doubt that the X-ray laser spatial coherence can be further improved in the future. Many characteristics of the beam are related to propagation through the plasma during amplification. Now the plasma has a refractive index gradient – due to the electronic density variation – perpendicular to the propagation direction. Optimization of the plasma density distribution for getting larger gain factors is in progress but less has been done so far for increasing spatial coherence. Let us mention that the double-passing configuration greatly enhances the fringe visibility with respect to the single-passing one. Spatial filtering of the beam could also be considered.

Comments

In addition to the coherence properties, the temporal and spectral properties (pulse length, repetition rate, peak power, energy per pulse, tunability, etc.) are additional elements that make the different sources more or less suitable for a specific application. For instance, it can be said that although the X-ray laser luminance is orders of magnitude above all other sources for one-pulse operation, its practicality is at the present time handicapped by its very low repetition rate, except when one pulse operation is practicable, or of particular interest. Similarly, although the HHG sources are still less luminant than the best SR sources, their much greater compactness might make them an interesting solution, not to mention its good temporal structure. In all cases, its high collimation makes it at least as luminant as the best plasma source, despite a much smaller flux. The major limitation of HHG sources for soft X-ray experiments is presently their spectral range, due to the plateau cut-off, but, as said above, there is no fundamental reason not to expect significant progress in that respect.

THE DESIGN AND REALIZATION OF INTERFEROMETRIC EXPERIMENTS

While the distinction between interference and diffraction is sometimes arbitrary, we shall stick to the usual simple concept that an interferometer should have at least two macroscopic channels, known as arms, that simultaneously send beams to the same detector.

Which Hardware?

Experiments based on interferometry always consist in measuring the relative phase of two wavefronts, and, whenever possible, their optical path difference. As is well known, the scale for estimating the performance and the accuracy requirements of such systems is the wavelength of the light used in the interference system. This basic remark leads to the most stringent constraint for X-ray interferometric devices. As a matter of fact, state-of-the-art X-ray optical components are far from delivering wavefronts with interferometric quality, except in one case: high-quality plane mirrors used under grazing incidence, possibly with a single metallic coating. As an example, a wavefront accuracy of $\lambda_X/10$ when $\lambda_X = 10$ nm requires a mirror accuracy of $(\lambda_{vis}/2 \sin i) \times (\lambda_X/10\lambda_{vis})$, with $\lambda_{vis} = 0.5$ μm as usual and i the grazing incidence angle. The factor 2 accounts for the reflection on the mirror surface, which doubles the profile defects. With a typical $i = 3°$, the corresponding mirror accuracy is $\lambda_{vis}/50$, which is well within the possibilities of top-level optical polishing of limited size mirrors. On the contrary, imaging components (spherical or toroidal mirrors, transmission Fresnel zone plates) are far from allowing a similar wavefront accuracy.

A second limitation occurs from the large absorption of materials in the spectral domain. If transmission beam splitters are required, they must consist of very thin membranes, which makes the shape accuracy requirement much more difficult to fulfil.

Finally, if flux is still a limiting factor, it is advisable to use as much spectral bandwidth as possible, and therefore to restrict the path difference just to what is necessary. In all cases, the limited monochromaticities that are practically available do not allow large path differences, e.g., not more than 40 μm at $\lambda = 4$ nm, for $\lambda/\delta\lambda = 10\,000$. From the technical point of view of the mechanical design and realization of the interferometer, this is a near-zero-path-difference interferometer.

This simple discussion leads us to consider that the best candidates for SXUV interferometry are *wavefront division interferometers*, with plane mirrors used at a grazing incidence, and working close to the zero path difference. This implies using beams having some spatial coherence.

In the next sections, we shall only consider such devices. Other solutions are possible, however [27, 25], but they lead to greater complexity in the design and/or difficulties in the manufacturing, in particular if beam splitting (for amplitude division) is necessary.

To end with, it is worth mentioning that a beautiful interferometric device working in the 10 keV ($\lambda = 0.1$ nm) range has been built and operated by Bonse *et al.* [28]. In contrast to what we just said, it is an *amplitude division* interferometer, namely a Mach–Zehnder, which includes two beam-splitters. But the difficulties we mentioned are very elegantly overcome by the use of the Bragg effect in a perfect monocrystal of silicon. The optical quality is therefore related not to the surface quality but to the regularity of the lattice, which is excellent even at the scale of a 0.1 nm wavelength. However, this principle is hardly applicable to much larger wavelengths, mainly because of the much larger absorption, and the lack of crystals with suitable lattice parameters.

...And What For?

Coherence is already routinely used in all scanning imaging systems that use a Fresnel zone plate (as a diffraction limited system, it accepts one coherence cell only), and also in holographic imaging experiments [29]. Despite the limitations set by the X-ray context, we believe that many new experiments can be performed, that bring new tools in various fields of the physical sciences. This belief is based on the extremely wide field of application of interferometry (as an experimental tool) in the optical domain. But it is also true that we certainly have no idea yet of all the possibilities that similar experiments offer in the X-ray domain. We certainly can *suggest* experiments, by analogy with the optical domain, but indeed, only the physicist community can help identify all the possible applications. However, progress is also linked to improvements in the instrument design, including the basic optical design, the mechanical design (stability, positioning accuracy of moving parts) and detection. This is the field of optical physicists.

New coherent applications are already being attempted, like the production and use of speckle patterns [30}, the interferometric diagnoses of X-ray producing plasmas [25], and the implementation of a differential interferential scanning microscope [31]. Others are definitely achievable, like the production of ultra-fine interference fringes (spacing of a few nanometres). We choose to illustrate the performance SXUV can actually reach, through the description of the following two experiments: (i) the interferometric determination of the real part of optical constants (already attempted), and (ii) high-resolution Fourier transform spectroscopy (planned).

The Direct Interferometric Determination of Refractive Indices

Generally speaking, the complex index of refraction, as a function of the wavelength, allows a complete description of the interaction of light with linear, isotropic materials. Its real part accounts for phase-sensitive phenomena (the optical path in matter, refraction), and its imaginary part accounts for absorption. The knowledge of the refractive index is therefore required as the basic parameter of all optical component design, and this holds for SXUV, for such components as phase diffractive optics and multilayer mirrors. Also, the complex index (in particular through its variations with wavelength) is a probe of the photon–matter interaction, and is therefore a useful tool for condensed matter studies.

In the SXUV spectral domain, up until recently, only the absorption part could be accessed experimentally in a direct way. Although in principle the refractive part may be computed from the absorption by using the well-known Kramers–Kronig (KK) relations, the determination process suffers from several difficulties, which in many cases make the results unreliable [32]. This is in particular true in the regions of absorption edges, where the index varies rapidly. The basic reason for this is the integral nature of the KK relations, which should take into account an unbounded continuous set of data. As this is obviously impossible, it is necessary to introduce physical assumptions on the interaction processes, and the practical completion of the numerical integration implies using numerical recipes, which may often locally change the final accuracy in an unknown manner. Other index determinations, based on reflectivity measurements, are also strongly affected by the knowledge of absorption, not to mention difficulties related to the perturbations of the reflectivity determination by surface roughness.

On the contrary, a direct determination of index can be performed in transmission, by interferometric methods, with practically no perturbation due to absorption or surface roughness. The fact that the refractive index is close to 1 in SXUV is merely an advantage (remember that the determination of small index differences is a basic application of visible interferometry). The experiment, which can be performed with a quite simple two-arm interferometer, consists in comparing the optical thickness of one arm that includes the sample to that of a reference arm, ideally the vacuum. The feasibility of such an experiment has been proven with a Fresnel mirror interferometer, on the Super-ACO SU7 undulator, in the carbon K edge region ($\lambda = 4.4$ nm).

As explained in more detail in refs. [*33–34*], a Fresnel bimirror interferometer was used. One quarter of the incident beam illuminated half of one of the mirrors, upon passing through the sample (here a thin polymer layer). This produced two mutually shifted fringe patterns, as only one of them was affected by the additional optical path introduced by the sample in one arm. The grazing incidence on to the mirrors was $3°$ in order to ensure good reflectivity. The tilt between the mirrors was $2'$ $15''$, producing a fringe spacing of about 3 μm (assuming the source at infinity). Fringes were recorded photographically on Kodak 1A high-resolution plates, under $6°$ grazing incidence in order to increase the recorded spacing by a factor of 10. The required spatial coherence was smaller than 150 μm. Figure 8 illustrates the fact that δ changes sign in the region of the carbon K edge, which was to our knowledge never directly demonstrated before this experiment. Further measurements allowed a precise determination of the large-wavelength side of the index peak, including the position of the sign change.

According to Chantler [*35*], such determinations are of primary importance to test and validate the theoretical calculations of index in the edge regions, and the underlying models. Also, the accurate determination of index might show 'index structures' similar to the well-known absorption fine structures (AFS) of the edge regions. Such structures may lead to AFS cross-tests or even bring additional information.

The same interferometric design might indeed be used for other measurements. As an example, all studies based on reflectometry use *intensity* reflectometry. There is no doubt that some of them should take advantage of the measurement of the *complex* reflectivity (intensity and phase), which basically could be implemented with the same Fresnel mirror setup.

SXUV Fourier Transform Spectroscopy

Absorption spectroscopy is a basic tool of atomic and molecular physics, for gaining an insight into the structure of matter. Resolution is an important parameter, as a higher resolution allows a finer analysis of the interactions and energy structures. The performances, however, are presently limited by instruments which hardly allow us to reach 20 000 in the SXUV domain. The

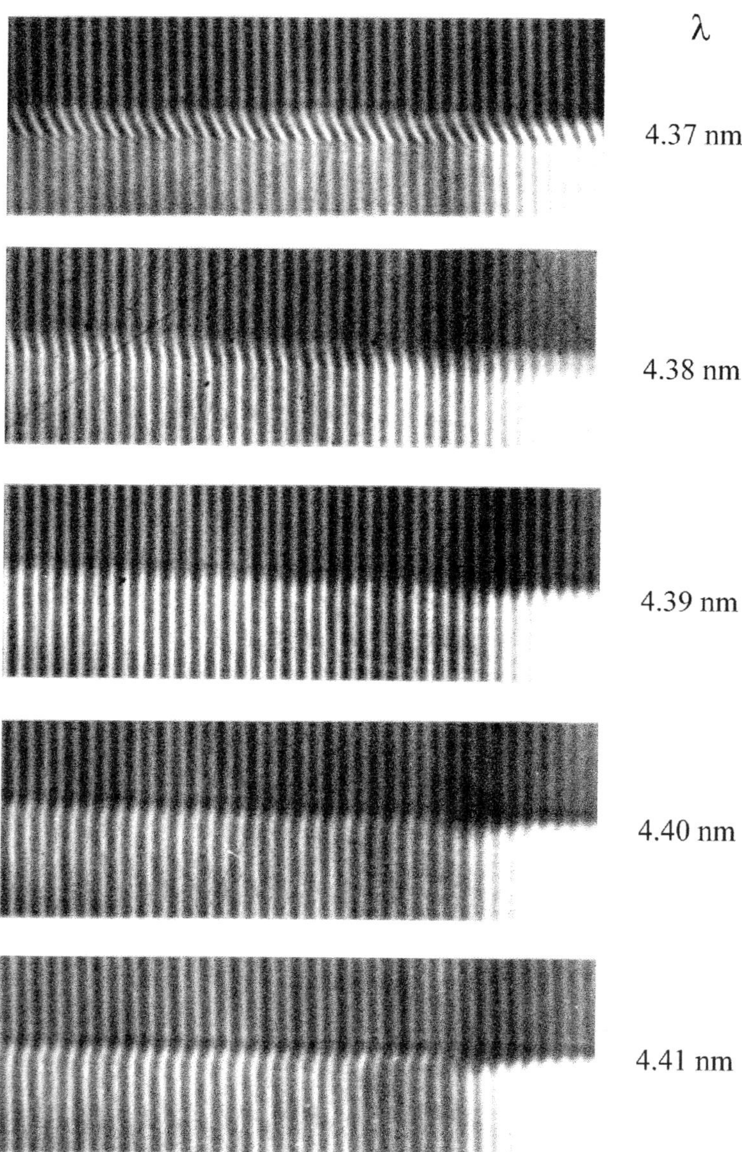

Fig. 8. This series of interferograms is a direct proof that the index of refraction of carbon passes through 1 at the K edge ($\lambda \approx 4.4$ nm). On each interferogram, the reference pattern is the top half, the sample pattern is the bottom half, and their relative shift is proportional to $\delta = n - 1$. The recorded fringe period is ≈ 25 μm. Only a part of each interferogram is shown, the actual recorded field being 1.5 mm wide by 6 mm high.

main reason lies in the basic design of classical spectrographs (grating monochromators), which are in this spectral domain always limited by geometric aberrations, and the manufacturing accuracy.

On the other hand, *Fourier transform spectroscopy* is now a classical technique in the optical domain, from IR to UV, which allows extremely high resolution. The instrument is basically a two-arm scanning interferometer, recording the interferogram intensity as a function of the path difference. As this signal is the cosine Fourier transform (in $\sigma = 1/\lambda$) of the source spectrum, the latter can be recovered by inverse Fourier transformation. Such a system is not limited by the performance of an imaging system, but merely by the performance of a scanning mechanical system, which must preserve the interferometer alignment during scanning, and allow the sampling of the interferometric signal at very precise spatial intervals. A simple calculation shows that resolution in excess of 100 000 can be obtained with interferometers having a reasonable size, scanning range and accuracy (here reasonable means: not out of reach of modern mechanical and optical techniques). Moreover, this kind of instrument is basically a wide-band instrument, in the sense that only geometrical parameters (mainly angles of incidence on to plane mirrors) have to be changed when the working wavelength is changed by large factors, e.g., from 5 to 80 nm.

The perspective of a gain of more than one order of magnitude in the spectral resolution in the XUV domain was exciting enough to generate two projects. The first (and most advanced) one is based on a scanning Mach–Zehnder configuration [27]. Its most delicate part is the necessary beam-splitters, for which impressive progress has been reported. The other one [31] is based on a wavefront division interferometer, which is depicted in Fig. 9. The system uses two pairs of plane mirrors, roof shaped with a large roof angle in order to maintain good reflectivity. Each path uses one roof, and one of them can be translated along the angle bisector in such a way that only the path length is changed, not the position of the emerging ray. By tilting slightly one roof sideways toward the other, rays are made to interfere at some distance downstream. The spatial coherence requirement can be not greater than 100–200 μm, and is constant throughout the field, and during scanning. It is easy to show that a resolution of 4×10^5 at $\lambda = 10$ nm is achieved with only 2 mm optical path difference, using a 6° grazing incidence angle. This requires an accuracy of 2.5 nm in the sampling position. Such specifications are reasonable in the sense explained above.

Theoretical studies in various fields predict the existence of many energy structures, the separation of which requires a resolution of several ten thousands, or even more. Among others, the Rydberg series of He at $\lambda \approx 15$–20 nm is a good example. New structures were recently observed [36] at a resolution of 10 000, and compared to theoretical prediction, and it is believed that a much larger resolution would bring valuable information. However, as stated above, only the physicist community interested in the

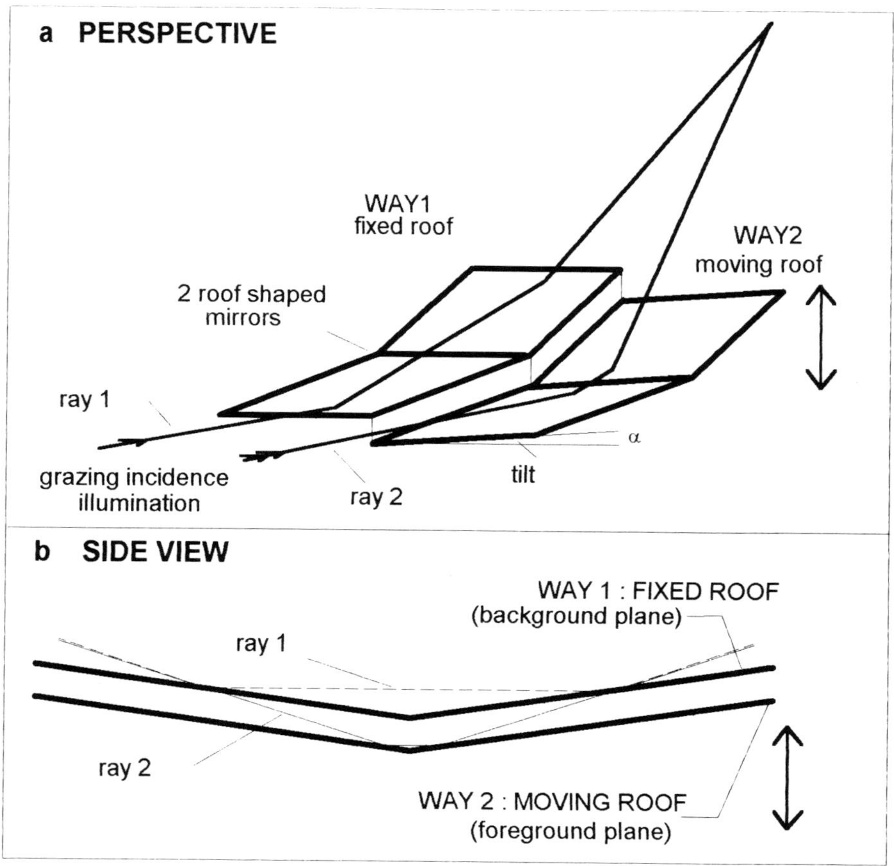

Fig. 9. A two-way, wavefront division interferometer for XUV Fourier transform spectroscopy. The side view (b) shows that the position of the emerging rays is unchanged as the mirror is translated along the roof bisector.

exploitation of high-resolution spectra can help identify all those cases where a large gain in resolution can generate a breakthrough.

CONCLUSION

The increase of luminance (rather than flux) in modern soft X-ray sources opens the way to coherent optics, a new research field in that spectral domain. Some experiments requiring coherence have been known for more than ten years, mostly related to imagery, such as scanning imaging with diffractive optics and holography. But now, the whole toolbox of 'classical' interferometry should be considered for doing physics in the SXUV range, which makes the optical phase a parameter that can be accessed macroscopically. Although some examples have been presented or mentioned in this chapter, it is most likely that we have so far only a faint idea of the new experimental capabilities that will be brought by these techniques. What to do precisely is an open question, which must be answered by the whole community.

REFERENCES

1. J.W. Giles Jr. *J. Opt. Soc. Am.* **59**, 1179–1188 (1969).
2. S. Aoki, S. Kikuta, *AIP Conf. Proc.* **47**, 49 (1986).
3. M. Born, E. Wolf, in *Progress in Optics*, p. 508 (Pergamon Press 2nd ed., New York, 1964).
4. See for instance the *Handbook on Synchrotron Radiation*, Vols I to IV, E.E. Koch, T. Sasaki, H. Winick, series editors, North Holland.
5. M. Ferray, A. L'Huillier, X.F. Li, L.A. Lompré, G. Mainfray and C. Manus, *J. Phys. B.* **21**, L31 (1988).
6. A. McPherson, G. Gibson, H. Jara, U. Johann, T.S. Luk, I. McIntyre, K. Boyer and C.K. Rhodes, *J. Opt. Soc. Am. B*, **4**, 825 (1987).
7. A. L'Huillier and Ph. Balcou, *Phys. Rev. Lett.* **70**, 774 (1993).
8. J.J. Macklin, J.D. Kmetec, and C.L. Gordon III, *Phys. Rev. Lett.* **70**, 766 (1993).
9. J.L. Krause, K.J. Schafer, and K.C. Kulander, *Phys. Rev. Lett.* **68**, 3535 (1992).
10. P.B. Corkum, *Phys. Rev. Lett.* **73**, 1995 (1993).
11. A. L'Huillier, M. Lewenstein, Ph. Balcou, P. Salières, M.Y. Ivanov, J. Larsson and C.G. Wahlström, *Phys. Rev. A* **48**, R3433 (1993).
12. A. L'Huillier *et al.*, *Int. J. Nonlin. Opt. Phys. and Mat.* **4**, 7467 (1995) and references therein.
13. T. Ditmire, J.K. Crane, H. Nguyen, L.B. DaSilva and M.D. Perry, *Phys. Rev. A* **51**, R902 (1995).
14. P. Saliéres, A. L'Huillier and M. Lewenstein, *Phys. Rev. Lett.* **74**, 3776 (1995).
15. H. Eichmann, S. Meyer, K. Riepl, C. Momma and B. Wellegehausen, *Phys. Rev. A* **50**, R2834 (1994).
16. A lot of literature exists on this topic. The reader may consult:
    (a) *Proceedings of International Colloquium on X-Ray Lasers*, ed. P. Jaeglé and A. Sureau, *Journal de Physique, Colloque C6*, **47** (1986);
    (b) *X-Ray Lasers 1990*, ed. G.J. Tallents, Institute of Physics Conference Series N° 116, IOP Publishing, Techno House, Redcliffe Way, Bristol BS1 6NX;
    (c) *X-Ray Lasers 1992*, ed. E.E. Fill, Institute of Physics Conference N° 125 (same address).
    (d) R.C. Elton, *X-Ray Lasers*, Academic Press, San Diego, 1990.
17. See for instance: D. Matthews, M. Rosen, S. Brown, N. Ceglio, D. Eder, A. Hawryluk, C. Keane, R. Londion, B. MacGowan, S. Maxon, D. Nilson, J. Scofield, I. Trebes, in *The Generation of Coherent X-UV and Soft-X-Ray Radiation. J. Opt. Soc. Am. B* **4** (1987) p. 575.
18. P. Jaeglé, A. Carillon, P. Dhez, P. Goedtkindt, G. Jamelot, A. Klisnick, B. Rus, Ph. Zeitoun, in *X-Ray Lasers 1994*, ed. D. Eder and D.L. Matthews, AIP Conference Proceedings 332 (1995) p. 25.

19. L.B. Da Silva, B.J. MacGowan, S. Mrowka, J.A. Koch, R.A. London, D.L. Matthews, J.H. Underwood, *Opt. Lett.* **18**, 1174 (1993).
20. B. Rus, C.L.S. Lewis, G.F. Cairns, P. Dhez, P. Jaeglé, M.H. Key, D. Neely, A.G. MacPhee, S.A. Ramsden, C.G. Smith, A. Sureau, *Phys. Rev.* **51**, 2316 (1995).
21. J.E. Trebes, K.A. Nugent, S. Mrowka, R.A. London, T.W. Barbee, M.R. Carter, J.A. Koch, B.J. MacGowan, D.L. Matthews, L.B. Da Silva, G.F. Sone, M.D. Feit, *Phys. Rev. Lett.* **68**, 588 (1992).
22. R.E. Burge, G. Slark, X. Cheng, M.T. Browne, D. Neely, C.L.S. Lewis, A.G. MacPhee, *Studies of the spatial coherence of an X-ray laser* Int. Symp. on Optical Science, Engineering and Instrumentation (Soft X-ray Lasers and Applications), San Diego, July 1995, to appear in *SPIE Proc.* Vol. 2520, 257–262 (1995).
23. H. Daido, M.S. Schultz, K. Murai, R. Kodama, G. Yuan, J. Goto, K.A. Tanaka, Y. Kato, S. Nakai, K. Shinohara, T. Honda, I. Kodama, H. Iwasaki, T. Yoshinobu, M. Tsukamoto, M. Niibe, Y. Fukuda, D. Neely, A. MacPhee, G. Slark, *J. X-ray Sci. Tech.*, **5**, 105 (1995).
24. B. Rus, Thèse n° 3549 (1995), Université Paris-Sud, Orsay, France.
25. B DaSilva *et al.*, *Phys. Rev. Lett.*, **74**, 3991 (1995).
26. F. Albert, LSAI internal report, Université Paris-Sud, Orsay, France (1995).
27. M.R. Howells, K. Frank, Z. Hussain, E.J. Moler, T. Reich, D. Möller, D.A. Shirley, *Nucl. Instrum. Methods*, **A347**, 182 (1994).
28. U. Bonse, H. Lotsch, A. Henning, *J. X-ray Sci. Technol.* **1**, 107, (1989).
29. See for instance: *X-Ray Microscopy III*, ed. A.G. Michette, G.R. Morisson, C.J. Buckley, Springer (Berlin, 1992).
30. G. Grübel, D. Abernathy, G.B. Stephenson, S. Brauer, I. McNulty, S.G.J. Mochrie, B. McClain, A. Sandy, M. Sutton, E. Dufresne, I.K. Robinson, R. Fleming, R. Pindak, and S. Dierker, *Intensity fluctuation spectroscopy using coherent X-Rays*, in *ESRF Newsletter* **23**, March 1995, p. 14.
31. F. Polack, D. Joyeux, *Rev. Sci. Instrum.* **66**, 2 (1995).
32. E. Spiller, *Appl. Opt.* **29**, 19 (1990).
33. J. Svatos, D. Joyeux, D. Phalippou, F. Polack, *Opt. Lett.* **18**, 1367 (1993).
34. J. Svatos, D. Joyeux, F. Polack, D. Phalippou, in *X-ray Microscopy IV*, ed. A.I. Erko and V.V. Aristovi (Bogorodski Pechatnik, Chernogolovka, Moscow, 1994).
35. C.T. Chantler, *Radiat. Phys. Chem.* **41**, 759 (1993).
36. M. Domke, G. Remmers, G. Kaindl, *Phys. Rev. Lett.* **69**, 1171 (1992).

21 Axially symmetric multiple mirror optics for soft X-ray projection microlithography

Sang Soo Lee, Cheon Seog Rim and Chang Hee Nam
Department of Physics, Korea Advanced Institute of Science and Technology (KAIST), Yusong-ku, Taejon 305–701, Korea

Young Min Cho
Korea Aero-Space Research Institute, Yusong-ku, Taejon, 305–600

Dong Eon Kim
Department of Physics, Pohang Institute of Technology, Pohang, 790–784

INTRODUCTION

In this chapter the results of the recent investigations carried out in several laboratories, and particularly in this laboratory, on the axially symmetric soft X-ray projection systems for high density (one gigabit) microlithography are briefly reviewed and the merits of the optical designs are compared. Our approach to deriving a soft X-ray projection system is the same as that which we developed previously for the projection system using UV beams from KrF and ArF excimer lasers [1, 2, 3, 4], however as the wavelength of soft X-rays is shorter (typically $\lambda = 13$ nm) the designed system should be closer to being optically aberration free, so that the optimization process during the system designing process usually takes much longer and extended elaboration is required. We have finally succeeded in obtaining several four-mirror systems with resolution beyond 3000 cycles mm^{-1}. These systems all have a depth of focus greater than or equal to 0.7 μm. The high-reflection quantum well coatings for $\lambda = 13$ nm are not discussed in this chapter, because the current development of high-reflection quantum well deposition technology for soft X-ray optics already seems satisfactory for our purposes.

The current optoelectronics industry such as the semiconductor DRAM industry and the digital image processing industry demand increasingly higher pixel densities for the electronic elements, and have started using ArF laser beams for 64M DRAM and 256M DRAM, and recently they are concentrating on the development of 1G DRAM chips. This requires a resolution of the order of $0.1-0.18$ μm [5], which corresponds to 5000 cycles mm^{-1}–3000 cycles mm^{-1} resolution in the MTF (Modulus transfer function) curve at the level of 0.4. In order to acquire that resolution in the optical imaging system, we have to use a shorter λ than the ArF excimer laser line ($\lambda = 1930$ Å) because the diffraction limited resolution $\varepsilon_R = k_1 \lambda / \text{NA}$ (Rayleigh's criterion) where $k_1 = 0.61$ for a circular aperture, is proportional to the wavelength employed for a given NA (numerical aperture). Together with the high resolution, the microlithographic system should have reasonable depth of focus (DOF) which is $k_2 \lambda / (\text{NA})^2$ ($k_2 = 1$). DOF is essentially required as the mask has relief structure. It can be smaller in the soft X-ray region where $\lambda \approx 100$ Å, which is about 1/20 of the ArF excimer laser line. The DOF required in the soft X-ray region is usually greater than or equal to 0.5 μm [6].

If we give DOF a value of 0.5 μm in the above expression thereby eliminating NA from ε_R and DOF, then we get $\varepsilon_R = \sqrt{0.5 \times (k_1^2/k_2)} \sqrt{\lambda}$ which means ε_R, the Rayleigh criterion of resolution, becomes smaller as a shorter wavelength λ is employed for microlithography. Thus the use of soft X-rays, typically $\lambda = -130$ Å, from the focused laser plasma becomes useful. In this wavelength region, material absorption is severe, so that we must use a reflection system, which has recently become feasible as a consequence of recent development in the high-reflection, multi-layer quantum well system such as the Si–Mo system (reflectance = 66% for 40 bilayers) [7].

As the wavelength of the soft X-rays from the laser plasma is typically 130 Å, one tenth of a wavelength in wavefront aberration in the ArF laser line becomes 1.5λ for the soft X-ray wavelength ($\lambda = 130$ Å), which is an unacceptably large amount of aberration for soft X-ray microlithography. This situation indicates that one has to make very precise attempts to eliminate optical aberration in designing the optical system using soft X-rays.

In our work, the primary attention is on axially symmetric systems, because the inherent off-axial aberration in an axially non-symmetric system such as an off-set three-mirror system [8, 9] creates difficult problems in eliminating minute aberration, which can be of the order of 10 Å in soft X-ray lithography optics. Also, the famous ring-field optical system [10, 11] is not our main interest, because the system is a scanning type and the perfectly correlated scanning mechanism in object space and image space is difficult, particularly in industry. Furthermore, this scanning system requires a stable, stationary soft X-ray source which is currently only at the stage of development. Non-imaging X-ray optics such as Kumakhov optics using a hollow capillary [12] does not give the high resolution required in high-density lithography. The most realistic high-performance soft X-ray imaging system reported so far is one developed by V. K. Viswanathan and Brian N. Newman in Los Alamos National Laboratory (LANL), USA [13], which is a reflective two-mirror system where reflections take place four times at four different annuli.

This reflective two-mirror system (the V-N system) is an inverse Cassegrain system and has achieved the impressive resolution of 0.1 μm or 5000 cycles mm^{-1}. The system has Petzval sum $4(c_1 - c_2)$, $c_1(>0)$, c_2 (>0) being the curvatures, so that in order to eliminate the field curvature aberration, the astigmatism in the opposite sign of the Petzval sum term, which is $-(1/2)\mathrm{H}^2 \cdot 4(c_1 - c_2)$, where H is Lagrangian invariant, should be retained in the system. $c_1 - c_2$ could be made small, but it is not zero in the case of the V-N system, so that the system should have a sizeable amount of the third-order astigmatism. Further, the beam has to move on the two mirror surfaces, so that the size of the mirror tends to become larger. Our final choice is to use an axially symmetric four-mirror system, such as a Cassegrain-inverse Cassegrain system. As the system has to have a resolution higher than 3000 cycles mm^{-1} for a soft X-ray wavelength of 130 Å for gigabit microlithography, the task of the optical design is immediately seen to be difficult. The four-mirror system we are working on is more versatile as the system has a reasonably large number of design parameters $(2 \times 4 = 8)$. In this chapter, however, the discussion will start with the two mirror system with four design parameters $(2 + 2 = 4)$ in order to obtain a better perspective of the four-mirror system which is described in the later sections.

The following table gives the currently required resolution of the optical system and wavelength employed for high-density microlithographic practice [5].

D-RAM	1M	4M	16M	64M	256M	1G
The resolution required (μm)	1.0	0.7	0.5	0.35	0.25	0.18~0.15
Wavelength employed (μm)	0.436 Hg-g	0.365 Hg-i		0.193 ArF excimer laser (DUV)		0.013 Soft X-ray (XUV)

It must be pointed out that the above discussion is based on third-order wavefront aberration theory, which is often called Seidel first-order aberration theory. The higher-order aberration theory, say fifth-order theory, is complex and beyond the scope of this chapter.

TWO-MIRROR SOFT X-RAY IMAGING SYSTEM

In the holosymmetric system which has the simplest optical configuration among the two-mirror systems, and shown in Fig. 1, c_2 is equal to $-c_1$ and inherently has unit magnification ($M = -1$). In Fig. 1, C is the concentric centre of the curvature c_1 and c_2 and F'_1 and F'_2 are the second Gaussian focal points of the two mirrors. O is the axial object point and O' is the image at the conjugate point and soft X-rays from a laser plasma source illuminate the object along the $-z$ direction. Note that the sine condition and the Hershel condition can be simultaneously fulfilled. This is an important advantage of the holosymmetric system.

In this system, the Petzval sum P is given by

$$P = \Sigma\left[-c_i\Delta\left(\frac{1}{n_i}\right)\right] = -c_1(-1 - 1) - c_2(1 - (-1))$$

$$= 2(c_1 - c_2) = 2(c_1 + c_1) = 4c_1, \qquad \text{as} \quad c_2 = -c_1 \tag{1}$$

and the astigmatism with the opposite sign of $(1/2)H^2 \times 4c_1$, which is the first term of the field curvature aberration, should be retained in order to make the field curvature aberration (third-order) zero. This remaining astigmatism gives a basic disadvantage to the two-mirror system.

This is a general statement. However, the holosymmetric system inherently has a solution which gives zero coma and zero distortion. This is an additional advantage, thus the aberrations (third-order) finally present in the system are spherical aberration and astigmatism. These aberrations may be, but are not certain to be, eliminated by using conic and/or aspherical surfaces c_1 and c_2 at the stage of optimization with respect to the finite aberration. As mentioned above, the field curvature aberration can be made zero only by providing astigmatism with the opposite sign of the Petzval sum aberration H^2P, so that there is always a limit in reducing the astigmatism. Spherical aberration and

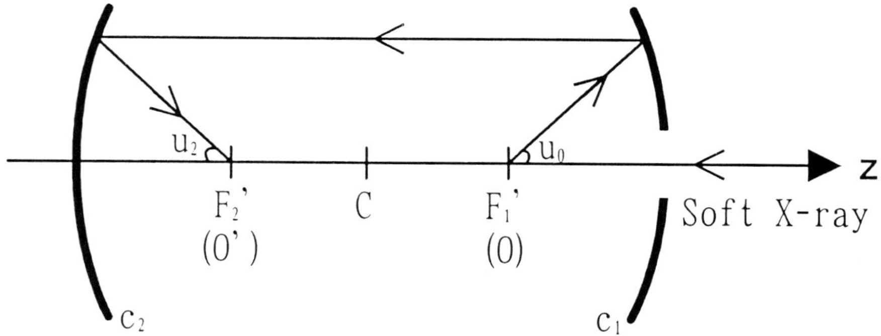

Fig. 1. Holosymmetric two-mirror system ($M = -1$), $c_1 = -c_2$.

that remaining astigmatism should be the two aberrations to be eliminated, possibly simultaneously during optimization.

Table 1 is an example of the optical design data of the holosymmetric two-mirror system shown in Fig. 2. The system has a resolution of only 2000 cycles mm^{-1} for $\lambda = 13$ nm and may be used in 256M DRAM lithography (image size = 0.4 cm). Remember that the system should have quite large astigmatism remaining.

The holosymmetric system (Fig. 1) with unit magnification discussed above is a special case of Gregory mirror optics. The system has inherently zero coma and zero distortion, but it has finite Petzval sum in the field curvature aberration. In the Cassegrain and the inverse Cassegrain systems, the Petzval sum P is $-2(c_1 - c_2)$ and c_1 and c_2 have the same sign, so that P can be made

Table 1. The design data of the aplanatic flat-field holosymmetric two-mirror system. The system has a focal length $f' = -60$ cm and the stop lies at the first mirror.

	Mirror No.			
	Object	Stop	I	II
Distance (cm)	60.000000	0.000000	−60 000000	
Radius of curvature (cm)			−120 000000	+120 000000
Conic constant			−1.000000 (parabolic)	−1.000000 (parabolic)

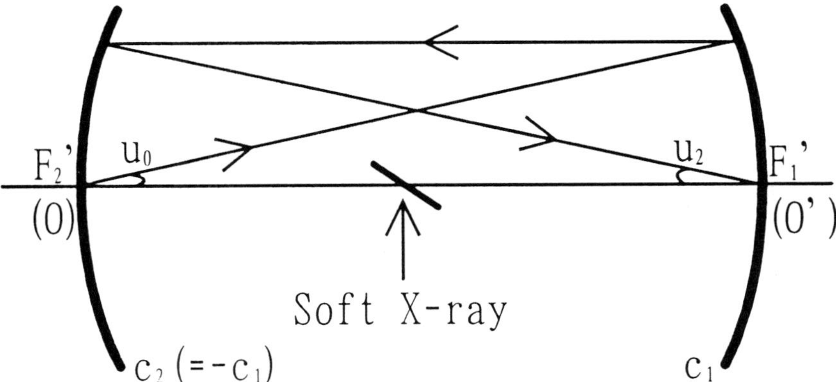

Fig. 2. Configuration of the holosymmetric two-mirror system ($M = -1$). This system has an NA of 0.032, the TCL(O–O', total conjugate length) of 60.0000 cm, and the largest clear aperture diameter is 4.4000 cm. The mask and wafer are at O and O', respectively.

Table 2. Diameters of clear apertures and holes of the mirrors for the system given by Table 1 (unit: cm).

	Mirror No.		
	Stop	I	II
Clear aperture	4.0000	4.0000	4.4000
Hole diameter	0.0000	0.6000	0.6000

small. The field curvature aberration (third-order) can be made zero by providing an astigmatism with the opposite sign of the Petzval sum. However, coma and distortion remain in the system and, further, in the case of unit magnification, holosymmetry does not exist in the Cassegrain or in the inverse Cassegrain system. Only the Gregory system can have the holosymmetry shown in Fig. 1 and discussed above. A holosymmetric Gregory system, however, does have a finite Petzval sum as shown in Eq. (1), and has zero coma and zero distortion, so that the system carries only astigmatism and spherical aberration. The system shown in Tables 1 and 2 is the optical design data obtained as the result of extensive optimization on the finite astigmatism and spherical aberration of the holosymmetric system shown in Fig. 2.

In imaging optics with reduction capability, Cassegrain and inverse Cassegrain systems seem advantageous in reducing astigmatism, as the Petzval sum can be made small (c_1 and c_2 have the same sign) or even zero, but considerable effort needs to be made in eliminating all other aberrations. In the astronomical telescope and the X-ray spectroscopic system, the object field angle is usually very small, so that the elimination of the spherical aberration is of primary importance. In imaging optics, particularly when high-resolution performance is required, two-mirror systems are not recommended.

AXIALLY SYMMETRIC FOUR–MIRROR SYSTEM WITH REDUCTION CAPABILITY

The reported two-mirror and four-reflection system [13] has an excellent performance of 5000 cycles mm^{-1} at the level of MTF = 0.4; however, in this system the beam (soft X-ray) has to walk on surfaces of the two mirrors during four reflections. Secondly, the system has only four design parameters which does not seem enough if one considers the five Seidel aberrations (third order), the given magnification and the scaling factor. One may think that the aspherization coefficients (a_4, a_6, a_8, a_{10}) [14] are the design parameters, but they are used only for small corrections to finite aberrations made on the spherical surfaces. It is more appropriate to exclude the aspherization coefficients from the design parameters.

With these considerations, we have started investigating the four-mirrors, four-reflections, axially symmetric system, i.e., the Cassegrain-inverse Cassegrain system (C-iC system) shown in Fig. 3.

In the C-iC system, there is ample opportunity for making the Petzval sum (P) zero, and due to the symmetric layout (like the Gauss-type camera lens), aberrations, including distortion and coma, are inherently small. The Petzval sum of the system P is given by

$$P = -c_1(-1-1) - c_2(1+1) - c_3(-1-1) - c_4(1+1)$$
$$= 2(c_1 - c_2 + c_3 - c_4) \tag{2}$$

and the number of design parameters is $2 \times 4 = 8$, which is larger by one than the total number of constraints which are zero, five Seidel aberrations (third order), a given magnification and a reference scale ($c_1 = -1$).

P of Eq. (2) can be made zero when

$$c_1 + c_3 = c_2 + c_4. \tag{3}$$

With this design condition, we first derive the four-mirror system from the third-order aberration theory which consists of four spherical surfaces. The five

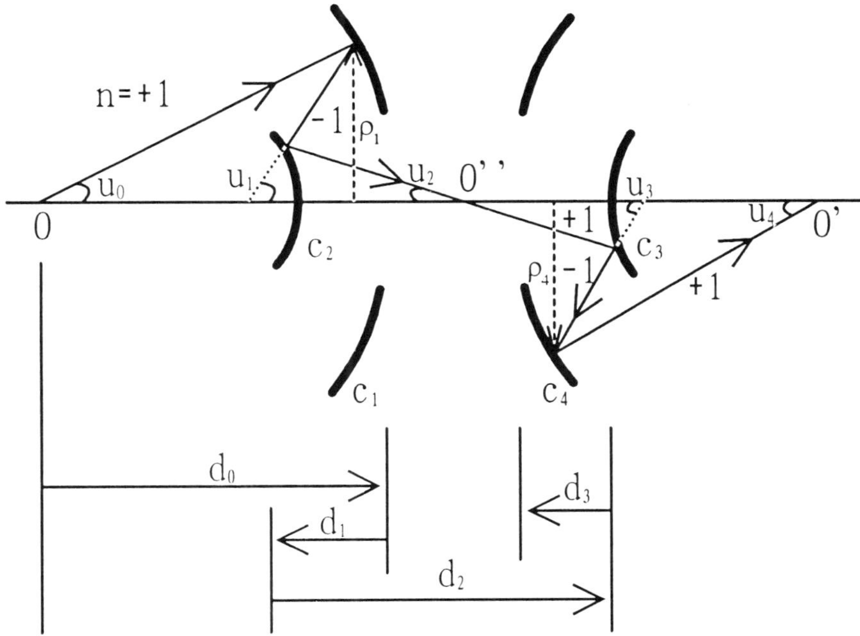

Fig. 3. Schematics of the Cassegrain-inverse Cassegrain system.

wavefront aberrations [*15, 16*] are expressed as follows:

$$\text{Spherical aberration} = \tfrac{1}{8} S_{\mathrm{I}}$$
$$\text{Coma} = \tfrac{1}{2} S_{\mathrm{II}} \cos\phi$$
$$\text{Astigmatism} = \tfrac{1}{2} S_{\mathrm{III}} \cos^2\phi \tag{4}$$
$$\text{Field curvature} = \tfrac{1}{4}(S_{\mathrm{IV}} + S_{\mathrm{III}}) = \tfrac{1}{4}(S_{\mathrm{IV}} + 2 \times \text{Astigmatism})$$

$$\text{Distortion} = \Sigma \frac{B}{A}(H^2 p + S_{\mathrm{III}});$$

$$S_{\mathrm{I}} = \Sigma A^2 \rho \Delta\left(\frac{u}{n}\right), \qquad S_{\mathrm{II}} = -\Sigma AB\rho \Delta\left(\frac{u}{n}\right), \qquad S_{\mathrm{III}} = \Sigma B^2 \rho \Delta\left(\frac{u}{n}\right),$$

$$S_{\mathrm{IV}} = H^2 \Sigma p = H^2 P, \qquad S_{\mathrm{V}} = \Sigma \frac{B}{A}\left[H^2 p + B^2 \rho \Delta\left(\frac{u}{n}\right)\right], \tag{5}$$

where H is Lagrange invariant, $p = -c\Delta(1/n)$, $A = ni$ (Snell's law for an axial ray), $B = n_p i_p$ (Snell's law for the principal ray), c_1 is the reference scale, and the given magnification is M; u and ρ for the paraxial ray are indicated in Fig. 3.

The simultaneous equations, $S_i = 0$ or minimal for $i = \mathrm{I}, \mathrm{II}, \ldots, \mathrm{V}$ and given magnification give the c_i ($i = 2, 3,$ and 4, $c_1 = -1$) and d_i ($i = 0, 1, 2$ and 3). Upon obtaining the layout from the third-order theory, one has to proceed to optimize with respect to finite aberration. In this process of optimization, aspherization may be included. Aspherization gives the small deviation from the spherical surface and we usually use the four aspherical constants (a_4, a_6, a_8, a_{10}), but in this work we use only the two aspherized constants (a_4 and a_6). Extensive computer calculation is required in order to arrive at the final aspherized system. The optical transfer function (OTF) or the modulus of the OTF and resolutions are worked out for the final system with some residual aberrations remaining.

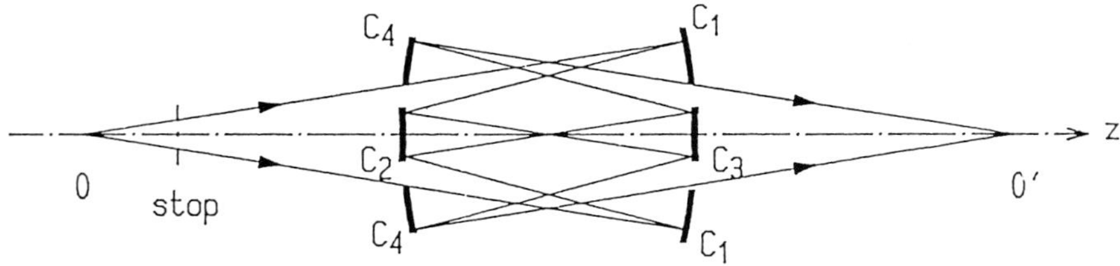

Fig. 4. The configuration of the soft X-ray four-mirror system ($M = +1$). This system has NA of 0.15, TCL of 98.8854 cm, and largest clear aperture diameter 22.0800 cm. The system is holosymmetric ($c_1 = -c_4$, $c_2 = -c_3$).

Table 3. The design data of the aspherized holosymmetric four-mirror system ($f' = -10$ cm) using a soft X-ray wavelength of 13 nm. The stop lies at the first focal point of the four-mirror system.

				Mirror No.		
	Object	Stop	I	II	III	IV
Radius of curvature (cm)			-49.442719	-49.442719	49.442719	49.442719
Distance (cm)	10.000000	54.721360	-30.557281	30.557281	-30.557281	
a_4 (cm^{-3})			0.294172×10^{-10}	0.000000	0.000000	$-0.294172 \times 10^{-10}$
a_6 (cm^{-5})			$-0.480050 \times 10^{-11}$	0.000000	0.000000	0.480050×10^{-11}

Table 4. Diameters of clear apertures and holes of the mirrors for the system given by Table 3 (unit: cm).

			Mirror No.		
	Stop	I	II	III	IV
Clear aperture	3.0350	22.0800	5.5200	5.5200	20.0000
Hole diameter	0.0000	10.6900	0.0000	0.0000	11.1040

Table 5. The design data of the aspherized four-mirror system (Cassegrain-inverse Cassegrain) with $M = +1/10$ ($f' = -20$ cm). The stop is at the first focal point (F).

				Mirror No.		
	Object	Stop	I	II	III	IV
Curvature (cm^{-1})			-0.0239187	-0.0165339	0.0112054	0.0038207
Distance (cm)	200.000131	58.375396	-11.241300	40.609442	-240.551504	
Conic constant			-0.0000001	0.000000	0.000000	0.000001
a_4 (cm^{-3})			1.522156×10^{-11}	0.000000	0.000000	$-2.504246 \times 10^{-14}$
a_6 (cm^{-5})			$-4.802285 \times 10^{-11}$	0.000000	0.000000	3.334372×10^{-16}

Table 6. Diameters of clear apertures and holes of the mirrors of the system in Table 5 (unit: cm).

			Mirror No.		
	Stop	I	II	III	IV
Clear aperture	6.0007	7.9467	4.1185	4.8000	85.0000
Hole diameter	0.0000	3.3399	0.0000	0.0000	44.1022

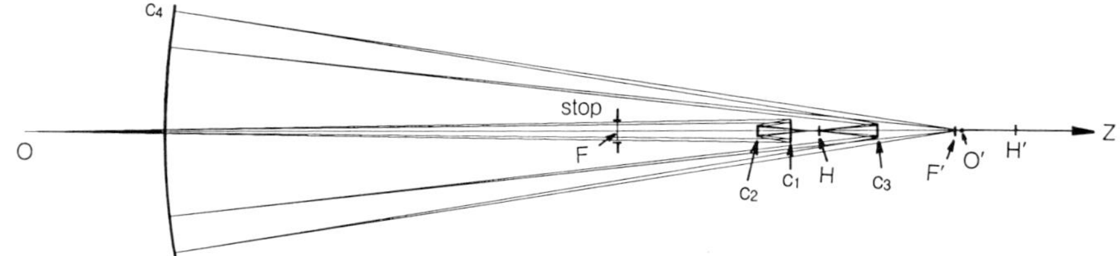

Fig. 5. The configuration of the aspherized four-mirror system with reduction capability ($f' = -20$ cm) using $\lambda = 13$ nm. This system has NA of 0.15, TCL of 3.15 m, and largest clear aperture diameter 85.0 cm. The stop lies at the first focal plane of the system.

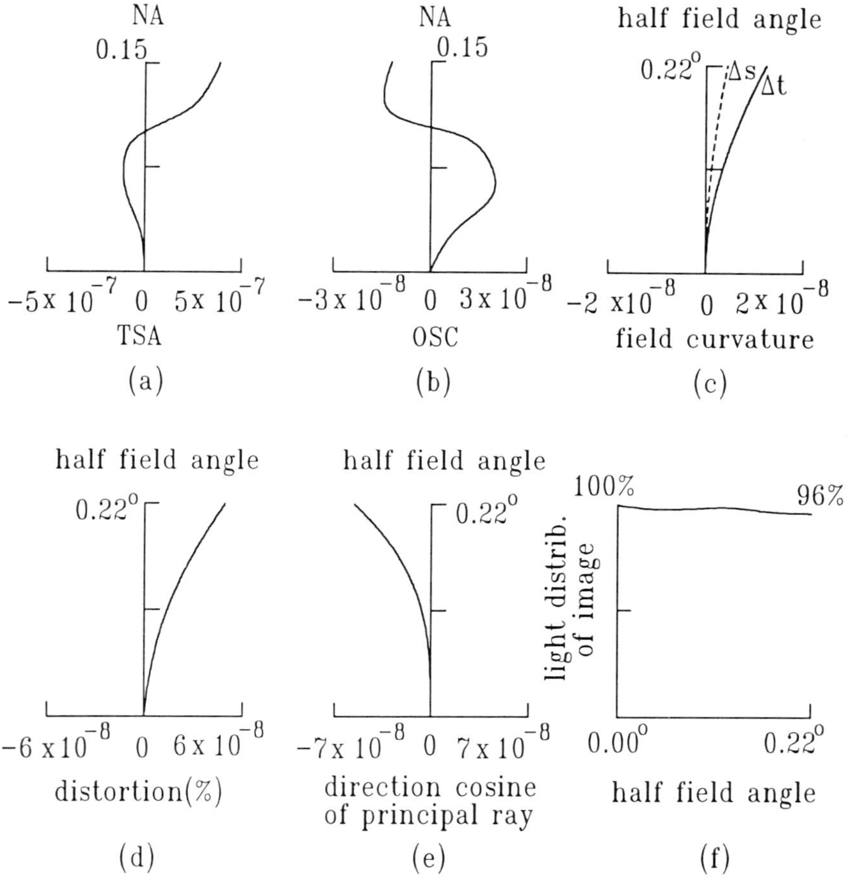

Fig. 6. (a), (b), (c), (d) are the residual finite ray aberrations of the aspherized four-mirror system ($M = +1/10$, $f' = -20$ cm) using $\lambda = 13$ nm; (e) is the direction cosine of the principal ray in image space; (f) is the relative amount of light in the image plane (TSA: transverse spherical aberration, OSC: offence against the sine condition, Δt: tangential ray aberration, Δs: sagittal ray aberration).

In Fig. 4, the four-mirror holosymmetric system [17] ($M = +1$) is shown and the system parameters thus obtained are listed in Tables 3 and 4.

A general four-mirror system with reduction capability ($M = +1/10$) is also found as shown in Tables 5 and 6, and Fig. 5. This system has resolution close to 4000 cycles mm^{-1} and is again usable in gigabit microlithography using 130 Å soft X-rays from the laser plasma. Characteristics of the performances are shown in Figs 6–8.

Fig. 7. The spot diagrams of the aspherized four-mirror system ($M = +1/10$) using $\lambda = 13$ nm for three half field angles at five defocused image positions. The circle shows the size of the Airy disk (radius $R = 0.053$ μm) for the given wavelength.

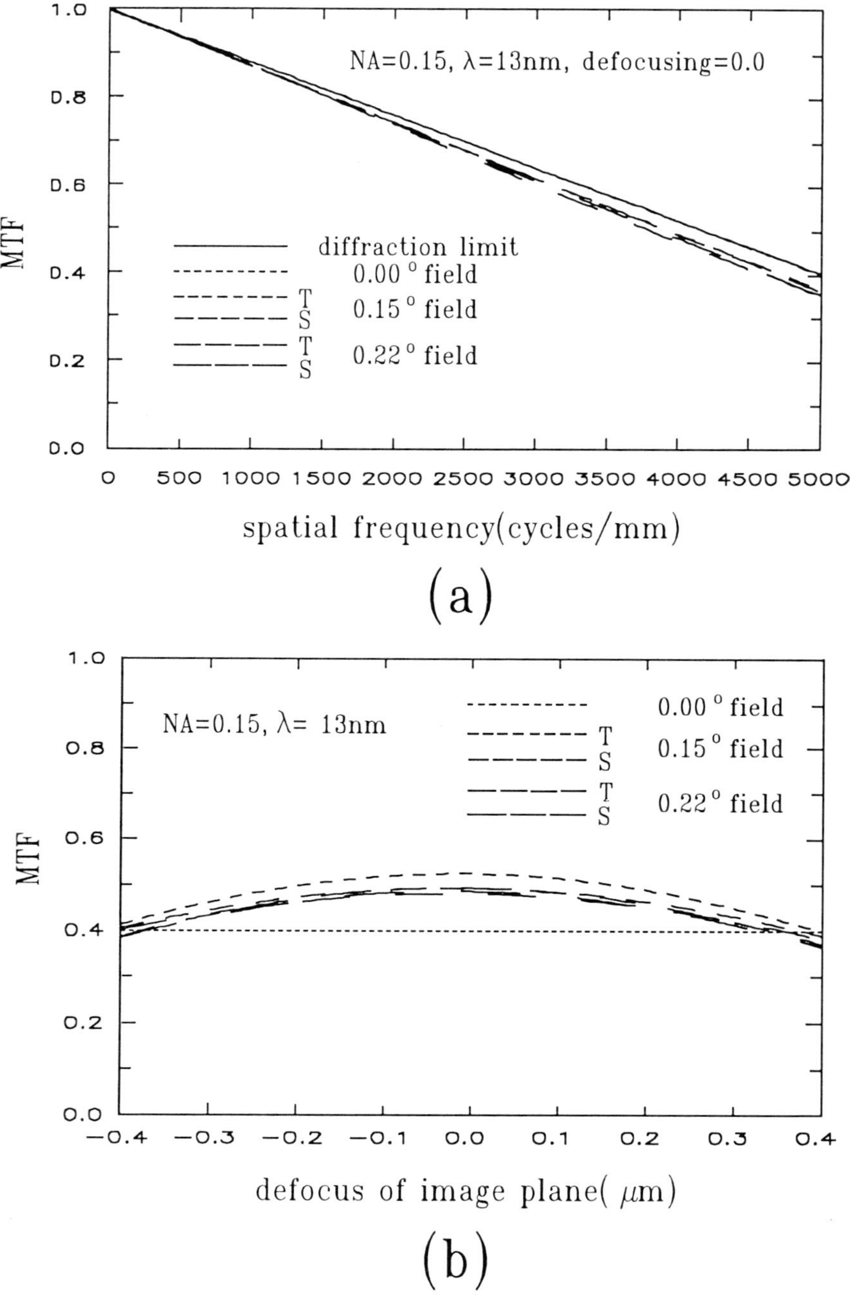

Fig. 8. (a) MTF of the aspherized four-mirror system ($\lambda = 13$ nm) with a ratio of 0.56 central obscuration. The diffraction limited MTF curves are plotted for the case with 0.56 obscuration. (b) Through focus MTF of the aspherized four-mirror system (13 nm) with 0.56 obscuration at the spatial frequency of 3950 cycles mm^{-1} (resolution = 0.127 μm, object size = 1.5 cm) for three half field angles.

MAGNIFICATION AND OBJECT SIZE DEPENDENT RESOLUTION

In the four-mirror system we have carried out quite a large volume of calculations to find the magnification ($M = +1/2$, $+1/3$, ..., $+1/10$, $+1/15$, and $+1/20$) dependent change of resolution, and have found that the resolution improves asymptotically (see Fig. 9). However, as the system becomes longer and larger, it takes the important disadvantage in mechanical stability. The total conjugate length (TCL), of the order of 3×10^2 cm could be about the limit. For that TCL, $M = +1/10$ gives the best performance of resolution 0.127 μm (3950 cycles mm^{-1}) for the object size of 1.5 cm.

Furthermore the resolution becomes poorer as the object size increases. We have checked numerically this aspect in the $M = +1/10$ system for the object sizes 1.5 cm, 2.0 cm, 2.5 cm, 3.0 cm, and 3.5 cm and found the largest object size of 3.0 cm is obtained, which gives a resolution of 0.139 μm (3600 cycles mm^{-1}).

FIVE−MIRROR SYSTEM

Although we are continuing to find the four-mirror system superior over those discussed about in the previous sections, we have not yet found a system with

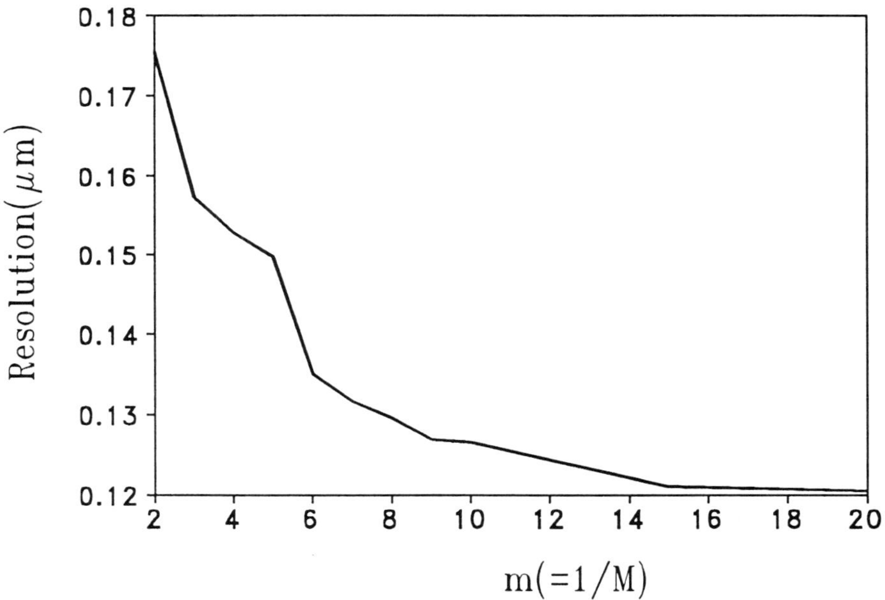

Fig. 9. Resolution behaviour of the aspherized four-mirror system according to the variation of the reduction magnification when NA is 0.15, DOF is 0.7 μm, and object size is 1.5 cm.

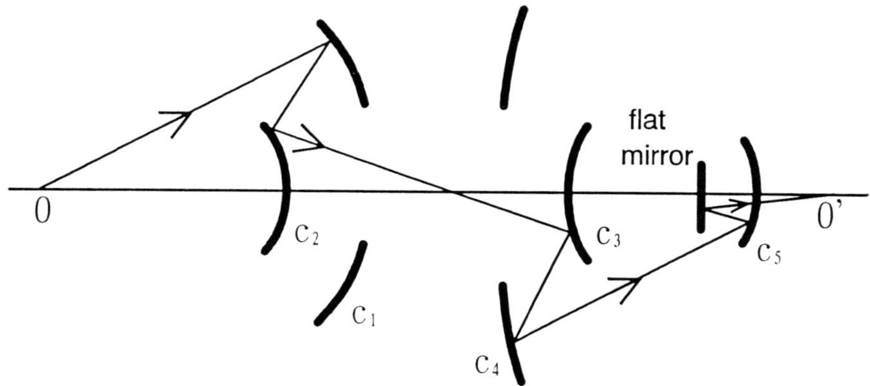

Fig. 10. Schematics of the five-mirror system.

resolution better than 4000 cycles mm^{-1}, and we now think that the five-mirror system may give us an improved system. In this laboratory we have experience of designing a five-mirror system [18] using an ArF excimer laser beam which has shown a much improved performance over the four-mirror system. The five-mirror system has $5 \times 2 = 10$ design parameters and there are two parameters remaining free (Fig. 10). We can make more detailed analyses in obtaining the higher-resolution optical design of the system.

CONCLUSION

Mirror systems with a multilayer quantum well reflection coating for soft X-ray ($\lambda = 13$ nm) lithography are reviewed. The holosymmetric two-mirror Gregorian type ($M = -1$) is found to be not very satisfactory as it has quite a large astigmatism remaining. It has resolution of 2000 cycles mm^{-1} at the MTF $= 0.4$ level and is useful only in 256M DRAM lithography for a small image size of 0.4 cm. Both Cassegrain and inverse Cassegrain two-mirror systems (four design parameters and two reflections) have severe coma aberration. A four-mirror system which has eight design parameters is far superior to the two-mirror system. A holosymmetric four-mirror system ($M = +1$) and Cassegrain-inverse Cassegrain system ($M = +1/10$) with resolution suitable for 1G DRAM lithography is obtained. A further improvement seems to require use of five mirrors which gives 10 design parameters. We are making extensive calculations for the final goal of a resolution of 5000 cycles mm^{-1} (0.1 μm), in the four- and/or five-mirror system. At this point, we want specifically to describe Prof. W. T. Welford's work at Imperial College, London on the four-mirror system used in the laser fusion target chamber. It is a folded Cassegrain-inverse Cassegrain system with minimized obscuration as shown in Fig. 11, and our four-mirror systems in Figs 4 and 5

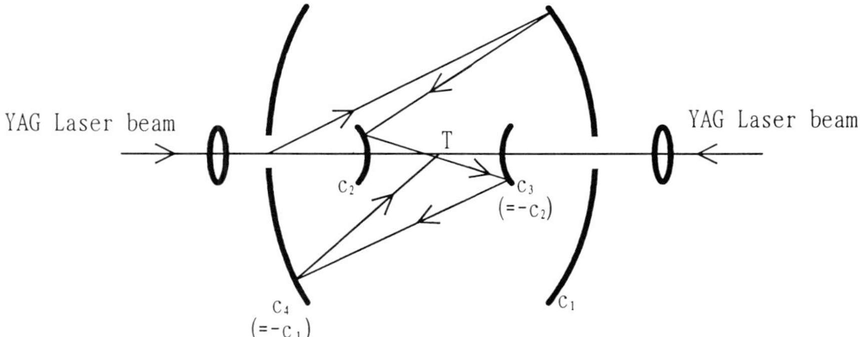

Fig. 11. Schematics of Welford's four-mirror system for high-power laser beam focusing in the laser fusion experiment.

are simply an unfolded Cassegrain-inverse Cassegrain system for imaging purposes. It is of interest and worth noting the similarity of his work for the YAG laser beam and our work for soft X-ray projection optics.

The work on the four-mirror system described in this chapter is a C(Cassegrain)-iC(inverse Cassegrain) system, and we have not yet fully explored C-C, iC-C and iC-iC systems, which may give us a further improved soft X-ray projection system.

ACKNOWLEDGEMENT

The work described in this chapter has been supported partially by the Korea Advanced Institute of Science and Technology (1993, 1994 and 1995), the Academic Research Promotion Foundation (1994, Ministry Education), and the Industrial Research Institute (1994, Pohang Steel Company, Korea). The authors are grateful for their supports.

REFERENCES

1. S.C. Park and S.S. Lee, 'Four-mirror optical system for UV submicrometer lithography', *Opt. Eng.* **30**(7), 1023–1027 (1991).
2. J.T. Kim and H.J. Kong, S.S. Lee, 'Improved four-mirror optical system for deep-ultraviolet submicrometer lithography', *Opt. Eng.* **32**(3), 536–541 (1993).
3. Young Min Cho, Hong Jin Kong and Sang Soo Lee, 'Cassegrainian-inverse Cassegrainian four-aspherical mirror system (magnification = +1) derived from the solution of all zero third order aberrations and suitable for deep ultraviolet optical lithography', *Opt. Eng.* **33**(7), 2480 (1994).
4. Cheon Seog Rim, Young Min Cho, Hong Jin Kong, Sang Soo Lee, 'Four mirror

imaging system (magnification + 1/5) for ArF excimer laser lithography', *Opt. Quant. Elect.* **27**(5), 319–325 (1995).

5. Michael W. Powell, 'IEEE Workshop on Submicrometer Lithography', *Solid State Technol.* **34**(12), 54–55 (1991).

6. Andrew M. Hawryluk and Natale M. Ceglio, 'Wavelength considerations in soft-X-ray projection lithography', *Appl. Opt.* **32**(34), 7062–7067 (1993).

7. D.M. Tennant, L.A. Fetter, L.R. Harriott, A.A. MacDowell, P.P. Mulgrew, J.Z. Pastalan, W.K. Waskiewicz, K.L. Windt, and O.R. Wood II, 'Mask technologies for soft X-ray projection lithography at 13 nm', *Appl. Opt.* **32**(34), 7007–7011 (1993).

8. Cheng Wang and David L. Shealy, 'Differential equation design finite-conjugate reflective systems', *Appl. Opt.* **32**, 1179 (1993).

9. Tanys E. Jewell, J. Michael Rodgers and Kevin K. Thompson, 'Reflective systems design study for soft X-ray projection lithography', *J. Vac. Sci. Technol.* B **8**(6), 1519–1523 (1990).

10. A.A. MacDowell, J.E. Bjorkholm, K. Early, R.R. Freeman, M.D. Hilmer, P.P. Mulgrew, L.H. Szeto, D.W. Taylor, D.M. Tennant, O.R. Wood II, J. Bokor, L. Eichner, T.E. Jewell, W.K. Waskiewicz, D.L. White, D.L. Windt, R.M. D'souza, W.T. Silfvast, and F. Zernike, 'Soft X-ray projection lithography using ring field optical system', *J. Vac. Sci. Technol.* B **9**(6), 3193 (1991).

11. H. Kinoshita, K. Kurihara, T. Mizota, T. Heya, H. Takenaka, and Y. Torii, 'Large-area, high-resolution pattern replication by the use of a two-aspherical-mirror system', *Appl. Opt.* **32**(34), 7079–7083 (1993).

12. M.A, Kumakhov, The Institute for Koetgen Optical System, the World Laboratory, Moscow CIS.

13. V.K. Viswanathan and Brian E. Newman, 'Development of reflective optical systems for XUV projection lithography', *OSA Proceedings on Soft X-Ray Projection Lithography*, **30** (1991).

14. D. Malacara, *Geometrical and Instrumental Optics*, Academic Press, 1988 (London).

15. W.T. Welford, *Aberrations of Optical Systems*, Adam Hilger Ltd., 1986 (Bristol).

16. H.H. Hopkins, *Wave Theory of Aberration*, Oxford, Clarendon Press, 1950 (London).

17. Young Min Cho and Sang Soo Lee, 'Holosymmetric 4 mirror optical system (mag. = +1) for soft X-ray projection lithography', *J. National Academy of Science, Korea*, **33**, 1 (1994).

18. Dong Hee Lee, Hong Jin Kong and Sang Soo Lee, 'Five mirror system derived from the numerical solutions of all zero 3rd-order aberrations and zero 5th-order spherical aberration for deep-ultraviolet optical lithography, *Kogaku J. Optical Society of Japan* **23**, 198–206 (1994).

22 Olmec mirrors: an example of archaeological American mirrors[*]

José J. Lunazzi
*Universidade Estadual de Campinas, Instituto de Física,
13084–970, Campinas, SP, Brazil*

INTRODUCTION

This chapter is not intended to give all available information on the subject of the title, but just a simple description that may be of interest to the optical community. The author believes he has consulted most of the available scientific literature that can be traced through cross-referencing from the most recent papers.

Olmec mirrors are the most ancient archaeological mirrors from Mexico and constitute a very good example of ancient American mirrors. The oldest mirrors found in America are from the Incas, made about 800 years before the Olmecs, dated from findings in archaeological sites in Peru. How this technology would have been extended to the north, appearing within the Olmecs, later within the Teotihuacan civilization, a few centuries before the Spanish colonization, is an interesting matter. Mirrors are important also within the Aztec civilization that appeared in the vicinity of the Olmec and Teotihuacan domains at about the time of their extinction.

The extent of the geographic area where these mirrors were employed is not entirely well known. Those made of pyrite, the material of highest reflectivity, deteriorate very easily, making it difficult to recognize them as mirrors. Besides

---

[*] The colour plate section for this chapter appears between p. 304 and p. 305.

this inconvenience, for civilizations living in the jungle, the extent and difficulties of the territory make it hard to obtain archaeological material. More mirrors are expected to be found in future excavations and this may help to better understand their utilization. Their existence can be associated with the skill of polishing stones with good quality and sphericity. A fundamental property of the iron ore minerals employed is the good reflectivity they have when thoroughly polished.

Most available Olmec mirrors were obtained at sites discovered during this century and studied by some authors. Many of the specimens are merely reported as belonging to unidentified 'private collections'. In many cases the archaeological context of their origin is unknown. They were dated by the radiocarbon method as being between 3125–2130 years old. Their use, as some archaeologists suggest, ranges from fire-making, self-contemplation, medicine, and divination, to imaging and astronomy. The range of focal lengths ranges from 5 cm to more than 80 cm for concave surfaces, while convex mirrors, although rare, are known.

Examples of their appearance in iconography are well known and their meaning is the subject of interesting studies. The knowledge the makers had of the laws of reflection and imaging is a very intriguing subject.

APPEARANCE OF THE MIRRORS

Mirrors are widely spread over South America, the oldest being found at Huaca de los Reyes and Gramalote, on the Peruvian coast [1] and in the Andes at Shillacoto and Kotosh [2]. They also existed in a later period at Chavin de Huántar, contemporary with the Olmecs. At later times in the Andes, polished copper and bronze mirrors were employed by the Incas. In Central America they appeared at a high-status burial ground in Sitio Conte on the Rio Grande de Coclé in Panamá.

One of the first general descriptions reported was published in 1926 [3], but most excavations have occurred since 1940 in a region to the south of Mexico City named La Venta, in Tabasco state, constituting at present the main site for Olmec mirrors. La Venta, San Lorenzo, Guerrero, Michoacan, Nayarit, Las Bocas-Puebla, San José Mogote-Oaxaca and Chalcatzingo-Morelos, are Mexican sites where Olmec mirrors were found, while Palenque is reported as an example of a site with mirrors from the Mayas. In many cases, the origin of the material and technology employed could not be identified, from which it can be assumed that mirrors were imported from other places. The use of broken mirrors in some offerings has been suggested as an indication that the technology could have disappeared, making the existing elements even more valuable at that time.

HOW TO FIND THEM

Although two mirrors are reported at the collection of the American Museum of Natural History in the United States and some at museums in Europe (the British Museum, 'Musée de l'Homme'), perhaps the best way of seeing them is at the 'Museo de Antropologia' in Mexico City. On my first visit to it, I saw in the Pre-classical section two small feminine ceramic figurines with brilliant mirrors between their breasts (Fig. 1).

They are referred to by Carlson [4] as findings of the Formative site of Tlatilco, and close to them there is one stone of irregular shape, smaller than 5 cm, but very well polished. The observer's face can be seen very clearly reflected, demonstrating the high quality and convexity of the surface [5]. This situation is shown in Fig. 2.

Its possible use for self-contemplation becomes evident, constituting a small, very portable and strong element. When I saw this unexpected example of archaeological optical techniques, I went excitedly to ask archaeologists for more information. I was told then that there were more Olmec mirrors at the 'Costa del Golfo' section. They belong to the findings at La Venta and are concave mirrors laying horizontally on the shelf, but there is one inclined toward the ceiling. In September 1992, there was a lamp on the ceiling located in a position that made it possible to see its image floating in front of that mirror. Once again, the quality of the image is astonishing. We returned in 1995 to produce the image of a hand that is shown in this chapter; this quality

Fig. 1. Two feminine figurines using mirrors, and a mirror sample close to them. Photograph by the author.

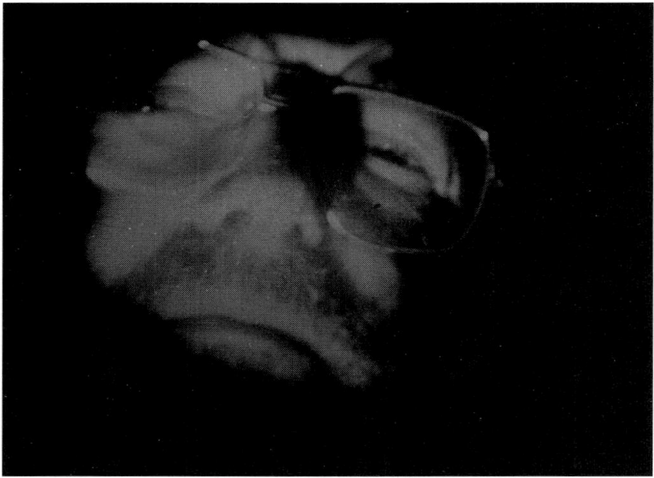

Fig. 2. Reflected image of a human face in a convex Olmec mirror. Photograph by the author.

of image is repeated in Plate 22.1. Six mirrors, five being concave and one convex, can be seen at the permanent exhibit of the 'Museo'.

THEIR SIGNIFICANCE IN THE CULTURAL CONTEXT OF THE OLMECS

The iconography shows that mirrors had an important presence in Pre-Columbian civilizations, giving even the name to an Aztec divinity: 'Tezcatlipoca', which means 'the smoking mirror'. This god is represented as having a smoking mirror instead of one of his feet. According to a transcription [6], there is an Aztec myth that says that Tezcatlipoca, one of the minor deities, proposed to the other deities a visit to the main god Quetzalcoatl, bringing a mirror draped in cotton. When Quetzalcoatl saw his face on the mirror, he cried because he thought himself to be a god but, having a human face, his destiny would also be human. Then he left the country announcing his future return. The remaining part of this myth has to do with the belief that the arrival of the Spanish conquerors represented the returning god, but it suffices for us to know that mirrors were an important element in Aztec mythology. Those mirrors were certainly a legacy from the Olmecs.

In many representations the mirrors are related to the sun god being fixed to his forehead [4] and with internal curved lines. This inner figure was interpreted as the curvature of straight lines distorted by an oblique angle of viewing. It appears displaced, occupying different positions in a way that makes us think about the image of the sun being visible on the mirror. The very well-known sculptures of a feathered serpent, very common at the Teotihuacan site, represent the serpent traversing the mirror [7]. To give an example of the capability of generating real images, we made the photograph of the image of one hand reflected in a concave Olmec mirror (Fig. 3).

These are good images although we could not work in ideal conditions: the focal length of a 105 mm objective was doubled by means of an additional diverging lens; also we were not allowed to remove the protective glass, which was traversed twice by the light at an angle of about 70°. We employed an aperture for the diaphragm of the objective on the camera of about 5 mm, to obtain an image similar to the one obtained by the naked eye. We employed KODAK TMAX 400 ISO film. Although there was ordinary glass between the mirror and the camera, and some deterioration may have occurred on the mirror surface, changing its original conditions, the image is very clear and appears floating in the air in front of the mirror. None of the bibliographic references we consulted indicate the possibility of images from the mirrors being present on the iconography as inverted images, which would certainly be an interesting finding.

We can imagine what strong feelings these images would have caused in the Olmec people from the existing records of other nearby civilizations. Drawings associating mirrors with serpents, human faces, eyes, cotton, water, flowers, butterflies and others have been described [7] for the Teotihuacan civilization. The representation of mirrors with a human-like eye inside could be the consequence of directly looking into a convex mirror of short focal length, but

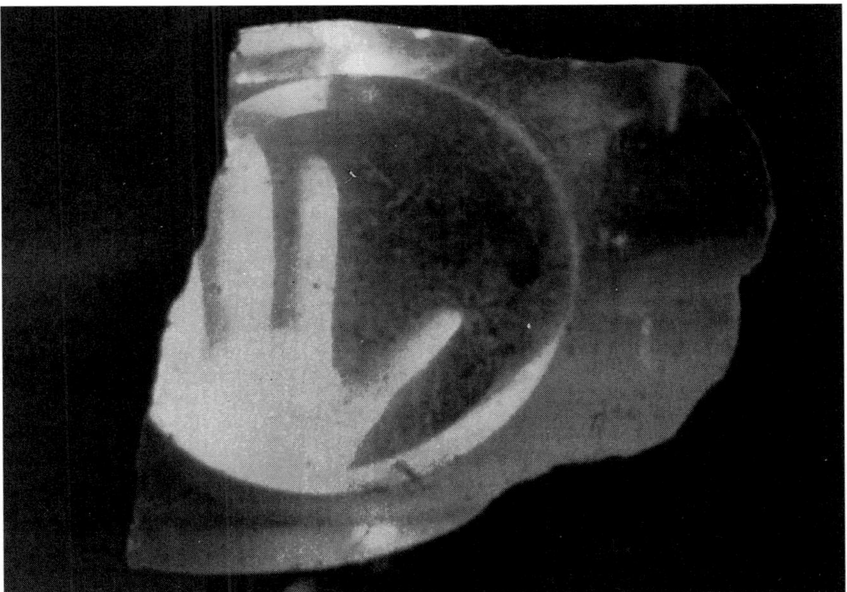

Fig. 3. Image of a hand obtained by means of a concave Olmec mirror. Photograph by the author.

it is also known that looking at mirrors was interpreted as getting in contact with a magic domain. Mirrors were associated also with the jaguar's eyes because of the reflections that can be seen on them. In many native languages, the Amerindians have the same or very similar words for designating 'mirror' and 'eye'. Regarding the Aztecs, for example, in Book 10 of the *Florentine Codex*, both the eye and the pupil are described as 'tezcatl' (mirror) [7]. The word for mirror in the central Maya domain was 'nen' derived from 'lem', which means something bright, gleaming and reflective. In the Yucatec phrases, 'nen' denotes rulers or persons of pre-eminent social status as 'the reflection of the world' [8].

Mica and stones were used to represent the eyes of divinities in Teotihuacan sculptures, while other reflective materials, such as mercury or pearls, could have had similar importance in relating their properties to spiritual worlds.

Making sacred fire by means of the sun is the most probable use of some mirrors, mainly for those which are spherical, although we know very little about successful experiences of firing with the original mirrors. The association of mirrors with cotton [7] was suggested by us [5] as a possible consequence of the Inca technique of making fire with mirrors described by the Inca Garcilaso [9] in the sixteenth century since it would be a more efficient way of starting fire than with dry wood or leaves.

When fire was obtained in this way in ritual sacrifices, this was interpreted as being the people in the grace of the gods, as reported by Garcilaso. Mirrors were used by the priests and the nobles, as a pectoral or on the back, probably being a symbol of high social status. We can get this impression from the photograph in Fig. 4.

It has been suggested that the Olmec élite identified themselves with a mythical jaguar ancestor. Even now some Amerindians preserve this identification, dressing as man-jaguars with a complete jaguar hide. One Olmec monument shows a jaguar-man figure apparently copulating with a human female, and a possible mirror pendant is seen suspended on the chest of the jaguar-like figure. The use as a pectoral is seen not only in the iconography but at the chest position of skeletons at burials.

The most probable uses of mirrors are divination, introspection, firing, imaging and medicine. Real images can be observed not only floating in the air but also projected on a white, partially covered surface [10]. It was mentioned that in Chinese culture, a 1000-year-old technique was reported where mirrors were applied for cauterizing in sunlight. Plane mirrors are not reported for the Olmecs, which is an intriguing fact due to the utility they could have had in light communications.

I would like to mention that many of us were taught at school that when the Spanish conquerors came to America they exchanged mirrors for many valuables and goods with the natives. This led us to think that they were fascinated by the images on the mirrors, which must be true, but also that they were not familiar with those images. Now we can suggest a different history;

Fig. 4. Olmec representation where a reflecting element is located at the chest. Jade figurine, 8 cm high. Photograph by the author.

that the mirror images were considered to be divine or highly important. That the people were taught to obey the priests and authorities who carried mirrors within their vestments. A natural transference of obeisance could have happened when seeing the mirrors the newcomers brought with them.

TYPES OF MIRRORS

Mirrors made in a single piece were ground mainly from iron ore minerals, which have the remarkable property of good reflectivity when very well polished. The reflectivity of the mirrors is not reported in the references but the same minerals have reported reflectivities of 21% for magnetite, 28% for hematite, 55% for pyrite [*11*].

Mica and obsidian are low-reflecting materials also employed. In the case of pyrite, the presence of 0.1% of gold within its structure was demonstrated, but not visible in the optical microscope. Being the most brilliant within the iron-ore materials, it is the easiest to deteriorate under humid conditions, to the point of leaving just a residual deposit. Pyrite seems not to have been available to the Olmecs but was reported to be found within the Mayas. The grinding materials could have been any of the elements available at that time, such as emery (aluminium oxide), hematite powder, or perhaps sand. Ochre (hematite jeweller's rouge) was used for polishing, a very laborious task. These materials are no different from the ones employed nowadays for making glass mirrors. It has been indicated that the polish of the specimens is so good that it represents the limit of perfection that the material will allow [*10*]. No trace of abrasion marks was found and the microstructure of the mineral was revealed. This result can only be obtained in modern technology by combining polishing and etching. When grinding, sphericity is a consequence of the lapidary process. A pair of complementary concave–convex surfaces is obtained, but it is clear that the preference was given to the concaves because convex surfaces were reshaped into concave. It may also be an indication of not having too much material available. Minute, local irregularities in the curvature of the surfaces suggest that the work was done on small areas at a time. In order to better analyse the procedure of fabrication, a site at San José Mogote-Oaxaca is mentioned as a probable artisan workshop area, deserving further study.

A great variety of curvatures can be found, generating focal lengths from -50 cm [*5*] to more than 80 cm. It is probable that the shorter focal lengths (about 5 cm) were the most appropriate for burning purposes. The existence of many mirrors whose vertical and horizontal curvature do not coincide may be an indication of a different use, not precluding that of imaging [*5*]. A parabolization of the surface intended to better concentrate the light is mentioned [*10*], although not characterized in detail. Plane mirrors are not explicitly referred to but it seems very probable that they existed. The author could see one, reported as belonging to a private collection, and also received a suggestion of its possible use in making signals by lighting remotely toward the observer.

Regarding the diameters of the mirrors, they corresponded to circular or elliptical shapes, the maximum reported diameter being 14 cm. Mosaic mirrors and the use of pyrite are not reported for the Olmecs but for the Teotihuacan and Kaminaljuyu findings. They correspond to civilizations appearing in the Mexican and Guatemalan regions, later than the Olmecs. For them, the assembling of pieces may have allowed mirror diameters of up to 50 cm. No good images were reported for mosaic mirrors. The quality of Olmec mirrors, although variable, can be very high, as can be seen from the images they generate at the short distance of a few millimetres [*12*] or even at the large distance of some metres [5]. Observations in a microscope showed good quality because no scratches were visible, although smoothness, roughness, reflectivity or aberration coefficients of the surfaces were not reported.

ON THE QUALITY OF REFRACTIVE ELEMENTS

Very well polished quartz objects are known to belong to the Olmec culture. Among them there is a small sphere, smaller than 10 mm in diameter, within the group of offerings of the E tomb at La Venta, Tabasco, in the permanent exhibit of the 'Museo Nacional de Antropologia' at Mexico City (Fig. 5, second from left on the upper line).

Although we could not have the opportunity of seeing an object through it, its appearance made us think that this element may function as a magnifying lens and that the knowledge the Olmecs had about optics could include some refractive properties. The focusing of the illuminating lamp is clear at the upper left side of the sphere.

CONCLUSIONS

From the information we obtained we can recognize the Olmec mirrors as very valuable cultural elements of unusual quality for ancient cultures. The precise

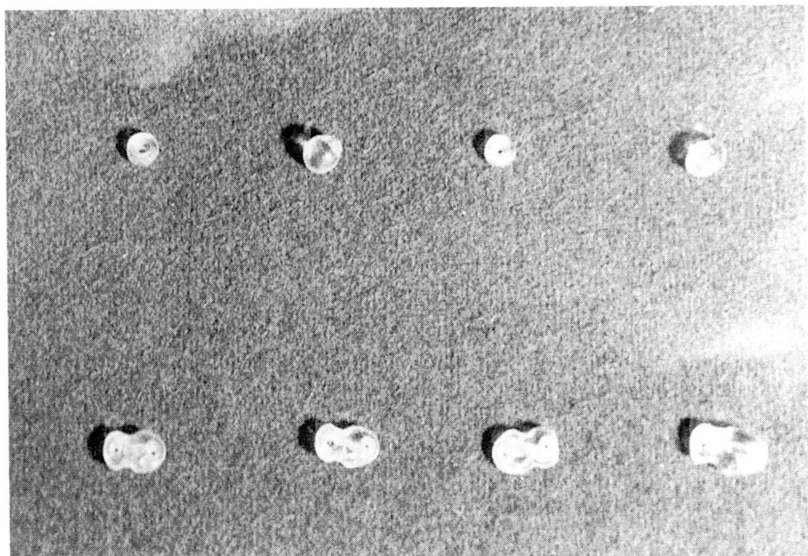

Fig. 5. Some small well-polished Olmec objects made of quartz, including one sphere. Photograph by the author.

characterization of their optical parameters seems not to have been completely reported, so that its measurement and analysis can still be an interesting subject of research in an interdisciplinary field involving archaeology and optics in close relationship.

ACKNOWLEDGMENTS

The author wishes to express his acknowledgment to the Organizing Committee of the 'II Reunión Iberoamericana de Óptica', Guanajuato – GTO, Mexico, 18–22 September 1995 for their invitation to present this work at that meeting, particularly to its General President Dr. Daniel Malacara Hernandez whose broad way of thinking and acting has allowed the realization of many important activities in Latin America.

Financial support from FAEP – Campinas State University, FAPESP – Foundation for Assistance to Research of the Sao Paulo State, and CNPq – National Council of Research made this work possible. I am also grateful to Jesus Najera who, representing the Esperanto world community, helped with equipment and personal work to obtain the photographs. The 'Museo Nacional de Antropologia' of Mexico City is acknowledged for allowing us to introduce the photographic and video cameras, their mechanical holders and a lamp into their exhibiting hall to make the photographs. Archaeologist Marcia L. Castro is acknowledged for the information reported at that museum in October 1992. Alexsandra Siqueira is acknowledged for helping at the photographic laboratory.

BIBLIOGRAPHY

1. Fung Pineda, R. (1987), 'The Late Pre-ceramic and Initial Period', in: R.W. Keatinge ed. *'Peruvian Prehistory'*, Cambridge University Press, 67–98.
2. Burger, R.L. (1984), *'The Prehistoric Occupation of Chavin de Huántar, Peru'*, University of California Publications in Anthropology, V. 14. University of California Press, Berkeley, 203–204.
3. Nordeskiold, E. (1926), 'Mirroirs convexes et concaves en Amérique', *Journal de la Societé des Americanistes de Paris*, n.s. tome **XVIII**, Paris, 102–110.
4. Carlson, J.B. (1981), The University of Maryland, 'Olmec concave iron ore mirrors: the aesthetic of a litic technology and the Lord of the Mirror', Dumbarton Oaks Conference on the Olmecs, Dumbarton Oaks, Washington, 117–147.
5. Lunazzi, J.J. (1995), On the Quality and Utilization of Olmec Mirrors, in: *Proc. of the II Reunión Iberoamericana de Óptica, Guanajuato – GTO, Mexico*, 18–22 September 1995, *SPIE V* **2370**, 2–7.
6. Fuentes, C. (1992), 'El espejo enterrado', Fondo de Cultura Económica, Mexico.
7. Taube, K.A. (1993), University of California at Riverside, 'The iconography of mirrors at Teotihuacan' in *'Art, Ideology and the City of Teotihuacan'*, J.C. Berlo, ed., 169–204, Dumbarton Oaks Research Library and Collection, Washington.

8. Saunders, N.J. (1988), 'Chatoyer, anthropological reflections on archaeological mirrors', in *'Recent Studies in Pre-Columbian Archaeology'*, Vol. I, N.J. Saunders, O. de Montmollin, eds, I–39, BAR International Series 313, Oxford, 1–37.

9. de la Vega, Garcilaso (1550?), *'Libro Sexto de los Comentários Reales de los Incas'*, ch. XXII.

10. Gullberg, J.E. (1959), 'Technical notes on concave mirrors' in *Excavations at La Venta, Tabasco, 1955*, Smithsonian Institution Bureau of American Ethnology, Bulletin 170, United States Government Printing Office, Washington, USA, 280–283 and pl. 62.

11. Craig, Vaughan (1981), *'Ore Microscopy and Ore Petrography'*, J. Wiley and Sons.

12. Heizer, R.F., Gullberg, J.E.(1981), University of California, Berkeley, in 'Concave mirrors from the site of La Venta, Tabasco: their occurrence, mineralogy, optical description, and function', Dumbarton Oaks Conference on the Olmecs, Dumbarton Oaks, Washington, 109–116.

23 Galileo Galilei: Research and development of the telescope

Giuseppe Molesini and Vincenzo Greco
Istituto Nazionale di Ottica, Largo E. Fermi 6, Firenze 50125, Italy

INTRODUCTION

The personality and the work of Galileo Galilei have fascinated generations of scientists engaged in the advancement of knowledge. In his life, one can find curiosity in natural phenomena, passion for research, an intuitive genius, and a consciousness of the importance of the findings. While the scientific merits of Galileo, along with their philosophical implications, are celebrated and fully acknowledged today, little is known about the engineering aspects of his inventions and discoveries. In the case of the telescope, in particular, acquaintance with and mastery of optical engineering skills play a fundamental role in its development into a successful instrument. Unfortunately, it is impossible to reconstruct the steps of such a development in detail: almost all of Galileo's optical instruments have been lost. It appears that only two telescopes and a single objective lens in the Science Museum of Florence (Italy) remain. They are part of the collection of scientific instruments handed down by the Medici and Habsburg-Lorraine families that ruled Tuscany in the past centuries.

Galileo's research and development of the telescope proceeded from a rumour of feasibility heard in May–June 1609, to a working device presented to the Senate of Venice in the following August, and then to a fully engineered instrument used for astronomical observations, offered to Noblemen and sold all throughout Europe. To document such progress, we have the reports of Galileo himself in his scientific works and in his letters. All these are collected in a comprehensive National Edition of Galileo's writings [1]. Various historical studies are also available in the literature [2]. As surviving evidence, we have the pieces now at the Science Museum of Florence. These optics were examined in 1923, both by visual sky observations and in the laboratory by the newly introduced Ronchi test [3,4]. Further inspections have been made recently with state-of-the-art equipment [5]. Here we summarize the main results of such studies, outlining the engineering achievements that Galileo was able to obtain as related to the technological knowledge of his time.

THE OPTICAL LAYOUT

It is known that Galileo first heard rumours of Dutch spyglasses (the name 'telescope' was given at a later time) during a visit to Venice; the news was confirmed by a letter from Paris [6]. Back in Padua, where he was appointed as a mathematics professor, within 24 hours Galileo had a prototype at work.

The description he was given was basically that of a landscape (non-inverting) device, capable of bringing a far scene closer; the layout implemented is that in Fig. 1, now referred to as Galilean, and still in use. The 'approaching' impression is actually due to angular magnification. A simple arrangement may be obtained with two spectacle lenses at the ends of a tube, one for the far-sighted viewers, the other for the near-sighted viewers. However, to perform as intended, the power of the lenses and their spacing must be properly matched. According to his own accounts, Galileo succeeded by speculating on the 'science of perspective', or on the 'doctrine of refraction' [6]. According to historians, nothing of such disciplines was actually known to Galileo, neither did he become familiar with them at later times. *Dioptrice*, the basic work explaining the use of lenses, was only written by Johannes Kepler in 1611, after a careful examination of Galileo's telescope. It is most likely that Galileo succeeded by intuitive feeling and experimentation, using lenses available in his shop or brought from Venice for the occasion.

Although Kepler, in his *Dioptrice*, also proposed a design with a positive

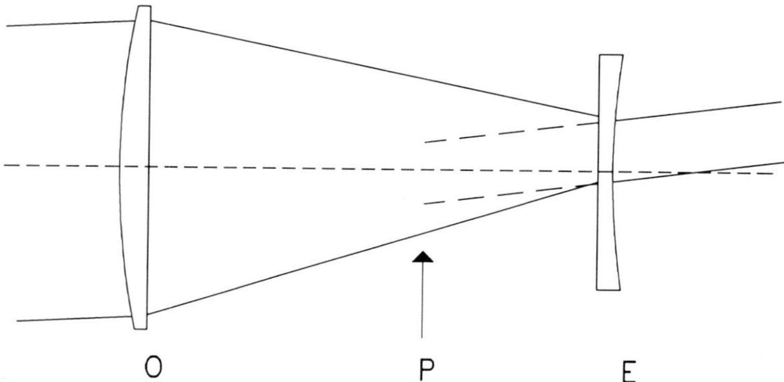

Fig. 1. Schematic layout of the Galilean telescope: O, objective; E, eyepiece; P, exit pupil.

eyepiece (the 'astronomical', or 'Keplerian' telescope, constructed by Christoph Scheiner a few years later) [7], Galileo always used the layout of Fig. 1, with a positive objective (quite exclusively, a plano-convex lens) and a negative eyepiece (a plano-concave lens). Incidentally, it is mentioned that in Germany an antique telescope with the signature of Galileo on its tube has been found recently. The layout is that of a terrestrial telescope, with a positive objective, a 1 : 1 erector made of two nearly equal positive lenses, and a positive eyepiece. The instrument has now been inspected and tested in the laboratory [8]. The signature of Galileo is considered to be authentic. The complexity of the layout and its technological implementation, however, throw serious doubts on the attribution of the instrument as a whole to Galileo.

Galileo's telescopes, referred to in documents, are ~0.5 to ~2 metres long. (Shorter devices, ~10 cm long, resembling today's tubes for opera glasses, appeared in Venice about 1618. Probably following Galileo's inquiry, objectives of focal length 15–30 cm were actually sent to him in Florence, where he had moved in 1610. Lens-making capabilities up to ~3.4 m focal length are also occasionally mentioned [9].) The magnification of the first prototype [6] was ~3, increased to ~8 for the instrument presented to the Senate of Venice in August 1609 [10]. In November he had a telescope with ~20 magnification; this was the one used by Galileo for most of his basic discoveries. In *Sidereus Nuncius* he also mentioned a 30-powered instrument, but he might never have used it [7]. Reference to performance improvements can be found in different letters. It appears that most of the development of the telescope occurred within a year of the first prototype, assessing the optimum magnification at about 20. Efforts were then devoted to obtain lenses of better quality.

To make a 1 metre long, 20 magnification telescope, the objective lens should have a focal length of ~1 metre. For spectacle-makers, such a lens is said to have ~1 dioptre of corrective power, which is about the limit of usefulness for far-sighted viewers. Lenses intended as magnifying glasses are also unsuited to the purpose, because their focal length is typically less than 25 cm. Further, the eyepiece should have a focal length of ~ -5 cm, or -20 dioptres, which is outside the range of interest for spectacle-makers. Unprecedented requirements on the quality compared to the existing technology were also demanded. In particular, it should have been clear from the very beginning that the major quality constraint affecting the optical performance was posed by the objective (at 20 magnification, the eyepiece useful diameter per field angle is only 1/20 of the clear aperture of the objective). Even Galileo's mother was aware of it, as she wanted some for her own business, specifying object lenses. Eyepieces, or 'acute' lenses, are rarely mentioned in writings.

While the opticians could accommodate the new power specifications of the lenses by extending the range of radii of curvature to be worked, the quality requirements posed serious technological problems in terms of glass purity and figure regularity. In the absence of fabrication techniques that could guarantee the end result, Galileo purchased great quantities of lenses, and then chose the best performing ones. The selection process was quite severe: a letter of 1616 relates that of some 300 objectives made by one of the best artisans in Venice, 22 were selected, but only three passed, and even these were not deemed perfect [*11*].

The demand for lenses of improved figure quality and in great quantities soon resulted in a technological breakthrough in the optical production process. In particular, Ippolito Francini, optical shop master in Florence and supplier of astronomical lenses for Galileo, is given credit for having introduced the polishing machine with vertical spindle, as is still in use today; Francini's lenses were unquestionably superior to those which Galileo had acquired from the opticians of Venice [*12*]. In 1644, two years after the death of Galileo, Evangelista Torricelli was able to produce two good objectives out of six polished in eight days of work [*13*].

Galileo also made a lens-polishing apparatus of his own. Indeed, he was known to be familiar with the grinding and polishing process. He reported, for example, having made oversize lenses to better control the surface figure, and then used diaphragms. In some cases Galileo suggested oval apertures because of defective optics. This also proves his awareness of the adjustments that would perfect the optical performance of his telescopes and his ability to carry out some optical testing procedures. The actual size and shape of the diaphragm should have been the result of a trade-off between image quality and brightness, depending on the features of the lenses being used.

A different problem was that of glass purity. Glass making had been known in Italy for a long time. By 1400, Venice had become a leading centre for glass manufacturing. The glass industry was located on the island of Murano, a separate quarter of the city, where the secret of special production techniques could better remain under control. By 1600, however, glass manufacture had spread from Venice to several places, including Florence. The basic components of glass were soda lime and silica, as are most of today's bottle and window glass compositions; the purity of the end-product depended on the quality of the components. The Florentine glass was acknowledged as superior, apparently because it used as a source of silica the pebbles of the River Arno (for Venetian glass, the pebbles of the River Ticino were used instead). In 1612, a sort of glass manufacturing handbook became available, also describing the use of lead oxide and of borax, which did not become common glass constituents until later [*14*].

What Galileo needed most were glass homogeneity, transparency, and freedom from bubbles and inclusions. Bubbles could be eliminated from the glass melt by increasing the furnace temperature. Wood fires used at the time could not succeed completely in a reasonable time. The use of coal instead of wood for fuel was introduced in 1615 in England. Galileo thus had to rely on the care of glassware masters in selecting the glass components and in their ability to control the production process as best they could. In a letter of

October 1610, just back from Padua to Florence, he reports of a special glass batch made on purpose for him with the finest materials in the foundry of the Grand Duke of Tuscany; this letter is considered to be the first citation of 'optical glass' ever made [*15*]. In addition, close friends of Galileo in Venice, Girolamo Magagnati and Giovanfrancesco Sagredo, were active in glass-making. A rich correspondence with Sagredo, in particular, reports on continuous efforts to improve the glass quality, including experimentation on new products and recipes.

Sagredo also acted as a trade correspondent in Venice for Galileo's supply of objectives, dealing with artisans, looking after their work, soliciting deliveries, shipping lenses to Galileo and taking good care of his business. Lenses were typically given a plano-convex shape by grinding on master tools, and then polished. The lens-makers' formula, relating the radii of curvature and the refractive index to the focal length, was not yet known. Instead of referring to the radius of curvature, the master tools were named by the focal length of the plano-convex lenses they served to produce, which ended up being approximately the length of the telescope. Each artisan had just a few masters of some quality. To obtain more values of focal length, soon they started to work biconvex lenses by combining the masters available. In a letter of Sagredo to Galileo of 1613, this practice is referred to as being known to both. In 1615, Sagredo returned to the same subject, becoming himself curious as to the way the powers combined. A week later he wrote again to Galileo and gave the thin lenses combination rule as it is given in today's textbooks [*16*].

In the first place, the nominal shape of convex surfaces was specified as spherical [*6*]. The problem arose, however, whether the sphere would be the best choice, due to intuitive ideas about ray aberrations. The parabola in particular was considered, and recommended by different persons at different times. Although Galileo leaned towards the hyperbola, as did Kepler (the Cartesian oval and Snell's law were still to come; the hyperbola in fact is the shape of the Cartesian oval in the case of focus-infinity conjugation), he noted that, within the small lens aperture, all such profiles differ by negligible amounts [*17*].

The lens apertures used by Galileo were in fact typically small. The telescope first presented to the Venetian Senate is described as a tube ~60 cm long, ~4 cm in diameter [*10*]. Considering the ~8 magnification of the instrument, the focal length of the objective should have been ~70 cm; assuming (arbitrarily, by guesswork) a lens diameter of ~2 cm, a relative aperture of ~f/35 would have resulted. The three objectives surviving to our day have relative apertures ~f/45 and higher. At such apertures the lenses work in nearly paraxial conditions, so that the Seidel aberrations are not expected to show up, neither are there appreciable differences between spherical shape and aspherics. Besides chromatic effects, the only features affecting the optical performance are the regularity of the surfaces and the glass homogeneity. The alignment and mounting tolerances of the lenses on the tube are very loose; the lenses could even be turned face to face without significant degradation of the imagery. The telescope assembly was therefore not critical at all. Galileo used to ship his instruments in the simple form of a kit case, containing a pair of matched lenses, tied up with a piece of twine that measured the exact distance between the lenses when properly mounted in the tube. The latter could be constructed by the consignee using cardboard, wood (also ebony), lead or iron plate, eventually covered with paper, cloth or leather, and decorated according to the circumstances [*12*]. The diaphragm of the objective was provided in the form of a lead plate attached to it, with a hole selecting the useful portion of the lens. Only as rare exceptions did Galileo undertake also the care of telescope building, having the tube prepared by artisans and artists of his time.

Intrinsic problems of the Galilean telescope are the size and location of the exit pupil, which make the observation similar to 'peering through a small keyhole at a distance' [*18*]. For the $f/35$, 70 mm focal length, 8 magnification telescope as above, the exit pupil would be 2.5 mm in diameter, located in between the objective and the eyepiece, at the distance of $(-)78$ mm from the latter. Assuming for the eye's pupil a diameter of ~5 mm, the instantaneous field of view (IFOV) would be ~30 arcmin (one Moon diameter). As a further indicative example, a 20 magnification telescope with a 1 m focal length, $f/50$ objective would have an exit pupil of ~0.94 mm in diameter, at a distance of $(-)47$ mm from the eyepiece, with ~15 arcmin (half a Moon) IFOV. Reference to similar IFOVs is found in Galileo's correspondence.

Galileo was also aware of the difficulties with the handling of his telescope, having to deal with unsteady hands and steam condensation on the eyepiece; insights on focusing by varying the telescope length were also given to users. As a matter of fact, the Moon maps that Galileo was able to draw would have been the result of laborious work, putting together the partial views available with his instruments during long observation sessions [*19*].

Chromatic effects are also intrinsic to the configuration of Fig. 1 with single lenses. To overcome colour problems, however, more than a century had to elapse.

OPTICAL TESTS

According to documents and biographies, Galileo made a considerable number of telescopes. In a letter of early 1610, seven months after the first demonstration in Venice, he mentioned having already constructed more than 60 instruments. The majority of the telescopes was for sale; some were reserved for his royal patrons. Mention of lenses broken during Galileo's lifetime is found in letters. Probably the same has occurred since then to the near totality of Galileo's production, with the fortunate exception of the pieces in the Science Museum of Florence [*20*]. These are two complete telescopes made of wood, plus a single objective lens. The single lens had been broken and rebuilt from several fragments. For identification purposes, and in accordance with the literature, the telescopes referred to here are as follows:

Telescope I, which is paper coated and has a longer tube length.
Telescope II, which is leather coated and has a shorter tube length.

The authenticity of the tubes is almost certain; there are some doubts, however, concerning the lenses. In particular, the eyepiece of telescope II is missing, and a non-original equiconcave lens has been substituted. Only the single lens can be attributed to Galileo with certainty. The lens was cracked when Galileo was alive and later given to the Grand Duke of Tuscany since it was the lens used to discover the Medicean stars. The Grand Duke had it placed in an ivory frame by Vettorio Croster in July 1677, and it has been preserved there.

Until Galileo, astronomical observations were made with the naked eye, for which it is generally assumed that the resolving power in white light is of the order of 1 arcmin, or 3×10^{-4} rad (in monochromatic light at the centre of the visible spectrum the diffraction limited resolving power of an aperture 5 mm in diameter as the eye's pupil is $\sim 1 \times 10^{-4}$ rad). Astronomical observations with Galileo's instruments made in 1923 led to estimations of a resolving power of 20 arcsec for telescope I and 10 arcsec for telescope II, both over a field angle of 15 arcmin; the single lens appeared to be of even superior quality [3, 4]. On the occasion of the 1992 Galilean celebrations (the 400th anniversary of the move of Galileo from Pisa to Padua and the 350th anniversary of his death), the same optics were partially dismantled and again underwent optical testing [5]. Whenever possible, the optical components were investigated individually, the focal length and the radii of curvature were measured, and the instrument layout was studied. The optical finish was observed with a Nomarski microscope, which revealed the surface characteristics. The optical quality of the surfaces and the overall performance of the two complete telescopes have been evaluated interferometrically by phase-shift techniques. The relevant data are reported here, and the major results are discussed.

The telescopes consist of three sections that fit together: the objective mounting, the tube, and the eyepiece mounting. For testing purposes the three parts were taken apart, but the lenses were left in place. In Fig. 2 the objective mounting of telescope I is shown as an example of the lens-holding technique. A paper diaphragm is placed on the lens, and a metal ring presses both on to the wooden frame. The single lens was removed from its case (Fig. 3), which made it easier to examine. The fragments had been held together and preserved by the glues used to restore the lens. Glass defects such as bubbles and inclusions are clearly visible in the figure.

Table 1 summarizes the results of the measurements carried out on the optics of Galileo. Some data that complement the table are just quoted from the literature, because new measurements would have required taking the lenses apart, causing serious risks of damage. The focal lengths are measured near the

    Giuseppe Molesini and Vincenzo Greco

Fig. 2. Objective lens of telescope I in its mounting.

Fig. 3. Single lens, out of its frame, reconstructed from fragments after it was broken.

Table 1.   Optical specifications of the lenses of Galileo (dimensions in mm).

Optics element	Front radius	Back radius	Central thickness	Full diameter	Aperture diameter	Focal length
Objective I	950	−2700	2.5	51	26	1330
Eyepiece I	plane	48.5 (*)	3.0	26	11	−94.0
Objective II	535 (*)	plane	2.0	37	16	980
Eyepiece II	−51.5 (*)	51.5 (*)	1.8	22	16	−47.5
Single lens	940	−12000	4.0	58	38	1710

(*) Data from Ref. 5.

centre of the visible spectrum (550 nm). For the positive lenses a calibrated target was projected, and the magnification was then accurately measured. The focal length of the negative lenses was calculated after the axial location of both the virtual object and the real image; the magnification was also measured. The radii of curvature were measured only on surfaces that could be inspected with an optical profile transducer. After the lens geometries and focal lengths were calculated, it was determined that the refractive index of the glass ranged from 1.510 to 1.546. Dispersion measurements reported in the literature [4] allow for Abbe numbers between 60 and 50. The configuration parameters of the telescopes are given in Table 2; the optical layouts are presented in Fig. 4, drawn to scale.

The typical appearance of the optical surfaces, observed with a differential interference-contrast (Nomarski) microscope, is shown in Fig. 5. Besides dust particles, scratches, stains, and contaminants related to age and use, there are marks, digs, and pits that are most likely a result of the raw materials and the polishing process. They would have caused serious problems in terms of stray light, particularly for the observation of extended sources. As a matter of fact, the inner part of the telescope tubes appears to be smoke darkened, which reduces the effects of diffuse light. The use of a tube extension to shadow the objective in daylight observation is also reported in letters.

Table 2.   Configuration parameters of the telescopes.

Tested optics	Magnification	Objective aperture
Telescope I	14	f/51
Telescope II	21 (*)	f/61
Single lens	–	f/45

(*) Based on the non-original eyepiece.

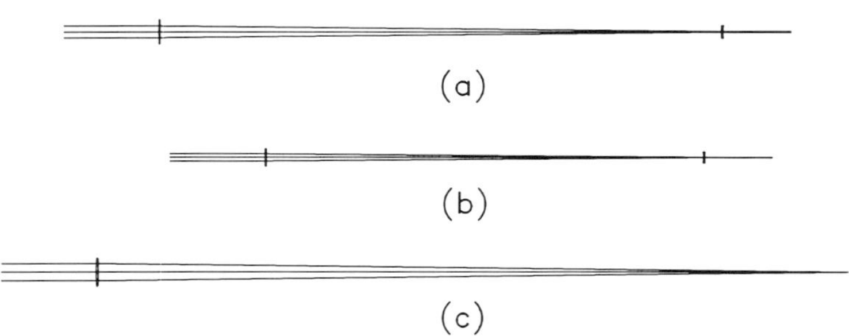

Fig. 4. Optical layout to scale of Galileo's instruments: (a) telescope I, (b) telescope II, (c) single lens.

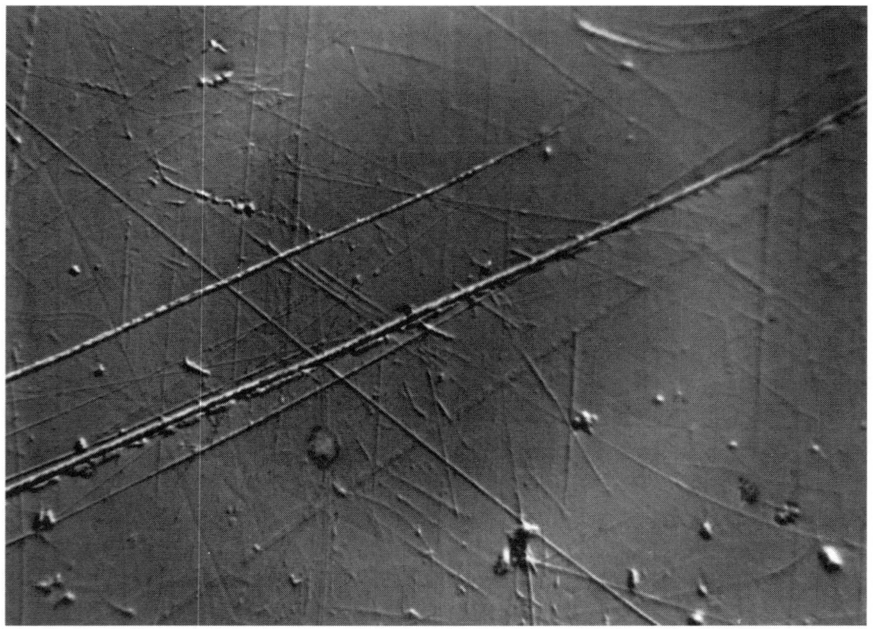

Fig. 5. Typical Nomarski micrograph of the lens surface. The area displayed measures 0.28 mm × 0.20 mm.

The optical shape of the surfaces and the wavefront distortion in transmission have been observed with a digital phase-shift interferometer at the wavelength of 633 nm (He–Ne laser source). The interferometric configuration used is of the Fizeau type, operating in double pass on the test optics. Static interferograms are presented in Fig. 6. As far as the shape of the optical surfaces is concerned, the quality of the objective lenses is far better than the quality of the eyepieces; as already noted, the effect of eyepiece imperfections is, however, far less important than that of objective defects. It is surprising that the plane surfaces of eyepiece I and objective II (the interferogram of Fig. 6c, taken against a reference plane without any power being removed) are really plane to a fraction of a wave, although flatness of these surfaces is not required in terms of image quality; polishing a surface to such absolute flatness

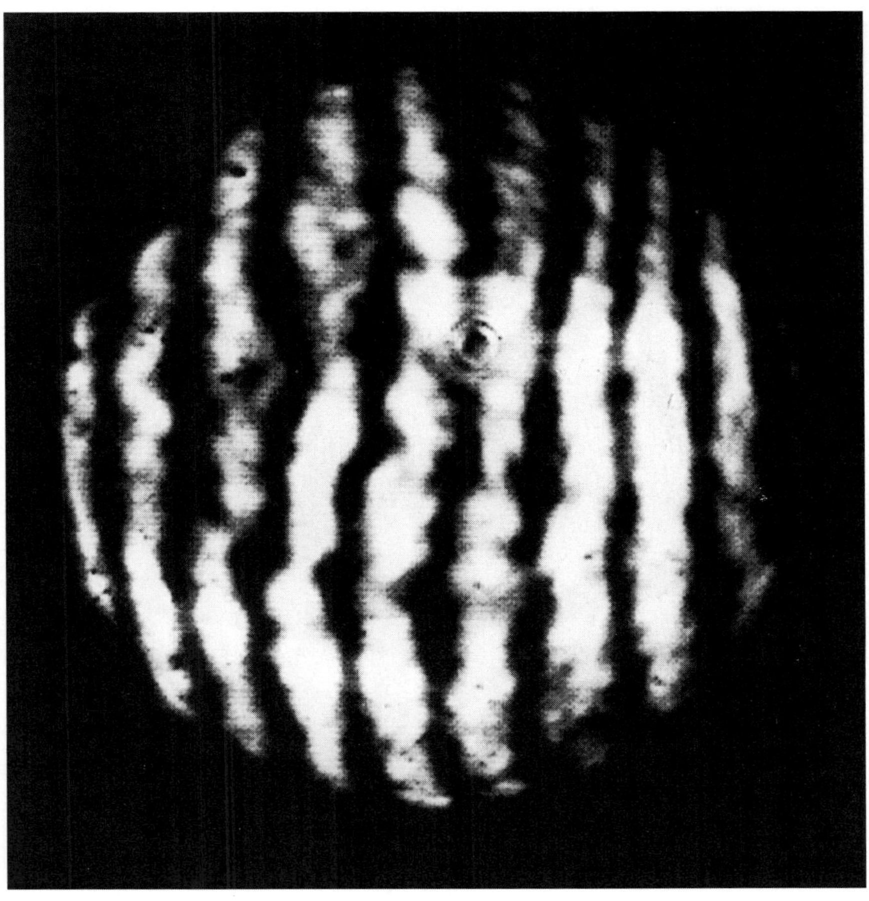

(a)

Fig. 6. (*Continued*)

is not trivial even for today's shops. A further notation comes from the appearance of the concave surface of eyepiece I (Fig. 6b): besides the interference fringes, a pattern of ring shadows is made visible, as if the surface had traces of a turning process. The wavefront distortion in transmission for objective I is very small (Fig. 6a). The best quality however belongs to the single lens, which can be considered to be nearly diffraction limited. The fringes appear to be fairly straight and uniformly spaced (Fig. 6d); the different orientation of the two parts accounts for a residual mismatch of the main lens fragments after restoration. Considering the lens useful aperture (38 mm), at the centre of the visible spectrum the monochromatic resolution would be ~3 arcsec.

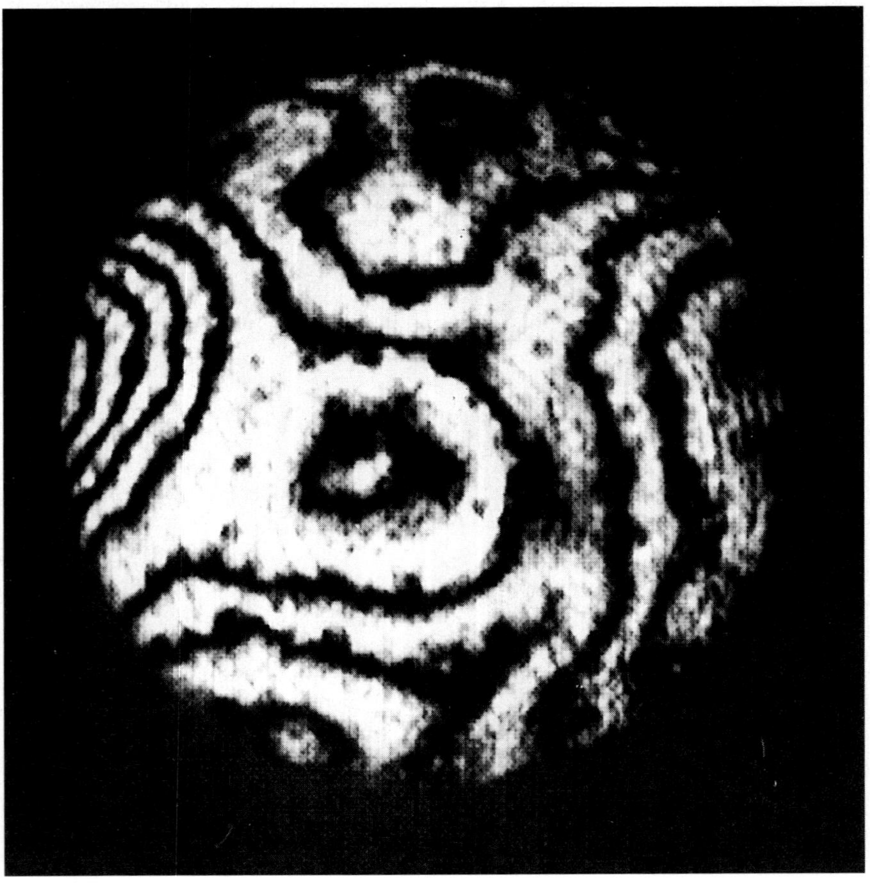

(b)

Fig. 6. *(Continued)*

The actual performance of Galileo's telescopes was necessarily degraded by chromatic aberration. Computer simulations have been made on the basis of the optics data of telescope I. For ray tracing purposes, a crown glass of type 522.595 was chosen, because of its close similarity to the lens material. The spectral range and the responsivity were selected according to the characteristics of the human eye. The interferometric data of Fig. 6a were also attached to the optics, although their effect on modulation transfer function (MTF) losses is much smaller than that of chromatic aberration. The resulting MTF curve so obtained drops to ~10% modulation at ~10 cycles mrad^{-1}, corresponding to a resolution of ~20 arcsec, in fair agreement with the observations of 1923.

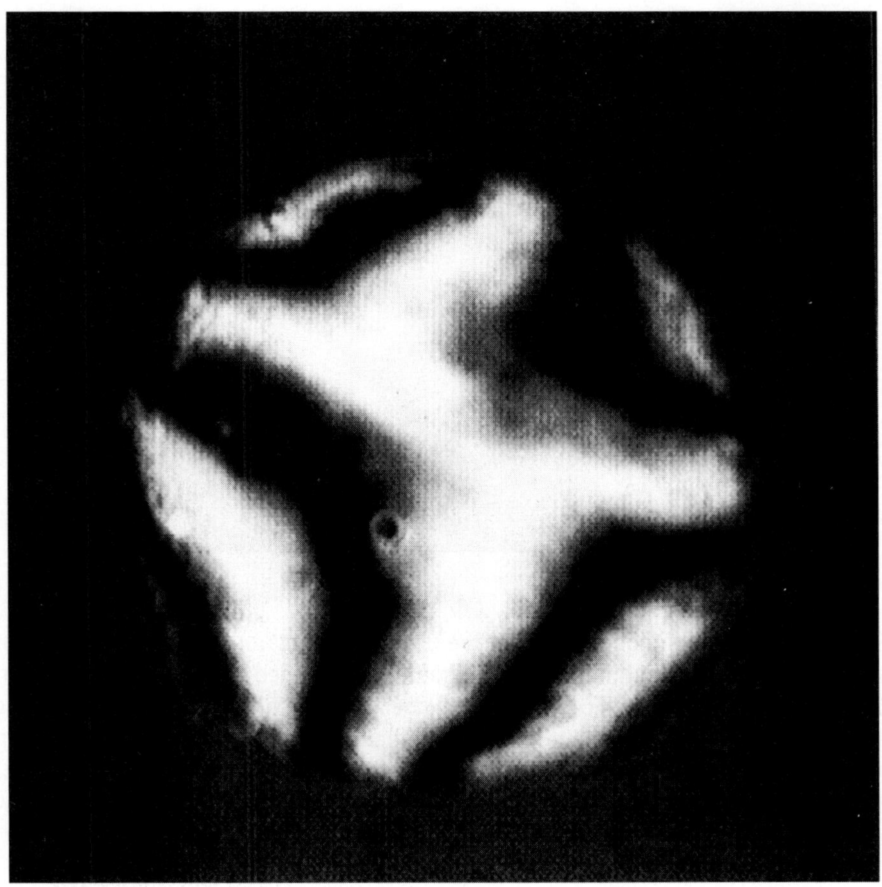

(c)

Fig. 6. *(Continued)*

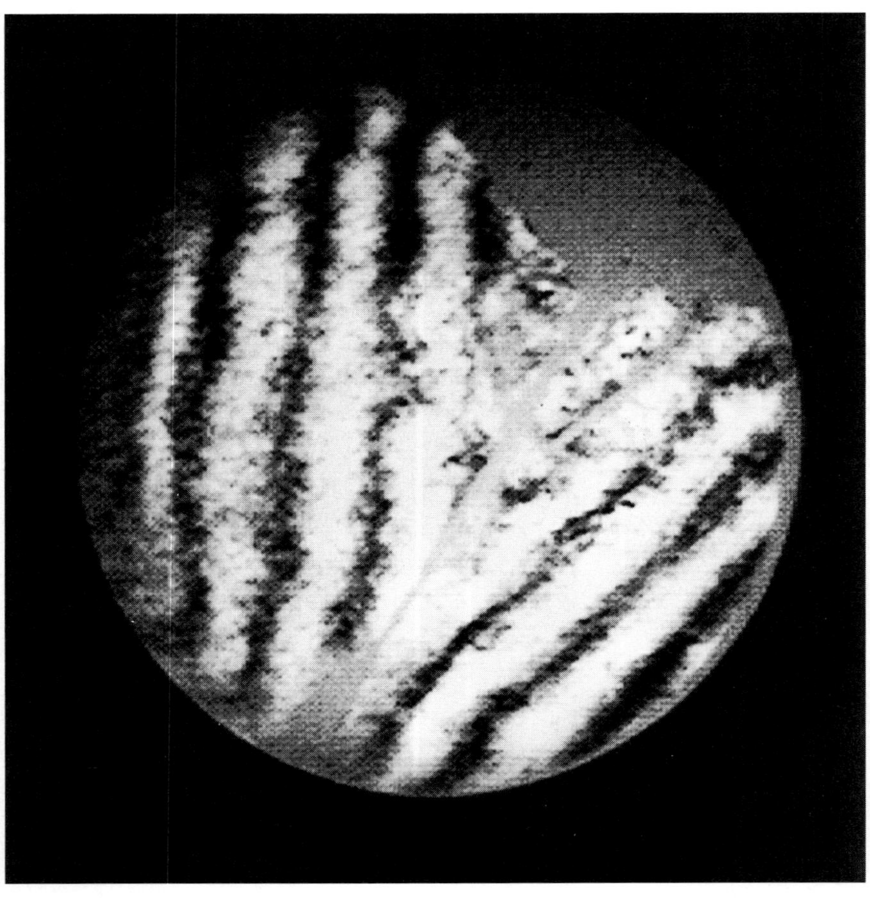

(d)

Fig. 6. Static Fizeau fringe patterns of the optical elements of Galileo's telescopes at 633 nm. (a) Double-pass interferogram of objective I, folded with a reference mirror. Deviations from straight fringes are of the order of half a pitch, meaning a departure from the ideal wavefront of the order of a quarter of a wavelength. (b) Reflection interferogram of the concave surface of eyepiece I. Fringes are highly irregular. (c) Reflection interferogram of the plane surface of objective II. The quadrant fringes show astigmatism. (d) Double-pass interferogram of the single lens.

CONCLUDING REMARKS

With the help of Galileo's writings and the study of the instruments remaining, some aspects of early telescope making have been highlighted. Although the information collected is largely incomplete, the telescopes of Galileo appear as an extraordinary technological achievement over the previous knowledge of optics and optical engineering.

It is particularly interesting that all the development was probably made empirically, using practical techniques of optimization, in the absence of a theory. The end result was a well-engineered instrument, in which optical fabrication, mounting tolerances, and testing procedures were all consistently integrated to obtain the best possible performance. Under such premises, the telescope could become a powerful instrument of investigation, as well as a commercial success.

It is also worth noting the effect that the promotion of telescope-making had on the advancement of the optical sciences and optics-related technologies. A key role had been played by the quality of the optics required for good performance of the telescope, stimulating efforts to understand the fabrication processes and improve the engineering approaches. In the documents that remain, no mention is found of the testing methods used by Galileo. There is no doubt, however, that the success of his telescope was based largely on the superior quality of the optics.

ACKNOWLEDGMENTS

The authors thank P. Galluzzi, A. Van Helden, L. Hoffer, M. Miniati and A. Righini for discussions and comments, and the Istituto e Museo di Storia della Scienza for permission to publish the material in this chapter.

REFERENCES

1. *Le opere di Galileo Galilei*, National Edition (Barbera, Florence, 1890–1909).
2. *Proceedings of the 1964 Symposium on Galileo* (Bemporad Marzocco, Florence, 1967).
3. G. Abetti, 'I cannocchiali di Galileo e dei suoi discepoli', *L'Universo* **4**, 685–692 (1923).
4. V. Ronchi, 'Sopra i cannocchiali di Galileo', *L'Universo* **4**, 791–804 (1923).
5. V. Greco, G. Molesini and F. Quercioli, 'Optical tests of Galileo's lenses', *Nature* **358**, 101 (1992).
6. *Sidereus Nuncius*, in Ref. 1, Vol. III.
7. A. Van Helden, 'The telescope from Galileo to today', in P. Mazzoldi, ed., *From Galileo's 'occhialino' to optoelectronics* (World Scientific, Singapore, 1993), pp. 318–331.

8. M. Miniati, V. Greco, G. Molesini and F. Quercioli, 'Examination of an antique telescope', *Nuncius* **9**, 677–682 (1994).
9. Letter of Giovanfrancesco Sagredo to Galileo, 2 June 1612, in Ref. 1, Vol. XI, p. 313.
10. *Cronaca di Antonio Priuli*, in Ref. 1, Vol. XIX, p. 587.
11. Letter of Giovanfrancesco Sagredo to Galileo, 23 April 1616, in Ref. 1, Vol. XII, p. 257.
12. S.A. Bedini, 'The makers of Galileo's scientific instruments', in Ref. 2, pp. 89–115.
13. Letter of Torricelli to Raffaello Magiotti, 6 February 1644, in *Opere di Evangelista Torricelli* (Lega, Faenza, 1919), Vol. 3, p. 165.
14. A. Neri, *L'arte vetraria distinta in libri VII* (Giunti, Firenze, 1612).
15. Letter of Galileo to Giuliano de'Medici, 1 October 1610, in Ref. 1, Vol. X, p. 440.
16. Letter of Giovanfrancesco Sagredo to Galileo, 17 October 1615, in Ref. 1, Vol. XII, p. 199.
17. Letter of Bartolomeo Imperiali to Galileo, 7 December 1624, in Ref. 1, Vol. XIII, p. 236.
18. We are indebted to Alberto Righini for suggesting this neat expression.
19. Note of Paolo Galluzzi, Istituto e Museo di Storia della Scienza (Firenze).
20. M.L. Bonelli, 'Note about Galileo's instruments', in Ref. 2, pp. 125–127.

24 GRIN optics: practical elements

Carlos Gómez-Reino and Jesús Liñares-Beiras
Departamento de Física Aplicada, Area de Optica. Escola Universitaria de Optica e Optometría e Facultade de Física, Universidade de Santiago de Compostela, E-15706 Santiago de Compostela, Galicia, Spain

INTRODUCTION

An inhomogeneous medium is one in which the refractive index profile varies from point to point within the medium. Currently the terms Gradient Index or Graded Index (GRIN) are often used to describe such a medium [1]. There are three basic gradient index types: axial, radial and spherical. Axial gradient has an index of refraction that varies only in a continuous way along the optical axis z: the iso-indicial surfaces (surfaces of constant index) are planes perpendicular to the optical axis. Radial gradient has an index of refraction that varies only as a function of the distance r from the optical axis: the iso-indicial surfaces are concentric cylinders about the optical axis of the medium. In the special case of the index varying quadratically with distance in the transverse direction, the medium is referred to as lens-like or self-focusing (selfoc). Spherical gradient may be thought of as a symmetric index around a point whose iso-indicial surfaces are concentric spheres.

In a GRIN medium the optical rays follow curved trajectories, instead of straight lines as in a homogeneous medium. By an appropriate choice of refractive index profile, a GRIN medium can have the same effect on light rays as a conventional optical component, such as a prism or a lens.

The possibility of using GRIN media in optical systems has been considered for many years, but the manufacture of materials has been the limiting feature in implementing GRIN optical elements until the 1970s. In the last 20 years, however, many different materials with gradient index have often been fabricated by adding impurities (dopants) of controlled concentrations, and the revival of GRIN optics has not been casual and is connected to a considerable degree with the enormous development of optical communications (for instance GRIN fibres are of special interest for transmission of information in broad-band systems), optical sensing and optical imaging systems.

Various methods of producing GRIN materials have been developed but these processes are limited by the small variation of the index, the small depth of the gradient region and the minimal control over the shape of the resultant index profile, as well as the cost. The processes most widely used to fabricate glass GRIN material have been ion exchange [2], chemical vapour deposition [3], ion-stuffing and sol–gel methods [4]. Plastic GRIN materials have been manufactured by copolymerization and monomer diffusion [5]. GRIN glasses with large variation of the index and large size have been fabricated by fusing together thin layers of glasses of progressively different indices of refraction [6].

Historically, Maxwell was among the first to consider inhomogeneous media in optics, when, in 1854, he described a GRIN lens, now known as Maxwell's fish eye, of spherical symmetry with the property that points on the surface and within the lens are sharply focused at conjugated points. In 1905, Wood constructed a lens by a dipping technique whereby a cylinder of gelatin is produced with refractive index of axial symmetry. Thin transverse sections, with flat end-faces, act like converging or diverging lenses depending on whether the index is a decreasing or increasing function of the radial distance. Nearly forty years later, Luneburg analysed ray propagation through inhomogeneous media, and described a variable index, spherically symmetric refracting structure, performing perfect geometrical imaging between two given concentric spheres on each other [7].

More recently, light propagation in cylindrical or planar GRIN media with a wide variety of gradient profiles has been analysed with emphasis on the waveguiding properties of importance for optical communications and optical sensing.

BASIC GRIN ELEMENTS

The basic GRIN optical elements are waveguides (slab, strip or cylinder) and rod lenses. A GRIN planar waveguide is a slab of dielectric material with variable refractive index surrounded by homogeneous media of lower refractive indices. A GRIN channel waveguide is a strip of dielectric material with variable refractive index embedded in a homogeneous medium of lower refractive index. The planar waveguide confines light in one transverse direction (the y direction, for instance) while guiding it along the z direction. A channel waveguide confines light in the two transverse directions (the x and y directions). The most widely used of these waveguides is the optical fibre which is made of two concentric cylinders (core and cladding) of low-loss dielectric material such as glass. In all types of waveguide light propagates in the form of modes. Each mode travels along the waveguide with a distinct propagation constant and phase velocity. The core of a GRIN fibre has a variable refractive index, highest in the axis and decreasing gradually to its lowest value at the core-cladding boundary (see Fig. 1). The phase velocity of light is therefore minimum at the centre and increases gradually with the radial distance. Guided rays of most axial modes travel the shortest distance at the smallest phase velocity. Rays of the most oblique mode travel a longer distance, mostly in a region where the phase velocity is high. Thus the differences in distances are compensated by opposite differences in phase velocities. As a consequence, the differences in the travel times of the modes are expected to be reduced so that the modal dispersion is minimized.

A GRIN fibre lens and rod lens, are, in essence, a segment of a cylindrical GRIN medium that can be manufactured as a single rod or as linear or planar array. A basic characteristic of GRIN components is the numerical aperture (NA), a quantity specifying the light-gathering capability of an optical system For a GRIN component with parabolic profile:

$$n^2(r) = n_0^2[1 - g_0^2 r^2] \tag{1}$$

and semi-aperture a, the NA, in the paraxial region, is given by the following expression

$$\text{NA} = \sin\,\theta_e \simeq n_0 g_0 a = (2n_0\,\Delta n)^{1/2} \tag{2}$$

provided that the GRIN component is surrounded by a vacuum and where θ_e is the maximum angle for which a ray trajectory is confined within the GRIN component and g_0 is the gradient parameter (see Fig. 2). Δn is the difference

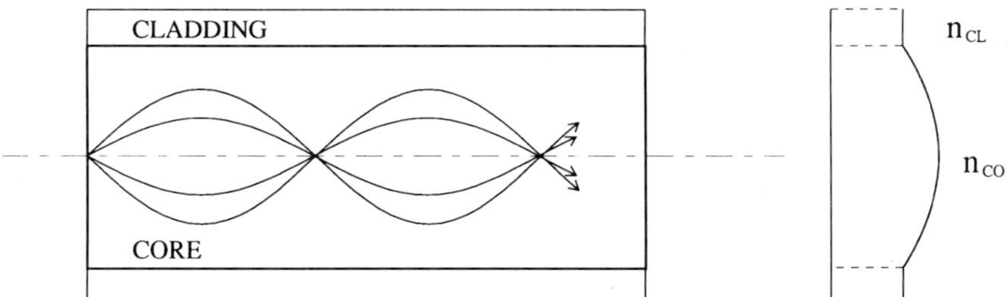

Fig. 1. Refractive index profile and trajectories of rays in a multimode GRIN fibre.

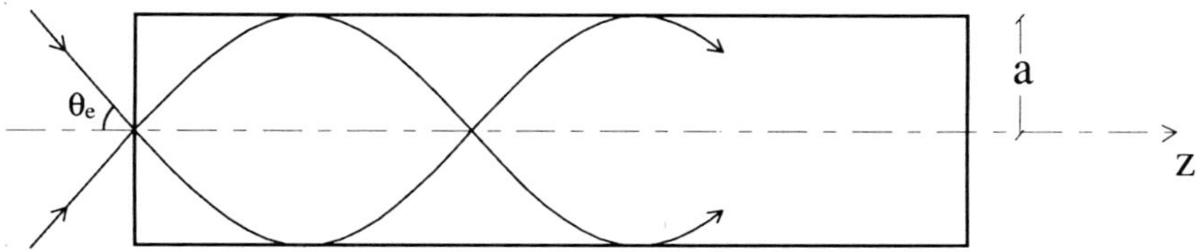

Fig. 2. Acceptance angle of a GRIN component.

between the refractive index at the optical axis n_0 and at the semi-aperture a. For instance, 0.2 NA means that at the input face of a GRIN component the maximum acceptance angle of guided light is $11.5°$.

On the other hand, a GRIN component may be used as a lens and the imaging rules are, in principle, the same as those for cylindrical or spherical lenses. In conventional lenses imaging is a result of discrete refraction occurring at the boundaries of homogeneous media of different refractive index. By using materials in which the refractive index varies in some controlled way it is possible to form images by continuous refraction. Index variation may be axial, radial or a combination of both variations. In particular, a tapered GRIN lens considered as a GRIN component in the form of a rod (inmersed in the vacuum) limited by plane parallel end-faces combines discrete refraction at plane end-surfaces with continuous refraction within the lens. Light propagation through a tapered GRIN lens (or fibre) of radius a, thickness d whose refractive index is given by

$$n^2(r, z) = n_0^2\{1 - [g(z)]^2 \, r^2\} \tag{3}$$

can be described, in the paraxial approximation, by the matrix representation (ABCD law):

$$q_2 = M_{ij} \, q_1 \tag{4}$$

where

$$M_{11} = A; \qquad M_{12} = B; \qquad M_{21} = C; \qquad M_{22} = D \tag{5a}$$

$$A = H_2(d); \qquad B = H_1(d)/n_0; \qquad C = n_0\dot{H}_2(d); \qquad D = \dot{H}_1(d) \tag{5b}$$

and $q_1^T = (r_1, n_0\dot{r}_1)$, $q_2^T = (r_2, n_0\dot{r}_2)$, where the superscript T indicates transpose and dot denotes the derivative with respect to z. Equation (4) relates the ray position r_1 and ray slope $\dot{r}_1$ at the input face of the GRIN lens to the ray

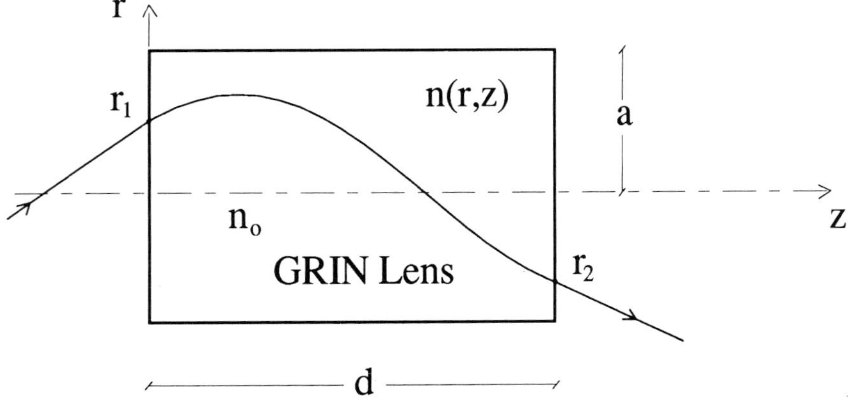

Fig. 3. Arbitrary ray trajectory through a GRIN lens.

position r_2 and ray slope $\dot{r}_2$ at the output face (see Fig. 3). Likewise H_1 and H_2 are the axial and field rays respectively [7]. The axial ray is a paraxial ray which originates at the point of the axis on the input face of the lens and the field ray is a paraxial ray leaving the input face parallel to the axis (see Fig. 4).

For a selfoc lens ($g(z) = g_0 = \text{constant}$), the ray oscillates about the axis of the lens with a period $2\pi/g_0$, known as the pitch, an important parameter to determine the lens length of a selfoc lens for practical applications.

When a collimated beam impinges on a GRIN lens, the point where the emerging light crosses the axis is the focal point. As illustrated in Fig. 5, the focal distance is measured from the principal plane of the GRIN lens and is

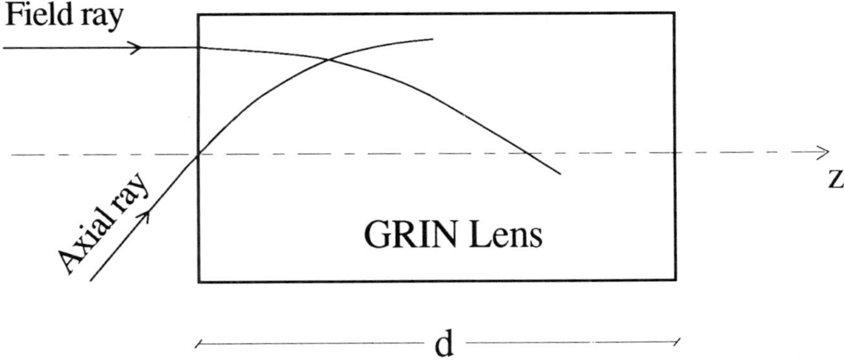

Fig. 4. Axial ray: $H_1(0) = 0$ and $\dot{H}_1(0) = 1$; and field ray: $H_2(0) = 1$ and $\dot{H}_2(0) = 0$, for describing light propagation through a GRIN lens (any other paraxial ray can be represented in terms of a linear combination of these rays).

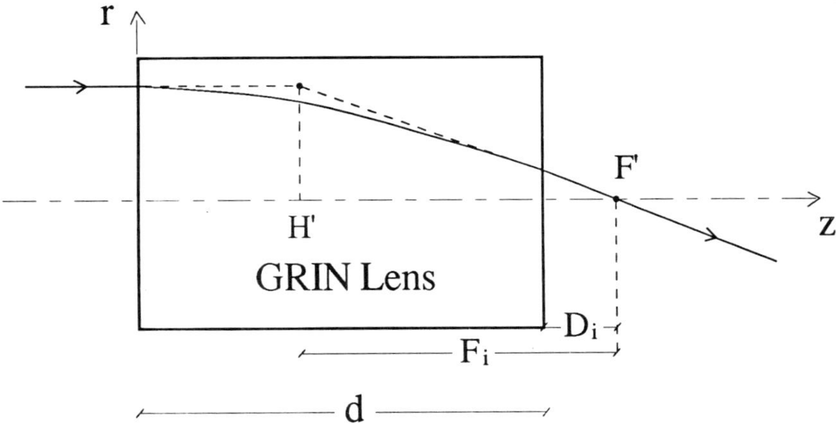

Fig. 5. Focal and working distances of a GRIN lens: F' is the focal point and H' is the principal point.

given by

$$F_i(d) = - [n_0 \dot{H}_2(d)]^{-1}. \tag{6a}$$

However, the working distance D_i is the length from the output face of the lens to the focusing point. It is more convenient to use the working distance and it is given by

$$D_i(d) = -H_2(d)/n_0\dot{H}_2(d) = H_2(d)F_i(d); \tag{6b}$$

both parameters depend on the thickness of the lens.

In the particular case of a selfoc lens, the focal and working distances are given by

$$F_i(d) = [n_0 g_0 \sin(g_0 d)]^{-1} \tag{7a}$$

$$D_i(d) = \cotan(g_0 d)/n_0 g_0 \tag{7b}$$

where the axial and field rays for selfoc lens are given by $H_1 = \sin(g_0 d)/g_0$ and $H_2 = \cos(g_0 d)$.

From Eq. (7b) it follows that $D_i > 0$ when $0 < g_0 d < \pi/2$ and $\pi < g_0 d < 3\pi/2$ and $D_i < 0$ when $\pi/2 < g_0 d < \pi$ and $3\pi/2 < g_0 d < 2\pi$. For a quarter-pitch lens ($g_0 d = \pi/2$) or three-quarter-pitch lens ($g_0 d = 3\pi/2$) $D_i = 0$ and the focal point is formed on the output face of the selfoc lens. When $D_i < 0$ the focal point is made inside the lens and the output beam diverges. The focal and working distances, for instance, of a 0.23 pitch lens with $n_0 = 1.602$ and $g_0 = 0.192$ mm^{-1} are 3.277 mm and 0.42 mm, respectively.

Finally, the lens law for a simple optical system composed of a GRIN lens (Fig. 6) is given by

$$d_2 = \frac{-[n_0 d_1 H_2(d) + H_1(d)]}{n_0[n_0 d_1 \dot{H}_2(d) + \dot{H}_1(d)]} \tag{8}$$

and the linear or transversal magnification m of the system is given by the following equation

$$m = n_0 d_2 \dot{H}_2(d) + H_2(d). \tag{9}$$

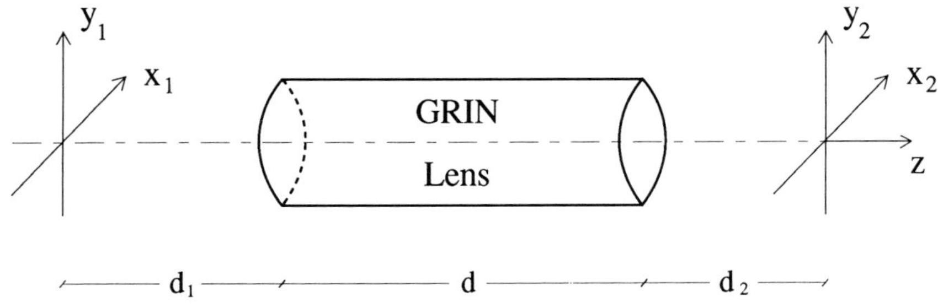

Fig. 6. Simple optical system composed of a GRIN lens.

Equation (8) specifies the distance where the ray, emanating from a single axial point located at d_1, will again cross the axis in an image point located at d_2.

GRIN OPTICS FOR OPTICAL CONNECTIONS

Within the communications area, most optical devices for manipulating and processing the optical signals in multimode fibre communication systems include lenses, and for most of these devices GRIN rod lenses have a number of advantages over conventional lenses, such as small size, light weight, low cost, short focal distance, easy mounting and adjustment and so on. A wide variety of such devices using GRIN rod lenses has been developed and reported to carry out typical functions as collimation and focusing, optical coupling and off-axis imaging [8].

Figure 7 shows how the output from a small light source such as a diode laser, LED or a single-mode optical fibre, can be collimated with a one-quarter pitch lens or how a collimated beam can be focused in another optical fibre or on a detector. A connector is the simplest application of a GRIN rod device for fibre–fibre coupling, source–fibre coupling [9] and so on. As an example, Fig. 8 shows a NA conversion using a selfoc lens in optical fibre lines. The light beam coming from a fibre (or a laser diode) is coupled to another fibre by a selfoc lens. The feature of the lens is the conversion of the NA of the beam.

The problem of both dividing the beam from a fibre into two or more beams and coupling those beams to two or more output fibres can be solved by

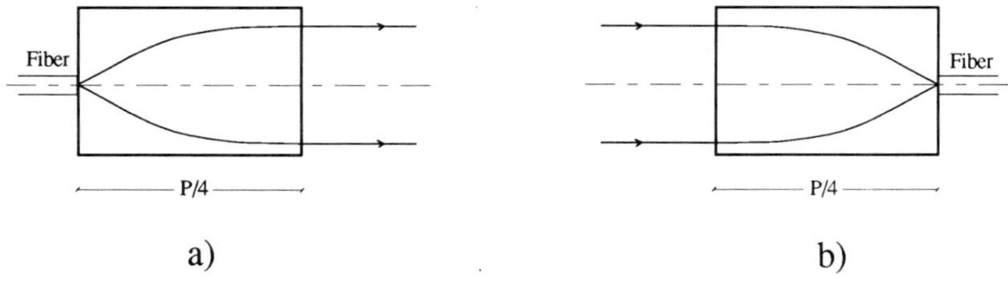

Fig. 7. A quarter-pitch lens collimates (a) or focuses (b) incoming light.

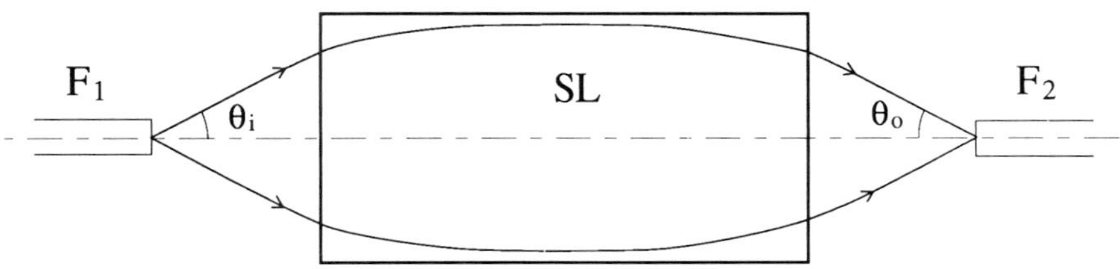

Fig. 8. Fibres coupling by a selfoc lens.

directional couplers. Figure 9 shows a four-port directional coupler using two GRIN lenses. The device uses a beam-splitter film between two selfoc lenses of 0.25 pitch arranged in series. The optical configuration is based on the off-axis imaging properties of the GRIN lenses so that each lens serves two fibres, and only two lenses are required. Finally, if the beam-splitter film in a directional coupler is substituted for an interference filter, the coupler serves as a multiplexer or demultiplexer in a wavelength division multiplexed system.

On the other hand, GRIN fibre lenses (GFLs) are being used in single-mode fibre (SMF) communication systems for obtaining low-loss connection between SMFs with different spot sizes. The GFLs are manufactured with multimode GRIN fibres of small diameter (diameters of about 100 μm) and strong focusing (gradient parameter values about 6 mm^{-1}). Figure 10 shows an optical connector by GFL, fused and aligned to monomode fibres. This optical connector reduces misalignment sensitivities and provides free space between single mode guiding structures to insert optical components [10]. As will be

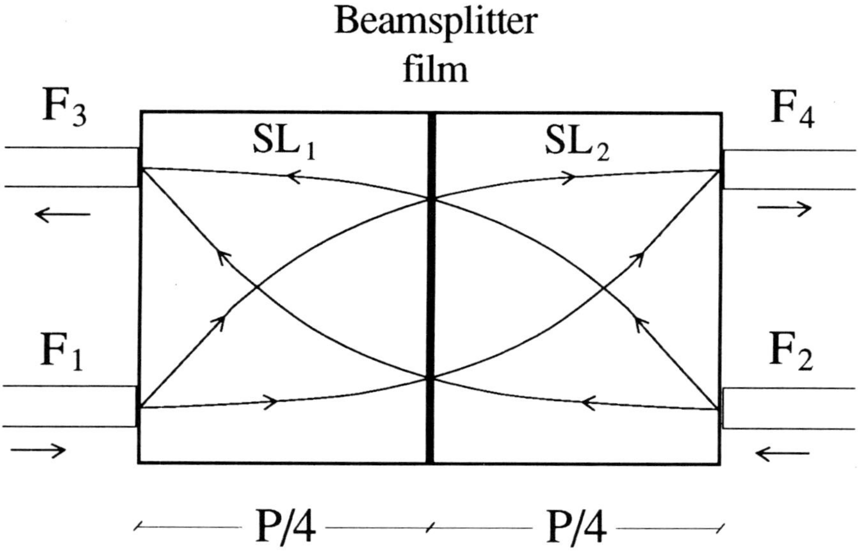

Fig. 9. A four-port directional coupler using two selfoc lenses.

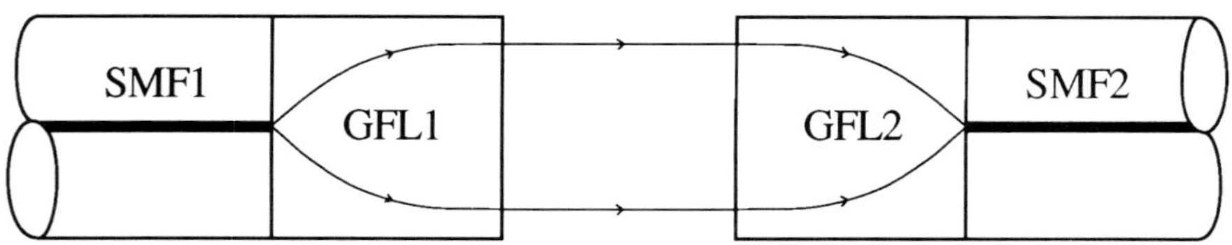

Fig. 10. Arrangement for GFLs fused to the end of single-mode fibres.

shown in the next section this kind of configuration with GRIN rod and/or fibre lenses is also useful for optical sensing purposes.

GRIN OPTICS FOR OPTICAL SENSING

Intensity-modulated fibre-optic sensors have been described in the literature for the measurement of a variety of environmental parameters such as pressure, fluid flow, temperature and so on. The inherent advantages of using fibre optics in these and other applications have been detailed by many authors [11, 12]. Many of these sensors require the direct measurement of a physical displacement which is used to infer the parameter of interest.

Many of these sensors employ GRIN rod lenses (GRL) and/or GRIN fibre lenses (GFL) in such a way that the sensing characteristics are significantly improved.

An important example of this kind of sensor has been recently proposed [13]. The optical sensor consists of a pair of optical fibres which are adjacent to a quarter-pitch GRL having a reflecting coating at one end. The two fibres are maintained at a fixed position relative to each other, symmetrically located with respect to the GRL axis. Optimum coupling efficiency is achieved by positioning the fibres very closely to the input plane of the quarter-pitch GRL, which has the property of being able to expand and collimate a point source of light (here by a small-diameter fibre) as was shown in Fig. 7(a). The sensor can be visualized as in Fig. 9, when the second lens is removed and the film is a totally reflecting coating: light coming from fibre F_1 (source fibre) is directed into the GRIN lens, reflected by the mirrored back surface, and then is collected by the fibre F_3 (the receiving fibre in Fig. 9).

Intensity modulation results from the relative transverse displacement between the lens and the pair of fibres. The transverse displacement technique can also be used with a single optical fibre acting as both source and receiver (located in the centre of the GRL), and therefore by using a beam-splitting element between the light source and the input plane of the fibre. Displacement sensitivities of 2.5 nm have been measured using 50-μm core diameter multimode fibres; likewise, theoretical sensitivities of the order of 0.01 nm are predicted with the use of single-mode fibres.

Many variants of this sensing structure can be implemented; thus, when a single-mode fibre is fused to a GFL (such as is shown in Fig. 10) and the second fibre SMF2 is replaced by a target such that its position or state is a function of some physical parameters to be sensed, then physical property sensors become possible. Targets such as a bimetallic element (for temperature sensing) or a flexible pressure-sensitive diaphragm with a reflective inner surface [14], varying its distance from GFL2 in response to a pressure input (or another physical input) can be used for intensity modulation sensing; note that the reflecting light by the target must be again coupled in SMF1, but the coupling efficiency is very sensitive to axial or lateral displacement [*10, 15*]. Usually the coupling efficiency can be evaluated by using a Gaussian mode approximation in such a way that the design of the sensor based on optical fibres and GRIN lenses is greatly simplified.

We must stress that one of the main advantages of using GRIN lenses (or GFL) in the configuration of a modulated intensity sensor is that only a single fibre need be used as source and receiver; moreover, depending on the configuration, the GRIN lenses can be used for increasing the sensor sensitivity or in other cases for improving the coupling efficiency (by reducing misalignments and therefore providing stable sensing structures). This last possibility has been taken into account in a single-crystal optical fibre high-temperature thermometer [*16*]; likewise, in many invasive catheters for sensing pH, pO_2, pCO_2, and so on, in blood (biosensors) there are GRIN lenses for refocusing the light coming from the blood in the vicinity of a second fibre (receiver fibre), that is, for ensuring the collection of the light emitted for sensing.

Up to now we have mainly used the reflective concept (unlike biosensors which are related to the absorption concept) which is especially attractive for broad sensor use, thanks to its accuracy, simplicity and potentially low cost. Generally, it was shown that the sensor comprises a pair of single fibres (or one fibre) connected to a GRL (or GFL) in such a way that one fibre transmits light to a reflecting target and the other one traps and transmits the light to a detector. The intensity of the detected light depends on the relative state between fibres and GRL (when the lens works as a target) or the system fibre-GRIN lens and target.

Another important and fundamental concept in sensing is microbending [*14*]. If a fibre (GRIN fibre lens) is bent, a small amount of light is lost through the wall of the fibre. If a transducer bends the GFL due to a change in some physical property, then the amount of receiving light is related to the value of this physical property. Following this concept, many sensors can be designed based on both the configuration given by Fig. 10 (when the free space distance between the GFLs is taken to be equal to zero) and the diffraction-limited coupling efficiency concept.

The diffraction-limited coupling efficiency has been recently taken into account for analysing the coupling between two Single Mode Fibres (SMF) [*15*]. It was shown that the coupling efficiency between two SMFs is very sensitive when the device shown in Fig. 10 (with SMF1 = SMF2 and GFL1 = GFL2) works under the condition of strong diffraction: it can be achieved when the diffraction parameter $\delta = kn_0g_0r_0w_0$ is close to unity (the radiation concept) [*15*], where k is the wavenumber in free space ($k = 2\pi/\lambda_0$), r_0 is the radius of the GFL and w_0 is the beam waist size of the Gaussian mode (approximately) of the single-mode fibres. Coupling efficiency η is given by the following expression:

$$\eta = [\exp(-\delta^2) - 1]^2. \tag{10}$$

To a first approximation the bending of a GFL can be considered as an effective change in the radius of the GFL, therefore if the coupling device works near $\delta = 1$, any bending will produce a drastic change in the intensity coupling efficiency to the second fibre. Figure 11 shows the normalized coupling efficiency versus the δ-parameter: note the high sensitivity of the device around $\delta = 1$. Moreover, since the intensity is a monotonic function of the δ-parameter, the sense of the variation can also be determined. Obviously, physical parameters such as pressure, temperature and so on can be sensed by this bending technique. Another important application of this type of GFL sensor is the study of structural defects. Numerous methods have been developed for detecting cracks in components after manufacturing or during their lifetime, nevertheless, optical fibre sensors have been used successfully, which suggests that the proposed GFL sensor can be used for improving the present sensors.

Finally, we will consider another important concept in optical sensing: absorption [*17*]. In this case, taking into account that $F(\omega)$ is the spectrum of the light source, we consider the total coupling efficiency as an integral over all frequencies. If $F(\omega)$ is centred at a central frequency ω_0, for which the diffraction parameter takes values close to unity, then the total coupling

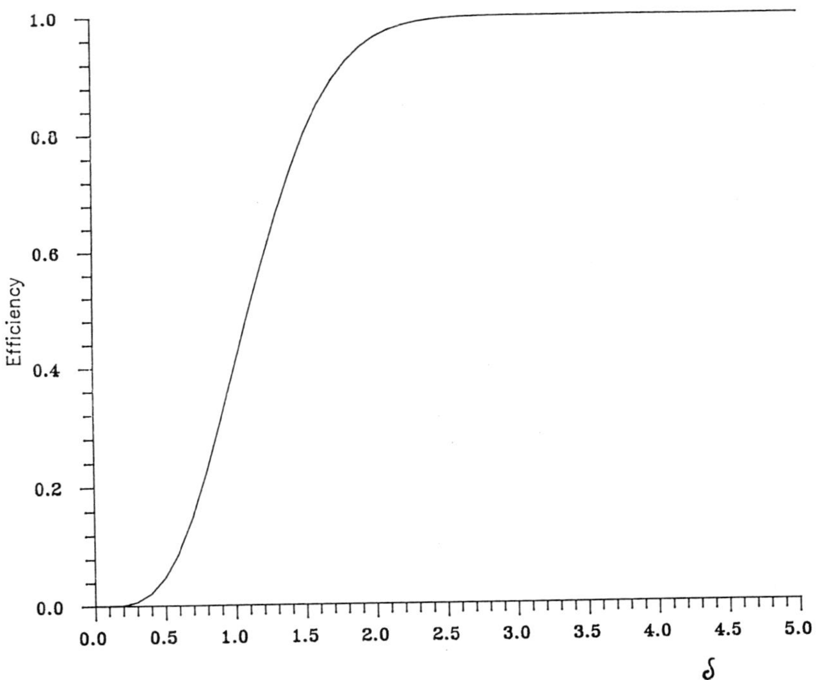

Fig. 11. Diffraction-limited coupling efficiency versus the δ-parameter.

efficiency will be very sensitive to absorption. Let us consider the case in which SMF2 in Fig. 10 is removed from the configuration and the end-face of GFL2 is coated by a reflecting material (obviously the device can be inserted in a catheter when biological applications are required); moreover, the free space between GFL1 and GFL2 is filled by the substance to be sensed. If the spectrum is changed by absorption the total coupling efficiency is also changed and therefore the presence of a determined substance in this space can be sensed (for instance, oxygen in blood). In short, GFLs combine the advantages of both the usual optical fibres and the optical properties of selfoc lenses. These advantages together can improve in a remarkable way the optical sensing devices.

GRIN OPTICS FOR OPTICAL IMAGING SYSTEMS

GRIN optical materials have remained a novelty for over 100 years. Only recently have applications beyond photocopy and facsimile imaging arrays, fibres couplers and GRIN sensors begun to emerge. From 1970 to 1975 several papers were published by P.J. Sands [18], E.W. Marchand [1] and D.T. Moore [19] that developed the aberration theory for optical imaging systems using GRIN optical materials. In 1974, Moore developed one of the first optical design programs for GRIN imaging systems. Today, most optical design programs include some form of support for GRIN material based on a Runge–Kutta method developed by A. Sharma in 1982. During the 1980s a number of optical designs were developed using both radial and axial GRIN materials. These included microscope objectives, telephoto, wide angle and zoom camera lenses, binoculars, laser diode collimators, and even a light-weight lens for satellite mapping applications. Due to a lack of GRIN optical materials, few of these were built. The main obstacle to the realization of practical commercial optical systems using gradient index has been a lack of good GRIN materials. They have been very expensive and have had undesir-able chromatic properties. At present, better and better GRIN optical materials are being developed, although only ion-exchange has been successfully commercialized.

The usual optical designs are either radial GRIN (RG) or axial GRIN (AG) materials. The index profile for RG is usually very close to parabolic with a maximum index at the centre of the rod resulting in a positive or converging lens (see Eq. (1)). Optical elements using RG may also have a power contribution from curved surfaces that are included to give additional control of the optical aberrations of the system or increase the numerical aperture of the element. Optical elements using AG materials have all of their power contributed by the curved surfaces of the lens element. The AG is used as an aspheric surface, to control the aberrations of the optical system [20]. To third order of ray tracing, aspherics and AG are equivalent [18]. An AG generally begins at the vertex of one surface and continues through the SAG sagittal line of the surface becoming homogeneous before reaching the SAG of the second surface of the element. These AG elements can be successfully fabricated by fusion of glasses [6]. In short, many optical imaging systems can be designed by RG and AG elements: imaging arrays in the input scanner of photocopy and facsimile machines, endoscopes and industrial boroscopes, and so on. We must stress that two of the most important advantages of using GRIN elements are that it reduces the number of elements required in a conventional imaging system [19] and it serves for taming optical aberrations [1,21].

As a particular imaging system we can consider the human eye. Briefly, we will indicate the main connections between GRIN optics and vision. First of all we must stress that the analysis of tapered parabolic GRIN rods is useful for modelling the tapered section in the cone-photoreceptor of the human eye [22]. Likewise, a model of the human eye based on an index distribution within the crystalline lens of the eye has been developed; in this way, novel studies based on a theoretical eye with a GRIN crystalline lens model can be made. The model corresponds to a representation of the refractive index profile given by the following function (parabolic index profile with axial graded index parameter):

$$n(x, z) = 1.42 - gx^2 - hz^2 - j\,|\,z\,| \tag{11}$$

where z is the optical axis, and the axis origin is at the centre of the crystalline lens. Three sets of values have been considered for the previous gradient index parameters: g, h, j. The first model is: $g = h = j = 0$, which is obviously a homogeneous lens (it is exactly the theoretical eye proposed by Le Grand [23] without accommodation). A second model is g, h different from zero and $j = 0$, which corresponds to the model developed by M.C.W. Campell and E.M. Harrison [24]. Finally when the three parameters are different from zero it corresponds to the model proposed by J.W. Blaker [25] for human lenses.

In short we must stress that a deeper knowledge of the behaviour of human lenses is being achieved by using GRIN modelling which can become an important goal for optometric research and vision sciences.

REFERENCES

1. E.W. Marchand, *Gradient Index Optics*, Academic Press, New York, 1978.
2. S.N. Houde-Walter and D.T. Moore, *Appl. Opt.* **25**, 3373 (1986).
3. M.A. Pickering, R.L. Taylor and D.T. Moore, *Appl. Opt.* **25**, 3364 (1986).
4. S.N. Houde-Walter, *Laser Focus World* **25**, 151 (1989).
5. Y. Koike, OSA Annual Meeting Technical Digest Series, 261 (1992).
6. R. Blankenbecler and G.E. Rindone, OSA Annual Meeting Technical Digest Series, 253 (1992).
7. R.K. Luneburg *Mathematical Theory of Optics*, University of California Press, Berkeley (1964).
8. W.J. Tomlinson, *Appl. Opt.* **19**, 1127 (1980).

9. C. Gómez-Reino, *Int. J. Optoelec.* **7**, 607 (1992).

10. J. Liñares and C. Gómez-Reino, *Appl. Opt.* **29**, 4003 (1990).

11. B. Calshaw, *Radio Electron. Eng.* **52**, 283 (1982).

12. C.M. Davis, *Opt. Eng.* **24**, 347 (1985).

13. P.J. Murphy and T.P. Courselle, *Appl. Opt.* **29**, 544 (1990).

14. D.A. Krohn, *Photonics Spectra*, January, 59 (1987).

15. J. Liñares and C. Gómez-Reino, *J. Mod. Opt.* **38**, 597 (1991).

16. W.H. Hu, C.F. Wang, W.Z. Tang, W. Zhow and J.H. Zhow, *Int. J. Optoelec.* **7**, 121 (1992).

17. G. Stewart, J. Norris, D.F. Clark and B. Culshaw, *Int. J. Optoelec.* **6**, 227 (1991).

18. O.J. Sands, *J. Opt. Soc. Am.* **60**, 1086 (1970).

19. D.T. Moore, *Appl. Opt.* **19**, 1035 (1980).

20. A.K. Ghatak and K. Thyagarajan, *Contemporary Optics*, New York, Plenum Press (1977).

21. Y.A. Carts, *Laser Focus World*, January, 142 (1994).

22. A. Sharma and I.C. Goyal, *Appl. Opt.* **18**, 1482 (1979).

23. Y. Le Grand and S.G. El Hage, *Physiological Optics*, Springer Verlag Series in Optical Sciences, N°13, Springer Verlag (1980).

24. M.C.W. Campell and E.M. Harrison, OSA Annual Meeting Technical Digest Series, 235 (1990).

25. J.W. Blaker, *J. Opt. Soc. Am.* **70**, 220 (1980).

25 Photorefractive fibres: fabrication and hologram construction

Francis T. S. Yu and Shizhuo Yin
Department of Electrical Engineering, The Pennsylvania State University, University Park, PA 16802

INTRODUCTION

The use of photorefractive (PR) materials for data processing and storage has offered promising applications in data storage and signal processing [1–8]. The salient features of the PR materials are the large storage capacity, the high diffraction efficiency, and the high angular wavelength selectivities. On the other hand, the growth of PR fibres [4, 9–11] has also provided an alternative approach to a variety of applications.

Although there have been several articles reported in this field, due to page limitation, we shall concentrate our effort on some of the more recent work done, and apologize for not being able to include all the appropriate references. In this connection, we shall briefly introduce a laser heated pedestal growth (LHPG) technique for single-crystal fibre fabrication. A phase conjugate fibre holographic construction architecture is presented, which has the advantage of minimizing the intramodal distortion [12]. The angular and the wavelength selectivities of PR fibre holograms, obtained with the coupled wave theory of [13] are provided. Since the reflection-type wavelength-multiplexed fibre hologram offers a higher and more uniform selectivity than the transmission-type, we would suggest that the reflection-type fibre holograms would be a better choice. Cross-talk noise between the wavelength-multiplexed channel is evaluated, and we have shown that the narrower the laser linewidth the lower the cross-talk would be. Finally, fibre hologram construction and its applications are given.

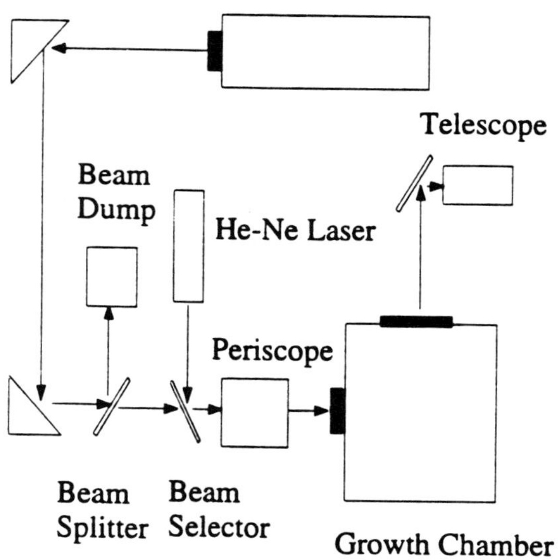

(a)

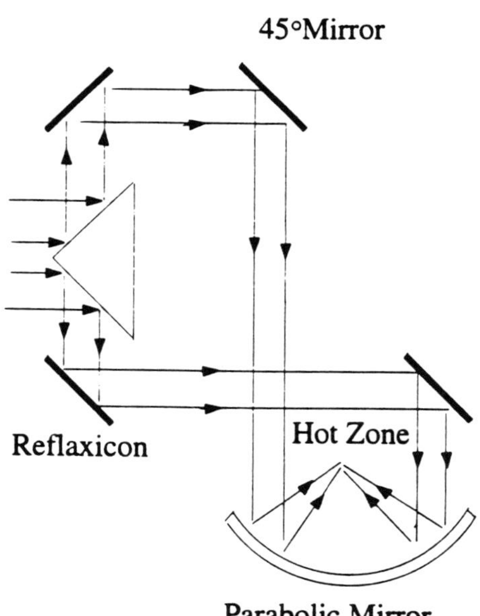

(b)

Fig. 1. *Continued.*

PR FIBRE GROWTH

The laser heated pedestal growth (LHPG) technique is a relatively new crystal fibre growth technology that can fabricate 50–1000 μm diameter and up to 5–10 cm long fibres [*14, 15*]. The groups at Stanford University and The Pennsylvania State University have independently grown single-crystal fibres by using this technique [*7, 9*]. The major advantages of growing single-crystal fibre using the LHPG system are as follows.

(1) It does not require any holding device (e.g., a crucible) so that the contamination from the holder can be avoided.
(2) Some growing limitations, such as the melting point, shape, etc., can be alleviated.
(3) Since the LHPG system used a CO_2 laser as the heating source, the growing process is much easier to control. For example, the laser beam has an annular taper form, such that the crystal fibre can be grown in the vertical direction, which is crucial for fibre growing.

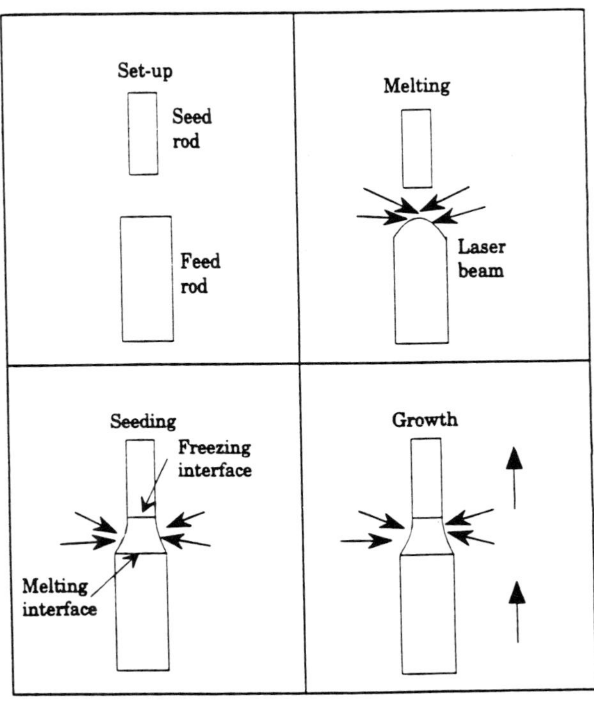

(c)

Fig. 1. The laser heated pedestal growth (LHPG) system. (a) Schematic diagram; (b) the growth chamber; (c) crystal fibre growth.

(4) The LHPG technique can also be used in a variety of applications, for example, as applied to high-temperature superconductor wire-rod and other novel devices.

A schematic diagram describing this system is shown in Fig. 1(a), in which a preformed PR crystal about 0.9×0.9 mm Ce: Fe: doped $LiNbO_3$ is placed in a holder that maintains a precise vertical orientation. A 50 watt CO_2 laser is brought to focus by first forming an annular beam with a reflaxicon. Then by steering the annular beam with a 45° flat mirror, a highly intense focusing beam can be formed on the preformed PR rod using a concave parabolic mirror, as shown in Fig. 1(b). When a seed rod is brought in contact with the molten tip of the preform, the fibre growth can be initiated by slowly pulling the seed rod upward with a computer control driver, as illustrated in Fig. 1(c). Thus we see that a single-crystal fibre can be drawn with the LHPG system.

PHASE CONJUGATE FIBRE HOLOGRAMS

We now describe a method for recording an axial hologram in a PR fibre, in which the intermodal distortion can be partially compensated [*12*]. Let us assume that a linearly polarized monochromatic plane wave is coupled into a PR fibre with a condensing lens. A cone of light beams having different polarizations [*16*], caused by the anisotropy and Fresnel reflection inside the fibre will eventually exit from the other end of the fibre. Thus, we see that as the light beam propagates along the fibre, the spatial information content of the incident light beam will be totally scrambled at the exit end. However, if the incident light beam (called an object beam) interferes with a reference beam, which is assumed to be coupled into the fibre at the opposite end, a hologram (similar to that of a reflection hologram) can be formed due to the induced PR effect. If the length-to-diameter ratio of the PR fibre is large in the order of hundreds (e.g., $l/d > 100$), the object beam cannot be faithfully read-out by merely using the reference beam. This is primarily due to modal scrambling as described earlier. Nevertheless, if the fibre hologram construction is made by using a conjugate reference beam, which is derived with a polarization preserving conjugating mirror shown in Fig. 2, a transmission-type (instead of a reflection-type) fibre hologram can be recorded. We note that the writing reference beam R has been made to exceed the coherence length of the object beam O, so that the returned conjugate reference beam R^* is coherent with the object beam O. Thus we see that the recorded object beam can be read-out by the reference beam R.

In view of the preceding fibre hologram construction, an object transparency is launched into the PR fibre with an imaging lens. This forms an object beam O that propagates along the fibre. Since the reference beam R is launched from the other end of the fibre, the polarization preserving phase conjugate mirrors

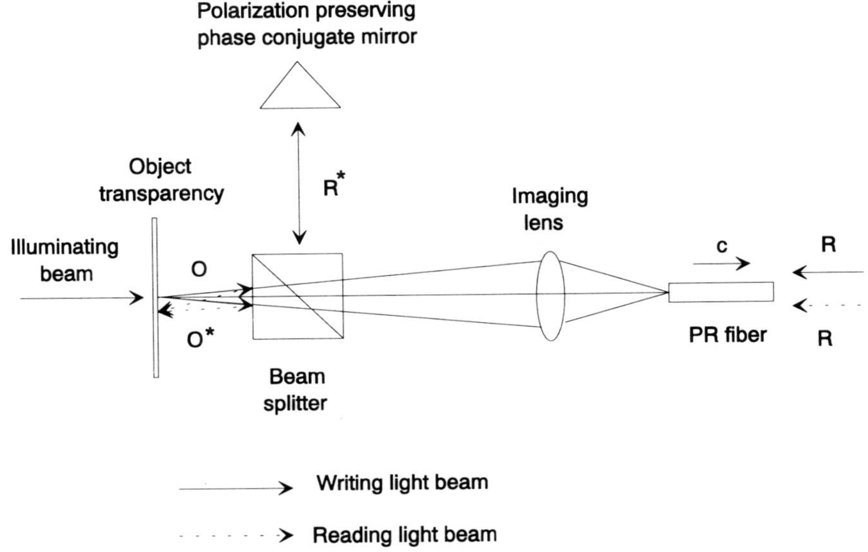

Fig. 2. Construction of a phase conjugate PR fibre hologram.

would reflect a phase conjugate reference beam R^* that is within the coherence length of the object beam. Then a transmission-type fibre hologram can be constructed, as given by

$$H = |O|^2 + |R|^2 + O^*R^* + OR. \tag{1}$$

If the recorded photorefractive fibre is illuminated by the reference beam R, a phase conjugate object beam O^* will be reconstructed as shown by the dashed lines in Fig. 2, in which no phase conjugate mirror is needed in the read-out process. Since the read-out conjugate object beam O^* retraces back within the fibre, in principle the modal distortions can be partially compensated. Thus we see that a faithful conjugate object beam O^* can be reconstructed.

However, if the length-to-diameter ratio is adequately small (e.g., $l/d \leqslant 10$), the PR fibre can be treated as a bulk crystal. Then a reflection-type PR fibre hologram can be directly constructed. This is particularly true for a direct counter-propagation construction, as will be seen in the following sections.

ANGULAR AND WAVELENGTH SELECTIVITIES

The PR fibres we considered are high-order mode fibres, by which the length-to-diameter ratio is assumed sufficiently small (e.g., $l/d < 10$). Thus, the PR fibre we will discuss can be treated as bulk materials for simplicity instead of thin fibres. This assumption is particularly true for reflection-type fibre holograms. By referring to the unslanted transmission and reflection type holograms shown in Fig. 3, the normalized diffraction efficiencies under weak coupling conditions, i.e., $|\nu| \ll |\xi|$, can be shown as [2],

$$\eta_t \sim \nu_t^2 \, \text{sinc}^2 \xi_t, \tag{2}$$

$$\eta_r \sim \nu_r^2 \, \text{sinc}^2 \xi_r. \tag{3}$$

where subscripts t and r denote the transmission-type and the reflection-type holograms, respectively,

$$\nu_t = \frac{\pi \Delta n d}{\lambda \cos \theta}, \qquad \xi_t = \frac{2\pi \, n d \sin \theta}{\lambda} \Delta\theta,$$

$$\nu_r = \frac{\pi \Delta n d}{\pi \sin \theta}, \qquad \xi_r = \frac{2\pi \, n d \cos \theta}{\lambda} \Delta\theta,$$

λ is the recording wavelength, n and Δn are the refractive index and the amplitude of its variation, d is the length of the fibre, and θ and $\Delta\theta$ are the internal construction angle and its variation. Since the typical values of n and Δn for a LiNbO$_3$ crystal are about $n = 2.28$ and $\Delta n_{max} = 10^{-3}$ to 10^{-5} [17], Eqs (2) and (3) can be considered as good approximations.

By considering Snell's law of refraction, the (external writing) angular

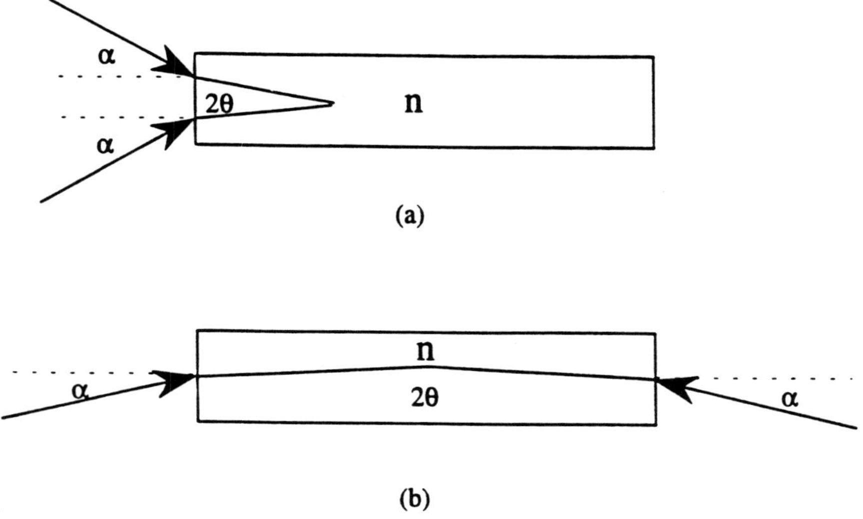

Fig. 3. Fibre holographic constructions: (a) transmission-type fibre hologram; (b) reflection-type fibre hologram.

selectivities of the filter holograms can be shown to be [*18*]

$$\left\{\frac{1}{\Delta\alpha}\right\}_t = \frac{\sin\alpha\cos\alpha}{\sqrt{n^2-\sin^2\alpha}}\,\frac{d}{\lambda} \tag{4}$$

$$\left\{\frac{1}{\Delta\alpha}\right\}_r = \frac{\sin\alpha\cos\alpha}{\sqrt{n^2-\cos^2\alpha}}\,\frac{d}{\lambda} \tag{5}$$

where the subscripts t and r represent the transmission- and reflection-type fibre holograms.

In view of the k-vector diagram of Fig. 4(a) (i.e., for the transmission-type 2), we have

$$(|k+\Delta k|)\sin\theta + (|k+\Delta k|)\sin(\theta-\Delta\theta) = 2|k|\sin\theta, \tag{6}$$

in which the readout wavelength is deviated from the recording wavelength by $\Delta\lambda$, and $|k| = 2\pi n/\lambda$. Thus, the wavelength selectivity for a transmission-type

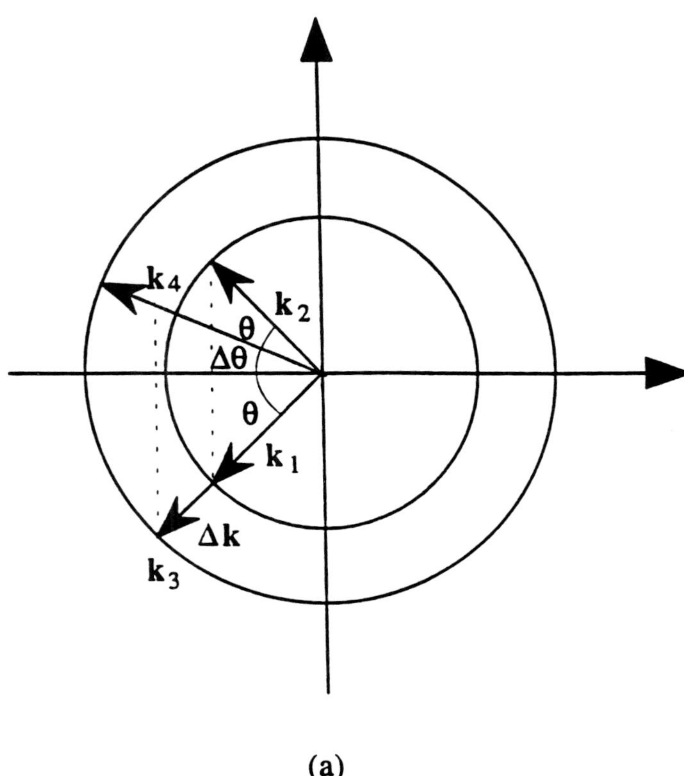

(a)

Fig. 4. *Continued.*

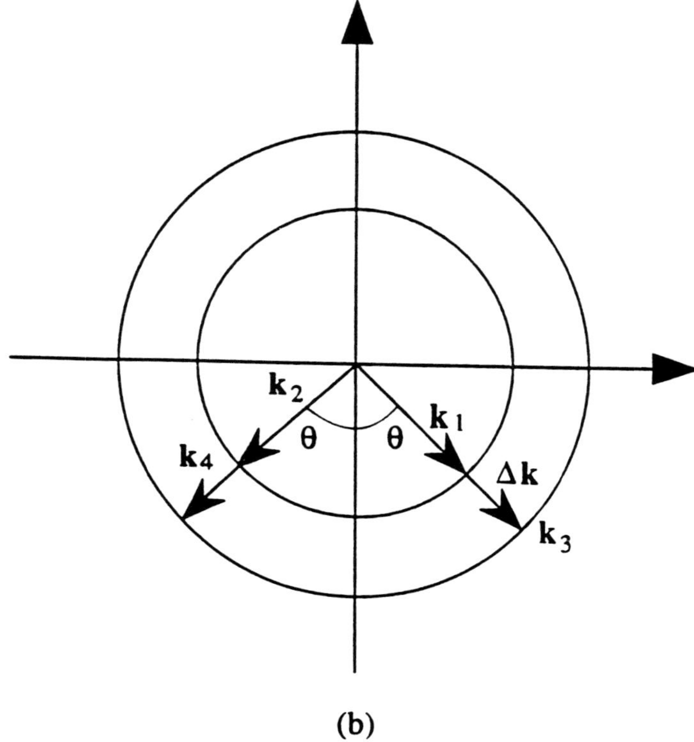

(b)

Fig. 4. Momentum conservation principle. k_1, k_2 are writing wave vectors; k_3 is the readout wave vector; k_4 is the diffracted wave vector; and Δk is the reading wavelength deviation. (a) Transmission-type hologram; (b) reflection-type hologram.

fibre hologram can be written as

$$\left\{\left|\left|\frac{\lambda}{\Delta\lambda}\right|\right|\right\}_t = 1 + \frac{2}{\dfrac{\sin(\alpha/n)}{\sin\left[\sin^{-1}\left(\dfrac{\sin\alpha}{n}\right) - \dfrac{\lambda}{d\sin\alpha}\right]} - 1}. \tag{7}$$

Similarly, by referring to the k-vector diagram of Fig. 4(b), the wavelength selectivity for the reflection-type fibre hologram can be shown to be

$$\left\{\left|\left|\frac{\lambda}{\Delta\lambda}\right|\right|\right\}_r = \frac{\sqrt{n^2 - \cos^2\alpha}}{\lambda}\, d. \tag{8}$$

From the preceding Eqs (7) and (8), the angular and the wavelength selectivities as a function of external writing angle are plotted in Figs 5(a) and (b). In these plots, we see that both selectivities increase as the fibre length increases. However, there is a limited usable-range angular selectivity, i.e., $15° \leqslant \alpha \leqslant 75°$. On the other hand, the wavelength selectivity for the reflection-type fibre hologram is several-fold higher than the transmission-type and it is relatively uniformly distributed over the entire range of recording angles. Thus, the reflection-type wavelength-multiplexed PR fibre hologram would be a better choice for holographic data storage. Mention must be made that Rakuljic *et al.* [*19*] and Curtis *et al.* [*20*] have subsequently shown that the cross-talk in a wavelength multiplexed volume hologram is less severe than in an angular multiplexed bulk hologram. To get a feeling of magnitude, we assume that the length of the PR fibre is $d = 7$ mm and the construction wavelength is $\lambda = 0.7$ μm; then the wavelength selectivity for a reflection-type fibre hologram can be as high as about 10^4.

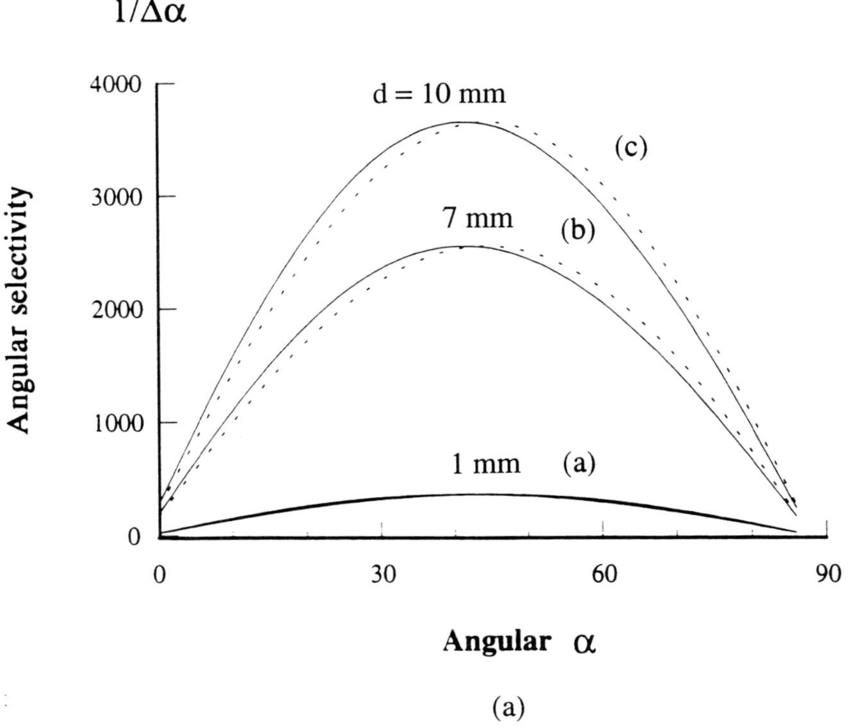

(a)

Fig. 5. *Continued.*

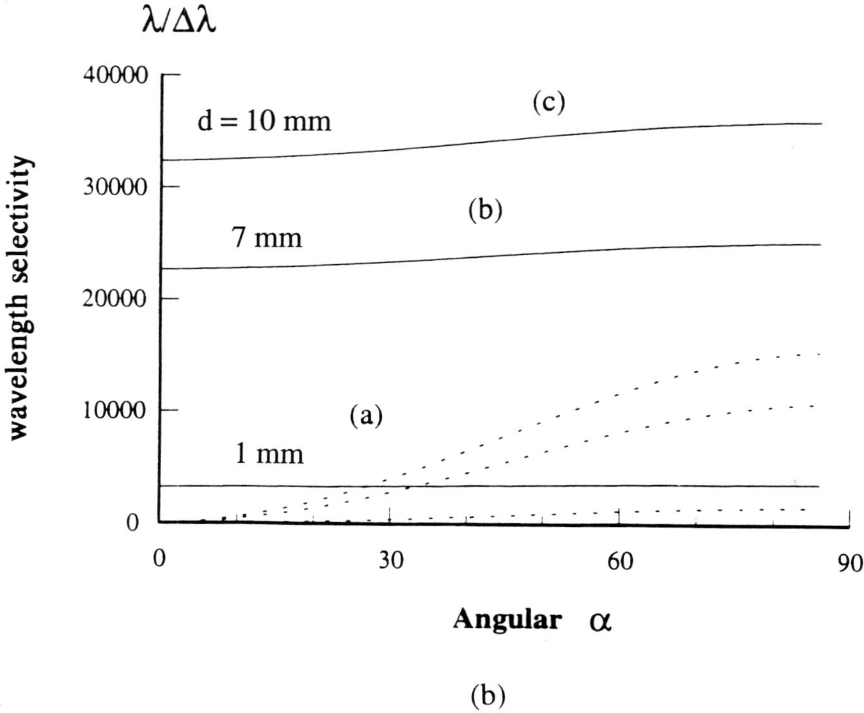

(b)

Fig. 5. Angular and wavelength selectivities; dashed line for transmission-type and solid line for reflection-type, recording wavelength $\lambda = 0.7$ μm. (a) Angular selectivity; (b) wavelength selectivity.

CROSS-TALK NOISE

Although fibre holograms have promising applications to high-density storage, there are however practical limitations that prevent it from attaining this upper limit. Certainly, cross-talk noise is one of the major limiting factors among them. Since the wavelength multiplexed reflection-type fibre hologram has been shown to be a better choice for data storage application we shall evaluate its cross-talk as follows.

Let us refer to the reflection-type fibre hologram construction process shown in Fig. 6, in which the object beam and the reference are assumed directly counter-propagated. Since the PR fibre is assumed to be sufficiently short (e.g., in the order of 0.5–1 mm in diameter and 3–20 mm in length), the length-to-diameter ratio is adequately small (e.g., ~10). The writing light beams would be reflected only a few times within the boundary wall of the fibre. By using the bulky crystal approximation, the diffraction efficiency (assuming no losses) can

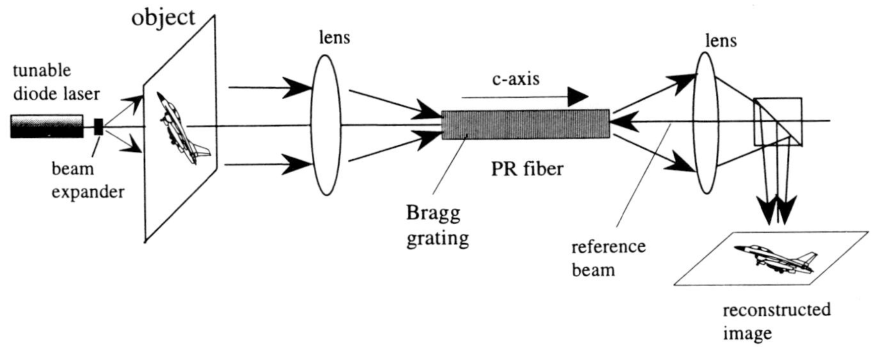

Fig. 6. Wavelength multiplexed reflection-type fibre holographic construction.

be shown to be [21]

$$\eta = \left[1 + \frac{1 - \xi^2/v^2}{\sinh^2(v^2 - \xi^2)^{1/2}} \right]^{-1}, \tag{9}$$

where

$$\xi = \frac{2\pi n d \Delta\lambda}{\lambda_0(\lambda_0 + 2\Delta\lambda)}, \qquad v = \frac{\pi \Delta n d}{(\lambda_0 + \Delta\lambda)\sqrt{1 + 2\Delta\lambda/\lambda_0}},$$

d is the fibre length, n is the refractive index of the fibre crystal, Δn is the induced refractive index due to the writing beam intensity, and $\Delta\lambda$ is the mismatched wavelength of the reading beam with respect to the writing wavelength λ_0. The diffraction efficiencies as a function of mismatched wavelength $\Delta\lambda$ are plotted in Fig. 7, in which we see that η decreases rapidly as $\Delta\lambda$ increases. We also see that as the fibre length increases, the diffraction efficiency improves rapidly and that spectral bandwidth becomes narrower. The peak efficiency can be shown to be $\eta_{\text{peak}} = \eta(\Delta\lambda = 0) = \tanh^2(\pi\Delta n d/\lambda_0)$, which is a monotonically increasing function as the coupling strength $\pi\Delta n d/\lambda_0$ increases. In other words, for a higher peak efficiency, it requires a higher refractive index Δn, longer fibre length, and a shorter writing wavelength λ_0. However, to have a large dynamic storage capacity, as pointed out by Hong *et al.* [22], Δn should be maintained as small as possible. Thus by imposing the weak coupling condition, then letting $\eta(\Delta\lambda) = 0$, the spectral bandwidth of the multiplexed channel can be shown as

$$\Delta\lambda_{\text{min}} \approx \frac{\lambda_0^2}{2nd} \tag{10}$$

which is proportional to the square of the construction wavelength and inversely proportional to the refractive index and the fibre length.

In order to gain a feeling of magnitude, we assume that $\lambda_0 = 670$ nm,

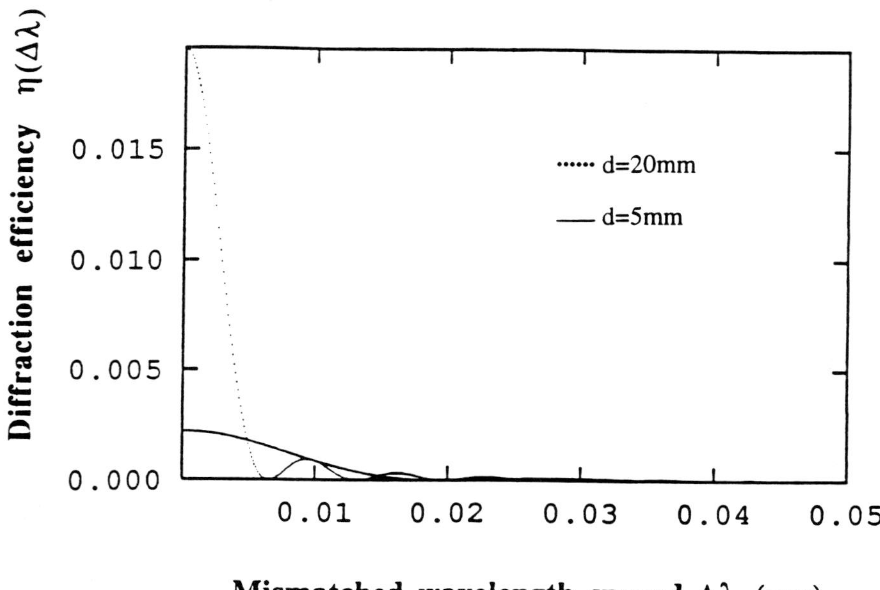

Mismatched wavelength spread Δλ (nm)

Fig. 7. Diffraction efficiency η as a function of mismatched wavelength $\Delta\lambda$. Solid line: $\eta_{\text{peak}} = 0.2\%$ ($\Delta n = 2 \times 10^{-6}$, $\lambda_0 = 670$ nm, $d = 5$ mm); dotted line: $\eta_{\text{peak}} = 0.2\%$ ($\Delta n = 2 \times 10^{-6}$, $\lambda_0 = 670$ nm, $d = 20$ mm).

$n = 2.28$ and the fibre length $d = 5$ mm; then the spectral bandwidth of the channel would be $\Delta\lambda_{\text{min}} = 0.02$ nm. We note that Eq. (9) is derived under the assumption of zero linewidth for the reading and the writing beams. It can be seen that if the writing beam has a finite spectral width $\Delta\lambda_l$, the profile of η would be broader, i.e., $\Delta\lambda_{\text{min}} + \Delta\lambda_l$. Thus, to have a fine tuning channel, a narrower spectral width of the laser is needed for the construction process.

If one regards the reconstructed hologram image from a given channel as the signal and the wavelength spread over into this channel as noise, then the signal-to-noise (cross-talk) ratio can be shown to be [21].

$$\text{SNR}_{\text{CT}}^{(j)} = \frac{\int \text{rect}\left(\frac{\Delta\lambda}{\Delta\lambda_l}\right) \eta \, d(\Delta\lambda)}{\sum_{\substack{m = -(M-j) \\ m \neq j}}^{M+j} \int \text{rect}\left(\frac{\Delta\lambda - m\delta\lambda}{\Delta\lambda_l}\right) \eta \, d(\Delta\lambda)}, \tag{11}$$

where $2M + 1$ is the total number of holographic channels, $\delta\lambda$ is the separation between holographic channels, and $\Delta\lambda_l$ is the spectral width of the light source. For simplicity, we assume that the recording linewidth is negligibly small, so that the cross-talk effect would be primarily due to the reconstruction linewidth. Notice that similar results can also be obtained when the recording linewidth is not assumed negligibly small. In order to minimize the cross-talk noise between the holographic channels, we let the channel separation ($\delta\lambda$) equal the spectral bandwidth of the channels $\Delta\lambda_{\min}$, that is

$$\delta\lambda = \Delta\lambda_{\min}. \tag{12}$$

For example, we assume that the modulated refractive index $\Delta n = 2 \times 10^{-5}$, the length of the fibre $d = 5$ mm, the central wavelength $\lambda_0 = 670$ nm, the wavelength multiplexed channel separation $\delta\lambda = \Delta\lambda_{\min} = 0.02$ nm, and the total number of multiplexed holographic channels $2M + 1 = 501$. Then the $\mathrm{SNR_{CT}}^{(j)}$ can be plotted as shown in Fig. 8, in which $\Delta\lambda_l = 1.5 \times 10^{-4}$ nm (~ 100 MHz) and $\Delta\lambda_l = 7.5 \times 10^{-3}$ nm (~ 5 GHz) have been used. In view of this figure, we see that the $\mathrm{SNR_{CT}}$ are relatively uniformly distributed over the jth channel, and the one using a narrower spectral width has a (≈ 10 dB) higher $\mathrm{SNR_{CT}}$.

Since the cross-talk in a reflection-type wavelength multiplexed fibre hologram is determined by the spectral bandwidth of the channel, it imposes a constraint on the storage capacity. For example, if the tuning range of the laser is assumed to be in the order of 10 to 10^2 nm, the cross-talk-limited storage capacity (C_{CT}) of the fibre hologram can be calculated by simply dividing the tuning range of the light source by the channel separation:

$$C_{\mathrm{CT}} = \frac{\text{tuning range}}{\text{channel separation}}. \tag{13}$$

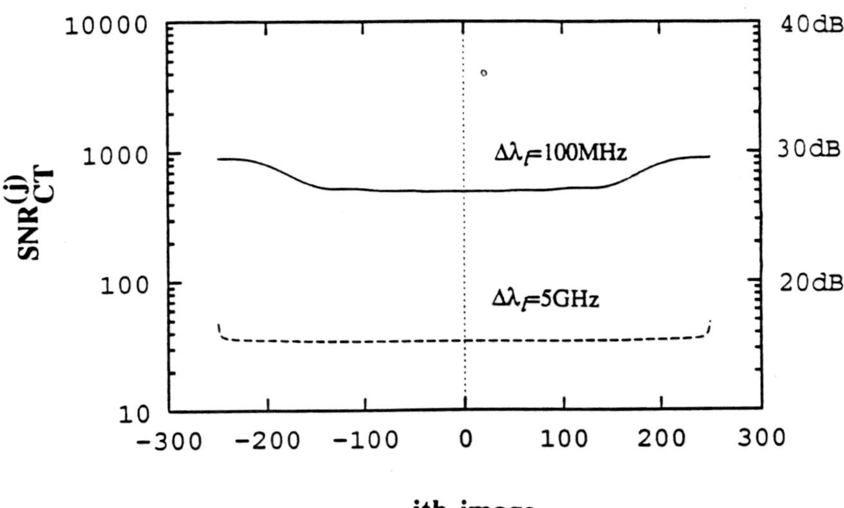

jth image

Fig. 8. Distribution of signal-to-noise (cross-talk) ratio $\mathrm{SNR}_{\mathrm{CT}}^{(j)}$ with respect to the jth multiplexed channel.

Let us assume that $\Delta\lambda_{min}$ is in the order of 10^{-2} to 10^{-3} nm; then C_{CT} would be in the order of 10^3 to 10^5 pages of holographic data. The corresponding read-out SNR_{CT} would be in the range of 10 to 20 dB.

FIBRE HOLOGRAM CONSTRUCTION AND ITS APPLICATIONS

Fibre Hologram Construction

We shall now show the construction of a wavelength multiplexed reflection-type PR fibre hologram as depicted in Fig. 9(a). A specially doped Ce(300 ppm): Fe(500 ppm): LiNbO$_3$ crystal fibre is used for the hologram construction. The specially doped fibre has a high photosensitivity in the red-light region, which is suitable for using a low-power visible-light tunable diode laser [*18–23*].

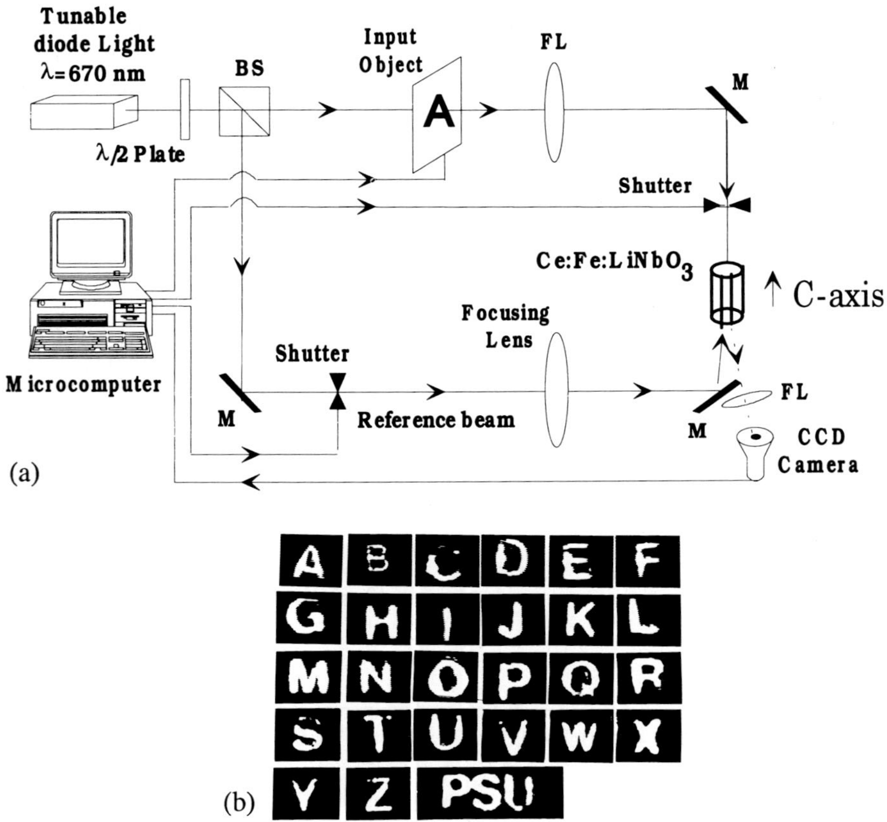

Fig. 9. Construction of a wavelength multiplexed reflection-type fibre hologram: (a) experimental set-up; (b) fibre hologram images.

A 5 mw tunable laser diode is used as the light source for the fibre hologram construction. The object beam is measured to be about $2\ \mathrm{mw\,cm^{-2}}$ with an external writing angle $2\alpha \approx 180°$. By using a wavelength tuning step of $\Delta\lambda = 0.1$ nm per step with about 5 seconds exposure time, a wavelength multiplexed PR fibre hologram is constructed. If the fibre hologram is read by tuning the laser diode, a sequence of holographic images can be sequentially reconstructed as depicted (in part) in Fig. 9(b). We note that the set of PSU letters was recorded as an input object. Aside from the TV scanning lines, the fibre hologram images suffer some fidelity distortions, which is primarily due to the non-smoothness of the boundary surface of the fibre. We believe that these fidelity distortions can be minimized if the PR fibre had been smoothly polished. In our experiment, we have actually recorded over a hundred holograms in a 7 mm length $\mathrm{Ce:Fe:LiNbO_3}$ fibre.

Applied to Fibre Sensor

Fibre-optic sensors have been used as one of several kinds of very important sensing devices [24]. However, most of the fibre sensors exploit the temporal content for sensing. We have recently developed a new type of fibre sensor called a fibre specklegram sensor (FSS) [25, 26], which exploits the spatial content instead of the temporal content for sensing. Instead of using a single-mode sensing fibre, we have used a multimode fibre for sensing, in which the complex speckle field (due to modal phasing) can be exploited for sensing [27]. One of the applications of the PR hologram is shown in Fig. 10(a) in which a reflection-type PR fibre specklegram is constructed by the interference of the counter-propagated speckle field and the reference beam. To introduce the transversal displacement on the sensing fibre, a micro-bending device is applied on the sensing fibre. The detected (normalized) correlation peak intensity (NCPI) as a function of transversal displacement is plotted in Fig. 10(b), in which the dynamic range for the transversal measurement and sensitivity are measured to about 1.3 µm, 0.05 µm, respectively. We note that the sensitivity can be further improved if a higher-mode sensing fibre is used. The major advantage of using the PR hologram for this application is direct-coupling, robustness against misalignment, and the compact size.

As Applied to a Tunable Filter

Narrow-band optical tunable filters potentially can be used for a variety of applications, such as spectroscopic solar astronomical imaging, lidar, terrestrial sensing, as well as wavelength division multiplexing channels in optics communication, and others. Although a bulk-crystal narrow-band fibre has recently been shown by Rakuljic and Leyva [28], we shall use the high wavelength selectivity of the PR fibre to develop such a tunable filter [29]. One of the advantages of using the fibre hologram is the long length, in contrast

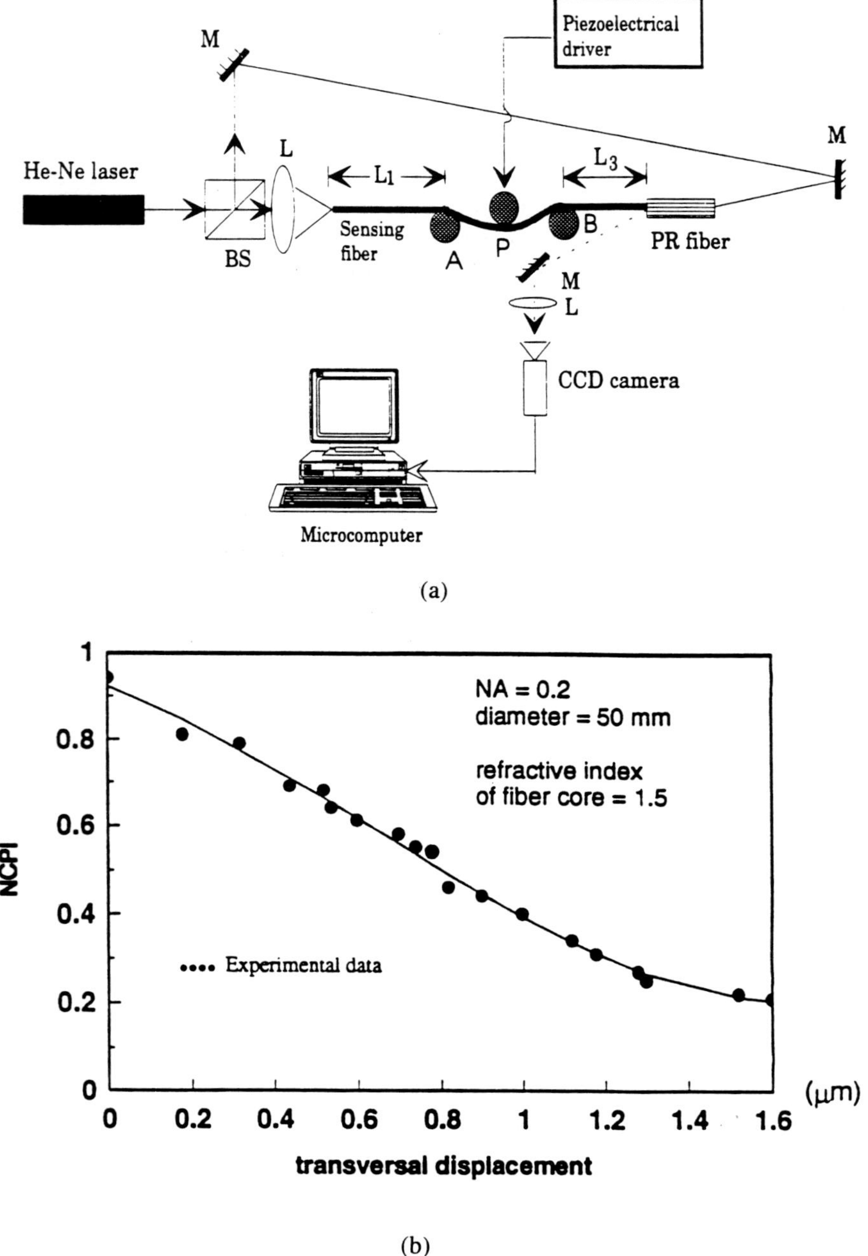

Fig. 10. Applied to fibre sensing: (a) experimental set-up; (b) normalized correlation peak value as a function of transversal displacement.

with bulk material, by which over 10 cm single-crystal fibres have been drawn in practice. This would be equivalent to a wavelength selectivity of order 10^5, which would have a spectral bandwidth smaller than 0.01 nm. In addition, the PR crystal fibre has a high-response electro-optic effect (the Pockels effect), in the order of 1 ns. Thus we see that the PR fibre can potentially be developed into a high-speed narrow-band tunable filter.

Since application of the fibre hologram would be used as an off-line operation, the single-crystal LiNbO$_3$ fibre is intentionally drawn along the *a*-axis, so that the (large electro-optic coefficient) *c*-axis is reserved for use of the applied field. In order to ensure a uniform applied electric field across the fibre, the drawn fibre is polished into a rectangular form, as shown in Fig. 11. By coating the external electrodes on the upper and the lower surface of the rectangular (fibre) waveguide, a uniform electric field can be applied in the *c*-axis direction. Thus we see that the refractive index of the fibre can be tuned by the induced electro-optic effect (i.e., the Pockels effect).

To improve the signal-to-sidelobe ratio, a Hamming (window) Bragg-grating (see Fig. 12) is used for the filter synthesis as given by

$$\Delta n(z) = \left[0.54 - 0.46 \cos\left(\frac{2\pi z}{L}\right)\right] \cos\left(\frac{4\pi}{\lambda} z\right), \tag{14}$$

which can be written as

$$
\begin{aligned}
n(z) &= n + n_1\left[0.54 - 0.46 \cos\left(\frac{2\pi z}{L}\right)\right]\cos\left(\frac{4\pi n}{\lambda} z\right) \\
&= n + n_1\left\{0.54 \cos\left(\frac{4\pi n}{\lambda} z\right) - 0.23 \cos\left[\left(\frac{4\pi n}{\lambda} + \frac{2\pi}{L}\right)z\right]\right. \\
&\qquad\qquad \left. - 0.23 \cos\left[\left(\frac{4\pi n}{\lambda} - \frac{2\pi}{L}\right)z\right]\right\}
\end{aligned} \tag{15}
$$

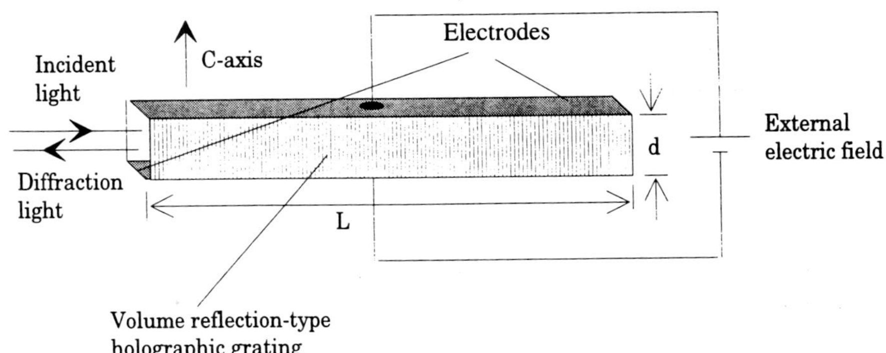

Fig. 11. The structure of a PR fibre based tunable filter.

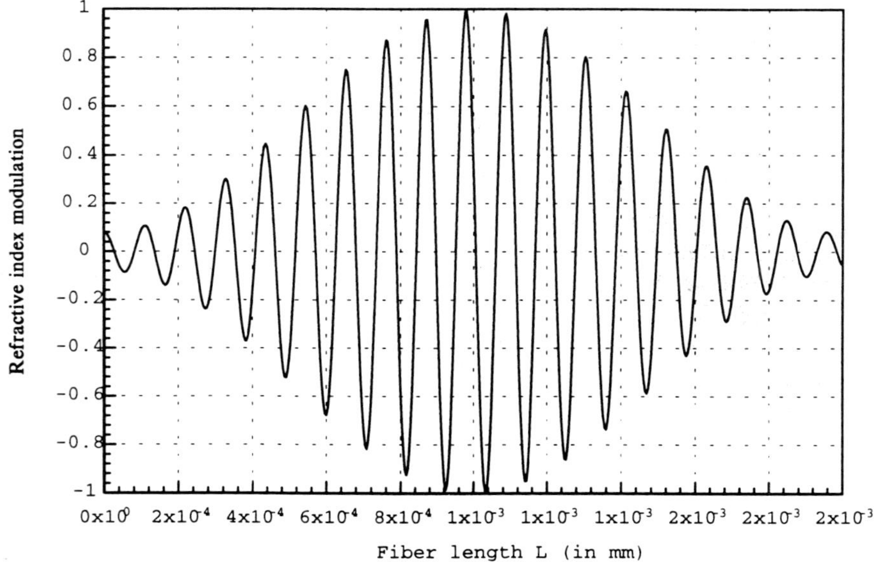

Fig. 12. Refractive index (Hamming) modulation profile.

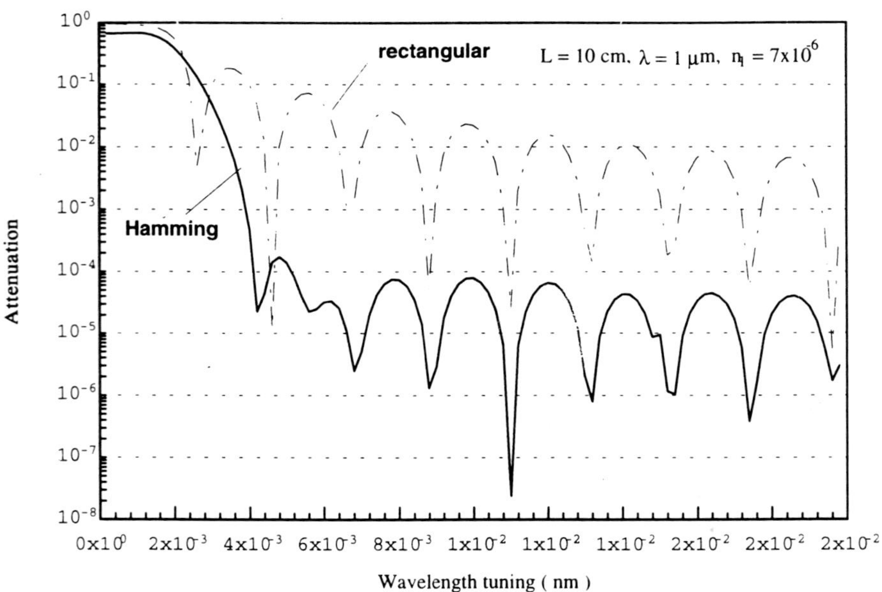

Fig. 13. Wavelength response.

where L and n are the length and the refractive index of the fibre, z is the axial coordinate, λ is the central wavelength and n_1 is the modulation depth of the refractive index. From the preceding equation, we see that the (Hamming) refractive index distribution can be obtained by superimposing three sinusoidal gratings of the same spatial frequency but different phase shifts. The wavelength responses for the rectangular and for the Hamming (Bragg) grading are plotted in Fig. 13, in which we see that the signal-to-sidelobe ratio (SNR) for the Hamming grating performs better. It has about 30 dB gain as compared with the rectangular grating. There is, however, a small price we pay by broadening the bandwidth of the filter. Nevertheless, the bandwidth can be made narrower by simply extending the length of the PR fibre. Thus, we see that a high-speed high-SNR narrow-band tunable filter, in principle, can be developed.

CONCLUSION

Some recent work on PR fibre holograms is discussed, in which the fabrication, angular and wavelength selectivities, cross-talk noise, and a couple of applications are given. In view of our findings, the reflection-type wavelength-multiplexed fibre holographic process would be the best choice for data storage. There are many promising applications of PR fibre holograms; for example, applied to optical interconnects, electro-optic switching, reconfigurable networks, and many others. However, the PR fibre devices are at the threshold of practical reality; there is much that remains to be done before they could become widespread applications.

REFERENCES

1. S. Wu, Q. Song, A. Mayers, D. Gregory, and F.T.S. Yu, *Appl. Opt.* **29**, 118 (1990).
2. F.T.S. Yu and S. Jutamulia, *Optical Signal Processing, Computing and Neural Networks*, Wiley-Interscience (1992) Chp. 7.
3. P.A. Yeh, A.E.T. Chiou, and J. Hong, *Appl. Opt.* **27**, 2093 (1988).
4. L. Hesselink and S. Redfield, *Opt. Lett.* **13**, 877 (1988).
5. F. Ito and K. Kitayama, *Opt. Lett.* **17**, 1152 (1992).
6. Y. Qiao and D. Psaltis, *Opt. Lett.* **17**, 1376 (1992).
7. L. Hesselink, *Int. J. Optoelectron.* **5**, 103 (1990).
8. K. Kitayama and F. Ito, *Multidimensional Systems and Signal Processing*, **2**, 401 (1991).
9. J.K. Yamamoto and A.S. Bhalla, *Mater. Res. Bldg.* **24**, 761 (1989).
10. H. Yoshinaga, K. Kitayama, and H. Oguri, *Appl. Phys. Lett.* **56**, 1728 (1990).
11. Y. Sugiyama, I. Yokohama, K. Kubogera, S. Yagi, *IEEE Photo. Tech. Lett.* **3**, 744 (1991).

12. S. Wu, A. Mayers, S. Rajan, and F.T.S. Yu, *Appl. Opt.* **29**, 1059 (1990).
13. H. Kogelnik, *Bell Sys. Tech. J.* **48**, 2909 (1969).
14. R.S. Feigelson, *Crystal Growth of Electronic Materials*, E. Kalkis, ed., p. 127. North-Holland, Amsterdam (1985).
15. M. Saifi, B. Dubois, E.M. Vogel and F.A. Thiel, *J. Mater. Res.* **1**, 452 (1986).
16. W.A. Gambling, D.N. Payne and H. Matsumura, *Appl. Opt.* **14**, 1538 (1975).
17. P. Gunter and J.-P. Huignard, *Photorefractive Materials and Their Applications I*, Springer-Verlag, Berlin (1988), p. 53.
18. F.T.S. Yu, S. Yin and A. S. Bhalla, *IEEE Photon. Tech. Lett.* **5**, 58 (1993).
19. G.A. Rakuljic, V. Leyva, and A. Yariv, *Opt. Lett.* **17**, 1471 (1992).
20. K. Curtis, C. Gu, and D. Psaltis, *Opt. Lett.* **18**, 1001 (1993).
21. F.T.S. Yu, F. Zhao, H. Zhou, and S. Yin, *Opt. Lett.* **18** 1849 (1993).
22. J.H. Hong, P. Yeh, D. Psaltis, and D. Brady, *Opt. Lett.* **15**, 344 (1990).
23. S. Yin and F.T.S. Yu, *IEEE Photon. Techn. Lett.* **5**, 581 (1993).
24. Udd, *Fiber Optic Sensors*, John Wiley & Sons, Inc. New York (1991).
25. S. Wu, S. Yin, and F.T.S. Yu, *Appl. Opt.* **30**, 4468 (1991).
26. S.Wu, S. Yin, S. Rajan, and F.T.S. Yu, *Appl. Opt.* **31**, 5975 (1992).
27. F.T.S. Yu, S. Yin, J. Zhang and R. Guo, *Appl. Opt.* **33**, 5202 (1994).
28. G.A. Rakuljic and V. Leyva, *Opt. Lett.* **18**, 459 (1993).
29. F.T.S. Yu, S. Yin, and B.D. Guenther, 'High SNR filter using volume holographic grating and the Hamming window', 1994 OSA Annual Meeting, Dallas, TX, 71 (1994).

26 Optical morphogenesis: dynamics of patterns in passive optical systems

F. T. Arecchi,[*] S. Boccaletti[*], E. Pampaloni, P. L. Ramazza and S. Residori
Istituto Nazionale di Ottica, 50125 Florence
[*]*also Physics Department, University of Florence*

INTRODUCTION

Before introducing the subject, we must define the terms used in the title. If we look at Webster's or the Oxford English dictionaries *morphogenesis* enters as a 'biological' term. In fact, in 1952 it was applied to chemical instabilities by Turing [1], who referred to it as the spontaneous birth of concentration shapes in a two (or more) component reaction–diffusion system. By spontaneous we mean not forced by an external agent but already built-in in the chemical dynamics.

By *shape*, the Oxford dictionary means *that quality of a material object which depends on constant relations of position, etc.*, thus implying long-range space (or space–time) correlations.

As for *patterns*, Webster provides the following definition (incidentally the word stems from the Latin *patronus*, which is related to *pater*, that is, *father!*): *a fully realized form, original, or model proposed for imitation.*

Thus the concept of shape, or form, can be associated with a specific indicator (correlations), whereas the concept of pattern implies the imitation or modelling (that is the fatherhood). Hence *pattern* is appropriate in a process whereby something is copied as in biology or in the textile industry.

However, we loosely speak of patterns as shapes. A recent review [2] has the title: *Pattern formation outside of equilibrium*, and here *pattern* stands for *shape*. From now on we will adhere to this use. Furthermore in optics spontaneous pattern formation has been recently called *optical morphogenesis* [3]. So much for the terminology introduced in the title.

In this chapter we plan to report on the dynamical interaction of a confined electromagnetic field in the optical domain, with a distributed medium whose refractive index depends upon the local field value.

We limit the discussion to passive media, that is, to media that have not been previously excited by an auxiliary pump source, such as lasers or photorefractive oscillators. In this latter case, which we call active, the energy is transferred from an external source to a self-generating patterned optical beam. On the contrary, in the passive case, it is the incident beam itself that acquires a spatial structure in the course of its interaction with matter.

Pattern forming instabilities are a rather general phenomenon in spatially extended systems driven out of thermodynamical equilibrium [2–4]. In general, patterns of different symmetry and/or characteristic scale can occur at different threshold values of some control parameter. Close to the lowest threshold, only one kind of structure can be observed, namely the one associated with the most unstable mode.

If the control parameter is increased to a value that allows for more than one pattern above threshold, some general scenarios have been identified. Among these we recall mode alternation in time (either periodic or chaotic) [5–7], domination of one mode over the others [8–9] and mode–mode coexistence [6, 10, 11].

Non-linear optical setups formed by a Kerr-like medium with feedback have become a popular subject of study for two-dimensional (2D) transverse pattern formation [12–19]. Here we present experiments on pattern formation and competition in a system formed by a liquid crystal light valve (LCLV) with optical feedback. As a first approximation, the LCLV can be considered as a defocusing Kerr-like medium [20].

The pattern-forming phenomena discussed here are based on the conversion of the phase fluctuations induced on an optical beam by the LCLV into intensity fluctuations. This occurs via diffraction of the beam [13]. When the diffracted output beam is sent on to the rear face of the LCLV the feedback loop is closed and this can give rise to dynamical instabilities.

It is worth emphasizing three main features of the patterns spontaneously formed from a homogeneous state, namely length scales, shape and dynamical behaviour.

The nature of the observed phenomena is strictly related on the one hand to the Kerr nature of the non-linearity and to the presence of diffusion in the medium, and on the other to the geometrical constraints that are imposed in the feedback loop. The experiments show that the length scale of the patterns is determined by the joint contribution of diffractive and diffusive phenomena occurring in the system.

As for the shapes of the structures that form, these can be determined by different factors. In the case in which the feedback loop is simply a propagation path, it is the quadratic nature of the Kerr non-linearity that dominates, leading to the formation of hexagons or rolls [13–16]. When a limiting aperture of size comparable to the intrinsic length of the pattern is added within the feedback loop, the role of the transverse boundary conditions overwhelms the other effects, resulting in the formation of polygon-like structures [17, 19]. In this case, for increasing input intensity, polygon rotation or alternation of different polygons in time is observed.

For large apertures, if a non-local interaction of the signal with itself is introduced in the feedback loop via a rotation in the transverse plane, new classes of symmetries arise in the patterns, corresponding to crystalline or quasi-crystalline structures [18]. For this case linear stability analysis of the problem gives rotation-dependent selection rules for the wavelength of the excited patterns. In the following sections we present phenomena related to the presence of such a non-local feedback.

Far above threshold, two-dimensional patterns having different spatial scales are shown to coexist in time, but segregated in spatially separated domains. The one-dimensional analogue of this phenomenon was demonstrated in hydrodynamical convection [21], and has been the object of a recent theoretical treatment [22].

The presence of topological defects in our system is closely connected both with pattern alternation and with pattern coexistence in different domains. A simple model where the role of defects is reduced to a noise term explains the main features of the experiments reported here.

EXPERIMENTAL RESULTS

The system used in the experiments consists of a liquid crystal light valve (LCLV) inserted in an optical feedback loop [12] as shown in Fig. 1(a). A schematic drawing of the LCLV is shown in Fig. 1(b). This device consists essentially of a nematic liquid crystal cell, a dielectric mirror and a photoconductive layer. A supply voltage V_0 is applied by means of transparent electrodes to the series of these three elements. The liquid crystal birefringence is a monotonic decreasing function of the voltage drop V_{LC} across it [23]. Since the liquid crystal and photoconductor are placed in series, the fraction V_{LC} of the total voltage V_0 will increase when increasing the writing illumination incident on the photoconductive layer. It follows that a writing beam impinging on the rear side of the LCLV induces a variation in the extraordinary index of refraction of the liquid crystals. This variation is of negative sign and in a first approximation is proportional to the writing intensity, so that the LCLV can be considered as a defocusing Kerr medium.

In our experimental setup a beam from a He-Ne laser operating at 632 nm is spatially filtered and expanded by means of the telescopic system formed by

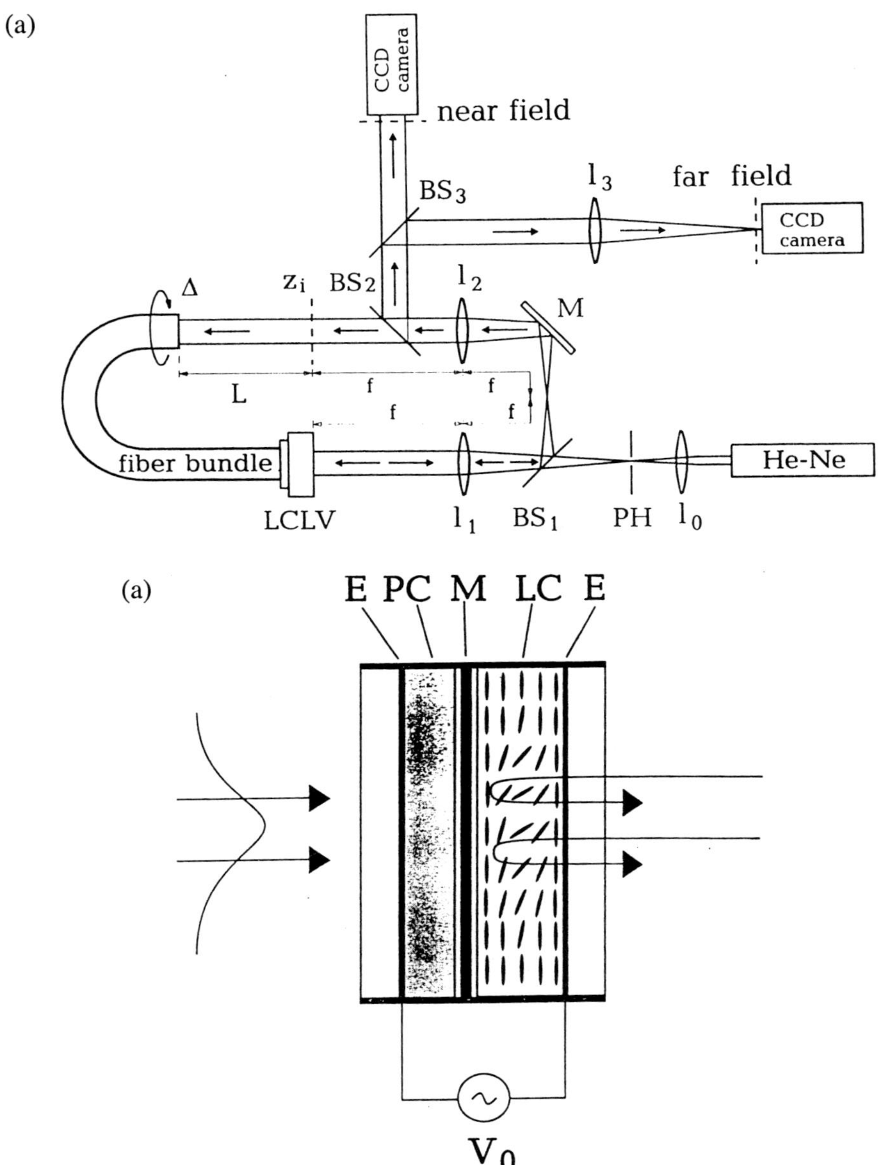

Fig 1. (a) Experimental setup; l_i = lens, M = mirror, BS$_i$ = beamsplitters, PH = pinhole, L = free propagation length, Δ = feedback rotation angle. (b) Liquid crystal light valve (LCLV): LC = liquid crystal layer, M = mirror, PC = photoconductor layer, E = transparent electrodes, V_0 = applied AC voltage. The liquid crystal molecules re-orient according to the write intensity incident on the PC.

lenses l_0 and l_1 and the pinhole PH. The resulting plane wave, which is polarized along the extraordinary axis of the liquid crystal, is sent on the front face of the LCLV. The wave reflected by the LCLV acquires a spatial phase modulation, which is determined by the distribution of the extraordinary index of refraction n in the liquid crystal layer. The front plane of the LCLV is imaged on to the plane z_i by means of lenses l_1 and l_2. Along the image-forming path, a fraction of the beam is extracted at the beam-splitter BS_2 and used for detection of the near and far field signal. From plane z_i to the input plane z_2 of the fibre bundle the wave undergoes a free propagation path of length L. At plane z_2 the diffracted wave enters an optical fibre bundle that just relays the intensity distribution from its input to its output plane. This last one is in contact with the rear face of the LCLV.

The origin of pattern formation in a system of this kind lies on the following mechanism [13]. The perturbations of the index of refraction n in the liquid crystal layer induce a phase modulation on the reflected beam. This phase modulation is converted into amplitude modulation by diffractive propagation along L. The light intensity reaching the rear side of the LCLV provides, via the Kerr effect, a feedback on the perturbations of n.

Here we review two sets of experimental results, namely, formation of orderly stationary patterns, which we call two-dimensional crystals and quasi-crystals [18] and the competition and coexistence of patterns with different symmetries [24] (see below).

Optical Crystals and Quasi-crystals

For an input intensity I_0 close to the pattern formation threshold, a circle of critical transverse wavevectors $|\vec{q}|$ ($\vec{q}$ being the two-dimensional Fourier transform of the transverse (x, y) dependence) becomes simultaneously unstable (see below). In this regime, the pattern is determined by linear superpositions of wavevectors belonging to this circle which has a radius $q_I = \sqrt{\pi k_0/L}$ for a focusing medium and $q_{II} = \sqrt{3\pi k_0/L}$ for a defocusing medium [13], $k_0 = 2\pi/\lambda$ being the optical wavenumber.

The free end of the fibre bundle is mounted on a rotation stage which allows for continuous angular positioning over a full 360° range with high resolution (readout to 0.2°). This way, the feedback image arriving at the photoconductive layer of the LCLV can be rotated of any angle Δ. It was shown theoretically [13] and experimentally [14–16] that for $\Delta = 0$ only hexagons are stable. On the other hand, for $\Delta = \pi$ it was shown that only rolls are stable and that a competition between hexagons and rolls can be achieved by inserting an attenuation filter in front of one half of the LCLV [16].

Initially, we fix the input intensity $I_0 = 4.5$ mW cm^{-2}, the r.m.s. amplitude (40 V) and frequency (9 kHz) of the voltage applied and the free propagation length $L = 75$ cm, and change the rotation angle Δ in the feedback loop. When the rotation angle is exactly commensurate to 2π, that is, when $\Delta = 2\pi/N$ with

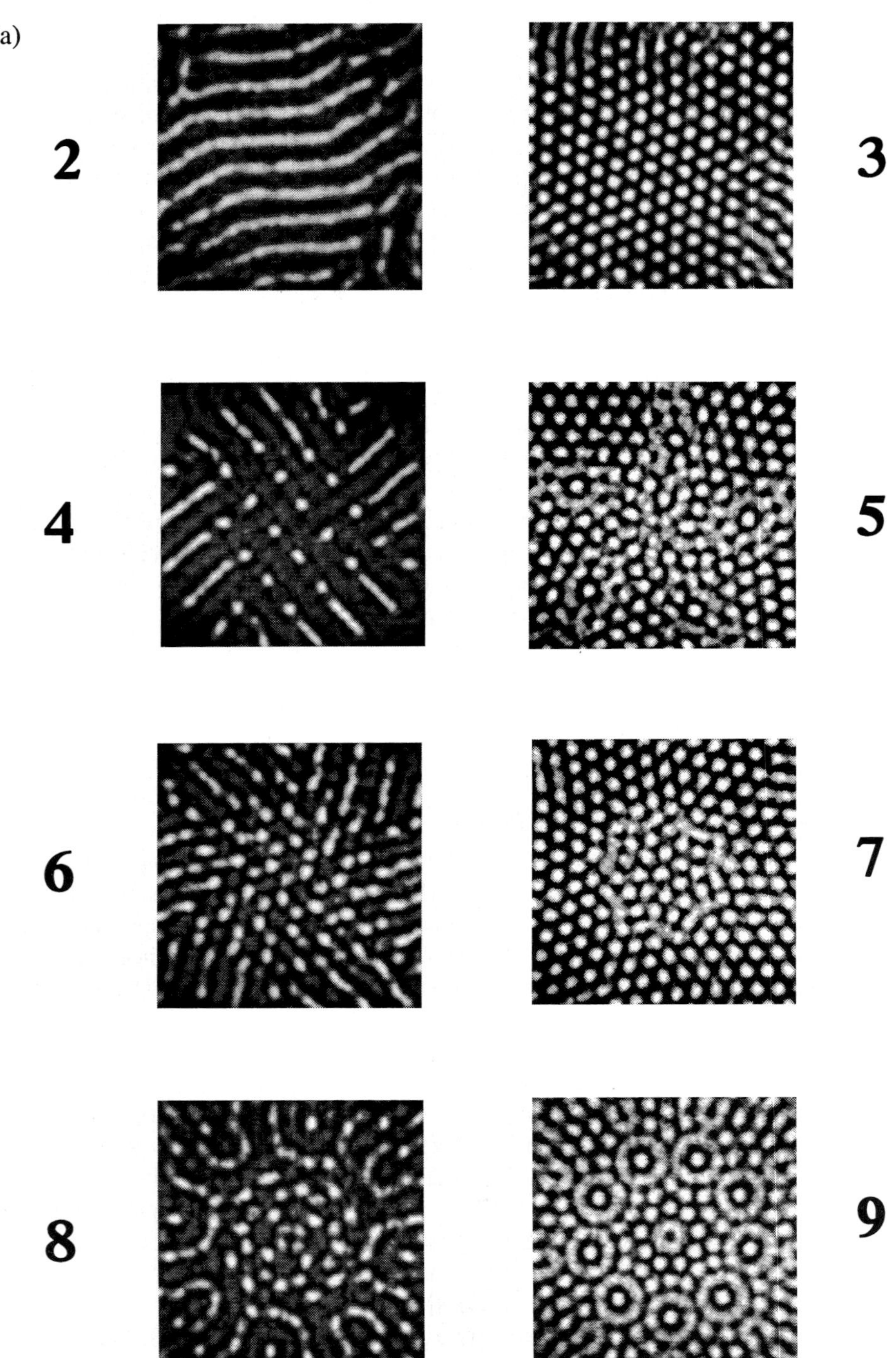

Fig. 2. *Continued.*

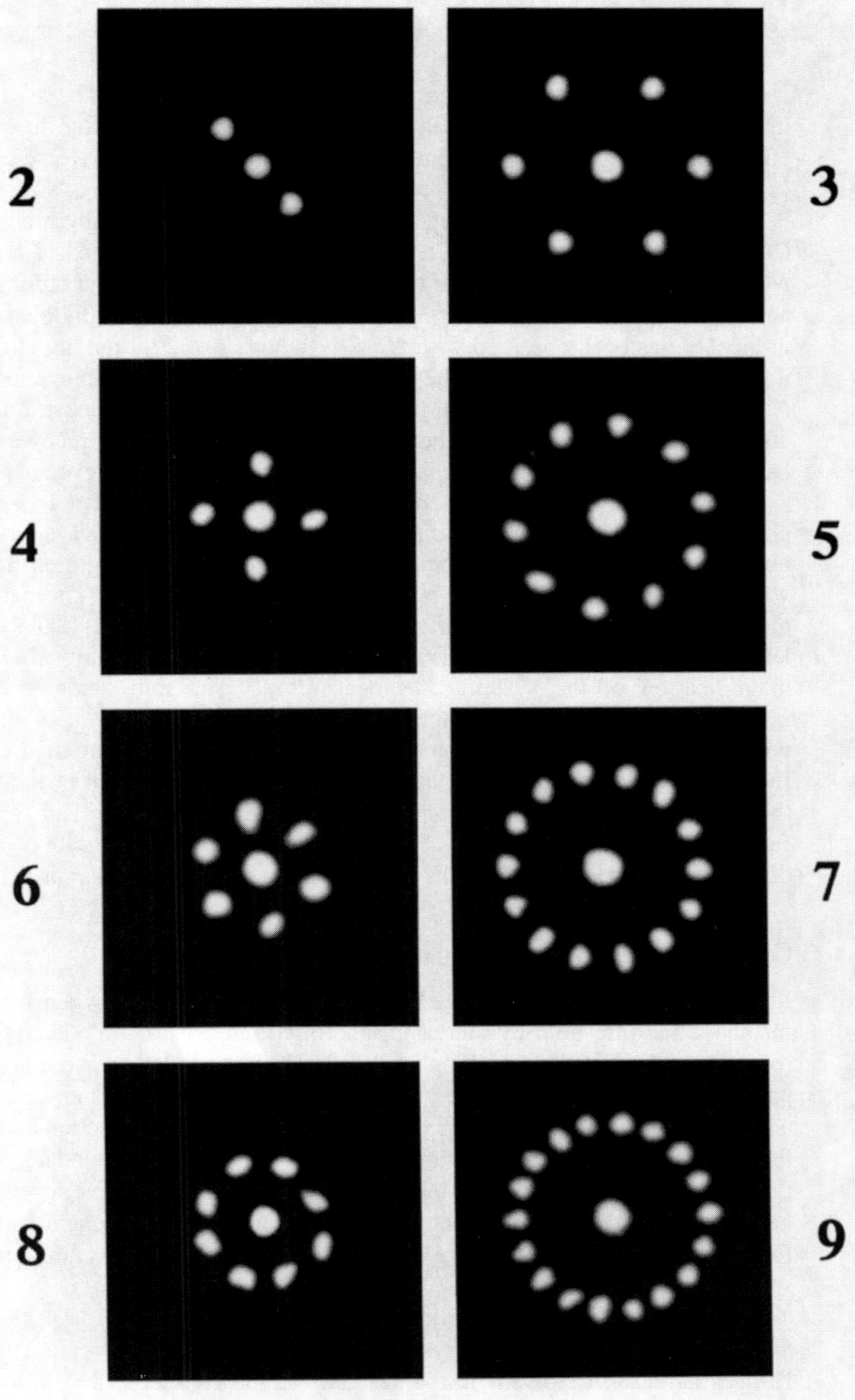

Fig. 2. Experimental patterns observed for feedback rotation angles $\Delta = 2\pi/N$. The N values are ind
close to each frame. (a) Near-field patterns; (b) far-field patterns; all frames correspond to the
magnification, thus the rings with N even and those with N odd are respectively proportional to q_1 a
(from Ref. 18).

$N = 2, 3, 4, \ldots$, a pattern with an N-fold symmetry develops. The symmetry is induced by the rotation angle in the feedback and it is not due to boundary effects that arise when the system is strongly limited in its transverse extension [*17, 19*].

In Fig. 2(a) we report different near-field patterns, obtained by imaging on a CCD camera the front of the LCLV, for various rotation angles $\Delta = 2\pi/N$ where N is changed from 2 to 9. Since the critical wavelength is of the order of 1 mm and the LCLV has a diameter of 3 cm, a central region with negligible boundary influences has been observed with N up to the order of 20. Looking at the edge, it can be seen that away from the centre the system tends to stabilize in rolls for N even and hexagons for N odd. In fact, the feedback rotation constraint is more efficient close to the centre, whereas at the edge the patterns recover the two basic symmetries corresponding to $\Delta = 0$ (N odd) and $\Delta = \pi$ (N even) [*16*].

The far-field patterns (Fig. 2b), collected on the focus of a lens, are the power spectra of the corresponding near-field images, and provide directly the number of modes involved in the pattern formation and the length of the critical wavevector (radius of the ring where the peaks are located). The patterns on the first and second column (respectively N even and odd) have different lengths of the critical wavevector. Taking into account the optical magnification of the system, we measure $q_{\mathrm{II}} = 12.5$ mm^{-1} for N odd and $q_{\mathrm{I}} = 7.3$ mm^{-1} for N even, so that the ratio $q_{\mathrm{II}}/q_{\mathrm{I}}$ is close to $\sqrt{3}$, in agreement with the critical wavenumber ratio of a defocusing to a focusing medium [*25*]. Indeed, even though the LCLV acts as a defocusing medium an instability with the q_{I} of a focusing medium can also be excited, as shown in Ref. [*16*]. The fact that for any pattern, each $\vec{q}$ component contributes as a pair of opposite directions, gives rise to $2N$ peaks for N odd, but only to N peaks for N even.

Competition and Coexistence of Optical Patterns

In this paragraph we investigate the behaviour of the system when it is driven far above the threshold by increasing the input intensity T_0. We keep the fixed rotation angle in the fixed loop at $\Delta = 2\pi/7$. Let us define a reduced pump parameter $\varepsilon = (I_0 - I_{\mathrm{th}})/I_{\mathrm{th}}$. For ε very small, a single q band is associated with a far-field made of $2N$ spots (fixed orientation of the wavevectors) and hence the near field shows mainly a single domain (besides some boundary perturbations), as discussed above. On the contrary, here (rather larger ε) even a single band is a collection of wavevectors with different orientations, and hence even for a single wavelength we have a many-domain pattern, with grain boundaries separating different orientations.

A gradual increase of ε starting from $\varepsilon = 0$ leads initially to an increase of the amplitude of the quasi-crystalline patterns, without a scale change. A further increase in ε results in the destabilization of a second band at $q_1 = 2\pi/\sqrt{2\lambda L}$ (Fig. 3b). In this situation the near-field signal does not appear as a uniform superposition of patterns at the two different wavelengths, but rather as

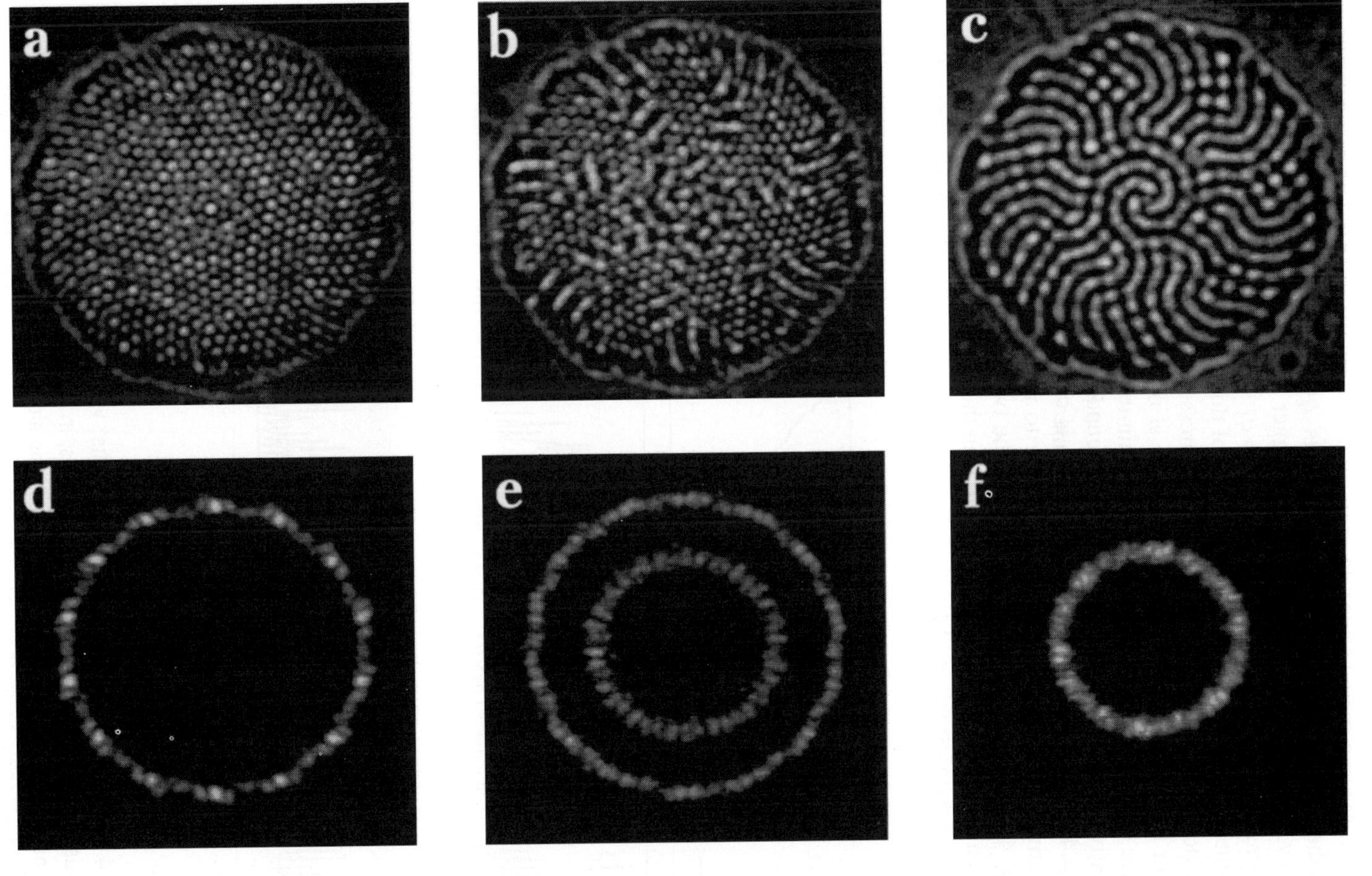

Fig. 3. Near-field (upper) and far-field (lower) patterns observed for $\varepsilon = 0.5$ (a, d), $\varepsilon = 2$ (b, e) and $\varepsilon = 4.2$ (c, f). The left (right) column corresponds to excitation of only the q_2 (q_1) band; in the middle column the two bands coexist. The single wavenumber cases (left and right) show the coexistence of many sets of $2N = 14$ vectors (from Ref. 24).

one of the two spatial scales. Domains with the smaller wavenumber q_1 emerge at the grain boundaries of the previous q_2 multiorientation patterns, thus showing that defects are sources that trigger the onset of the q_1 patterns. The average size of the domains with $q = q_1$ increases for increasing ε and eventually the whole wavefront is made of domains at this wavenumber, while the domains at $q = q_2$ are completely suppressed (Fig. 3c).

In Fig. 4 we show the local intensities at one point of the near field for the three cases described above. When wavenumber q_2 is excited ($\varepsilon = 0.2$) we have relatively slow drifts of the domain boundaries. When only q_1 is excited ($\varepsilon = 4.5$) the corresponding eigenvalue λ is complex [18] and thus we obtain rotating patterns. The rotation gives rise to a high frequency as observed in Fig. 4(c). Finally in Fig. 4(b) ($\varepsilon = 1.9$) the two wavenumbers coexist and at a given pixel we have an alternation between the two regimes.

Further information about the observed phenomena can be gained from the

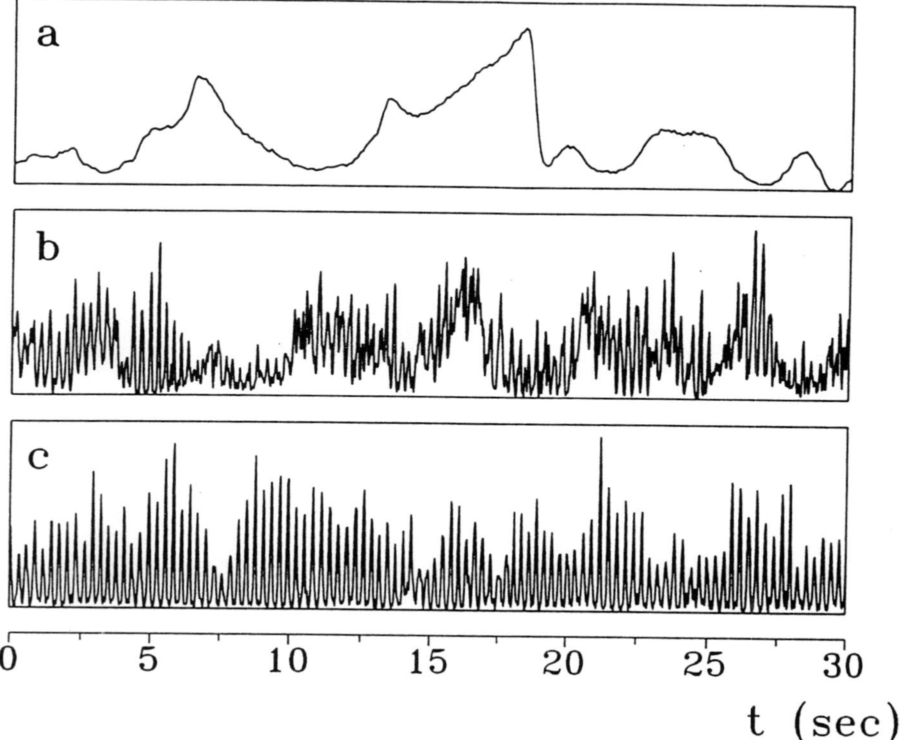

Fig. 4. Near-field local intensity (arbitrary units) vs. time. In (a) ($\varepsilon = 0.2$, q_2 band) the fluctuations are due only to domain dynamics; in (c) ($\varepsilon = 4.5$. q_1 band) there is also a fast oscillation due to the imaginary part of the eigenvalue; in (b) ($\varepsilon = 1.9$, both q_1 and q_2 bands) there is a superposition of the other two cases (from Ref. 24).

spatial power spectra of the signal, corresponding to the far field. Typical examples of these spectra are shown in Figs 3(d)–(f). In order to obtain global information about the temporal behaviour of the signal we define the quantity $\eta(t) = S_1(t)/(S_1(t) + S_2(t))$ as the fraction of the total power that instantaneously belongs to the first band. Here $S_j(t)$ ($j = 1, 2$) is the instantaneous power radially integrated in the Fourier space over a circular corona of radius q_j. A plot of $\eta(t)$ for three different values of ε is shown in Fig. 5. It is seen here that, when the system is dominated by one of the two competing bands, the time fluctuations of $\eta(t)$ are very small. On the contrary, the range of ε for which the two bands show coexistence corresponds to regions of high fluctuation for $\eta(t)$, meaning that neither the coexistence of the two bands nor the domination of one band over the other are stable phenomena.

A quantitative measurement of the transition from the band q_2 to the band q_1 dominated regime is given by the behaviour of the time average $\bar{\eta} \equiv \langle \eta(t) \rangle_t$ and of the standard deviation $\sigma \equiv [\langle \bar{\eta}^2 - \eta(t)^2 \rangle_t]^{1/2}$ of the quantity $\eta(t)$, versus the pump parameter ε. Plots of the results of these measurements are shown in Fig. 6 (left). These plots give a quantitative confirmation of the enhancement of fluctuations in the signal that accompanies the regimes of competition–coexistence between the two bands. In the next section we will explain the criteria that give rise to the theoretical plots reported on the right of Fig. 6.

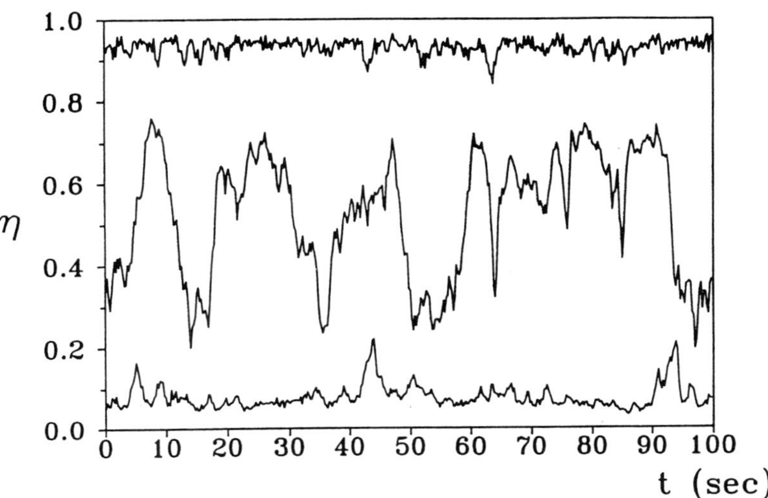

Fig. 5. Temporal evolution of the normalized spectral power η on the first ring: $\varepsilon = 1$, q_2 band (lower curve), $\varepsilon = 4.1$, q_1 band (upper curve), and $\varepsilon = 2.1$, both q_1 and q_2 bands (middle curve) (from Ref. 24).

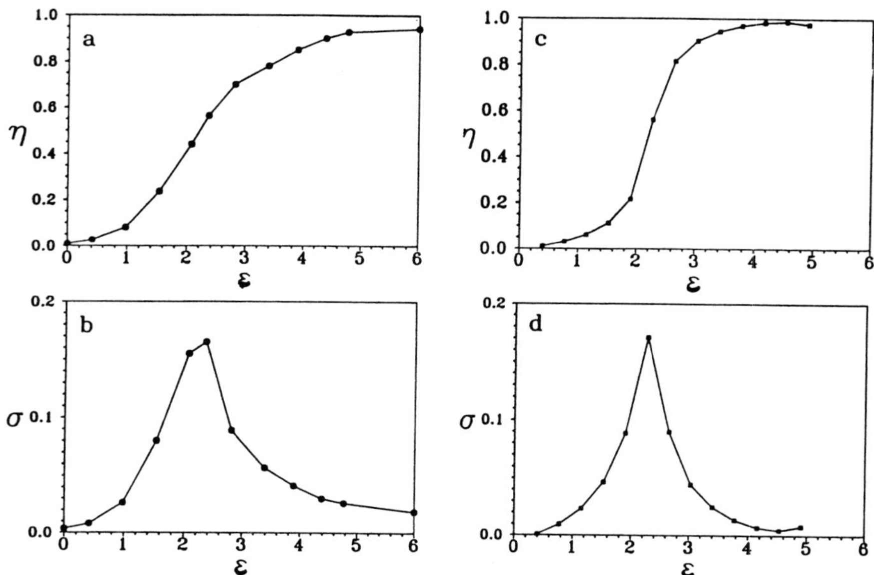

Fig. 6. Experimental (left column) and theoretical (right column) plots of $\eta(\varepsilon)$ (a, c) and $\sigma(\varepsilon)$ (b, d). Experimental error bars are within the size of the black circles. Theoretical points (black squares) are obtained from numerical integration of Eq. (3) with $\mu_1 = 1 - (I_0 - 5)^2$, $\mu_2 = 1 - (I_0 - 5.5)^2$, $\beta_1 = \beta_2 = 1.5$, $\gamma_1 = \gamma_2 = 2.4$ and noise addition. The x axis has been normalized to the reduced pump parameter ε. In all cases, lines are just a guide connecting points (from Ref. 24).

THE THEORETICAL MODEL

The coupled field non-linear medium dynamics is ruled by two partial differential equations (1) and (2). Equation (1) describes the dissipative–diffusive reorganization of the refractive index $n(x, y)$ in a thin cell (thickness d much smaller than the diffusive length $l_D = \sqrt{D\tau}$, where D and τ are defined by the same equation (1)) as

$$\left(\partial_t + \frac{1}{\tau} - D\nabla^2 \right) n = -\chi |E_w|^2. \tag{1}$$

Here ∇^2 is the (x, y) Laplacian, $\chi > 0$ (defocusing medium) is a constant and E_w (w for writing) is the field impinging on to the photoconductor back of the LCLV.

The electromagnetic equation in the eikonal approximation is given by

$$\left(\partial_z + \frac{i}{2k} \nabla^2 \right) E = inE \, \mathrm{rect}(z/d), \tag{2}$$

where $\mathrm{rect}(\xi) \equiv 1$ (0) for $\xi < 1$ (> 1). In Eq. (2) the non-linear r.h.s. is limited by the rect function to a length d around $z = 0$, z being the longitudinal coordinate along the axis of the optical system, and the time derivative is neglected since the transit time along the feedback loop (10^{-7} sec) is much faster than the characteristic time of the refractive medium ($\tau \cong 10^{-2}$ sec). The solution of Eq. (2) for an impinging field E_0 gives a field E_1 exiting the Kerr cell given by

$$E_1 = E_0 \mathrm{e}^{ind},\tag{3}$$

since diffraction plays no role over the distance d. This field freely propagates providing an input E_2 to the fibre bundle, which is symbolically written in terms of an operator corresponding to a formal integration of Eq. (2) as

$$E_2 = \exp\left(-\mathrm{i}\,\frac{L}{2k}\,\nabla^2\right)E_1.\tag{4}$$

The diffraction operator becomes a simple phase factor in the $\vec{q}$ space, where $\vec{q}$ is the Fourier transform of the (x, y) dependence.

The azimuthal rotation within the fibre bundle corresponds to applying a rotation operator $R(\Delta)$ to the field E_2. Thus we have

$$|E_\mathrm{w}|^2 = |R(\Delta)E_2|^2.\tag{5}$$

Combining Eqs (3)–(5) into Eq. (1) we arrive at a single closed, non-local and highly non-linear, equation for $n(x, y)$ with a source term

$$|E_\mathrm{w}(r, \theta)|^2 = I_0\left|\exp\left(-\mathrm{i}\,\frac{L}{2k_0}\,\nabla^2\right)\exp[-in(r, \theta + \Delta)]\right|^2,\tag{6}$$

where $I_0 \equiv |E_0|^2$, and (r, θ) are polar coordinates in the transverse plane.

Suppose that the homogeneous solution is perturbed by a phase modulation of wavevector $\vec{q}$ and that the feedback rotation of an angle $\Delta = 2\pi/N$ excites N vectors with the same length in N directions spaced by the angle Δ. Therefore, the phase perturbation can be expanded as $n = \sum_{j=1}^{N} a_j \cos \vec{q}_j \cdot \vec{r}$. The $\vec{q}$ vectors are related by the rotation operator in such a way that

$$R\vec{q}_j = \vec{q}_j + 1.\tag{7}$$

With this expansion, the propagation over a distance L of a field gives rise to the following feedback intensity

$$|E_2(L, \vec{r})|^2 = I_0\left(1 + \sin\frac{q^2L}{2k_0}\sum_{j=1}^{N} a_j \cos \vec{q}_j \cdot \vec{r}\right).\tag{8}$$

Using Eq. (5) and substituting this expression into Eq. (1), we evaluate the eigenvalues λ_j of the perturbation vector $(a_1, a_2, \ldots, a_N)$ as

$$\lambda_j = -(1 + l_D^2 q^2) + \chi I_0 \sin q^2 L / 2k_0 [e^{iN\pi}]_j^{1/N} \tag{9}$$

where $[\ ]_j^{1/N}$ denotes the jth Nth root of the unity. By selecting the eigenvalues with maximal real part, we derive two marginal stability curves:

$$\text{branches I:} \quad \chi I_{th} = \begin{cases} \dfrac{1 + l_D^2 q^2}{\sin \dfrac{q^2 L}{2k_0}} & \text{for } N \text{ even,} \\[3em] \dfrac{1 + l_D^2 q^2}{\cos \dfrac{\Delta}{2} \sin \dfrac{q^2 L}{2k_0}} & \text{for } N \text{ odd,} \end{cases} \tag{10}$$

$$\text{branches II:} \quad \chi I_{th} = -\dfrac{(1 + l_D^2 q^2)}{\sin \dfrac{q^2 L}{2k_0}} \quad \text{for } N \text{ even or odd.} \tag{11}$$

The positive branches of these two curves give the value I_{th} of the input intensity I_0 necessary to excite an instability with a spatial frequency q. In the diffractive limit, $l_D^2 \ll \lambda L$, all branches have their minima at $q^2 L / 2k_0 = (2n + 1)\pi/2$ with $n = 0, 2, \ldots$ (even integer) for branches I and $n = 1, 3 \ldots$ (odd integer) for branches II, so that, with respect to the first critical wavenumber $q_1 = \sqrt{\pi k_0 / L}$, the next is located at $q_{II} = \sqrt{3}\, q_1$ and the higher ones at $\sqrt{5}\, q_1$. $\sqrt{7}\, q_1, \ldots$. Close to threshold, only the first two branches of I and II (shown in Fig. 7a) are involved in the wavenumber selection process since they have the lowest minima. Furthermore, for N odd, branch I depends on N through Δ, hence its minimum can be above or below that of branch II. At threshold, the lowest local minimum determines the excited mode.

Comparing the magnitude of the real parts of the two unstable eigenvalues, it turns out that for N even the excited mode is always the one with q_1, independently of N. On the other hand, for N odd branch I has an enhancement factor $1/\cos(\Delta/2)$ (see Eq. (10)) which is larger for lower N, as shown for $N = 3$ in Fig. 7(b), and hence branch II is favoured, whereas for high values of N there exists a pair of eigenvalues closest to the real axis, for which branch I can still have the lower threshold.

To deal with competition we must introduce non-linearities. We do that by accounting for the leading non-linearities and truncating the Fourier spectrum by a Galerkin procedure.

Let us consider the Fourier expansion of the local optical field on the rear face of the LCLV corresponding to the observation plane:

$$E(r, t) = \int \mathrm{d}q a_q(\bar{r}) \exp(i\boldsymbol{q} \cdot \boldsymbol{r}) \tag{12}$$

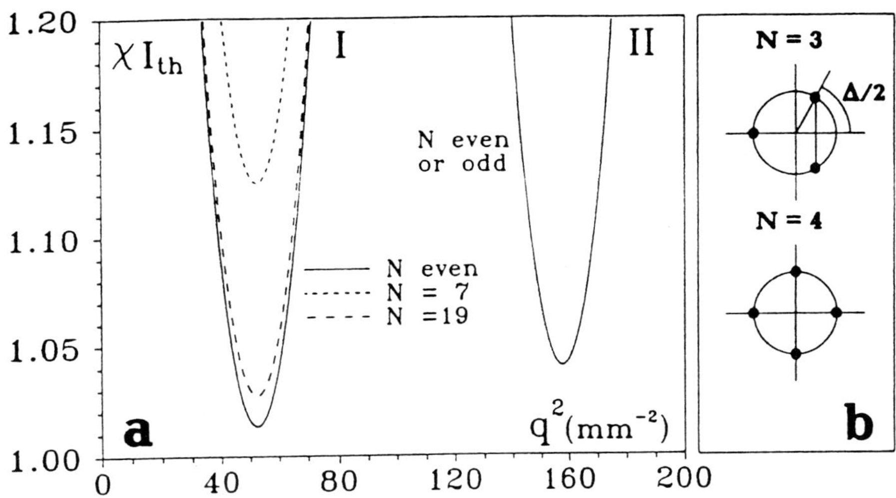

Fig. 7. (a) Marginal stability curves evaluated for $L = 75$ cm and $l_D = 15.5$ μm. The first branch has the lowest threshold for any N if N is even and also for N odd if N is large. For N odd and small the second branch has a lower threshold. (b) The N roots of unity in the complex plane, for $N = 3$ and 4. For $N = 3$ it is easy to realize the enhancement factor $1/\cos(\Delta/2)$ shown in Eq. (10) for N odd (from Ref. 18).

where

$$a_q(\vec{r}) = \sum_{n=1}^{2N} a_n \delta_{q - q_n(\vec{r})}. \tag{13}$$

The discretization in the set of the Fourier amplitudes is imposed in the experiment by the rotation introduced in the feedback loop and by the recursive relation (7). We are in the presence of two active rings in Fourier space, corresponding to the two unstable values of q.

The mode coupling within one ring (at constant q modulus) was treated in [25], and for $N \neq 3l$ (l being a positive integer) the quadratic non-linearity disappears by closure considerations, thus leaving a cubic mode coupling of the type considered in laser theory for population coupling in the absence of phase coupling [26]. This applies to our case since we have selected $N = 7$.

The data of Fig. 3 show that a situation of almost isotropic amplitude distribution on the two rings is easily reached. Even though the far field displays this isotropy, the closure relations in building the quadratic convolution term for the evolution equation of $a_q(\vec{r}, t)$ must be built with a unique set of $2N$ vectors. This rules out the possibility of having $q_i^I + q_i^{II} = q_j$ ($i \neq j$, i and $j = 1, 2$) since with $N = 7$ ($2N = 14$ points regularly spaced over each ring) and with the ratio $|q_2|/|q_1| = \sqrt{3}$, the above relations are never satisfied. Thus, also the inter-ring competitions are ruled only by cubic non-linearities.

displays this isotropy, the closure relations in building the quadratic convolution term for the evolution equation of $a_q(\vec{r}, t)$ must be built with a unique set of $2N$ vectors. This rules out the possibility of having $\boldsymbol{q}_i^{\mathrm{I}} + \boldsymbol{q}_i^{\mathrm{II}} = \boldsymbol{q}_j$ ($i \neq j$, i and $j = 1, 2$) since with $N = 7$ ($2N = 14$ points regularly spaced over each ring) and with the ratio $|q_2|/|q_1| = \sqrt{3}$, the above relations are never satisfied. Thus, also the inter-ring competitions are ruled only by cubic non-linearities.

We find it convenient to follow the evolution of the corresponding integrated spectral powers $S_i = 2\pi q_i |a_{q_i}|^2$ ($i = 1, 2$). The equations for S_1 and S_2 are

$$\dot{S}_1 = \mu_1 S_1 - \beta_1 S_1^2 - \gamma_1 S_1 S_2$$
$$\dot{S}_2 = \mu_2 S_2 - \beta_2 S_2^2 - \gamma_2 S_1 S_2. \tag{14}$$

We have thus arrived at general equations well known for competing populations [27] and already used in laser dynamics for two-mode operation [26].

Due to the saturating characteristic of the LCLV [20], the linear growth rates μ_i depend on the input intensity I_0. The function $\mu_i(I_0)$ is increasing for moderate I_0 and decreasing for high I_0, where saturation of the LCLV characteristic is effective. We choose as a functional form for $\mu_i(I_0)$ a parabola, that is $\mu_i = \alpha_i - (I_0 - \rho_i)^2$.

The system (14) admits the following four fixed points: $\mathrm{O} = (0, 0)$, $\mathrm{F1} = (\mu_1/\beta_1, 0)$, $\mathrm{F2} = (0, \mu_2/\beta_2)$ and $\mathrm{c} = (\mu_1\beta_2 - \gamma_1\mu_2)/(\beta_1\beta_2 - \gamma_1\gamma_2)$, $(\mu_2\beta_1 - \gamma_2\mu_1)/(\beta_1\beta_2 - \gamma_1\gamma_2)$.

The spatial interaction neglected in Eqs. (14) permits the birth of coherent F1 or F2 domain structures, nucleating from local defects. Indeed, when a single family locally displays a defect, this becomes a nucleation centre for the other family. Hence, the observed sharing process on the near field can be interpreted as a continuous nucleation and competition of the two coherent domains, and it can be modelled by adding $\mu_2\xi(t)$ and $\mu_1\xi(t)$ to the first and second of Eqs. (14) respectively, where $\xi(t)$ is a wide-band stochastic process with zero average. The noise contribution in the S_1 equation has been multiplied for μ_2 to account for the fact that the perturbation to S_1 arises from S_2 domains nucleating from local defects, hence it is proportional to the growth rate of the second family. Similar considerations hold for the S_2 equation.

In Fig. 6 (right) we show the plots of $\eta(\varepsilon)$ and of $\sigma(\varepsilon)$ extracted from the numerical solutions of Eqs. (14) with noise addition. For a suitable choice of parameters, they are in good qualitative agreement with experiment.

We conclude that the interband dynamics implies both competitive deterministic terms as well as a stochastic force modelling defect nucleation. In fact a better model, rather than relying on a noise source, should account for the deterministic character of the underlying dynamics. However it has become customary, in many-mode dynamics, to account for the perturbation that the modes below threshold induce on the modes already unstable in terms of a suitable noise source [5].

REFERENCES

1. A.M. Turing, *Phil. Trans. R. Soc. London B* **237**, 37 (1952).
2. M. Cross and P.C. Hohenberg. *Rev. Mod. Phys.* **65**, 851 (1993).
3. F.T. Arecchi, *Il Nuovo Cimento* **107A**, 1111 (1994).
4. *Nonlinear Dynamics and Spatial Complexity in Optical Systems* (edited by R.G. Harrison and J.S. Uppal) (SUSSP Publications, Edinburgh, and Institute of Physics Publishing, London, 1993).
5. K. Ikeda, K. Matsumoto, and K. Otsuka, *Progr. Theor. Phys. Suppl. n.* **99**, 295 (1989).
6. F.T. Arecchi, G. Giacomelli, P.L. Ramazza, and S. Residori, *Phys. Rev. Lett.* **65**, 2531 (1990).
7. P. Mandel, M. Georgiou, K. Otsuka and D. Pieroux, *Opt. Comm.* **100**, 341 (1993).
8. C. Benkert and D.Z. Anderson, *Phys. Rev.* **A44**, 4633 (1991).
9. M.A. Vorontsov and W.J. Firth, *Phys. Rev.* **A49**, 2891 (1994).
10. S. Ciliberto. E. Pampaloni and C. Perez-Garcia, *Phys. Rev. Lett.* **61**, 1198 (1988).
11. D. Dangoisse, D. Hennequin, C. Lepers, E. Lovergnaux and P. Glorieux, *Phys Rev.* **A46**, 5955 (1992).
12. S.A. Akhmanov, M.A. Vorontsov and V. Yu. Ivanov, A.V. Larichev and N.I. Zheleznykh, *J. Opt. Soc. Am. B* **9**, 78 (1992).
13. G. D'Alessandro and W.J. Firth, *Phys. Rev. Lett.* **66**, 2597 (1990). G. D'Alessandro and W.J. Firth, *Phys. Rev.* **A 46**, 537 (1992).
14. R. Macdonald and H.J. Eichler. *Opt. Commun.* **89**, 289 (1992).
15. M. Tamburrini. M. Bonavita. S. Wabnitz and E. Santamato. *Opt. Lett.* **18**, 855 (1993).
16. E. Pampaloni. S. Residori and F.T. Arecchi. *Europhys. Lett.* **24**, 647 (1993).
17. E. Pampaloni. P.L. Ramazza, S. Residori and F.T. Arecchi, *Europhys. Lett.* **25**, 587 (1994).
18. E. Pampaloni. P.L. Ramazza, S. Residori and F.T. Arecchi, *Phys. Rev. Lett.* **74**, 258 (1995).
19. F. Papoff, G. D'Alessandro, G.L. Oppo and W.J. Firth, *Phys. Rev.* **A48**, 634 (1993).
20. M.A. Vorontsov, M.E. Kirakosyan and A.V. Larichev, *Sov. J. Quant. Electron.* **21**, 105 (1991).
21. J. Hegseth, J.M. Vince, M. Dubois and P. Berge', *Europhys. Lett.* **17**, 413 (1992).
22. D. Raitt, and H. Riecke, *Physica* **D82**, 79 (1995).
23. P.G. De Gennes, *The Physics of Liquid Crystals* (Oxford University Press, Oxford, 1974).
24. S. Residori. P.L. Ramazza, E. Pampaloni, S. Boccaletti and F.T. Arecchi, Domain coexistence in two dimensional optical patterns, *Phys. Rev. Lett.* **76**, 1063 (1996).
25. B.A. Malomed. A.A. Nepomnyaschiĭ, and M.I. Tribelskiĭ, *Sov. Phys. JETP* **69**, 388 (1989).
26. W.E. Lamb jr., *Phys. Rev.* **A134**, 1429 (1964).
27. J.D. Murray, *Mathematical Biology* (Springer-Verlag, Berlin, Heidelberg, 1989).

27 High sensitivity molecular spectroscopy with diode lasers

Krzysztof Ernst
Institute of Experimental Physics, Warsaw University, Hoza 69, 00681 Warsaw, Poland

INTRODUCTION

Recent progress in semiconductor diode lasers [1] has been of great importance for their continuously increasing use in both pure and applied
spectroscopy. Because of their good spectral purity and low noise fluctuations,
diode lasers represent attractive sources for inexpensive, room-temperature
spectroscopic probing of a variety of molecular gases.

The tunability of the wavelength and large spectral coverage, including near-
infrared and part of the visible, give a wide range of applications. Presently,
AlGaAs diode lasers are available in the range 750–900 nm, InGaP lasers
operate between 630 and 690 nm, and InGaAsP lasers emit in the range
1300–1500 nm. The frequency tuning of all these lasers can be easily
performed by means of temperature and injection current variation. As a
consequence many absorption lines can be studied, allowing the determination
of such parameters as line strength, pressure broadening coefficient and
molecular constants.

Semiconductor diode lasers also have the advantage of small size, relatively
low cost, high speed, low power consumption, reliability, ease of use, and
compatibility with fibre optic technology. Their particular features such as
excellent quality single-mode operation and very low amplitude noise allow
one to perform measurements in the field of high-resolution spectroscopy and
high sensitivity detection. This leads to various applications such as, for
example, real time non-contact monitoring of the atmosphere.

As is well known, all fundamental vibrational transitions of molecules are in
the infrared region. At the same time only a few molecules have electronic

absorption bands in the visible. This situation is not favourable for performing molecular spectroscopy measurements in this range, which would otherwise be very convenient for several reasons (tunable laser sources, sensitive detectors). On the other hand, because of the transparency of molecular gases for visible radiation, sunlight can easily penetrate the atmosphere, which is of primary importance for all kinds of life on the Earth.

In order to find a way out of this difficulty we can move from the infrared fundamental vibrational transitions to the overtone and combination bands in the visible and near-infrared region. Absorption coefficients decrease even by several orders of magnitude, but in comparison with laser sources operating in the infrared, semiconductor diode lasers in the visible (and in the near-infrared) offer the advantage of much simpler operation and much better amplitude stability. Moreover, remote sensing of atmospheric species in these spectral ranges could be more useful and more convenient than in the infrared because of the reduced opacity of the atmosphere.

For the reasons mentioned above, the overtone and combination transitions of various molecules, as well as high sensitivity detection techniques, have recently attracted considerable interest. Both aspects are discussed in this chapter.

TECHNIQUES DEVELOPED TO DETECT VERY WEAK ABSORPTIONS

Several detection methods for high sensitivity absorption measurements have been proposed and successfully applied in the last few years. The simplest possible technique is a direct absorption measurement which consists of sweeping the frequency and detecting the signal against the constant background. The sensitivity of this method is certainly very low. It can be improved by modulating the amplitude of the light source, but we are still limited by the background contribution.

In order to understand the motivation for developing specialized techniques for direct absorption spectroscopy (i.e., the direct measurement of the optical attenuation of a light beam through an absorbing sample) let us consider some general requirements concerning the signal-to-noise ratio (SNR).

The signal is usually a detector photocurrent and may be written as $S = kI$, where k represents the net attenuation of the laser intensity I incident upon the absorption sample. Noise contributions may be separated into three terms: N_e – originating from the detection electronics and therefore independent of I; N_0 – detector shot noise proportional to $I^{1/2}$; and $N_s I$ – the contribution of the amplitude-fluctuation background proportional to I. The SNR may then be written [2]

$$\mathrm{SNR} = \frac{kI}{[N_e^2 + (\beta I^{1/2})^2 + (N_s I)^2]^{1/2}} .$$

As the intensity increases the SNR increases proportionally to I until the I-dependent terms in the denominator exceed N_e. As we can see, amplitude modulation of the laser cannot solve our problem, since it imparts a time dependence to I, which is common to both the numerator and the denominator in the equation for the SNR. However, by using frequency modulation, one can effectively modulate the signal kI without modulating the noise terms.

The ease with which diode lasers can be modulated forms the basis for a number of sensitive spectroscopic detection methods. Amplitude modulation of the diode laser injection current results in emission that is both amplitude and frequency modulated. The interaction between the absorbing sample and the spectrally modulated radiation field generates a signal that varies with modulation frequency. Thus, frequency- and phase-sensitive detection electronics eliminates noise contributions that do not fall into the detection frequency band. In the case of sinusoidal frequency modulation with modulation index M and frequency ω_m, the optical field with carrier frequency ω_c can be written as $E_0\exp[i(\omega_c t + M \sin \omega_m t)]$. In view of the interest in high sensitivity measurements, techniques operating with widely different values of the frequency ω_m and the modulation index M have recently been developed.

Wavelength Modulation (WM) Spectroscopy [3, 4, 5, 6]

The single-frequency laser is modulated at a relatively low frequency (10^3 Hz), small compared to the width of the spectroscopic feature of interest, so that the absorption is probed simultaneously by a number of sidebands. Usually only the first- and second-harmonic signals are recorded and they are proportional to the first and second derivatives of the absorption line shape. For weak absorption, the signals are also proportional to the concentration of the absorbing molecular species.

Wave modulation is a sensitive form of derivative spectroscopy. Unlike direct absorption methods in which the signal is detected as a change against a constant background, in WM spectroscopy the signal arises from the difference in the absorption of different sidebands. Therefore, the sensitivity and spectral resolution are greatly enhanced, provided that the laser shows no or very little variation of intensity with wavelength and that the response of the detector is independent of λ. Consider the current C from the detector, $C = IGT$, where I is the laser light intensity, T is the transmissivity of the sample, and G is the response of the detector. We can then write for the detector signal

$$\frac{1}{C}\frac{dC}{d\lambda}\Delta\lambda = \frac{1}{I}\frac{dI}{d\lambda}\Delta\lambda + \frac{1}{G}\frac{dG}{d\lambda}\Delta\lambda + \frac{1}{T}\frac{dT}{d\lambda}\Delta\lambda$$

where $\Delta\lambda$ is the wavelength modulation depth. Only the last term in the above expression should give a contribution to the signal. Unfortunately $dI/d\lambda$ is not negligible in the case of semiconductor diode lasers and gives an unwanted contribution to the noise level.

Frequency Modulation (FM) Spectroscopy [*6, 7, 8*]

FM spectroscopy is an extension of WM spectroscopy to much higher frequencies ($\sim 10^8$ Hz). Frequency modulation produces sidebands which are widely spaced in frequency so that the spectral feature of interest can be probed by only one sideband at a time. Viewed in frequency space, the spectral distribution of the modulated laser field consists of the strong carrier at ω_c and two sidebands of the same amplitude but 180° out of phase, displaced by the angular modulation frequency ω_m from the carrier. When there is no absorption present, the beat signal at ω_m between the carrier and the upper sideband cancels exactly with the beat signal between the carrier and the lower sideband. If, however, the laser frequency is tuned over an absorption, so that one of the sidebands is absorbed, the balance condition between the sidebands is disturbed and a signal at the modulation frequency appears in the detector photocurrent. Although detection is usually performed at the modulation frequency, it can also be performed at various harmonics.

Modulation and detection are performed in the RF region because the intensity noise in diode lasers (and in most lasers) is at a minimum, so that high signal-to-noise ratios can be achieved. Moreover, this condition is particularly simple to implement with diode lasers.

If the modulation frequency does not exceed the absorption linewidth, the technique is usually called high wavelength modulation (HWM). In fact, the difference between FM and WM is slight and they can be considered as two limiting cases of the same technique. In FM the modulation index is small, while the ratio of the modulation frequency to the absorption linewidth is large. As a result, the absorption feature of interest is probed by a single sideband. In WM, on the other hand, the ratio of the modulation frequency to the absorption linewidth is small, but the modulation index is large. The absorption feature is then probed with a large number of sidebands.

Two-Tone Frequency Modulation (TTFM) [*9, 10*]

A third method of diode laser modulation spectroscopy is a simple variation of FM. If we wish to investigate a broad spectral feature, such as absorption lines broadened by atmospheric pressure to 2–3 GHz, we must have correspondingly high modulation frequencies. The TTFM technique has been applied in order to reduce the detection bandwidth requirement since detectors with bandwidth in the GHz range are not easily available.

In the TTFM, the laser is modulated simultaneously at two distinct but closely spaced angular frequencies $\omega_1 = \omega_m + \Omega/2$ and $\omega_2 = \omega_m - \Omega/2$ ($\Omega/\omega_m < 10^{-3}$). The TTFM absorption signal arises from the difference between the absorption of the carrier and the sum of the absorptions of the sidebands. Thus, it is approximately analogous to a second derivative signal, whereas single-tone FM is analogous to a first derivative signal. In both cases, however, there is no background signal if there is no absorption. This method eliminates the need for high-speed detectors and has the advantage that arbitrarily large modulation frequencies can be applied to the laser maximizing the differential absorption experienced by the sidebands and making possible large linewidth detection.

Unfortunately a pure modulation frequency is rarely achieved. There is always some residual amplitude modulation (RAM) present, especially in diode lasers where any variation in the injection current changes not only the frequency of the laser but also its output power. It is evident that RAM is an undesirable feature in all kinds of frequency modulation spectroscopy. It gives rise to a background signal with accompanying noise, even when there is no absorption present, and thus limits the detection sensitivity.

It is worth mentioning here that a wide variety of laser techniques allowing high sensitivity detection of atomic and molecular species has been proposed and successfully applied. Let us give a few examples such as laser-induced fluorescence (LIF) [11], optoacoustic spectroscopy (OA) [12], optogalvanic spectroscopy (OG) [12], resonance ionization spectroscopy (RIS) [13], and laser-intracavity absorption [14]. All these techniques monitor some indirect effects of the optical absorption.

Any qualitative comparison of these methods would be extremely difficult since their applicability depends essentially on specific aims and experimental conditions of the desired measurements. However, two important advantages of extracavity direct absorption should be emphasized: simple calibration procedures and remote-sensing possibilities. Both are very important from the point of view of environmental studies.

SPECTROSCOPIC DATA AND EXPERIMENTALLY MEASURED SENSITIVITY LIMITS

The techniques described above have already been successfully applied for high sensitivity detection of various molecular species. It has allowed us to determine molecular parameters, as well as to establish and compare the sensitivity limits in different experimental conditions. Table 1 contains information concerning wavelengths of overtone and combination vibrational bands of selected molecular species in the visible and near-infrared spectral range covered by diode lasers.

As one can see, there are many candidates for spectroscopic studies with diode lasers. Most transitions presented in the table have already been investigated with diode and dye lasers, but some of them refer to reported numerical values and can be considered as potential candidates for future experimental studies.

Table 1. Overtone and combination molecular transitions in spectral ranges covered by diode lasers.

Molecule	λ (nm)	Molecule	λ (nm)	Molecule	λ (μm)
C_2H_2	640	HCN	791	CH_4	0.89
NH_3	647	NH_3	793	HCl	1.2
C_2H_2	670	H_2O_2	793	HF	1.33
CH_4	682	H_2O	796	HBr	1.34
C_2H_6	741	NO	800	H_2O	1.37
HCl	750	CO	804	HI	1.54
O_2	761	H_2O	818	NH_3	1.54
CH_4	782	HI	820	CO	1.57
CO_2	783	C_2H_2	848	CO_2	1.57
C_2HD	785	CH_4	856	H_2S	1.58
C_3H_4	787	CH_3D	861	CH_4	1.65
H_2O	789	CO_2	869	NH_3	1.65
CO_2	789	C_6H_6	869	HCl	1.75
NH_3	790	HF	877		

The results obtained for H_2O and CH_4 were used for comparing the sensitivity of different frequency modulation techniques. In the first experimental work [15] dedicated to water vapour the authors found that the difference in sensitivity between the FM and the TTFM was about two orders of magnitude. In later work [16], the minimum detectable absorption was measured using the WM (1 kHz), the FM (100 MHz), and the TTFM (390 ± 5 MHz) techniques by observing a test transition of methane – a gas of considerable interest for its environmental implications. A component of the combination band at 886 nm has been chosen mainly for its strength. Two recordings are shown in Fig. 1.

The minimum detectable absorption for the WM, FM and TTFM techniques was measured as 4.5×10^{-7}, 9.7×10^{-8}, and 6.4×10^{-8}, respectively. In the first case the detection limit is determined by the laser amplitude excess noise, while in the other two cases this noise contribution is limited with respect to detector-induced shot noise, and thermal and RAM noises.

The detection limits agree well with the calculated 'quantum limited' values based on measured laser power, modulation index, noise figure of the electronic components and other parameters of the apparatus. They are also consistent with the results obtained for N_2O absorption by using lead salt diodes [17, 18], while they differ from those for water vapour [15].

The absorption limits measured for methane are extremely low which is very promising for all kinds of trace-gas monitoring applications. It is also worth noting that the advantage of using high-frequency detection with respect to low-frequency detection is, in the case of diode lasers, not so pronounced as in the case of dye lasers. Very sensitive trace-gas measurements

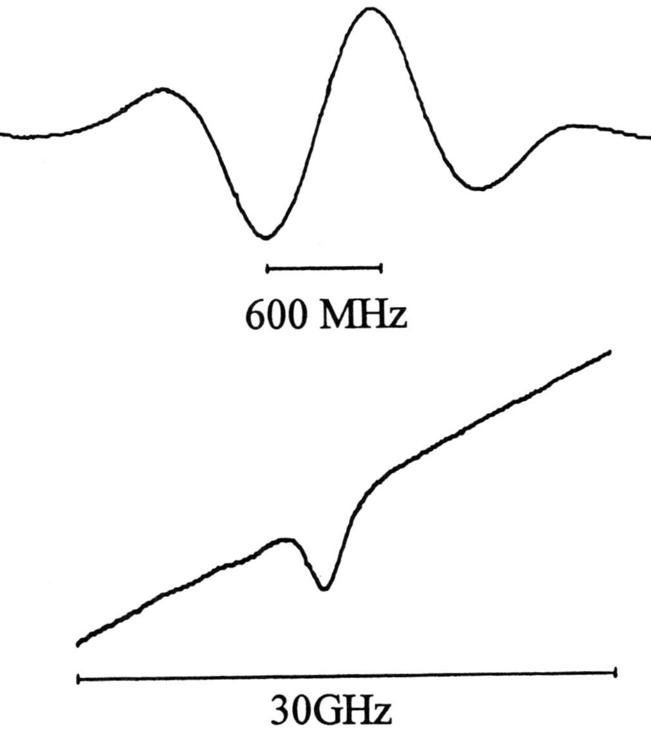

Fig. 1. Derivative lineshape of a two-tone recording at 500 mTorr of methane in 1.5 m pathway (upper curve), from Pavone and Inguscio (1993) Fig. 4; and pure absorption signal at 100 Torr (lower curve), from Pavone and Inguscio (1993) Fig. 9.

for environmental purposes with such compact sources can then be performed using a simple experimental configuration.

In most measurements with diode lasers the frequency modulation was produced by modulating the injection current. In the case of acetylene [19] the diode laser was used in an extended cavity configuration. The experimental arrangement for C_2H_2 absorption measurements was quite simple as is schematically shown in Fig. 2. Low-frequency modulation could be produced by both changing the injection laser current and/or varying the external cavity length (by means of the PZT (piezoelectric transducer)). The extended cavity configuration reduced the linewidth (to less than 1 MHz) and improved short-term stability. Another advantage of this configuration was that wavelength modulation achieved by varying the cavity length led to a lower level of RAM noise than direct injection current modulation.

An example of the derivative signal for the P(11) line at the 12646.966 cm^{-1} transition is shown in Fig. 3 for two different pressure values: 10 Torr (a) and 36 mTorr (b). The latter measurement gives a minimum relative absorption of 10^{-6}. Acetylene as a combustion gas is present in small amounts in both the urban atmosphere and in the troposphere. In view of environmental monitoring,

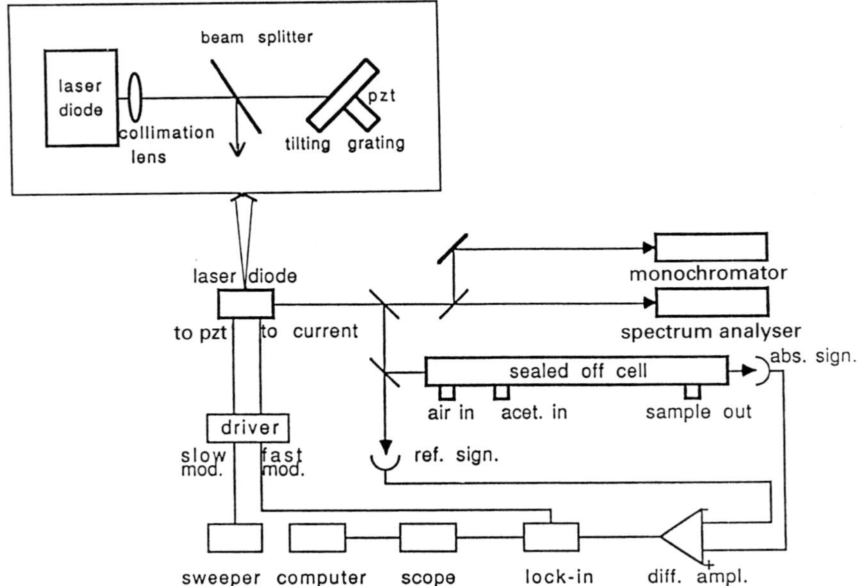

Fig. 2. Experimental apparatus for absorption measurement in acetylene, including the scheme for the extended cavity diode laser configuration, from Pavone *et al.* (1993) Figs 1 & 2.

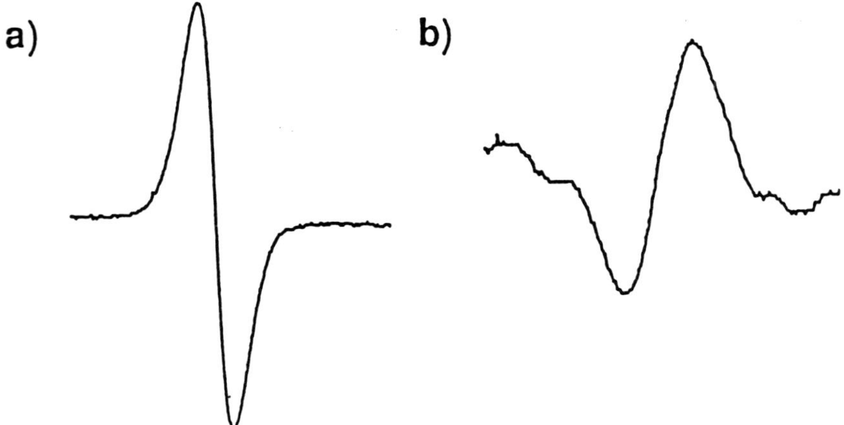

Fig. 3. Derivative signal for the P(11) component of acetylene at 10 Torr (a) and 36 mTorr (b), from Pavone *et al.* (1993) Fig. 4.

the minimum C_2H_2 pressure in the presence of air at atmospheric pressure has also been extracted. The minimum detectable concentration was estimated to be 0.1 ppm per km of acetylene in air.

In the spectral range around 1.6 μm, a new type of diode laser has been recently developed [20]. The device structures are strained-layer multiple-quantum-well separated-confinement distributed-feedback (DFB) lasers. Their main characteristics with respect to other diode lasers is given by the first-order Bragg grating layer growth inside the chip device. It induces the distributed feedback required for the laser action by resonant reflections of the confined optical mode off the grating. The distributed feedback mechanism avoids the random mode-hops frequently present in frequency tuning of semiconductor diode lasers with a simple Fabry-Perot guided cavity. The lack of mode-hops is a unique feature of DFB diode lasers and gives them a significant advantage in spectroscopic applications.

A sensitive detection study of NH_3 at 1.66 μm was performed by using a DFB diode laser absorption device [21, 22]. The high sensitivity of ammonia is interesting for its importance in industrial processes. Its concentration limit has to be controlled in order to avoid health hazards. Moreover, the 1.66 μm combination band of NH_3 is more convenient for detection than the stronger band at 1.55 μm because of the lack of interference effects from other gases. Two recordings for the 1.66454 μm component are shown in Fig. 4. The minimum detectable concentration for ammonia without another gas in the cell

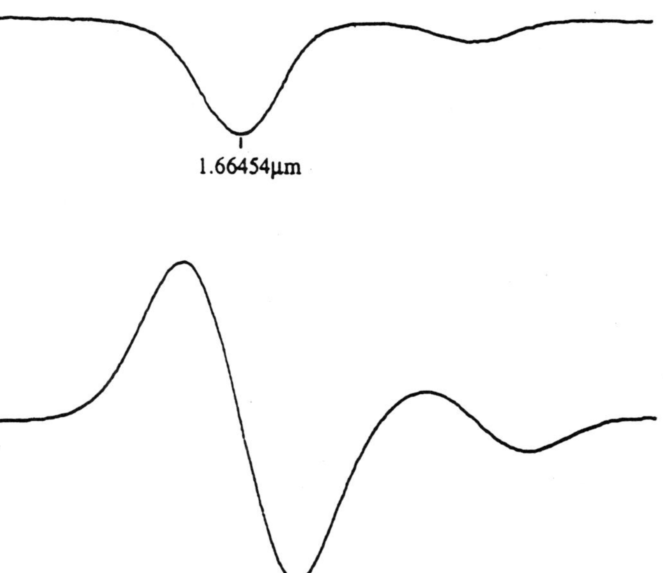

Fig 4. Pure absorption, first and second derivative signals of the 1.66454 μm line of NH_3 at 1 Torr pressure and 1.5 m absorption path, from Cancio (1996) Fig. 2.

Table 2. Pressure broadening parameters.

Molecule	Transition	Γ (MHz Torr^{-1}) self	air	Refer.
C_2H_2	789 nm $(v_1 + v_3)$			[19]
	P(11)	11	8	
	R(5)		9	
	862 nm $(v_2 + 3v_3)$			[23]
	R(7)	12.3		
	v_5			
	R(5)	7		[24]
	R(11)	6.2		
H_2O	822 nm $(2v_1 + v_2 + v_3)$		5.9	[25]
CH_4	860 nm $(2v_1 + 2v_3)$	3.7	3.2	[26]
NH_3	790 nm $(2v_1 + 2v_3)$	21.7		[26]
	792 nm $(4v_1)$	27.2		[26]
HCl	1.2 μm $(v = 3 \leftarrow 0)$		2.5	[27]

(1.5 m long) was as low as 1.2 mTorr which corresponds to the sensitivity limit of the apparatus of 2 ppm per m.

In many practical applications it is required to monitor gas-traces in an open-path configuration in the atmosphere. This is one of the reasons for which the analysis of line broadening with pressure and the knowledge of characteristic parameters can be useful. Unfortunately, in real atmospheric conditions the strong broadening due to the collisions of absorbing molecules with air molecules strongly affects the sensitivity achieved. Moreover, the overlapping of different broadened lines may occur and has to be taken into consideration.

In the case of ammonia the minimum detectable concentration decreased drastically in the presence of foreign gases (N_2 and O_2) added to the cell. At 76 Torr total pressure it was measured as 55 ppm per m (in N_2) and 59 ppm per m (in O_2). So large foreign gas broadening behaviour explains the fact that the ammonia absorption signal at atmospheric pressure is not detectable under 2 Torr of partial pressure in a 1.5 m measurement path.

The self-broadening and air-broadening coefficients have also been measured for several overtone and combination transitions for other molecules by using diode laser based devices. The broadening parameters Γ for selected molecules are shown in Table 2.

APPLICATIONS AND FUTURE PERSPECTIVES

The continuously increasing interest in molecular spectroscopy with semiconductor diode lasers in the last few years has been possible because of considerable progress in laser technology and the development of sensitive detection techniques. As a consequence, many molecular transitions in the visible and the near-infrared region, hardly accessible before, have been investigated, providing new spectroscopic data and/or revealing potential applications.

One of these applications is the high sensitivity remote sensing of the atmosphere including the monitoring of industrial emissions as well as various photochemistry processes in the atmosphere. The possibility of detection of air pollutants with tunable diode lasers was demonstrated in 1971 for ammonia [28]. Much effort has been made since then in order to build diode laser based monitoring sensors [29, 30, 31]. The latter system, called OPS (open path sensor), is a computerized, three-diode-laser instrument for monitoring oxygen, water vapour and carbon dioxide. Its detection sensitivity of 2×10^{-6} for oxygen has been reported.

The stabilization of a diode laser operating at wavelengths below 1 μm could also be important for use in efficient pulsed solid-state lasers, such as titanium:sapphire operating in the same region. This can be interesting in view of lidar measurements [32], where diode laser injection seeding could allow one to perform detection at larger distances with improved sensitivity and easier spectral control.

Sensitive and selective detection of molecular species have also found an interesting application in medical diagnostics [20]. By measuring the ratio of $^{12}CO_2$ to $^{13}CO_2$ in human breath it is possible to diagnose a number of disorders that result from the presence or absence of a specific enzyme. Since both isotopic CO_2 molecules have combination bands around 1.6 μm, their ratio can be determined with adequate precision by using a diode laser based sensor. It is already possible to measure the ratio of the two lines with a precision of 1%.

Looking into the future, let us conclude by indicating two important laser devices that, taking into consideration the present state of the art, should be available soon. The first is related to a commercially available, sensitive instrument for monitoring the atmosphere, able to detect a variety of molecular species present as trace constituents. In spite of attempts undertaken by a few companies, all operational systems are still limited to the laboratory.

The second is an extension of the operation of diode lasers toward the blue [33]. Various investigations of materials are now proceeding with quite promising results. Many interesting applications may come when this new generation of diode lasers will allow one to investigate electronic molecular transitions.

REFERENCES

1. C. Wieman and L. Hollberg, 'Using diode laser for atomic physics', *Rev. Sci. Instr.* **62**, 1 (1991).

2.	M. Gehrtz, G. Bjorklund and E. Whitaker, 'Quantum limited laser frequency modulation spectroscopy', *J. Opt. Soc. Am. B* **2**, 1510 (1985).

3.	J. Telle and C. Tang, 'New method for electrooptical tuning of tunable lasers', *App. Phys. Lett.* **24**, 85 (1974).

4.	E. Moses and C. Tang, 'High-sensitivity laser wavelength modulation spectroscopy', *Opt. Lett.* **1**, 115 (1977).

5.	P. Pokrowsky, W. Zapka, F. Chu and G. Bjorklund, 'High frequency wavelength modulation spectroscopy with diode lasers', *Opt. Commun.* **44**, 175 (1983).

6.	J. Supplee, E. Whitaker and W. Lenth, 'Theoretical description of frequency modulation and wavelength modulation spectroscopy', *Appl. Opt.* **33**, 6294 (1994).

7.	G. Bjorklund, 'Frequency modulation spectroscopy: a new method for measuring weak absorptions and dispersions', *Opt. Lett.* **5**, 15 (1980).

8.	G. Bjorklund, M. Levenson, W. Lenth and C. Ortiz, 'Frequency modulation spectroscopy: theory of lineshapes and signal to noise ratio', *Appl. Phys. B* **32**, 145 (1983).

9.	G. Janik, C. Carlisle and T. Gallagher, 'Two-tone frequency modulation spectroscopy', *J. Opt. Soc. Am. B* **3**, 1070 (1986).

10	D. Cooper and R. Warren, 'Two-tone optical heterodyne spectroscopy with a lead-salt diode laser: theory of line shapes and experimental results', *J. Opt. Soc. Am. B* **4**, 470 (1987).

11.	W. Demtroder, *Laser Spectroscopy*, Springer Verlag, 1981, p. 416.

12.	K. Ernst and M. Inguscio, 'Unconventional techniques in laser spectroscopy', *Rivista del Nuovo Cimento* **11**, no. 2 (1988).

13.	G. Hurst, M. Payne, S. Kramer and J. Young, 'Resonance ionization spectroscopy', *Rev. Mod Phys.* **51**, 767 (1979).

14.	S. Harris, 'Intracavity laser spectroscopy: an old field with new prospects for combustion diagnostics', *Appl. Opt.* **23**, 1311 (1984).

15.	L. Wang, H. Riris, C. Carlisle and T. Gallagher, 'Comparison of approaches to modulation spectroscopy with GaAlAs semiconductor lasers: application to water vapor', *Appl. Opt.* **27**, 2071 (1988).

16.	F. Pavone and M. Inguscio, 'Frequency- and wavelength-modulation spectroscopies: comparison of experimental methods using an AlGaAs diode lasers', *Appl. Phys. B* **56**, 118 (1993).

17.	J.A. Silver, 'Frequency-modulation spectroscopy for trace species detection: theory and comparison among experimental techniques', *Appl. Opt.* **31**, 707 (1992).

18.	D. S. Bomse, A. C. Stanton and J. A. Silver, 'Frequency modulation and wavelength modulation spectroscopies comparison of experimental methods using a lead-salt diode laser', *Appl. Opt.* **31**, 718 (1992).

19.	F. Pavone, F. Marin, M. Inguscio, K. Ernst, G. Di Lonardo, 'Sensitive detection of acetylene absorption in the visible by using a stabilized AlGaAs diode laser', *Appl. Opt.* **32**, 259 (1993).

20.	D.E. Cooper and R.U. Martinelli, 'Near-infrared diode lasers monitor molecular species', *Laser Focus World*, Nov. 1992, p. 133.

21.	F. Pavone, C. Corsi, P. Cancio, K. Ernst, S. Chudzynski, M. Pokora and L. Lundberg-Nielsen, 'Overtone molecular spectroscopy using semiconductor diode lasers in the 1–2 μm. European Quantum Electronic Conference, Amsterdam 1994.

22. P. Cancio, C. Corsi, F. Pavone, R.U. Martinelli and R.J. Menna, 'Sensitive detection of ammonia absorption by using a 1.65 µm distributed feedback InGaAsP diode laser', (to be published).

23. Y. Ohsugi and N. Ohashi, '0.85 µm diode laser spectroscopy of C_2H_2', *J. Mol. Spectrosc.* **131**, 215 (1988).

24. D. Lambot and G. Blanquet, 'Diode laser measurements of collisional broadening in the v_5 band of the C_2H_2 perturbed by O_2 and N_2, *J. Mol. Spectrosc.* **136**, 86 (1989).

25. A. Lucchesini, L.Dell'Amico, I. Longo, C. Gabbanini, S. Gozzini and L. Moi, 'Diode laser spectroscopy: water vapour detection in the atmosphere', *Nuovo Cimento D*, **13**, 677 (1991).

26. A. Lucchesini, I. Longo, C. Gabbanini, S. Gozzini and L. Moi, 'Diode laser overtone spectroscopy: a possible atmospheric monitoring technique', SPIE Conf. *'High Performance Optical Spectrometry'*, Warsaw 1992.

27. A.C. Stanton and J.A. Silver, 'Measurements in the HCl $3 \leftarrow 0$ band using a near-IR InGaAsP diode laser', *Appl. Opt.* **27**, 5009 (1988).

28. E.D. Hinkley and P.L. Kelley, 'Detection of air pollutants with tunable diode lasers', *Science*, **171**, 635 (1971).

29. D.T. Cassidy and J. Reid, Atmospheric pressure monitoring of trace gases using tunable diode lasers', *Appl. Opt.* **21**, 1185 (1982).

30. K. Uehara and H. Tai, 'Remote detection of methane with a 1.66 µm diode laser', *Appl. Opt.* **31**, 809 (1992).

31. H. Riris, C.B. Carlisle, L.W. Carr, D.E. Cooper, R.U. Martinelli and R.J. Menna, 'Design of an open path near-infrared diode laser sensor: application to oxygen, water and carbon dioxide vapor detection', *Appl. Opt.* **33**, 7059 (1994).

32. H.J. Kolsch, 'Probing the atmosphere: air pollution studies by LIDAR', Theses, Freie Universität Berlin, 1990.

33. H. Okuyama and A. Ishibashi, 'ZnMgSSe based blue laser diodes', European Quantum Electronic Conference, Amsterdam 1994.

28 Sub-micrometre optical metrology using laser diodes and polychromatic light sources

Christophe Gorechki and Patrick Sandoz
*Laboratoire d'Optique P. M. Duffieux (URA CNRS 214),
Université de Franche-Comté, 16, route de Gray,
25030 Besançon Cedex, France*

INTRODUCTION

Sub-micrometre optical metrology is a relatively new field for laser diodes. When only gas lasers were available, the He-Ne laser was the light source of choice in interferometric measurements. Since then, many interferometers have switched to near-infrared, tunable laser diodes. The semiconductor laser is a more compact and efficient device, usually less expensive, and more easily directly modulated at high rates. The wavelength stability and output beam quality of a laser diode are inferior to those of other lasers. However, as the output characteristics are refined, and the frequency stability improved, the overwhelming advantages of the laser diode make it the source of choice in the design of new interferometers. They have already been integrated in metrology systems initially dedicated to traditional gas or solid crystal lasers.

Monomode and multimode laser diodes, super-luminescent laser-diodes (SLDs) and light emitting diodes (LEDs) are widely used as light sources for high-precision displacement measurements, velocity evaluation and rough surface profilometry. The methods for interferometric range finding using the laser diode can be classified basically into two categories: multiple wavelength interferometry and techniques based on frequency chirping. In the case of rough surface profilometry or 3D shape measurements, several methods based on interferometry have been developed over the last decade using laser diodes in manufacturing technologies, and new measurement principles are proposed which profit from the specific spectrum or sensitivity of such light sources.

Laser diodes are similar in many ways to wide-spectrum light sources such as SLDs and LEDs. Laser diodes emit very intense coherent light at only a few discrete wavelengths, whereas the SLDs emit broad-band incoherent light over a relatively wide spectral range. The main difference in the use of laser diodes resides in the fact that the laser diode can be used mostly as an active light source in interferometric devices, where the intrinsic diode features play an important role (coherence, modulation properties, etc.). In contrast to monomode or multimode laser diodes, the wide-spectrum light sources such as SLD and LED act mostly as passive light sources, where only the emitted spectrum characteristics are used.

LASER DIODES FOR INTERFEROMETRIC USE

Single-mode or Multimode Laser Diode?

A laser is basically an oscillator working at fixed optical frequencies and requiring gain and feedback. In laser diodes, the gain is produced by injection of a current into the active region, and feedback is produced by reflection from the facets. In contrast to other lasers which are usually pumped either by an electrical discharge or optically, the pumping of a semiconductor laser is performed by passing a current through the laser structure.

Figure 1 shows the cavity structure of an AlGaAs laser. It is a double heterojunction which consists of an active layer of GaAs placed between two cladding layers, one n-AlGaAs doped, the other p-AlGaAs doped. The active medium, where lasing occurs, is of low band-gap than those of the two cladding layers. Under a bias voltage applied in the forward direction, electrons are injected into the active region from the n-type layer, and the holes are injected from the p-type layer. Since the band gap energy is greater in the passive region than in the active region, the injected carriers are prevented from diffusing across the junction by potential barriers formed between the active and passive layers. The carriers confined to the active region generate a state of population inversion, permitting the amplification of light by stimulated emission. The charge carriers are confined in a waveguide cavity having a rectangular cross-section of size dw, where d is the thickness of the active layer along the y-axis and w is the width of the active layer along the x-axis. Light propagation occurs along the z-axis because the refractive index of the active layer is larger than those of the clad layers. The cavity forms a waveguide of $w = 1-10\ \mu m$ wide, $d = 0.1-0.2\ \mu m$ thick, and $L = 200-500\ \mu m$ long along the z-axis. The stimulated emission is the basic factor for laser action. Both

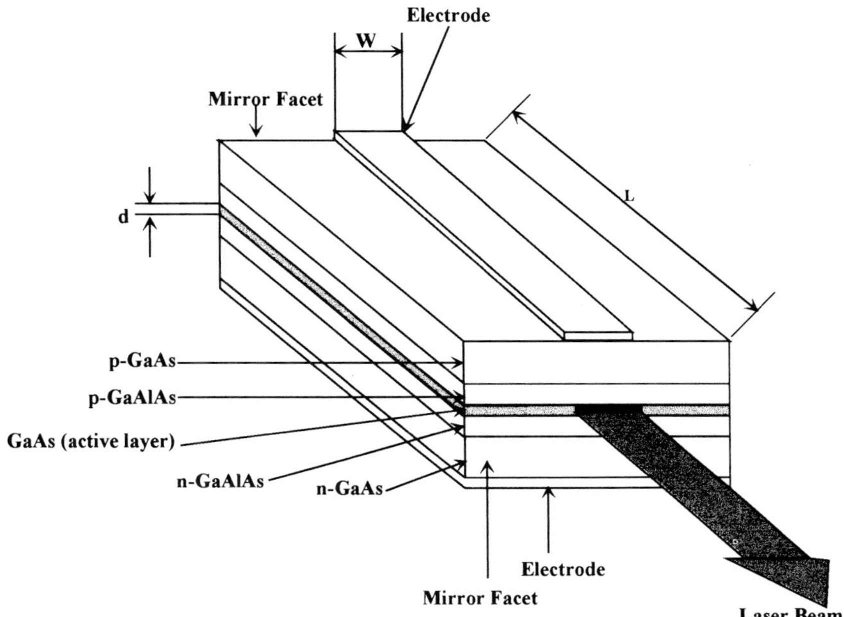

Fig. 1.

ends of the laser cavity are smooth facets which have been formed to act as mirrors forming a Fabry–Perot resonator. This Fabry–Perot cavity is realized by making flat mirrors on the waveguide facets by cleaving or chemical etching. Mirror facets provide optical feedback; this is the second critical element to make the device a laser. The Fabry–Perot cavity of AlGaAs laser supports several longitudinal modes. The cavity loss is the same for all longitudinal modes but all modes do not have the same gain at a given pumping level because the dominant lasing regime is reached when the peak gain is nearly equal to the loss. In the case of a laser diode with emission wavelength $\lambda_i = 780$ nm, refractive index $n-3.5$ (GaAs) and $L = 300$ μm, the wavelength separation between adjacent longitudinal modes is $\Delta\lambda_i = 0.29$ nm and the frequency separation is $\Delta v_i = 143$ GHz.

Laser diodes can be classified into two generic types depending on the mechanism of confinement of the lasing mode in the plane of the junction: gain-guided and index-guided lasers [1]. Gain-guided laser diodes have relatively wide spectral linewidths and usually emit multiple longitudinal modes and sometimes transverse modes. Index-guiding generally generates both single-transverse-mode and single-longitudinal-mode behaviour. For single-mode operations, the index-guided laser is clearly preferred.

Figure 2 shows the longitudinal mode spectra of a Sharp LT022 laser at several power levels [2]. In this index-guided laser diode, the main mode is

SINGLE LONGITUDINAL MODE
(Index Guided)

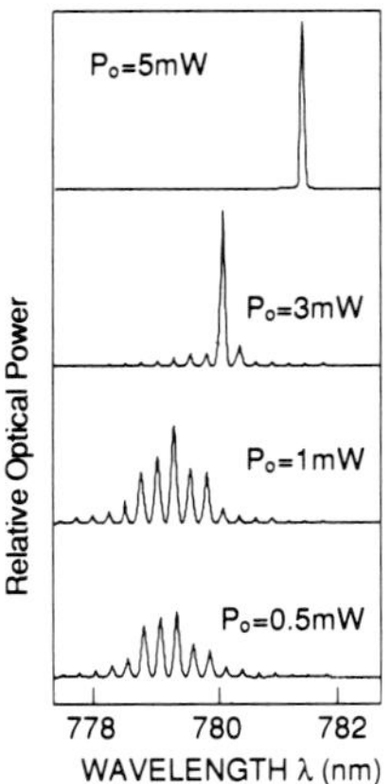

Fig. 2.

accompanied by several Fabry–Perot modes, whose amplitude decreases with an increase in the power of the main mode. At high output powers and for high operating currents much above threshold, the laser spectrum is a narrow peak with only one mode leading to a relatively long coherence length. Then, as output power and current are reduced, less stimulated gain in the laser results in more longitudinal modes with a coherence length reduction. These types of spectra are neither 'true single mode' nor a 'true multimode'.

It is quite different for the 'true multimode' spectra of gain-guided lasers, illustrated in Fig. 3 for the Sharp LT023 laser diode [2]. In these devices, when the current is reduced well below threshold, the laser output can be dominated by spontaneous emission and the source can be operating as an LED. As seen in Figs 2 and 3, a major property of laser diodes is that the wavelength changes with modifications in both injection current and temperature. Laser oscillations are generated at the wavelength corresponding to the maximum gain, which is determined by the band-gap energy. As the band gap varies with temperature, so too does the oscillation wavelength. For GaAlAs laser diodes emitting at 780 nm, the wavelength increases approximately 0.23 nm for 1 °C temperature increase.

Longitudinal mode hopping can be observed if the ambient temperature or injection current are varied. This phenomenon illustrated in Fig. 4 for a single-mode Sharp LT027 laser diode [2]. It presents seven narrowly spaced mode hops over a temperature range of 25–50 °C. This kind of laser diode is widely used in heterodyne interferometry [3] and two-wavelength holographic devices [4], where the capability of direct modulation of the laser by changing the injection current permits one to remove the frequency shifters with mechanical moving parts.

MULTI LONGITUDINAL MODE
(Gain Guided)

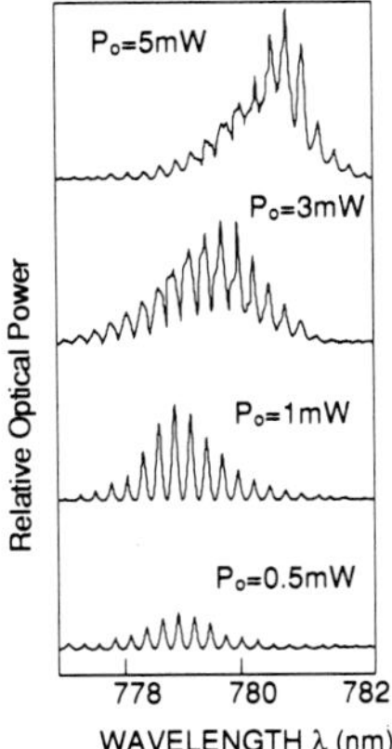

Fig. 3.

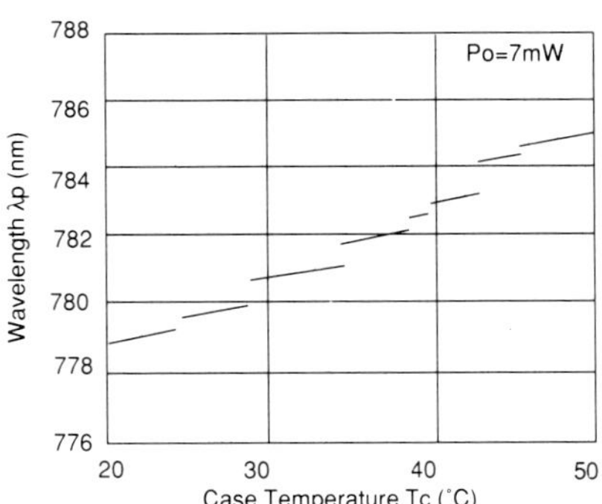

Fig. 4.

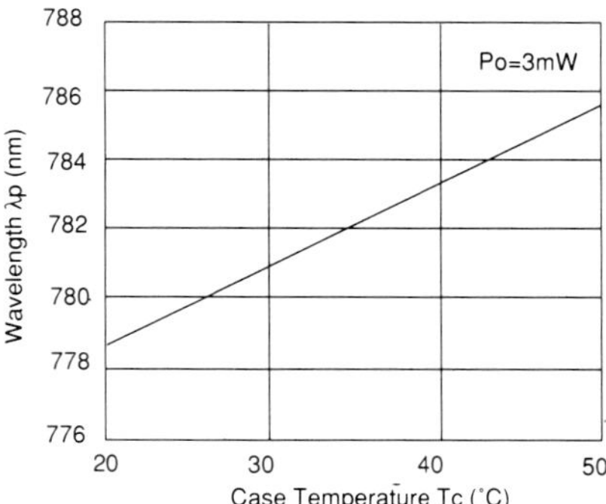

Fig. 5.

Longitudinal mode hopping does occur for multimode laser diodes. Figure 5 illustrates the output wavelength versus case temperature over a temperature range of 25–50 °C for a multimode Sharp LT023 laser diode [2]. In contrast with Fig. 4, this curve behaves as a linear function. The mode hops are not visible because of their large number and the small energy separating mode hopping. This kind of laser can be used as an alternative to the low coherence sources for interferometric-type applications, normally using white light sources [5]. The multimode laser diode is also very useful for multiplexed systems where light source power is coupled into an optical fibre [6].

Special Case: External Cavity Laser Diode

Sometimes conventional single-mode laser diodes have too wide spectral linewidth or too low coherence for interferometric-type applications that cannot tolerate mode hops. One example of an excellent source which improves these performances is the external cavity laser diode. A specific example of an external cavity is shown in Fig. 6, where the architecture consists of a laser chip and an external mirror. The coupling between the cavities is governed by an air gap of width X. The air gap itself acts as a third Fabry–Perot cavity, and the intercavity coupling depends on the loss and phase shift suffered by the optical beam while traversing the gap. In this architecture R_2 and R_3 represent the reflection coefficients for the left laser facet and the external mirror, respectively. R_2 includes all losses due to the coupling between the laser diode mode and external cavity mode. In this case the feedback level is defined in dB as $20 \log R_3$, and the effective reflection coefficient R_{eff} is [7]:

$$R_{\text{eff}} = \frac{1 + R_3/R_2 e^{-i\omega\tau}}{1 + R_2/R_3 e^{-i\omega\tau}} \tag{1}$$

where $\tau = 2X/c$ is the external cavity round-trip time and ω is the lasing frequency.

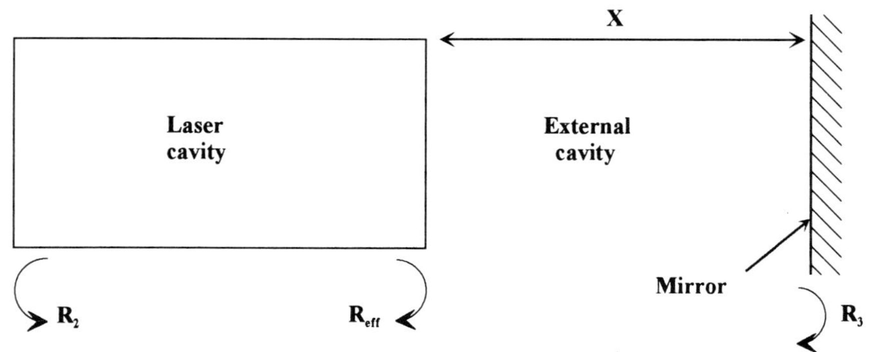

Fig. 6.

The main advantages of this configuration are that it has excellent wavelength stability and wavelength selectivity together with wavelength tunability. Single-mode operation is possible over a relatively wide current range. The external cavity generates the effective wavelength-dependent reflectivity R_{eff}. In contrast with a single Fabry–Perot cavity where the loss profile is flat, the resulting periodic loss profile permits one to selectively amplify the side modes that have the lowest cavity loss and are closest to the peak of the gain profile. Spectrally dependent reflectivity, induced by the external cavity and the effective longitudinal mode selection of the gain medium, produce the highly single-mode behaviour of this laser, especially when the laser is operated well above threshold. This is illustrated in Fig. 7 which represents the wavelength output versus case temperature from the short external cavity structure of the Sharp LT080 laser diode [2]. It has only one widely spaced mode hop over a temperature range of 25–50 °C. As will be demonstrated in Section 3, this commercially available external-cavity laser diode is the light source of choice for range-finding applications based on frequency chirping and backscatter-modulation velocimetry.

Operating Parameters and Error Factors with Laser Diodes

An important feature of the laser diode is its modulation property, very useful in interferometric applications. Intensity or amplitude modulation (AM) is always accompanied by phase or frequency modulation (FM). The mechanism of AM and FM is related to the index change that occurs when the optical gain is changed in response to modifications of carrier population. A carrier density decrease leads to an increase in the refractive index. An increasing refractive index means an increasing wavelength or decreasing frequency. Thus, the gradual shift of frequency is called 'chirping'. When we use the chirping technique for interferometric applications, the evaluation of stable regions between mode hopping where the laser diode can be linearly modulated is

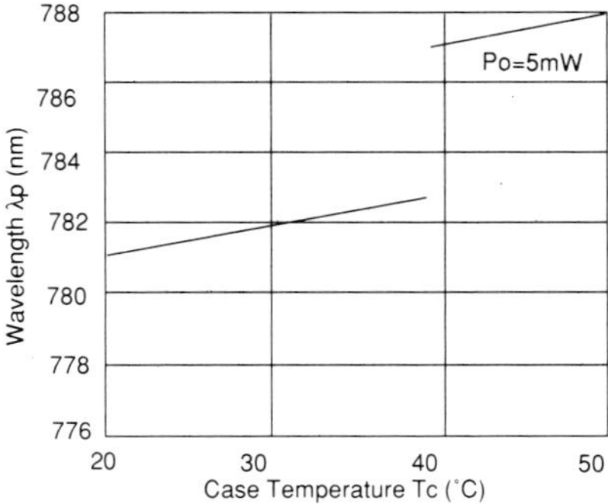

Fig. 7.

needed. This linear behaviour can be measured by the frequency deviation per unit of modulation current $\xi = \Delta v / \Delta I_{\mathrm{m}}$, called the modulation efficiency. This was measured for a wide frequency range in the case of two commercially available laser diodes (Sharp LT080 and Hitachi HL7802), as shown in Fig. 8 [8]. As we can see, the modulation efficiency gradually decreases between 10 Hz and 500 kHz, and presents a declining tendency from 500 kHz to 1 MHz. At frequencies up to 1 MHz, the FM effect is primarily due to the thermal effect in the active junction of the laser diode.

The main error factors with the laser diode are based on mode hopping, high sensitivity to optical feedback, spectral ageing, and dispersion of all the characteristics even in the case of laser diodes of the same kind. These factors are mainly responsible for the relative instability and inreproducibility of the central frequency and cause systematic measuring errors, reducing the use of laser diodes as well-calibrated-frequency light sources.

The more useful characteristics of tunable single-mode laser diodes for heterodyne detection are the narrow linewidth, associated with the ability to tune it continuously over a large wavelength range. In practice, the usable range of this tuning is limited by longitudinal mode hops which introduce discontinuities of the emitted wavelength and may influence the measurement accuracy. The discrete wavelength changes between mode hops are due to the changes in the optical length of the active junction produced by the modifications in the band gap of the lasing medium [9].

Optical feedback produced by any optical element of the optical system may seriously disturb the dynamic properties of the laser diode beam because the external reflector surface acts as an additional external mirror. In particular,

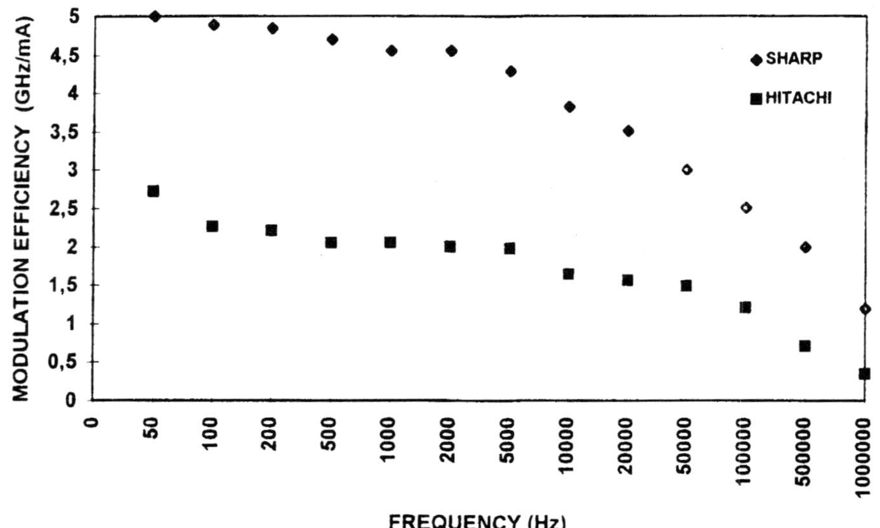

Fig. 8.

feedback changes the linear relationship relating output power to injection current. That can make the laser diode multistable, and hysteresis phenomena may occur. To reduce the feedback effect, light returning into the laser can be reduced by an adequate choice of mirror reflectivity on the laser output facets which permits the suppression of mode competition noise [10]. Thus, by coating these mirrors with high reflectivities $R_2 = 75\%$ and $R_3 = 96\%$, feedback-induced noise is suppressed by up to 1.4% optical feedback for a V-channelled substrate inner strip (VSIS) laser diode with a 460 µm cavity from Sharp. The instabilities produced in single-longitudinal-mode operation can be noticed even when the feedback reinjection rate is as low as 0.001%. This means that an optical isolator with attenuation of about 60 dB should be used. In practice, Chebbour *et al.* [11] demonstrated that the use of a 40 dB Faraday isolator inserted between the laser and the interferometer combined with two cube corner retroreflectors used as interferometric mirrors are sufficient to cut back the feedback effects.

In interferometry, based on a frequency-modulated laser diode, good control of stability either in short-term and long-term of both frequency and intensity characteristics of the ramp generator increases considerably the measuring accuracy. For a commercially available function generator, the frequency relative error is currently about 10^{-5} with a ramp linearity of about 1% peak-to-peak. Those frequency and linearity performances do not permit high-resolution measurements because the fluctuations of amplitude current and modulation frequency of the function generator may reduce considerably the measurement repeatability. For a given amount of feedback light, the relationship between emitted power and wavelength is influenced. For this reason, Chebbour *et al.* [11] developed a function generator using a cooled and highly stable oscillator with relative short-term stability of 10^{-11} and long-

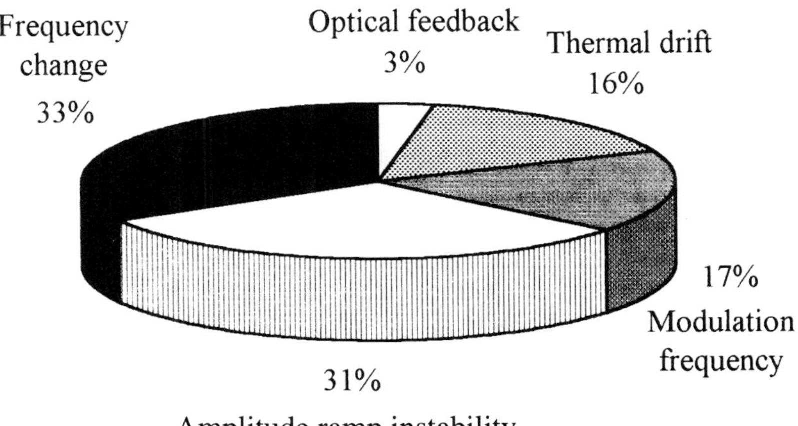

Fig. 9. Error factors in distance measurement for the sharp LTO80.

term stability of 10^{-9} with a ramp linearity of 0.01%. In a specific application of wavelength-shift interferometry for absolute distance and velocity measurements, the main systematic error sources are measured experimentally. Total interferometer system accuracy and repeatability are defined by summing the individual error terms listed in Fig. 9 [*12*].

RANGE FINDING WITH LASER DIODES

Distance Measurements by Two-wavelength Interferometry

Conventional interferometry has the disadvantage of having an ambiguity range depending on the wavelength of the laser used. To enlarge that ambiguity range, a second wavelength is used, which, in combination with the first wavelength, forms an equivalent wavelength called the synthetic wavelength Λ:

$$\Lambda = \frac{\lambda_1 \lambda_2}{|\lambda_1 - \lambda_2|}. \tag{2}$$

The synthetic wavelength is much larger than the original ones, resulting in an increase of the measuring range at the expense of resolution. Since laser diodes can be easily tuned, they are capable of generating a wide range of synthetic wavelengths, making them a good alternative to more expensive argon lasers or dye lasers. Thus, actual laser diodes can be easily tuned by several tens of GHz, by changing the junction temperature, and can provide Λ in the centimetre range.

When L is the optical path difference in the interferometer, the measured phase difference is:

$$\Delta\Phi = \frac{2\pi L}{\Lambda} = \frac{2\pi L(\lambda_1 - \lambda_2)}{\lambda_1 \lambda_2}. \tag{3}$$

Two-wavelength interferometry using laser diodes associated with superheterodyne detection is very effective in distance measurements, increasing the measuring resolution at an arbitrary controlled synthetic wavelength without the need for interferometric stability at the individual optical wavelengths [*13*]. For this purpose, two laser diodes operating at two different optical frequencies v_1 and v_2, corresponding to the wavelength λ_1 and λ_2, are used to illuminate a Michelson interferometer, as shown in Fig. 10. Each source generates two orthogonal polarized beams of slightly different frequencies, that have been controlled accurately by means of two acousto-optical modulators working at frequencies of $f_1 = 40$ MHz and $f_2 = 40.1$ MHz, producing a beat signal with carrier frequency $(f_1 + f_2)/2$ and a modulation frequency $(f_2 - f_1)/2$. The output signal, obtained from a superhetorodyne detector, as one of the

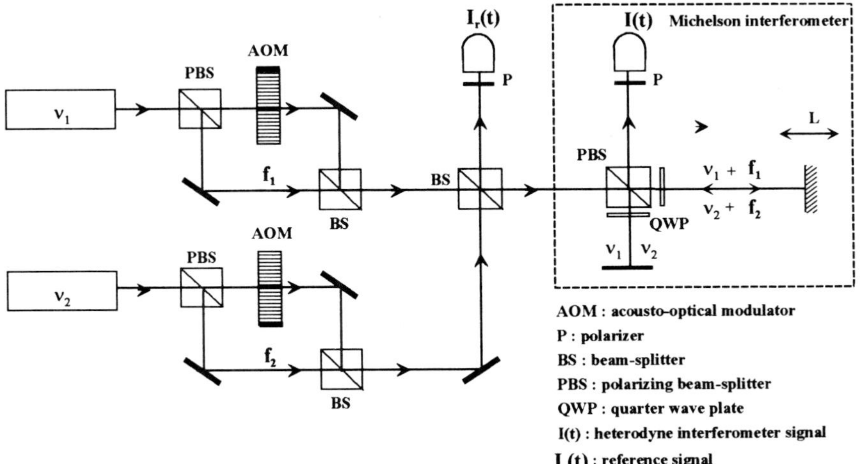

Fig. 10.

interferometer mirrors is moved, is squared, low-pass filtered, and processed in a computer to obtain the phase difference. In this case, with the demodulated frequency $f_2 - f_1 = 100$ kHz, phase measurement is obtained with an accuracy of 0.0063 rad. Using two laser diodes with a frequency difference of 15 GHz, the synthetic wavelength is $\Lambda = 20$ mm and distances up to 10 m can be measured with an accuracy of 1 part in 10^6.

For highly accurate distance evaluation, laser diodes have to be locked with respect to external wavelength references. This consists of locking each laser to its own external reference as a Fabry–Perot interferometer, an absorption atomic line, or a combination of both [*14*]. Dändliker *et al.* [*15*] applied these techniques for two laser diodes (Sharp LT027) operating at 780 nm. By generating between them a wavelength difference of 6 nm, a synthetic wavelength $\Lambda = 0.12$ mm was obtained. To achieve higher stability, one of the laser diodes was stabilized with the help of a rubidium absorption line, to give the 'master' frequency laser. The second 'slave' laser diode was then tracked by means of a Fabry–Perot etalon. The error signal measured in the stabilization control loop was estimated with an overall stability of the two laser frequencies of about ±150 kHz, corresponding to a relative measurement stability of 5 parts in 10^8 for a distance range of several metres.

Distance and Velocity Measurements by FM Laser Diode Interferometry

The FM technique has been used for absolute distance measurements with an approach similar to the frequency modulated continuous wave technique commonly used for ranging in radar and lidar systems [*16*]. In contrast with the technique previously described, frequency chirp interferometry can use only one frequency-modulated laser diode and is more closely related to the dependence of the operating frequency of the laser diode on the injection current. When the injection current is modulated by a ramp signal, the laser diode frequency is time dependent and the interferometer output intensity becomes a beat signal. When triangular current modulation (modulation frequency f_M, modulation amplitude ΔI_m) is added to the dc bias current (Fig. 11), the corresponding frequency change is given by

$$v(t) = v_0 + 2\,\Delta v f_M t \tag{4}$$

where v_0 is the central optical frequency, t is the time, and Δv represents the frequency deviation. Kubota *et al.* [16] developed an interferometer using a laser diode whose frequency is swept linearly with time for measurements of distances, as shown in Fig. 12. In this arrangement, interference takes place between the beams reflected from the front face of a fixed mirror and a movable mirror. For an optical path difference L, a time delay $2L/c$ is produced between the two beams, and the beams intefere to yield a beat signal f_b which is proportional to the distance L to be measured:

$$L = \frac{c f_b}{4\chi \Delta I_m f_M} \tag{5}$$

where c is the velocity of light.

Experiments have been carried out with the Hitachi HL 7802 laser diode operating at 790 nm. A 90 Hz triangular wave with 15 mA peak-to-peak current modulation was applied, introducing a modulation efficiency of $\xi = 4.1\ \text{GHz}\,\text{mA}^{-1}$. A displacement accuracy of 0.02 μm and a distance accuracy of 100 μm over a range of a few metres were obtained by means of this massive configuration.

This kind of system is not adapted for 'on-line' inspection in the micro-industry because of its size, relative complexity, and significant cost. To solve

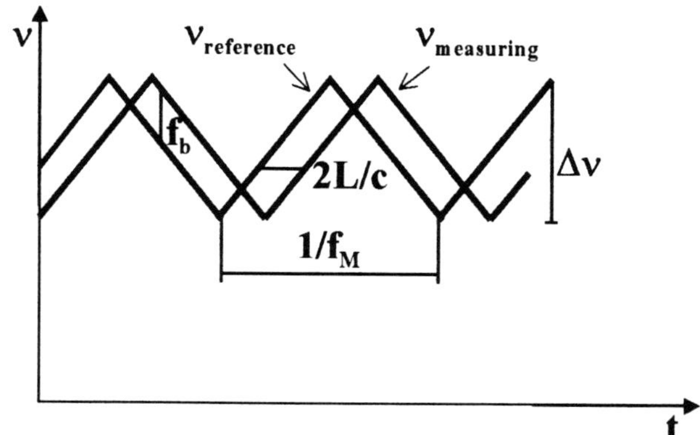

Fig. 11.

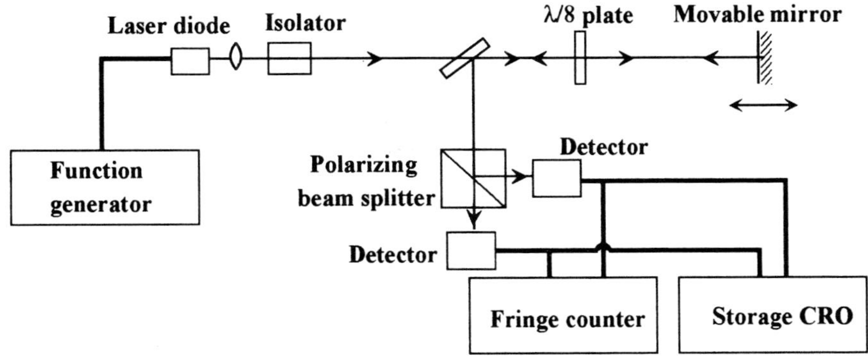

Fig. 12.

this problem and to increase the measurement accuracy in distance measurements, Chebbour *et al.* proposed a technique in which a double Mach–Zehnder interferometer is integrated on a monolithic optical circuit, as shown in Fig. 13 [*17*]. This has a $14 \times 23 \times 95$ mm^3 head sensor developed by CSEM (Neuchâtel, Switzerland), and one of the interferometers measures the displacement of a moving target retroreflector introducing an optical path difference L_m, while the second interferometer acts as a fixed-length reference with an optical path difference L_r. This permits a direct comparison between the phase read in the measuring interferometer with the phase given in the reference interferometer. The resulting distance measurement is independent of frequency modulation f_m and frequency shift Δv. It is also stabilized for external air-refractivity effects. In addition, by using a waveguide multiple imaging technique [*18*], four optical signals close to quadrature are obtained simultaneously in the output plane of each interferometer. All the heterodyned

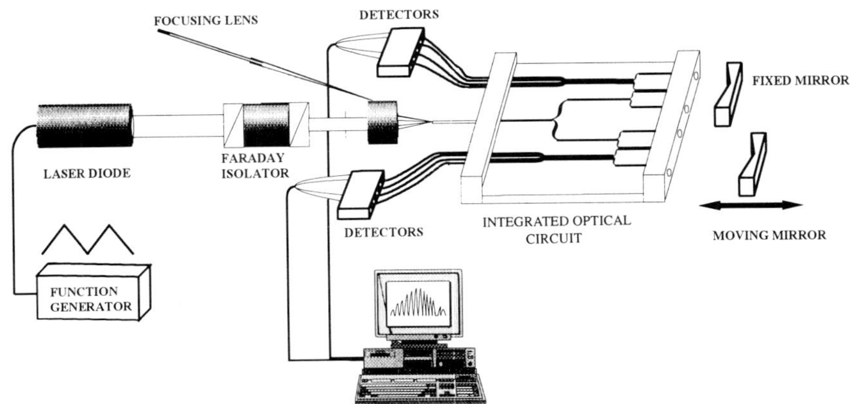

Fig. 13.

signals are collected by eight pigtailed detectors interfaced to a microcomputer via a high-speed interface board. The phase-stepping technique improves the phase evaluation. The laser is a short external cavity laser diode (Sharp LT080) with the temperature controlled at 0.05 °C by means of a thermoelectric cooler. When a linear current ramp is applied to the laser diode, the phase measurement in the output of the measuring interferometer is a linear function of time:

$$\Delta\Phi_{\mathrm{m}}(t) = 4\pi\,\frac{\Delta\nu}{c}\,f_{\mathrm{M}}L_{\mathrm{m}}t + \Phi_0 \tag{6}$$

where Φ_0 represents the initial phase, and c is the velocity of light.

The corresponding measuring beat frequency is then:

$$f_{\mathrm{bm}} = 2\Delta\nu f_{\mathrm{M}}\,\frac{L_{\mathrm{m}}}{c} \tag{7}$$

where f_{bm} and f_{br} are the beat frequencies corresponding to the measuring and reference interferometers, respectively.

Thus, the distance to be measured is defined as the ratio of two beat frequencies:

$$L_{\mathrm{m}} = \frac{f_{\mathrm{bm}}}{f_{\mathrm{br}}}\,L_{\mathrm{r}}. \tag{8}$$

When the laser diode was operated in single mode with a frequency deviation of $\Delta\nu = 30$ GHz, a relative distance accuracy of 2 parts in 10^6 for the maximum distance of 400 mm was obtained. One of the range limitations is due to microlenses used to collimate beams emerging from the waveguide which produce a significant beam divergence, decreasing the interference signal intensity at distances larger than 300 mm. The system is able to determine simultaneously the absolute distance and the velocity of a moving target with directional discrimination [17]. An example of wavefronts measured at a distance of 200 mm, corresponding to opposite directions, is shown in Fig. 14. Figure 14(a) shows the beat signal wavefronts obtained for a target translating with a velocity of $V = +5$ mm s^{-1}, and Fig. 14(b) represents the opposite case. It can be seen that the form of the beat frequency ramp reverses as the target changes its sense of translation. It was possible to determine the direction velocities from a few µm s^{-1} to some mm s^{-1}.

Laser Diode Feedback Metrology

Laser diode behaviour can be seriously affected by external optical feedback, which is due to the portion of the laser output injected back into the laser cavity from a diffuse reflector external to it. On one hand, unintentional external feedback in the laser cavity is a source of noise and degrades the spectral

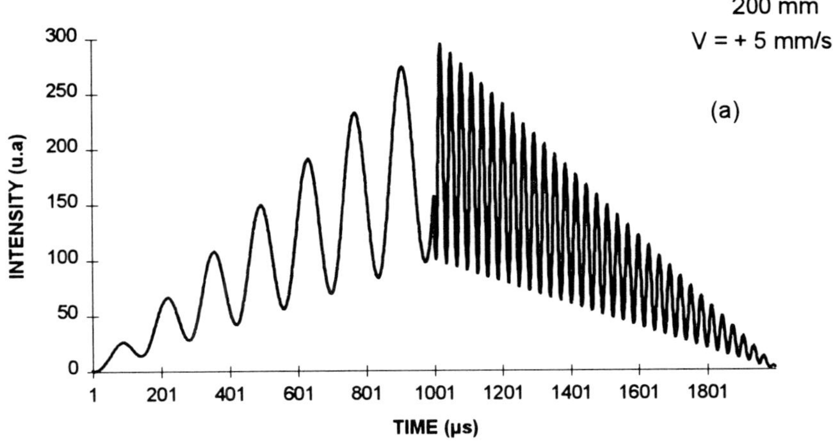

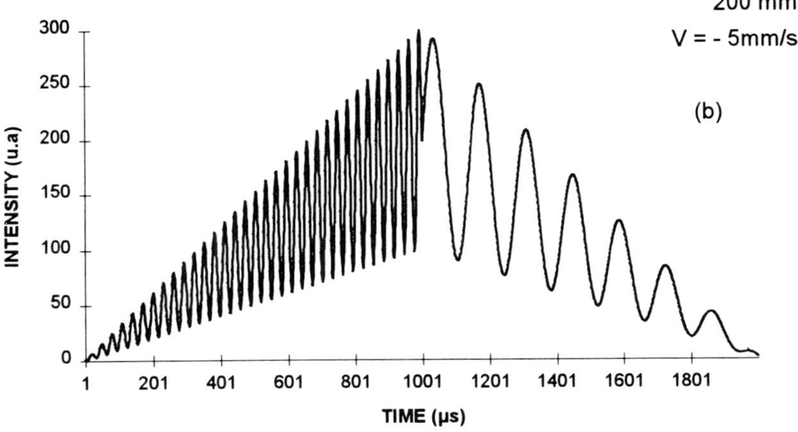

Fig. 14.

properties of the laser diode. On the other hand, controlled external feedback is potentially of practical use.

The typical configuration for demonstrating self-mixing laser ranging is shown in Fig. 15. The incident beam emitted from the laser diode is focused by a microlens placed in front of a laser diode and is scattered back by the rotating disk generating Doppler-shifted scattered light. A small part of it returns into the laser diode cavity mixing with the original oscillating wave. The resulting power modulation signal is detected using a photodetector mounted in the backside of laser package and recording the laser output power emitted from the rear end of the laser diode. In this technique the careful alignment required by conventional interferometry is not required because the only optical

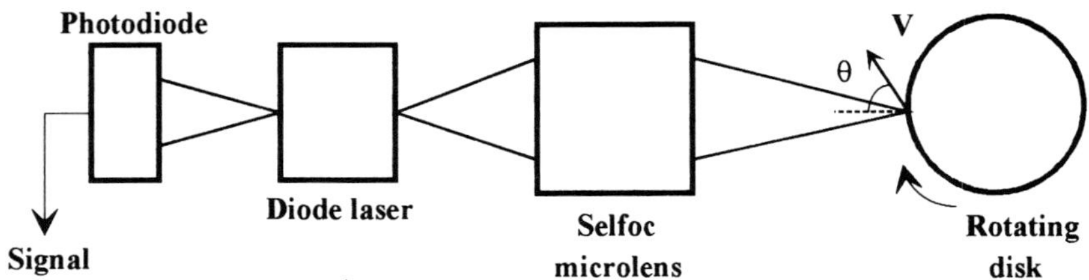

Fig. 15.

component needed is a microlens used to collimate the laser beam on the target disk. The observed Doppler beat signal is significantly different from that obtained by conventional interferometry because the beat signal spectrum consists not only of the fundamental Doppler peak but also its higher harmonics [19]. The corresponding beat waveform is now not sinusoidal, but is strongly distorted having the form of a sawtooth wave. The inclination of the sawtooth-like waveform is of practical use in directional discrimination of the target motion because the waveform is reversed when the target changes its sense of rotation. When the disk rotates with a uniform velocity V and the angle between the laser beam direction and the vector V is θ, the Doppler beat frequency can be written:

$$f_{\mathrm{d}} = \frac{2V}{\lambda} \cos \theta. \tag{9}$$

Since the moving object acts as a weakly reflective external cavity mirror we can model this system as a three-mirror Fabry–Perot cavity (see Fig. 6). Until now we have considered the situation in which the laser diode is frequency modulated and the external mirror has a fixed distance X_0 from the laser at time zero. When the target is in motion, the time-of-flight is given by:

$$\tau_x = \frac{2(X_0 + Vt)}{c}. \tag{10}$$

The beat Doppler frequency can be written as [20]:

$$f_{\mathrm{d}} = \frac{2V}{\lambda} + \frac{2X_0}{c}\left(\frac{\delta v}{\delta t}\right) \tag{11}$$

where $(\delta v/\delta t)$ is the frequency 'chirp' obtained by thermally tuning the laser diode via the pump current.

De Groot and Gallatin [21] demonstrated that the self-mixing modulation in the output power of a laser diode generated by a moving target can modify the lasing modes and threshold conditions of the resulting three-mirror cavity. The

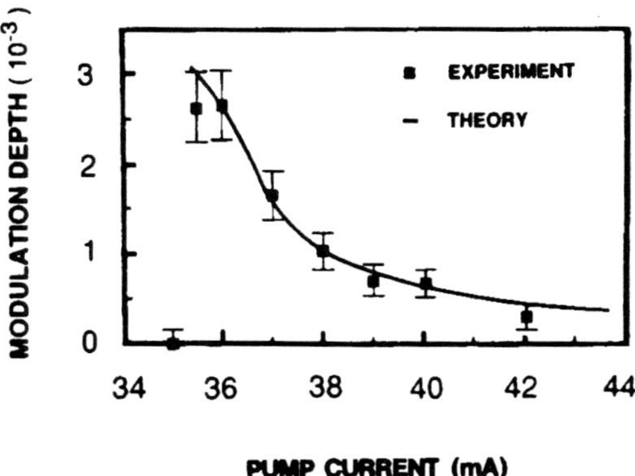

Fig. 16.

intensity modulation for a self-mixing laser diode can be written as:

$$W = \frac{(\Delta P_{out})_{max}}{P_{out}} \approx \left[\frac{J/J_{th}}{J/J_{th} - 1} \right] R_t \eta \tag{12}$$

where R_t is the reflectivity of the target, and η represents the coupling efficiency of the backscattered light into the laser diode.

As shown in Fig. 16 [21], the optimal modulation depth as a function of pump current was measured for a Sharp LT015 laser diode. In this curve the modulation depth is given by the ratio of the amplitude of the ac component to the dc component of the signal monitored by a photodetector mounted in the backside of the laser diode. The best modulation depth for self-mixing is obtained with the laser running near its threshold level (35 mA). The use of the external cavity allows for a good control of the near-threshold operation, where the modulation depth is better than that obtained with conventional homodyne systems. The consequence of this fact is to extend the operational range of self-mixed laser diode velocimeters from 2 m to 50 m.

SURFACE PROFILING BY USING POLYCHROMATIC LIGHT SOURCES

For many years, interferometry has been used as a standard instrument for surface inspection. Recent progress in semiconductor technology for two-dimensional light sensors, data acquisition boards and microcomputers has allowed the proposal of accurate optoelectronic systems to replace visual analysis of fringe patterns. The first developments were carried out with monochromatic light sources. In 1981, Sommargren [22] proposed a phase measurement technique based on heterodyne detection. His work gave birth to phase shifting interferometry (PSI) [23] which is widely used in the industrial world as a powerful surface inspection tool. The height sensitivity of commercially available systems is better than 1 nm, while lateral resolution is limited by the diffraction of the lens.

Further developments of low-coherence (or polychromatic) interferometry, also called white light interferometry (WLI), were motivated by the intrinsic limitation of monochromatic PSI techniques due to the well-known phase ambiguity problem. If the phase difference between two consecutive points is larger than π, we are unable to correctly unwrap the phase and an error occurs. The basic purpose of research carried out on low-coherence interferometric profilers is to remove this phase ambiguity problem while maintaining a good height resolution. The success of this research will allow the extension of the field of application of interferometric profile measurement to rougher surfaces. In these methods, a wide light spectrum is necessary to retrieve height information relative to the surface under inspection.

Before we describe the operating principles of these profilers, let us consider some theoretical background. The output intensity of a two-beam interferometer may be expressed as a function of the optical path difference (OPD) [24]:

$$I(z) = I_{\mathrm{ref}} + I_{\mathrm{surf}} + 2(I_{\mathrm{ref}}I_{\mathrm{surf}})^{1/2}\gamma^{(\mathrm{r})}(2z/c) \tag{13}$$

where I_{ref} and I_{surf} are intensities issued from the reference and measuring arms of the interferometer, respectively. $2z/c$ is the time delay between the two optical paths, and $2z$ is the round trip OPD. In Eq. (13), $\gamma^{(\mathrm{r})}(2z/c)$ is the real part of $\gamma(2z/c)$, that is the complex degree of coherence of the light source. $\gamma(2z/c)$ is the normalized form of the self-coherence function of the light source $\Gamma(2z/c)$. $\Gamma(2z/c)$ is the Fourier transform of the power spectral density function of the light source. It can be written:

$$\Gamma(2z/c) = \int_{-\infty}^{+\infty} |a(\sigma)|^2 \exp(-\mathrm{i}2\pi\sigma 2z)\,\mathrm{d}\sigma. \tag{14}$$

This formula shows that the interferometer output intensity is defined by the linear superimposition of the individual contribution of each spectral component of the light source.

For usual polychromatic light sources, coherence functions may be analytically described and the output intensity of the interferometer is [25]:

$$I(z) = I_{\mathrm{ref}} + I_{\mathrm{surf}} + 2(I_{\mathrm{ref}}I_{\mathrm{surf}})^{1/2}g(2z)\cos(4\pi\bar{\sigma}z + \phi_0) \tag{15}$$

where $g(2z)$ is the fringe envelope which defines the fringe contrast, $\bar{\sigma}$ is the mean wavenumber of the light source, $\phi(z) = 4\pi\bar{\sigma}z$ is the phase term due to the length difference z between the interferometer arms and ϕ_0 is an additional phase term due to different reflections on the beam-splitter and mirrors.

The width of the light spectrum appears through the term $g(2z)$. For a monochromatic light source, this term always equals one, and the z-response

$I(z)$ becomes perfectly periodic. We may write:

$$I(z) = I\left(z + \frac{k}{2\sigma}\right) \qquad (16)$$

where k is an integer. This periodicity is responsible for the phase ambiguity problem of monochromatic PSI which limits its field of application to height steps smaller than $1/2\sigma$ (or $\lambda/2$).

If a polychromatic light source is used, the interference fringes produced by the cosine term are modulated by $g(2z)$ which modifies the fringe contrast as a function of the OPD equal to $2z$. Figure 17 shows the curve of the experimental z-response of a Mirau interferometer illuminated by a halogen lamp. Due to the short coherence length of the light source, only a few interference fringes are present.

The term $g(2z)$ in Eq. (15) shows that low-coherence interferometry presents a better potential for optical profilometry than monochromatic interferometry. The width of the light source spectrum brings complementary information which is encoded into this term. Thus Eq. (15) provides two information sources about the OPD $2z$. One is given by the fringe envelope $g(2z)$, the other one by the phase term $\phi(z)$. Both terms may be used for profilometry. However, they vary simultaneously and height calculations are necessarily more complex than in the case of monochromatic interferometry. For instance, PSI techniques using OPD modulation (by using for example a piezoelectric transducer PZT) are no longer applicable.

Various profile measurement techniques using low-coherence interferometry have been proposed. They use different ways of reconstructing the surface profile but they are based on a single operating principle (similar to that for

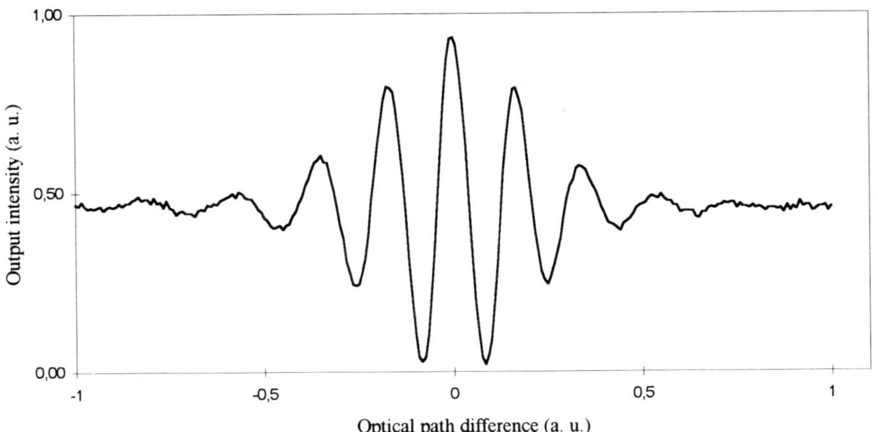

Fig. 17. *z*-response of a Mirau interferometer illuminated by low-coherence light source.

profilometry by monochromatic confocal microscopy). The surface is progressively shifted from side to side with respect to the optical contact along the z-axis, for instance by using a PZT. The z-response $I(z)$ of the low-coherence interferometer used is sampled. The recorded sets of data are processed to deduce the particular position z for which the interferometer arms match. Electronic detection units [26] or FFT based data treatments [27] are used for 'coherence peak' identification. In this case, we look for the amplitude maximum of $g(2z)$. Height resolution depends both on the z sampling rate and data treatment but height sensitivity is limited by the relatively low dependence of $g(2z)$ on the change in z.

The phase term $\phi(z)$ is much more sensitive to z variations. Therefore $\phi(z)$ can provide a greater accuracy on height measurements. Direct identification of the zero-order interference fringe is then possible. This easy-to-implement technique requires a null value for ϕ_0 and cannot be applied to every kind of surface. The surface profile presented in Fig. 18 was measured with a 3 nm resolution using this method [28].

A different phase measurement approach is based on FFT computation of the sampled z-response $I(z)$. In this way, the phase $\phi(z, \sigma) = 4\pi\sigma z$ associated with each spectral component σ of the light source is computed. The phase variation $\Delta\phi$ associated with a wavenumber change $\Delta\sigma$ is proportional to the OPD:

$$\Delta\phi/\Delta\sigma = 4\pi z \tag{17}$$

By computing the phase for different values of σ, we are able to extract the

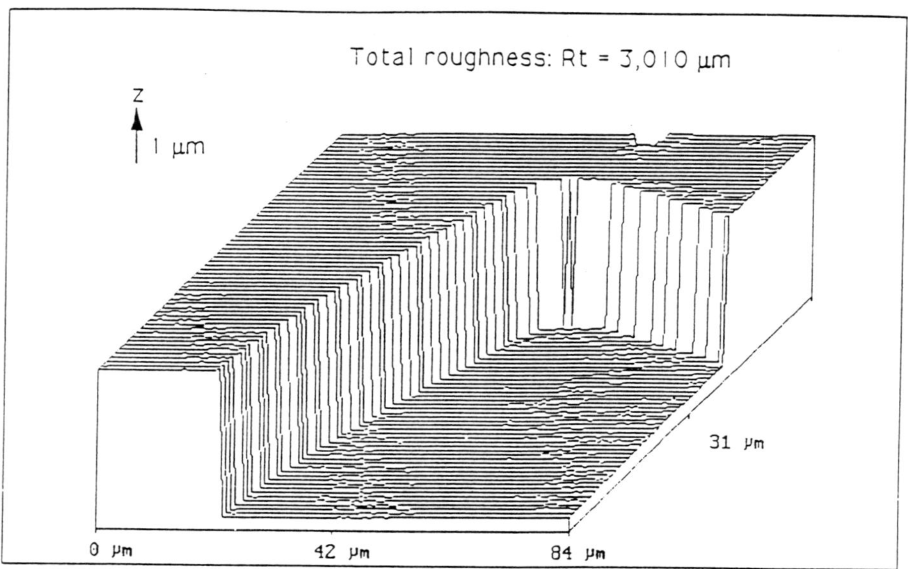

Fig. 18. Profile of a silicon sample measured by low-coherence interferometer.

surface height z. Available data to perform this FFT are recorded during z scanning of the OPD. The sampling distance between two consecutive positions of the surface must be carefully chosen. To respect the Nyquist criterion, the sampling distance must be shorter than a quarter of the mean wavelength $\bar{\lambda}$ of the light source. For $\bar{\lambda} = 0.5~\mu m$, at least 800 steps are necessary to inspect a depth equal to 100 μm. This frequency domain analysis performs the inverse Fourier transform of the self-coherence function $\Gamma(2z/c)$ (given in Eq. (14)) in order to retrieve the individual contribution of each spectral component. This technique allows the profiling of surfaces with a height repeatability of 0.5 nm [29]. To avoid calibrated OPD scanning and numerical FFT treatment of the set of data, this frequency domain analysis can be performed optically [30]. In this case, the white-light encoded interferogram is spectrally analysed by using a spectroscopic device. The absolute height distribution of one line of the surface is reconstructed from only one two-dimensional interferogram and with a one nanometre resolution.

CONCLUSION

For many unconventional dimensional measurement tasks, the wavelength tunability, multiple wavelength operation, good spatial coherence, small size and low cost of laser diodes greatly extend the range of applications of sub-micrometric metrology, especially in range finding and profilometry of polished and unpolished optical surfaces. There is a critical need to bridge the gap between conventional mechanical measuring techniques and high-resolution non-contact optical methods in the manufacture of advanced components. A new class of instruments is now available because of the rapid progress in laser diode technology. These instruments are more suitable for metrology applications in industrial environments because the laser diode may be the key for fulfilling various requirements such as resolution, measure rate, mechanical stability, cost or overall dimensions.

Using new laser diodes it should be possible to build instruments with nanometre resolution and no ambiguity interval. The measuring range and resolution can be adjusted to the desired application by simply tuning the wavelengths of the individual monomode laser diode (multiple wavelength interferometry). In the case of polychromatic light interferometers using broadband sources (LEDs, SLDs), it should be possible to generate a 'multiple wavelength effect'. Then, absolute profile or distance measurements are obtained by spectral analysis. The latter may be performed either optically by means of dispersive devices or numerically by frequency domain analysis. The measuring accuracy depends on the number of spectral components resolved by the spectral analysis.

In specific applications of velocimetry, self-mixing interference based on optical feedback inside the laser diode cavity can be used to avoid the coherence length limit of conventional interferometry and to permit directional discrimination.

REFERENCES

1. (1989), in *Handbook of Solid-State Lasers* (ed. P.K. Cheo, Marcel Dekker, Inc.), New York.
2. (1988), in *Laser Diode User's Manual* (ed. Sharp Corp.).
3. A.J. den Boef (1988), *Appl. Opt.* **27**, 306–311.
4. M. Yonemura (1985), *Opt. Lett.* **10**, 1–3.
5. A.S. Gerges, T.P. Newson and D.A. Jackson (1990), *Appl. Opt.* **29**, 4473–4480.
6. Y. Ning, K.T.V. Grattan, B.T. Meggit and A.W. Palmer (1989), *Appl. Opt.* **28**, 3657–3661.
7. D.R. Hjelme, A.R. Mickelson and G. Beausoleil (1991), *IEEE J. Quantum Electron.* **27**, 352–371.
8. R. Escalona (1991), Thèse de Doctorat en Sciences pour l'Ingénieur, Université de Besançon.
9. M. Ohtsu and Y. Teramachi (1989), *IEEE J. of Quantum Electron.* **25**, 31–38.
10. S. Matsui, T. Takiguchi, M. Taneya, O. Yamamoto, H. Hayashi, S. Yamamoto, S. Yano and T. Hijikata, *Appl. Opt.* **23**, 4001–4006.
11. A. Chebbour and C. Gorecki (1993), EOS Annual Meetings Digest, Zaragoza, 203–204.
12. C. Gorecki, A. Chebbour and G. Tribillon (1994), *Proc. SPIE* **2340**, 356–365.
13. R. Dändliker, R. Thalmann and D. Prongué (1988), *Opt. Lett.* **13**, 339–341.
14. P.J. de Groot and S. Kishner S. (1991), *Appl. Opt.* **30**, 4026–4033.
15. R. Dändliker (1993), EOS Annual Meetings Digest, Zaragoza, 243–246.
16. T. Kubota, M. Nara and T. Yoshino (1987), *Opt. Lett* **12**, 310–312.
17. A. Chebbour, C. Gorecki and G. Tribillon (1994), *Proc SPIE* **2248**, 176–186.
18. G. Voirin, L. Falco, O. Boillat, O. Zogmal, P. Regnault and O. Parriaux (1993), 6th European Conference on Integrated Optics, Neuchâtel.
19. E.S. Shimizu (1987), *Appl. Opt.* **26**, 4541–4544.
20. P.J. de Groot, G.M. Gallatin and S.H. Macomber 1988), *Appl. Opt.* **27**, 4475–4480.
21. P.J. de Groot and G.M. Gallatin (1989), *Opt. Lett.* **14**, 165–167.
22. C.E. Sommargren (1981), *Appl. Opt.* **20**, 610–618.
23. K. Creath (1988), in *Progress in Optics* vol. 26, (ed E. Wolf), 357–373.
24. M. Born, E. Wolf (1987), *Principles of Optics* (6th ed. Pergamon Press), 502.
25. P. Sandoz and G. Tribillon (1993), *J. Mod. Opt.* **40**, 1691–1700.
26. P. Hariharan and M. Roy (1994), *J. Mod. Opt.* **41**, 2197–2201.
27. P.J. Caber (1993), *Appl. Opt.* **32**, 3438–3440.
28. G.S. Kino and S.S.C. Chim (1990), *Appl. Opt.* **29**, 3775–3783.
29. L. Deck and P.J. de Groot (1993), *Appl. Opt.* **33**, 7334–7338.
30. P. Sandoz, G. Tribillon and H. Perrin (1996), *J. Mod. Opt.* **43**, 701–708.

29 A physical method for colour photography

Guo-Guang Mu, Zhi-Liang Fang, Fu-Lai Liu and Hong-Chen Zhai
Institute of Modern Optics, NanKai University, Tianjin 300071, China

INTRODUCTION

Until now, several different kinds of materials have been developed to record, store, and display colour images or information, among which photographic colour film has become the most popular and commercially available. The multilayered colour film made of organic dyes, however, suffers from slowly fading with the passage of time, which limits its application for long-term storage.

Much effort has been made to develop practical techniques of using silver halide black-and-white film, which is stable for long-term storage and is responsive, however, only to light intensity. The most common technique used is to decompose a colour image into three sub-images of different primary colours with the help of three filters, followed by the recording of the sub-image separately in three black-and-white films for storage. For displaying the original colour image, three projectors with different primary-colour light are needed to project the three sub-images and compose them precisely into the original image.

The drawback of this technique is that three times as much film is needed, compared with the usage of colour film. Besides, much attention has to be paid to precise adjustment during the composition process.

Ives reported for the first time a technique [1] for colour image retrieval

---

involving the composition of a colour image from a black-and-white transparency, in which a projector and gratings of different spatial frequencies or orientations were used to compose colour images based on the diffraction process. Since then, similar techniques for recording and storage of colour information or colour retrieval with black-and-white film were developed by Mueller [2], Macovski [3], Grousson and Kinany [4], and Yu [5].

In Yu's approach, the colour image was recorded on black-and-white film by three sequential encodings with a Ronchi grating through three filters of different primary colour.

Mu *et al.* [6] designed and fabricated an optical spatial modulator (tricolour grating encoder), with which colour encoding can be executed in a single process. During the recording, the tricolour grating is placed in contact with a panchromatic black-and-white film, and the color information is encoded and recorded simultaneously on the black-and-white film in a single white-light exposure.

For the retrieval and display of the recorded colour information, the developed film is placed in the input plane of a white-light processor, and its spatial frequency spectra are filtered through three corresponding primary colour filters at its Fourier plane. In this way, the original colour image can be retrieved and displayed at the output plane of the processor.

With this technique, real-time colour photography with black-and-white film using commercially available cameras becomes practical.

The principle of this technique is based on the physical solution of reducing the number of recording parameters and transforming them from three related chemical processes into physical parameters related to space frequency, which can be expressed in a grey scale distribution for storage on black-and-white film.

In this chapter, we outline this technique with the emphasis on the mathematical analysis and its physical meaning.

GREY SCALE ENCODING OF COLOUR INFORMATION

Suppose that a colour transparency is used as the object to be encoded on a black-and-white film. With proper design of the tricolour grating encoder, in the physical solution, the three primary colour sub-images of it will be spatially sampled with a certain specific sampling frequency and three directions, and then separated and recorded on a monochrome black-and-white film in the form of only a grey scale distribution. In this section, we describe the principle of this technique in detail.

Tricolour Grating

Figure 1 shows a schematic drawing of the tricolour grating, which is

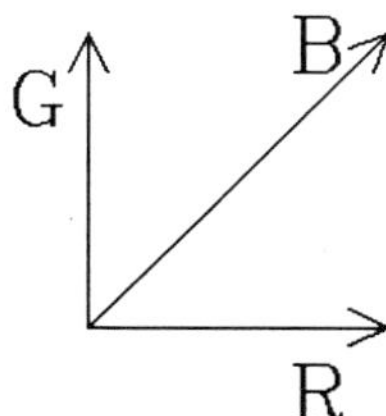

Fig. 1. Schematic drawing of the optical colour modulator.

composed of three primary colour gratings with the same spatial frequency and oriented in three different directions, namely red ruling in the x direction, green in x', and blue in x'', on an opaque background. Thus, the intensity transmittance from the grating can be given by the following expression:

$$T_{\mathrm{T}}(x,y) = [\tfrac{1}{2} + \tfrac{1}{2}\,\mathrm{sgn}(\cos p_0 x)]_{\mathrm{R}} + [\tfrac{1}{2} + \tfrac{1}{2}\,\mathrm{sgn}(\cos p_0 x')] + [\tfrac{1}{2} + \tfrac{1}{2}\,\mathrm{sgn}(\cos p_0 x'')]_{\mathrm{B}}$$

$$(1)$$

where p_0 is the spatial frequency of the ruling, and R, G, and B denote the transmittance of the three primary colour (red, green, and blue) gratings, respectively, and

$$\mathrm{sgn}(\cos p_0 x) = \begin{cases} 1, & \cos(p_0 x) \geq 0 \\ -1, & \cos(p_0 x) < 0. \end{cases}$$

A tricolour grating can be fabricated in the laboratory by superimposing a fresh colour slide film with a close-contact Ronchi ruling of angular spatial frequency p_0, and the recording will include sequential exposures of the slide film to the three primary collimated light illuminations. Between the exposures, the Ronchi ruling is rotated by a certain angle, say 45°. After the developing and fixing processes, in which the gamma of the slide film is maintained at unity, a colour transparency will be obtained, whose intensity transmittance can be described by Eq. (1). Proper design of the tricolour grating will greatly eliminate the appearance of moive fringes.

Encoding

The encoding of a colour image is performed in the same way as a normal recording process, except that a tricolour grating is used and superimposed with the black-and-white photographic recording film. This process can be carried out by using an ordinary camera or other commercially available instruments. The optical scheme of the colour encoding is shown in Fig. 2, where a tricolour grating is placed at the imaging plane of the imaging lens L, which is in contact with a panchromatic black-and-white film. Through the modulation of the tricolour grating, the colour image is imaged by the lens on the black-and-white film. The irradiance of the colour image can be expressed as

$$T(x, y) = [T_r(x, y)]_R + [T_g(xc, y)]_G + [T_b(x, y)]_B \tag{2}$$

where T_r, T_g, and T_b are the red, green, and blue primary colour components of the colour image irradiance, and subscripts R, G, and B denote the display colours. For simplicity of the formulation, the green and blue components are expressed with the coordinate systems (x', y') and (x'', y''), respectively. That is

$$T(x, y) = [T_r(x, y)]_R + [T_g(x', y')]_G + [T_b(x'', y'')]_B. \tag{3}$$

The intensity distribution in the grey scale of the panchromatic black-and-white film is the product of $T(x, y)$ and $T_T(x, y)$. For each primary colour sub-image, the other primary colour gratings seem to contain only opaque parts, for the other two transparent parts do not allow any passage of it. An optimal encoding result depends on a properly designed modulator. There is no need to balance the colour exposures as in the multiexposure encoding process. After

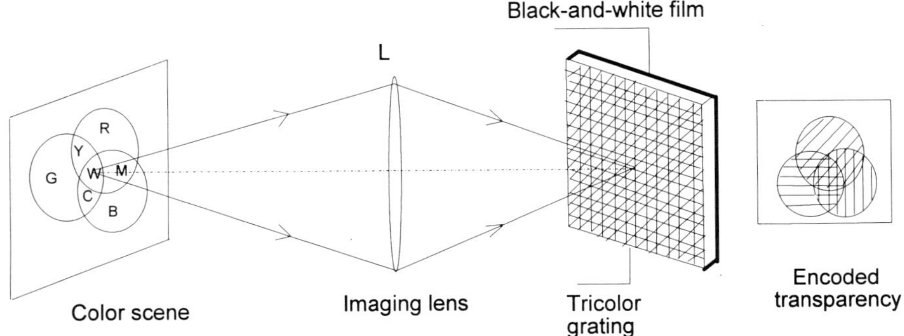

Fig. 2. Optical configuration of the encoding.

the exposure and development, a negative encoded transparency is obtained, whose intensity transmittance can be described as

$$
\begin{aligned}
T_n(x, y) &= [T(x, y)T_T(x, y)]^{-\gamma_{n1}} \\
&= \{T_r(x, y)[\tfrac{1}{2} + \tfrac{1}{2}\mathrm{sgn}(\cos p_0 x)] \\
&\quad + T_g(x', y')[\tfrac{1}{2} + \tfrac{1}{2}\mathrm{sgn}(\cos p_0 x')] \\
&\quad + T_b(x'', y'')[\tfrac{1}{2} + \tfrac{1}{2}\mathrm{sgn}(\cos p_0 x'')]\}^{-\gamma_{n1}}
\end{aligned}
\tag{4}
$$

where γ_{n1} is the gamma of this negative film. We will omit the proportional constant hereafter for simplicity. It is noticed that Eq. (4) is the same expression reported by Yu [5], where a triple exposure encoding process is introduced. In our case, however, the colour encoding is performed by only a single exposure, thanks to the usage of our novel tricolour grating.

Through a contact printing process with another black-and-white film of a gamma value γ_{n2}, a positive encoding transparency can be obtained, of which the intensity transmittance can be given by

$$
\begin{aligned}
T_p(x, y) &= [T_n(x, y)]^{-\gamma_{n2}} = \{T_r(x, y)[\tfrac{1}{2} + \tfrac{1}{2}\mathrm{sgn}(\cos p_0 x)] \\
&\quad + T_g(x', y')[\tfrac{1}{2} + \tfrac{1}{2}\mathrm{sgn}(\cos p_0 x')] \\
&\quad + T_b(x'', y'')[\tfrac{1}{2} + \tfrac{1}{2}\mathrm{sgn}(\cos p_0 x'')]\}^{\gamma_{n1}\gamma_{n2}}.
\end{aligned}
\tag{5}
$$

The films used in negative encoding and positive printing are chosen such that the product of their gammas is equal to 2, so that the intensity transmittance of the positive encoded transparency can be given by

$$
\begin{aligned}
T_p(x, y) &= \{T_r(x, y)[\tfrac{1}{2} + \tfrac{1}{2}\mathrm{sgn}(\cos p_0 x)] + T_g(x', y')[\tfrac{1}{2} + \tfrac{1}{2}\mathrm{sgn}(\cos p_0 x')] \\
&\quad + T_b(x'', y'')[\tfrac{1}{2} + \tfrac{1}{2}\mathrm{sgn}(\cos p_0 x'')]\}^2.
\end{aligned}
\tag{6}
$$

So we have the amplitude transmittance of this encoded transparency as

$$
\begin{aligned}
t_p(x, y) &= T_r(x, y)[\tfrac{1}{2} + \tfrac{1}{2}\mathrm{sgn}(\cos p_0 x)] + T_g(x', y')[\tfrac{1}{2} + \tfrac{1}{2}\mathrm{sgn}(\cos p_0 x')] \\
&\quad + T_b(x'', y'')[\tfrac{1}{2} + \tfrac{1}{2}\mathrm{sgn}(\cos p_0 x'')]
\end{aligned}
\tag{7}
$$

In this way the encoded positive transparency is obtained with a single exposure with the modulator. The three primary colour sub-images are modulated by gratings of different azimuth orientations. The rightmost part of Fig. 2 shows a schematic example.

This two-step colour encoding process can be replaced by a one-step encoding process through a different developing procedure. A positive encoded transparency obtained in this way will have an intensity transmittance

$$
\begin{aligned}
T_p(x, y) &= \{T_r(x, y)[\tfrac{1}{2} + \tfrac{1}{2}\mathrm{sgn}(\cos p_0 x)] + T_g(x', y')[\tfrac{1}{2} + \tfrac{1}{2}\mathrm{sgn}(\cos p_0 x')] \\
&\quad + T_b(x'', y'')[\tfrac{1}{2} + \tfrac{1}{2}\mathrm{sgn}(\cos p_0 x'')]\}^{\gamma}
\end{aligned}
\tag{8}
$$

where γ is determined by the parameters of the film used and by the related photochemical process. With certain development conditions the γ can be controlled to be equal to 2, so that an encoded positive transparency can be obtained directly. The positive linear amplitude transmittance of this encoded black-and-white transparency then becomes

$$t_{\mathrm{p}}(x, y) = [T_{\mathrm{r}}(x, y)[\tfrac{1}{2} + \tfrac{1}{2}\mathrm{sgn}(\cos p_0 x)] + T_{\mathrm{g}}(x', y')[\tfrac{1}{2} + \tfrac{1}{2}\mathrm{sgn}(\cos p_0 x')]$$
$$+ T_{\mathrm{b}}(x'', y'')[\tfrac{1}{2} + \tfrac{1}{2}\mathrm{sgn}(\cos p_0 x'')]]. \tag{9}$$

This is the same expression as Eq. (7). A second contact printing process will, however, not be necessary at all.

The key points of the technique described here are to reduce the number of recording parameters and to transform them from those related to chemical processes into physical parameters related to space frequency.

Equation (1), for example, describes the process of decomposing an original colour image into only three primary colour sub-images, based on the physiological vision effect. Every term in Eq. (1) corresponds respectively to a primary colour chemical process both in the recording and storage; Eq. (3), on the other hand, stands for a further treatment to transform the three image irradiances, T, from the three coordinate systems related respectively to the three higher space carrier frequencies, which in our case, depend on the orientation angles and the interval of the rulings in the tricolour grating. Thus, the information in lower space frequencies of all the three sub-images will be carried by the three space carrier frequencies respectively, and recorded on black-and-white film in the form of a grey scale distribution.

In principle, this information recorded in a grey scale distribution can be stored either in analogue form on a black-and-white film, or in digital form in a computer, if a proper sampling frequency is used.

RETRIEVAL OF THE COLOUR INFORMATION

The technique introduced above for recording and storing colour information in a grey scale distribution will find a successful application only if a joint technique for the retrieval of the colour information is accompanied with it.

In this section, we describe the corresponding technique for colour information retrieval.

Optical Decoding System

The optical decoding system used for retrieving and displaying the colour images stored on a black-and-white encoded transparency is shown in Fig. 3.

The principle of displaying the colour image is based on the Fourier transform technique. The stored colour image can be optically displayed by

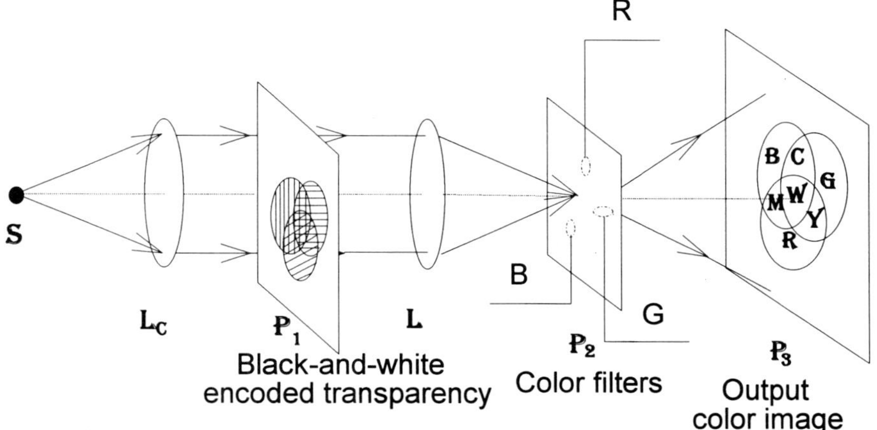

Fig. 3. Optical configuration of the decoding.

employing three primary colour filters on the Fourier spectrum of the encoded transparency. As shown in Fig. 3, the encoded positive transparency is placed at the input plane P_1. Under white-light illumination, its multichromatic Fourier spectra are formed on the Fourier plane P_2 by the Fourier transform of the first achromatic Fourier lens L. Red, green, and blue primary colour filters are inserted at the corresponding first-order of the Fourier spectra to modulate the first Fourier transform orders of the transparency, as shown in Fig. 4.

The complex light distribution for the wavelength λ at the Fourier plane P_2 is

$$E(p, q; \lambda) = \int t_p(x, y)\exp[-i(xp + yq)]\, dx\, dy \tag{10}$$

Evaluating this equation, we have

$$E(p, q; \lambda) = T_r(p, q) + \frac{1}{2}\sum_{n=1}^{\infty} a_n T_r(p \pm np_0, q)$$

$$+ T_g(p', q') + \frac{1}{2}\sum_{n=1}^{\infty} a_n T_g(p' \pm np_0, q')$$

$$+ T_b(p'', q'') + \frac{1}{2}\sum_{n=1}^{\infty} a_n T_b(p'' \pm np_0, q'') \tag{11}$$

where (p, q), (p', q') and (p'', q'') are the angular spatial frequency coordinate systems and $T_r(p, q)$, $T_g(p', q')$, and $T_b(p'', q'')$ are the Fourier transforms of $T_r(x, y)$, $T_g(x', y')$, and $T_b(x'', y'')$, respectively. Equation (11) can be rewritten in terms of the linear spatial coordinates of the Fourier

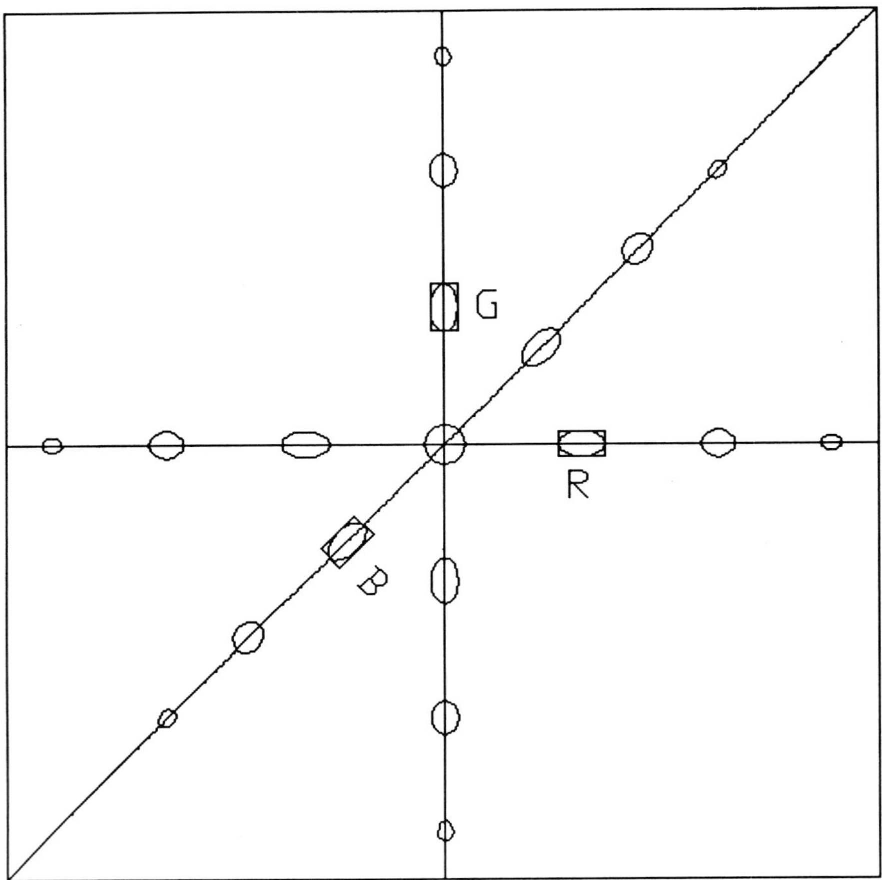

Fig. 4. Primary colour filters.

plane as

$$E(\alpha,\beta;\lambda) = T_r(\alpha,\beta) + \frac{1}{2}\sum_{n=1}^{\infty} a_n T_r\left(\alpha \pm \frac{n\lambda f}{2\pi}\,p_0,\beta\right)$$

$$+T_g(\alpha',\beta') + \frac{1}{2}\sum_{n=1}^{\infty} a_n T_g\left(\alpha' \pm \frac{n\lambda f}{2\pi}\,p_0,\beta'\right)$$

$$+T_b(\alpha'',\beta'') + \frac{1}{2}\sum_{n=1}^{\infty} a_n T_b\left(\alpha'' \pm \frac{n\lambda f}{2\pi}\,p_0,\beta''\right) \tag{12}$$

where $\alpha = (\lambda f/2\pi)p$, $\beta = (\lambda f/2\pi)q$, $\alpha' = (\lambda f/2\pi)p'$, $\beta' = (\lambda f/2\pi)q'$, α'' $= (\lambda f/2\pi)p''$, $\beta'' = (\lambda f/2\pi)q''$, and f is the focal length of the transform lens.

From that we know that the Fourier diffraction orders of different components of the encoded transparency are spatially separated except for their zero orders at the Fourier plane. If the spatial frequency of the tricolour grating is greater than twice the highest frequency of the encoded scene, there will be no overlap between the different Fourier orders. At the Fourier plane, the three first orders are filtered by a red, a green, and a blue transparent film, respectively, while the other orders are blocked. This filtering function can be described as

$$F_{\mathrm{T}}(\alpha,\beta) = \left[F\left(\alpha - \frac{\lambda_{\mathrm{R}}f}{2\pi}p_0,\beta\right)\right]_{\mathrm{R}} + \left[F\left(\alpha' - \frac{\lambda_{\mathrm{G}}f}{2\pi}p_0,\beta'\right)\right]_{\mathrm{G}} + \left[F\left(\alpha'' - \frac{\lambda_{\mathrm{B}}f}{2\pi}p_0,\beta''\right)\right]_{\mathrm{B}}$$

(13)

where

$$F(\alpha - \alpha_0,\beta) = \begin{cases} 1 & |\alpha - \alpha_0| \leq \alpha_0/2;\ |\beta| \leq \alpha_0/2 \\ 0 & \text{otherwise} \end{cases}$$

λ_{R}, λ_{G} and λ_{B} are the wavelengths for red, green, and blue light, respectively, and the terms in brackets [ ]$_{\mathrm{R}}$, [ ]$_{\mathrm{G}}$ and [ ]$_{\mathrm{B}}$ denote the colour components for which the transmittances act. As discussed above, the transmittance will become zero (opaque) for the colour components other than the one denoted by the subscript. The optical field at the Fourier plane immediately behind the colour filters will then be

$$E'(\alpha,\beta) = \tfrac{1}{2}a_1\left\{\left[T_{\mathrm{r}}\left(\alpha - \frac{\lambda_{\mathrm{R}}f}{2\pi}p_0,\beta\right)\right]_{\mathrm{R}} + \left[T_{\mathrm{g}}\left(\alpha' - \frac{\lambda_{\mathrm{G}}f}{2\pi}p_0,\beta'\right)\right]_{\mathrm{G}} \right.$$
$$\left. + \left[T_{\mathrm{b}}\left(\alpha'' - \frac{\lambda_{\mathrm{B}}f}{2\pi}p_0,\beta''\right)\right]_{\mathrm{B}}\right\}$$

(14)

Here, the Fourier spectrum for the three components of the original colour scene are represented by red, green, and blue light, respectively. Through an inverse Fourier transform by the following achromatic lens, the retrieved colour image with the original colours can then be displayed on the output plane P_3. The multiple colour field distribution is given by

$$E''(x,y) = [T_{\mathrm{r}}(x,y)\exp(ixp_0)]_{\mathrm{R}} + [T_{\mathrm{r}}(x',y')\exp(ix'p_0)]_{\mathrm{G}}$$
$$+ [T_{\mathrm{b}}(x'',y'')\exp(ix''p_0)]_{\mathrm{B}}.$$

(15)

As the primary colours are mutually incoherent, interference between different primary colour components will not occur at the output plane of the decoder. So the output image intensity distribution can be expressed as

$$I(x,y) = [T_{\mathrm{r}}^2(x,y)]_{\mathrm{R}} + [T_{\mathrm{g}}^2(x',y')]_{\mathrm{G}} + [T_{\mathrm{b}}^2(x'',y'')]_{\mathrm{B}}.$$

(16)

This retrieved colour image can be viewed directly by placing a sheet of ground glass at the output plane of the decoder or fed to a colour monitor by a portable video camera, as we did. So the colour image can be displayed directly on a monitor as the final display result.

Plate 29.1 shows the retrieved colour image produced by our display system with the encoded transparency.

Decoding with a Computer

According to the previous discussion, we know that the decoding process of the stored colour image is carried out by transforming the encoded grey scale distribution into its space spectrum, so that the information in lower space frequencies of the three sub-images can be spatially separated and retrieved with the help of three primary colour filters in the space spectral plane. The composition of the original colour image is finally completed by a reversed Fourier transformation.

These kinds of processes for Fourier transformation and reversed transformation from image space into its spectral space or vice versa can be easily undertaken by a computer in discrete form, whereby the final colour image can be composed in its inner standard colourity space. The construction of this computer-controlled decoding and display system with an opto-electronic detector is shown in Fig. 5, where the opto-electronic detector is used to detect the decoded grey scale information stored on the black-and-white transparency, which will be transferred into discrete digital data for further treatment in a computer by an image grab card.

Figure 6 shows a block diagram for decoding and for colour image composition processes in a computer. The final colour image will appear on the monitor as the output of the computer-controlled decoding and display system.

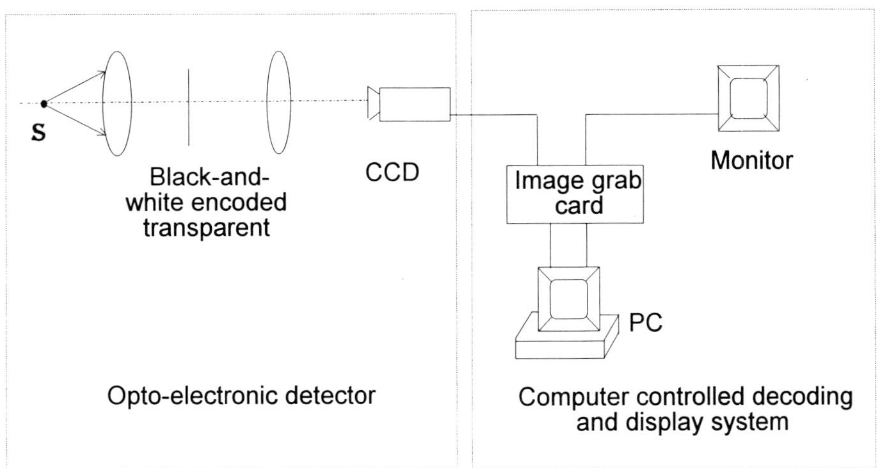

Fig. 5. Computer-controlled decoding and display system.

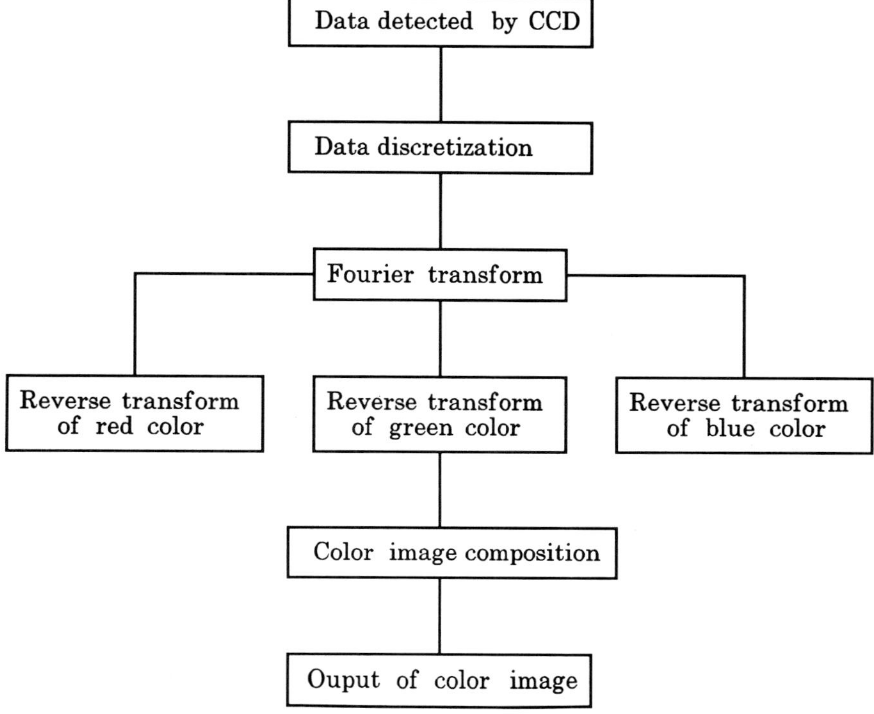

Fig. 6. Block diagram for decoding and colour image composition in a computer.

SOME TECHNICAL NOTATION

We have described the principle of the recording, storage and display of colour information with black-and-white film in detail. In this section, some technical points of this technique will be discussed, first about the tricolour grating, and then the colour-encoding camera and the colour image display system.

The Tricolour Grating

The tricolour grating is the key optical element for colour image encoding, and therefore its quality will be an important factor for the application of this technique, because the recorded resolution of the colour image and therefore the colour saturation of the retrieved colour image rely mainly on the specifications of the tricolour grating. Here are some technical considerations.

Spatial frequency of the tricolour grating

The tricolour grating acts as a sampling element encoding the sampled image components, therefore the spatial frequency of the tricolour grating determines the resolution of the encoded image. It is obvious that the higher the spatial frequency of the tricolour grating, the more sampling points there are and the higher the encoded image resolution is. According to the sampling theorem, the sampling rate should be greater than $2M$, provided that the highest resolution requirement of the photo image is M. For the resolution requirement in ordinary colour photography, a spatial frequency range of $301-501$ p/mm will be satisfactory.

Diffraction efficiency and intensity transmittance

Encoding a colour image with a tricolour grating can be understood as selectively printing the gratings on the black-and-white film. The quality and other specifications of the duplicated grating rely on the printing technique and the contrast, shape and transparent/opaque ratio of the grating used. Mathematical derivation shows that the diffraction efficiency is maximized for a transparent/opaque ratio of $1:1$. This implies that the higher the contrast of the grating, the higher the diffraction efficiency of the printed grating.

The brightness and other specifications of the retrieved colour image are directly related to the diffraction efficiency of the encoded black-and-white film, because during the decoding process the colour filtering is performed at certain diffraction orders of the encoded black-and-white transparency. Experimental results show that the increase in the diffraction efficiency of the encoded black-and-white film improves the brightness of the retrieved colour image and the output signal-to-noise ratio.

To eliminate the exposure loss introduced by the tricolour grating, the average intensity transmittance of the tricolour grating should be as high as possible. According to our results, the optimal value for the average transmittance ranges from 40% to 50%.

Orientations of the colour gratings

For the retrieval of the colour image, the colour filters are placed at the first orders of the Fourier spectrum of an encoded black-and-white transparency. In general, this spectrum has three series of diffraction orders corresponding to the three primary colours, respectively, in the form of 0, ±1, ±2, etc. Cross-orders will appear, if the gammas of the black-and-white films do not exactly coincide with the mathematical prediction. For high quality in the first-order filtering, the gratings should be oriented such that no overlaps of the first orders with other cross-orders exist on the spectrum plane and therefore no more fringes appear in the retrieved colour image. Mathematical calculations have been carried out to determine the optimal orientations of the colour gratings. According to our calculations, the orientations of the three colour grating should differ by $45°$, that is, the optimal orientations will be $0°$, $45°$, $90°$, respectively, for the three colours.

Colour-Encoding Camera

In fact, there is no difference between the encoding camera shown in Fig. 7 and an ordinary camera, except for the tricolour grating on the imaging plane. The tricolour grating is mechanically connected to the film advance system. During the film advance, it is pulled away from the film to prevent friction between them. As the next frame is in position, the grating is pushed to its duty position by a spring for a close contact with the film. The loss of the light energy due to the introduction of the tricolour grating is about 50%, which therefore doubles the exposure time. The tricolour grating is removable, and, without it, the encoding camera can be used as an ordinary camera.

Colour Image Decoder

The colour image decoder is the main body of the colour image display system. The optical system is schematically shown in Fig. 8. Figure 9 is a photograph of the structure of the decoder designed and manufactured by the Academia Sinica.

In this system, the white-light source, an indium arc, is imaged by the achromatic converging lens on to the pinhole filter to form a natural spot of the

Fig. 7. The colour encoding camera.

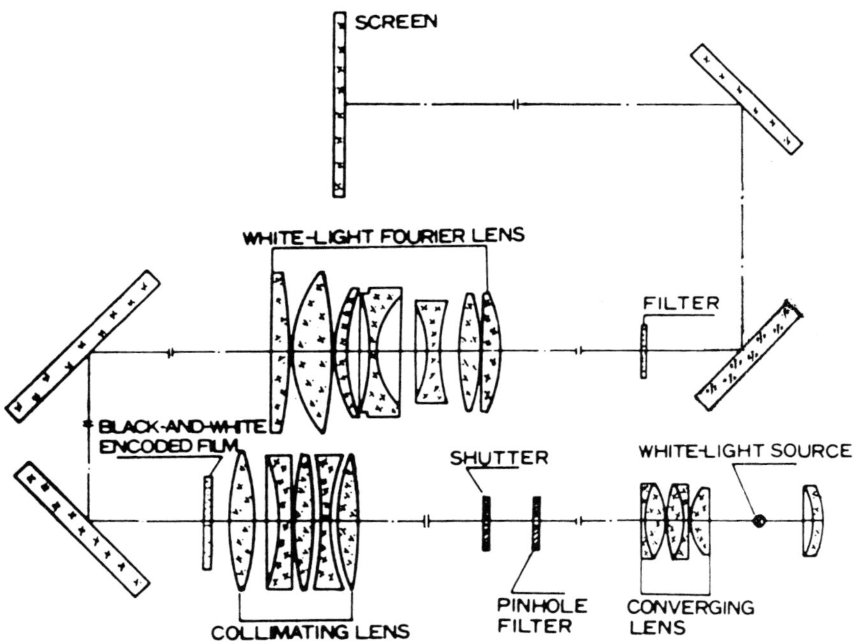

Fig. 8. Inner structure of the colour image decoder.

Fig. 9. A photograph of the colour image decoder.

indium arc. The following collimating lens is an achromatic lens of high aperture ratio (1 : 3). Therefore a large parallel white-light beam is available to illuminate the input encoded transparency. The input transparency is Fourier transformed by the white-light Fourier transform lens, which is a complex achromatic objective (semisymmetric). The three first orders of the Fourier spectrum are filtered by a red, green, and blue filter, respectively. At the output plane of the decoder, a faithful colour image is retrieved.

Professionally designed optical elements in the optical system ensure coaxial conditions in the whole system. The optical loss in reflection from the converging lens is minimized to increase the optical efficiency. Further reduction of the size of this system is possible.

CONCLUSIONS

In this chapter, we have outlined the technique of using black-and-white film to record and restore colour images or information. The technique has opened up the possibility of a physical solution to recording and restoring colour information in a grey scale distribution on black-and-white film, which is chemically stable for long-term storage.

Because of the improvement of the technique, optical encoding can be executed with a single exposure, and a special optical colour modulator (tricolour grating) can be used in an ordinary camera, which makes the encoding technique practical.

Some optical and computer-controlled decoding systems developed by us are discussed and presented as whole packages of instruments, which have been successfully employed in their practical applications.

ACKNOWLEDGMENT

The authors appreciate the contribution of Mr. Fu-Lai Liu and Mr. Lie Lin in this communication.

REFERENCES

1. H.E. Ives, Improvement in the diffraction process of colour photography, *Br. J. Phtog.* **609**, 1906.
2. P.F. Mueller, Color image retrieval from monochromatic transparencies, *Appl. Opt.* **8**, 2051, 1969.
3. A. Macovski, Encoding and decoding of colour information, *Appl. Opt.* **11**, 16, 1972.
4. R. Grousson, and R.S. Kinany, Multi-colour image storage on black and white film using a crossed grating, *J. Opt.* **9**, 333, 1978.

5. F.T.S. Yu, White-light processing technique for archival storage of colour films, *Appl. Opt.*, **19** 2457, 1980.
   F.T.S. Yu, *Optical Information Processing*, Wiley, New York, 313, 1983.
   F.T.S. Yu, G.G. Mu, and S.L. Zhuang, Colour restoration of faded colour films, *Optik*, **58**, 389, 1981.
6. G.G. Mu, J.Q. Wang, Z.L. Fang, and X.Y. Li, A white-light processing technique for colour photography with a black-and-white film and a tricolour grating, *Chin. J. Sci. Instrum.* **4**, 124, 1983.
   G.G. Mu, F.X. Wu, and Z.Q. Wang, Holographic image encoding and white-light image processing, Proc. Int. Conf. Lasers, Guangzhou, China, 376, 1983.

30 Multiwavelength vertical cavity laser arrays by molecular beam epitaxy

C. J. Chang-Hasnain, W. Yuen and G. S. Li

*EECS Department, University of California, Berkeley, CA 94720
and Stanford University, Stanford, CA 94305–4085*

INTRODUCTION

The capability of fabricating two-dimensional (2D) semiconductor diode laser arrays is a key and most important step towards making wafer-scale low-cost lasers. The recent emergence of vertical cavity surface emitting lasers (VCSELs) facilitates the fabrication of such large 2D arrays [1–15]. The VCSELs are promising for a large range of applications from optical interconnects, optical communications, and optical recording, to remote sensing. Due to their topology, the VCSELs will be especially important for applications requiring a laser array.

A multiwavelength laser array is a monolithic array of single-wavelength lasers emitting distinct wavelengths with uniform wavelength spacings. They are promising sources for communication systems using wavelength-division multiplexing (WDM). The VCSEL structure, having an ultrashort cavity and thus a single Fabry–Perot (FP) mode, offers an inherent advantage for the making of multiwavelength laser arrays.

In this chapter, recent progress on multiwavelength VCSEL arrays grown by molecular beam epitaxy (MBE) is reviewed. I will discuss the basic principle of the fabrication, namely, how to vary the emission wavelength of an array of VCSELs. Thereafter, I will discuss two techniques for realizing multi-wavelength VCSEL arrays and their corresponding results. Finally, the wavelength repeatability and controllability issues are addressed.

BASIC PRINCIPLE

A typical VCSEL consists of two highly reflecting mirrors with an active region sandwiched in between. The laser emits in the direction normal to the substrate, and thus is named a surface emitting laser by its pioneer, K. Iga [1]. Most of the present VCSEL designs utilize semiconductor epitaxial layers as mirrors, as shown in Fig. 1. The emission wavelengths of a laser are determined by the laser cavity round trip phase condition, known as Fabry–Perot (FP) modes. A VCSEL has a very short effective cavity length, ~1 µm, and thus only one Fabry–Perot mode within the ~100 nm Bragg reflector bandwidth. Therefore, this FP mode determines the VCSEL emission wavelength. By designing and changing the effective cavity length of an array of VCSELs, a multiwavelength laser array can be made.

The VCSEL effective cavity length can be changed by varying the layer thickness of the one-wave cavity layer or some Bragg reflector layers. In fact, the FP wavelength depends critically on the position and the number of layers that are varied. Figure 2 shows the calculated FP wavelength as a function of the amount of thickness variation. The wavelength varies nearly linearly and monotonically with the thickness variation. A thickness gradient across the wafer can therefore translate into a wavelength gradient of the lasers fabricated across the wafer. Figure 3 shows a schematic of a three-element laser array

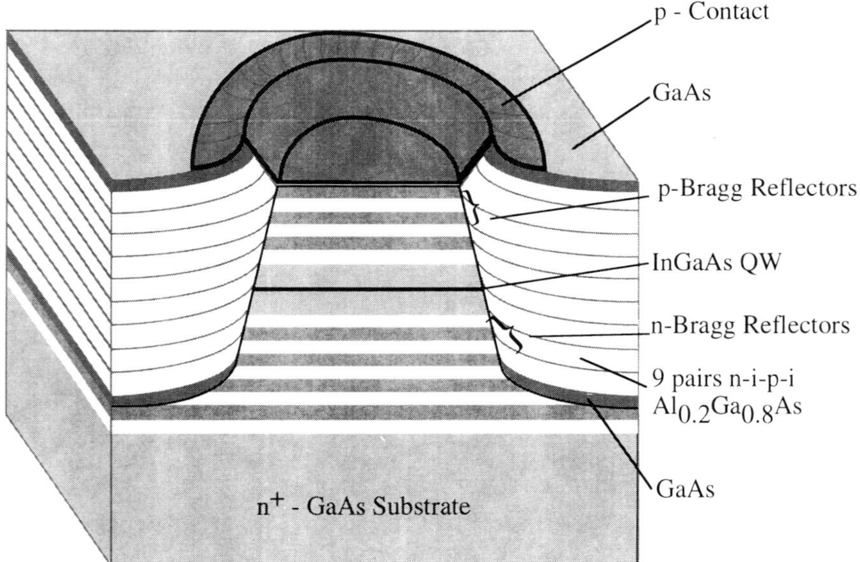

Fig. 1. Schematic of a typical VCSEL structure with epitaxial distributed Bragg reflectors (DBR) sandwiching a 1-λ cavity layer. The quantum well active region is placed at the centre of the 1-λ layer. The top and bottom DBR are p- and n-doped, respectively.

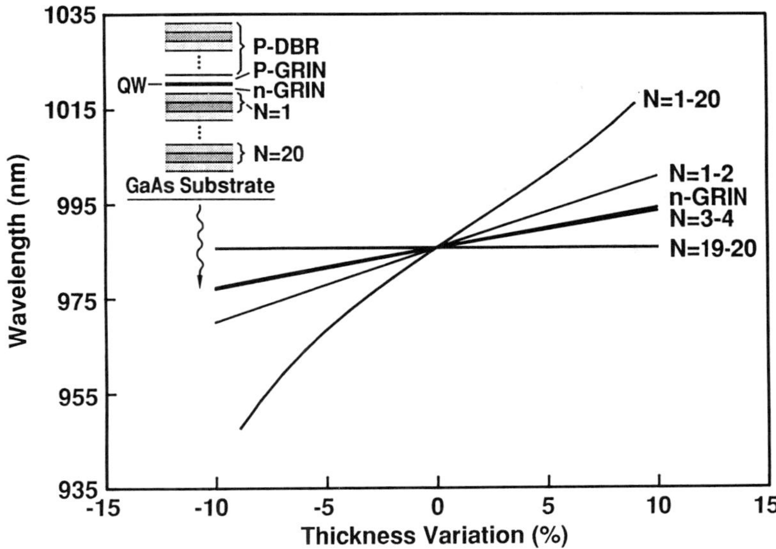

Fig. 2. VCSEL Fabry–Perot wavelength as a function of thickness variation in two pairs of DBR at different locations in the vertical cavity. For a given percentage variation, the closer the varied layer is to the active region, the larger the wavelength change.

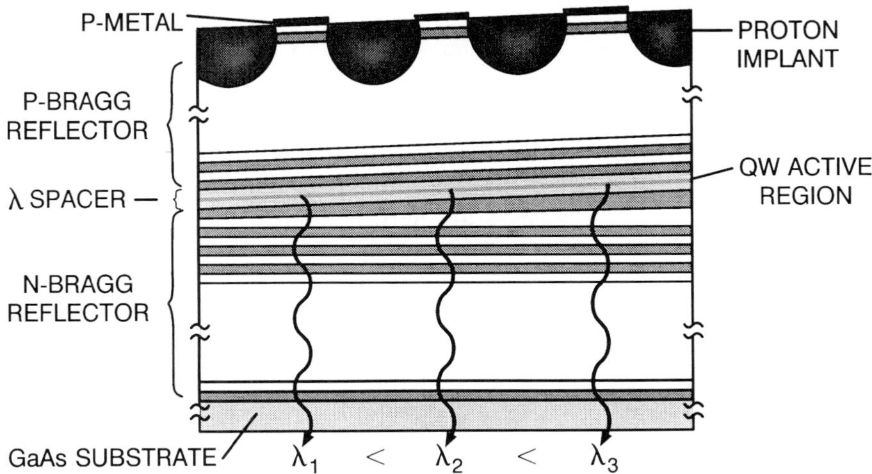

Fig. 3. Schematic of a three-element multiwavelength VCSEL array. A different wavelength is obtained by implementing graded layers in the VC.

based on this idea. With spatial thickness gradient being created in two layers closest to the central spacer, the lasing wavelengths for these three lasers are thus tailored according to the thickness variation.

There are several approaches to attaining a spatial thickness gradient at desirable parts of the VCSEL growth sequence. Serving as a proof of the concept, a demonstration of multiwavelength VCSELs was first made using the inherent beam flux gradient in a molecular beam epitaxy (MBE) system [2]. A 2D VCSEL array emitting 140 distinct wavelengths over a 43 nm wide wavelength span was reported. A limitation of this method is that cavity thickness is monotonically graded across the wafer, and thus the wavelength chirp and range are fixed. For manufacturing purposes, it is desirable to fabricate many identical multiwavelength arrays in a single growth. In applications such as parallel WDM architectures [16] where each array is one pixel, it is also necessary to be able to reproduce many identical arrays on the same wafer.

Recently, there have been several publications reporting various techniques to fabricate multiwavelength VCSELs [17–20]. In ref. 17, metal-organic chemical vapour deposition (MOCVD) growth on a front-side patterned substrate was used. Using the fact that the growth rate varies with the pattern size, VCSELs emitting at different wavelengths were obtained. Ref. 18, on the other hand, uses an oxide layer, which is part of the top reflector, with different thicknesses to attain different VCSEL wavelengths. Ref. 19 used a mask to perform shadow growth in MBE. Ref. 20 reported a technique using a lateral substrate–temperature profile during growth to create a varied thickness. The thickness variation is controlled by a lithographically defined pattern on a backing dummy substrate. By bonding the growth substrate to a patterned backing substrate and using a radiant heat source, a lateral surface temperature gradient is induced on the substrate. This is simply because better heat transfer is attained where the two wafers are bonded, resulting in a hotter surface temperature on that part of the growth wafer. On the other hand, for the parts where the wafers are not in contact, due to the poor thermal conductivity in a vacuum, the surface temperature is lower. This temperature variation thereby alters the GaAs growth rate across the wafer during the growth of the cavity. A multiwavelength VCSEL array with a 20 nm shift in emission wavelength was reported [20].

This patterned substrate technique is further modified by directly patterning the backside of the growth wafer and mounting it on a Mo block to obtain a better controlled thermal contact and therefore a more uniform wavelength shift. This resulted in a multiple-wavelength VCSEL array with the largest lasing wavelength span (62.7 nm) and excellent continuous wave (CW) lasing characteristics. A very sharp wavelength dispersion rate of 117.14 nm mm^{-1} with a very small deviation of 4.1% for 15 arrays is reported [21].

Repeatable wavelengths, nonetheless, cannot rely solely upon substrate temperature control. Accurate control of multiwavelength VCSEL growth can be obtained using a location-resolvable *in situ* optical calibration technique. In particular, this technique results in accurate high and low wavelength bounds of the arrays. Since the intermediate wavelengths are determined by the growth rate profile that is determined by the high and low bounds, multiple-wavelength VCSEL arrays with controllable and repeatable wavelengths can be obtained. With this technique, a wavelength standard deviation of 2.66 Å from 4×35 arrays was achieved [22]. In the following section, we will only discuss the MBE beam flux gradient and backside patterned substrate methods. This is followed by a discussion on wavelength repeatability control using a location-resolvable optical monitoring technique.

TECHNICAL APPROACHES AND RESULTS

Inherent Beam Flux Gradient in MBE

One way to create a thickness gradient during the desirable part of a growth sequence while leaving the rest of the growth spatially uniform is to simply keep the wafer stationary during that part of a molecular beam epitaxy (MBE)

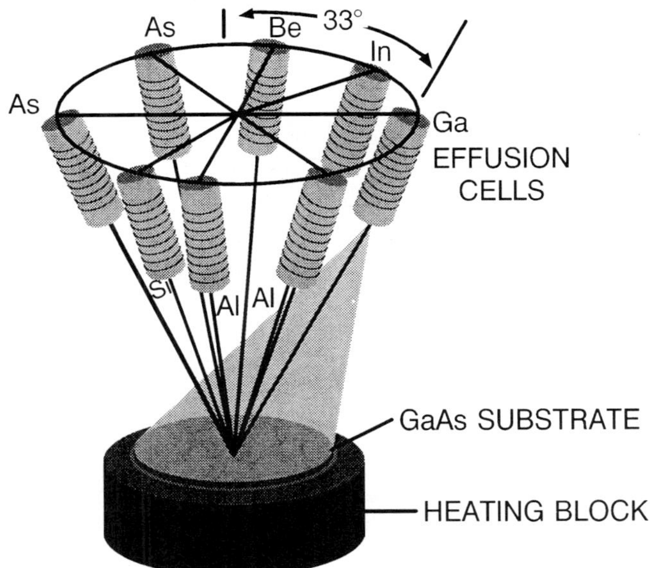

Fig. 4. Schematic of a typical molecular beam epitaxy system. The sources are positioned in a cone off-normal to the substrate resulting in an inherent non-uniform beam flux. The substrate is typically rotated through growth to attain growth uniformity.

growth. The thickness variation originates from the fact that the atomic sources in an MBE system are incident on the wafer at an angle off-normal (~33° for the Varian Gen II system we used) and hence the number of atoms arriving at the wafer varies monotonically in the direction parallel to the planes of incidence of the sources. Figure 4 shows schematically the arrangement of the atomic sources in an MBE system. Since the MBE material is grown in an As-rich environment, the thickness of the MBE growth is determined by the number of group III sources that arrive at the wafer. Thus, the directions of thickness variation for GaAs and AlAs layers are parallel to the directions of the Ga and Al sources, respectively. Therefore, we can obtain the desired small but definite thickness variation across the wafer by rotating the wafer for uniformity during the MBE growth of the VCSEL structure except for two pairs of AlAs and GaAs DBR layers during whose growth the wafer is kept stationary. Since the Ga and Al sources are placed next to each other in our MBE system, the direction of the resulting cavity thickness variation runs parallel to the line intersecting the directions of the two sources. Knowing the direction of increasing wavelength on the wafer, both 1D and 2D multi-wavelength laser arrays can be fabricated [2].

Figure 5 shows one example of results obtained with this technique. The measured wavelength distribution of a 7 × 20 multiwavelength VCSEL array is plotted. There are 140 distinct wavelengths with a well-behaved wavelength

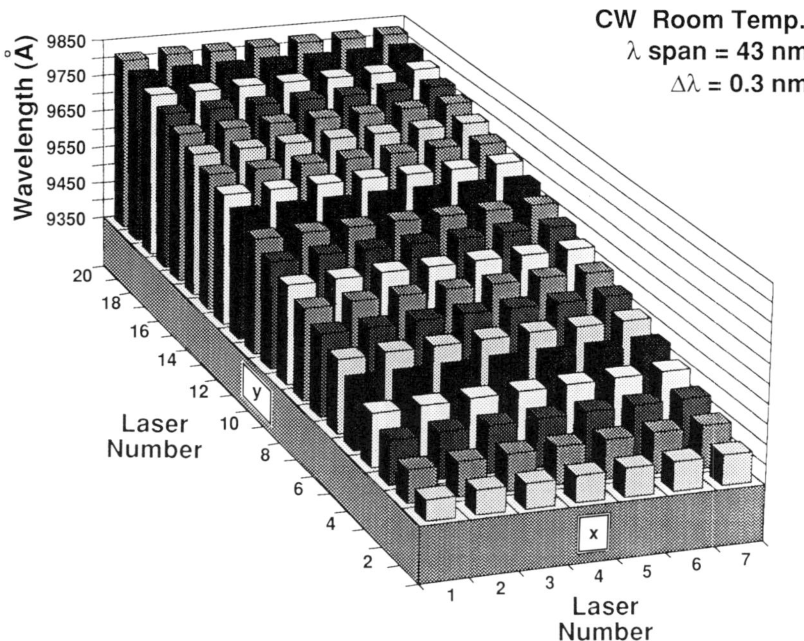

Fig. 5. 7 × 20 VCSEL array wavelength distribution.

sequence through the array as designed. The total wavelength range is as large as 43 nm, from 940 to 983 nm. The average wavelength separation between two neighbouring lasers on a row and a column are 0.3 and 2.1 nm, respectively. The laser spectra are measured under CW operation with the lasers emitting a single TEM_{00} mode. This was the largest monolithic multi-wavelength laser array reported.

Backside Patterned Substrate

The principle of the patterned-substrate growth technique is based on the dependence of GaAs growth rate on the surface temperature, T_s, in an MBE system. For the case when T_s is above ~640°C [23], a wafer surface temperature change of 20–30°C results in a growth rate change of as much as 20%; whereas for T_s well below 640°C, the same temperature change would result in virtually no growth rate change. Thus, given the surface temperature profile, by setting the nominal temperature range to below or above ~640°C, one can obtain either uniform growth or a spatially varied growth induced by the temperature profile, respectively. As mentioned in the previous section, a backside pattern on the substrate is used to first create a surface temperature profile. Previously, this effect has been used to grow laterally tapered GaAs quantum wells and AlGaAs waveguides [24, 25]. In that work, 1 mm grooves were machined in the molybdenum substrate holder in an MBE system, to create a non-uniform thermal contact to the substrate during growth. This technique is further extended to fabricate VCSELs by defining patterns lithographically in the backside of a GaAs substrate and use the selective heating to alter the cavity length in vertical cavity structures. The advantage of patterning GaAs substrates, rather than a molybdenum block is that arbitrary patterns can easily be defined using lithography. In the following, a detailed example of multiwavelength VCSELs fabricated with this method is described.

Figure 6 shows the schematic of the patterned-substrate growth. A 200 μm-deep trench on the backside of a GaAs wafer is patterned by wet etching. A single stripe is used in this study for simplicity of implementation. After a regular wafer cleaning procedure, the patterned wafer is directly mounted on to a Mo block. A substrate surface temperature profile is thus created by the patterned thermal contact which translates into a GaAs layer thickness profile and thereby the desired lasing wavelength distribution. In particular, the transition regions are used to fabricate multiwavelength VCSEL arrays.

The epitaxial structure consists of 21.5 pairs of n-DBR, 25 pairs of p-DBR and an active region, where two identical InGaAs quantum wells are centred in a 1-λ $Al_{0.5}Ga_{0.5}As$ cavity. We first grow the n-DBR at a regular growth temperature without Ga desorption. We then grow a 200 Å GaAs cap layer to protect the wafer surface, stop the growth and raise the substrate temperature. The wavelength-shifting layer is grown after the surface temperature is stabilized. After this layer is done the growth is interrupted again to allow the

                                          Connie J. Chang-Hasnain et al.

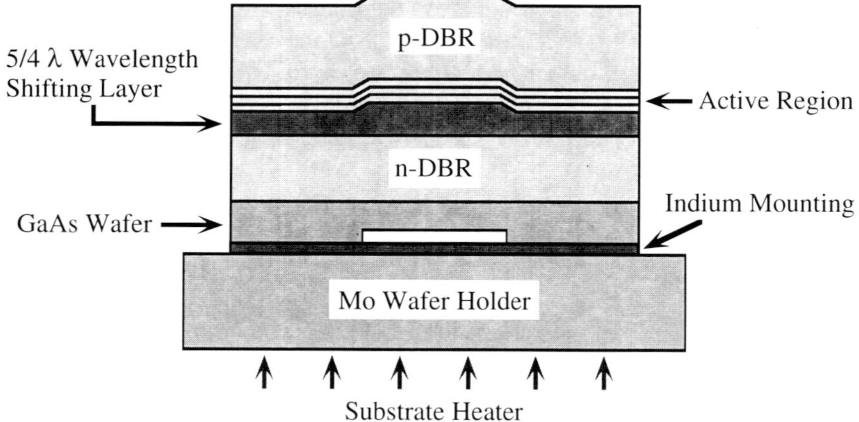

Fig. 6. Schematic of the patterned-substrate growth technique.

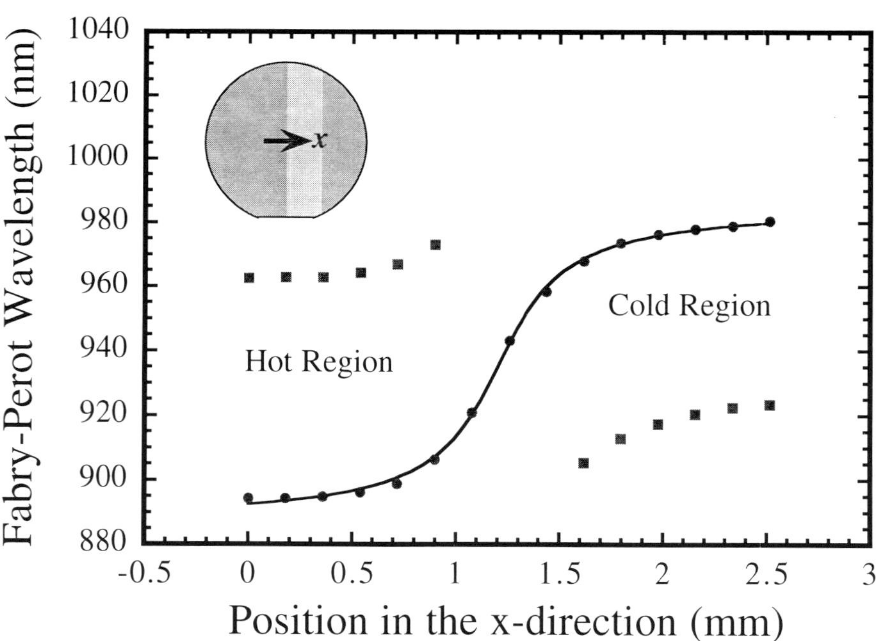

Fig. 7. Fabry–Perot wavelength versus laser position of an array, moving from the high-temperature region toward the low-temperature region. The solid line shows the best fitting by the tan^{-1} function, as predicted by the simple thermal model.

surface temperature to be ramped down to normal growth temperature. From here on, the growth sequence follows that of a regular VCSEL growth. The wafer was fully rotated during the entire growth.

The spontaneous emission wavelength of bottom-emitting VCSELs is first measured as shown in Fig. 7 as a function of position in the direction perpendicular to the stripe of the substrate pattern. The data are compared against the analytic solution of a step temperature profile [20].

$$\lambda(x) = \frac{(\lambda_{\text{hot}} + \lambda_{\text{cold}})}{2} + \frac{(\lambda_{\text{hot}} - \lambda_{\text{cold}})}{\pi} \tan^{-1}\left(\frac{x - x_0}{\Delta x}\right) \tag{1}$$

shown by the solid curve. The Fabry–Perot wavelength shifts more than 90 nm as the laser position varies from the cold region toward the hot region over a small distance of ~1.5 mm. The excellent agreement demonstrates the validity of the simple thermal model [4]. Due to the longer cavity in this particular design having a $(5/4)\lambda$ wavelength shifting layer, a secondary Fabry–Perot mode exists for both ends of the wavelength spectrum. However, since the wavelength spacing between the two Fabry–Perot modes is large (~65 nm), all the VCSELs lase in a single longitudinal mode, as will be discussed next.

Figure 8 shows the lasing wavelength vs. position in the direction per-

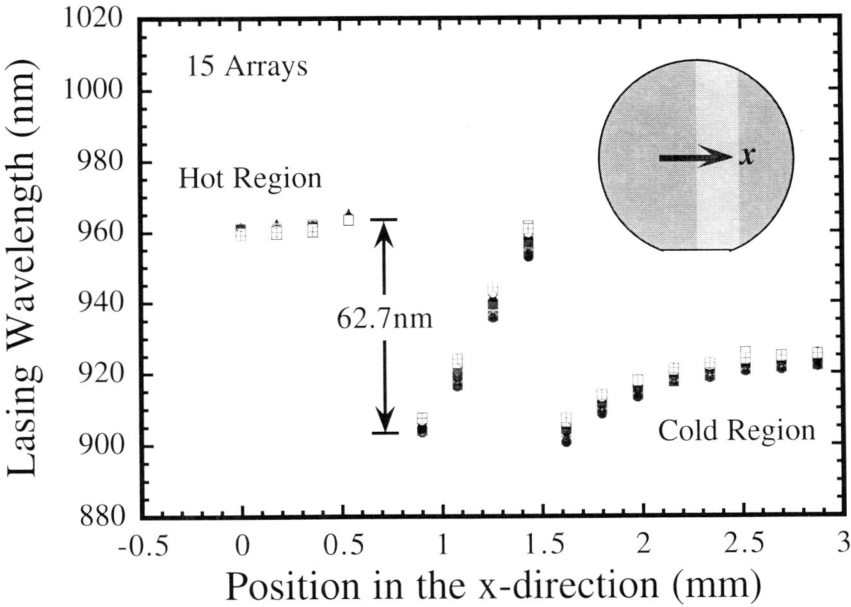

Fig. 8. Pulsed lasing wavelength versus laser position, moving from the high-temperature region toward the low-temperature region. 15 neighbouring arrays are shown.

pendicular to the patterned stripe for 15 neighbouring arrays. The pulsed lasing wavelength span we achieved is about 62.7 nm. This is the largest span achieved for a multiwavelength laser array to date, to the best of our knowledge. The lasing wavelength span is primarily limited by the Fabry–Perot mode spacing and the QW gain bandwidth. In this case, the lasing wavelength of the VCSELs on the two sides of the wavelength transition region is forced to be the one closer to the QW gain peak ~940 nm. The lasing wavelength distribution thus follows the shape of the FP distribution with two discontinuities. From Fig. 8 we also see that the lasing wavelength shift within the temperature transition region is very rapid and linear. The data is curve-fitted to Eq. (1) to study the uniformity of wavelength distribution of the 15 arrays. The 15 curves exhibit a slight shift in position (x) and in wavelength with respect to each other. The position shift was introduced by the misalignment of the photolithographic VCSEL pattern with the stripe on the backside, which can be improved. The wavelength shift was due to the non-uniformity of the MBE growth, which is dependent on the MBE system used. Both types of shifts can be corrected by a thermal electric cooler in the final package and thus are not of concern. However, it is the uniformity of wavelength spacing between lasers in a given array that is important. Therefore, the most significant parameter of

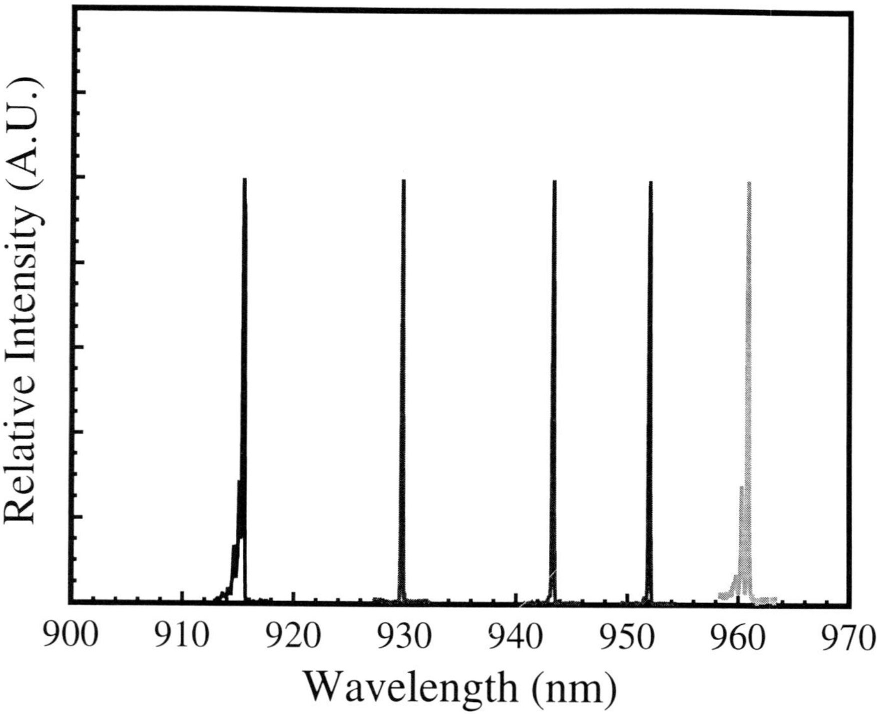

Fig. 9. Typical lasing spectra of an array in CW operation.

concern here is the slope (dispersion rate) uniformity among the arrays. We achieved a highly uniform slope of 117.14 nm mm^{-1} with a standard deviation 4.80 nm mm^{-1} corresponding to 4.1% variation. This is the lowest variation ever reported for multiwavelength laser arrays, to the best of our knowledge.

Figure 9 shows a typical CW lasing spectrum emitting from one array at 180 μm laser spacing. It is seen that CW lasing is achieved over 45 nm. The narrower CW lasing wavelength range compared with that for pulsed operation is attributed to the excess heating which reduces the gain at the ends of the wavelength span. This range can be expanded with a QW active region designed to have a broader gain bandwidth, e.g. by using two QWs with different thickness. It should be noted that a smaller λ spacing between lasers can always be obtained by aligning the VCSEL array at an oblique angle to the λ gradient, by using a thinner wavelength shifting layer and/or placing it farther from the active region, etc. Figure 10 shows the threshold current vs. lasing wavelength with an inset showing a typical $L–I$ curve. The threshold current is essentially flat within a 30 nm range with the minimum of 1.18 mA. The average and standard deviation are 2.16 mA and 0.81 mA, respectively. A

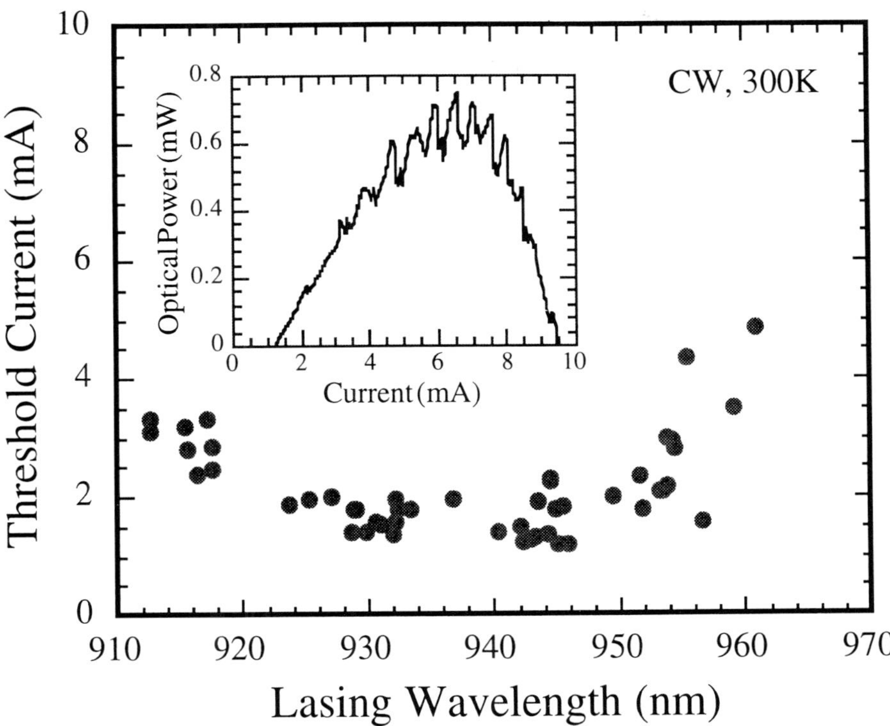

Fig. 10. Threshold current versus lasing wavelength in CW operation. Inset: a typical $L–I$ curve.

typical optical output level of ~0.8 mW is obtained as shown in the inset of Fig. 10.

WAVELENGTH REPEATABILITY CONTROL

Accurate VCSEL growth is very important to ensure wavelength precision. Recently, various optical monitoring techniques have been reported [26–29]. A diode laser reflectometry *in situ* calibration technique is of particular

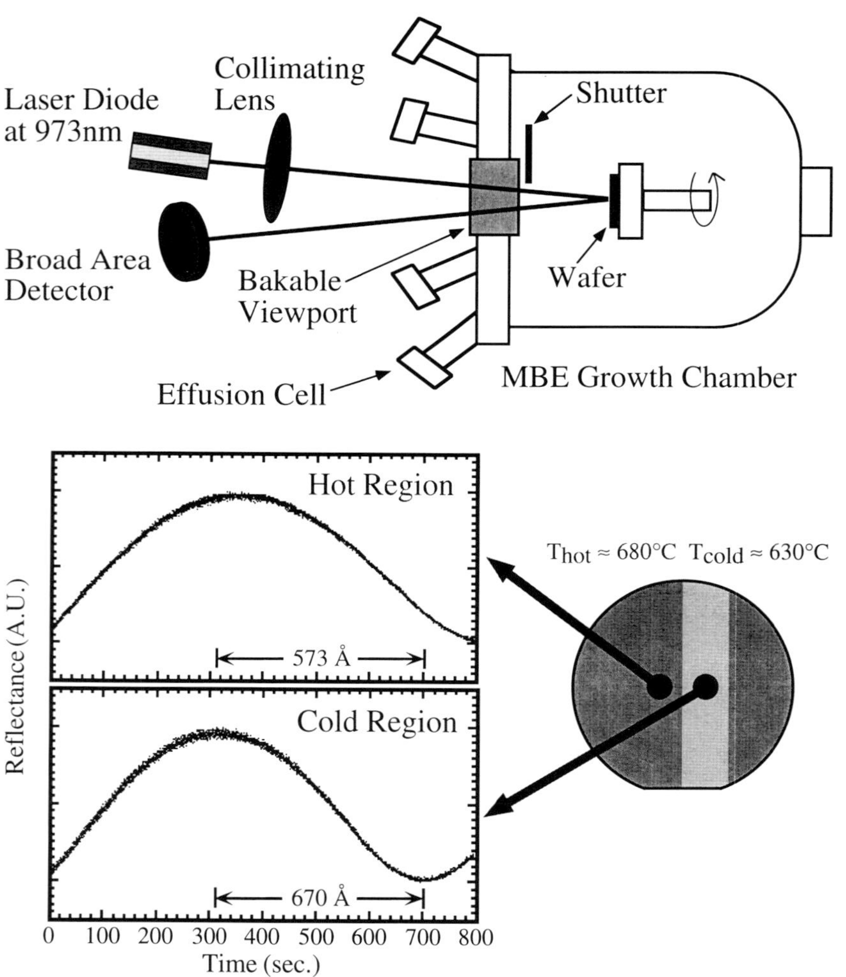

Fig. 11. Schematic of the location-resolvable optical *in situ* monitoring technique. The reflectance curves are those obtained for this particular laser wafer.

interest to monitor the patterned substrate growth because of its location-resolvable nature [29]. Thus, the growth of the wavelength shifting layer in both hot and cold regions can be monitored simultaneously. This allows one to obtain accurate high and low wavelength bounds of the arrays. Since the intermediate wavelengths are determined by the growth rate profile fixed by the two bounds, controllable and repeatable wavelengths can be obtained for the entire array.

An example of a recent implementation of this method along with results is described in this section. Figure 11 shows the schematic of this optical *in situ* calibration technique with the reflectance curves obtained for this laser wafer. It is implemented by directing the output beam of a diode laser on to hot and cold regions of a patterned substrate where the control of growth rates is desirable. The reflectance curves of both regions are recorded sequentially as epitaxial material for calibration is being grown. With the information of the GaAs refractive indices at growth temperatures and the period of the measured reflectance curves, GaAs growth rates of both regions can be accurately calculated. Since the laser beam can be focused down to a spot size as small as 1 mm × 1 mm, good spatial resolution can be obtained.

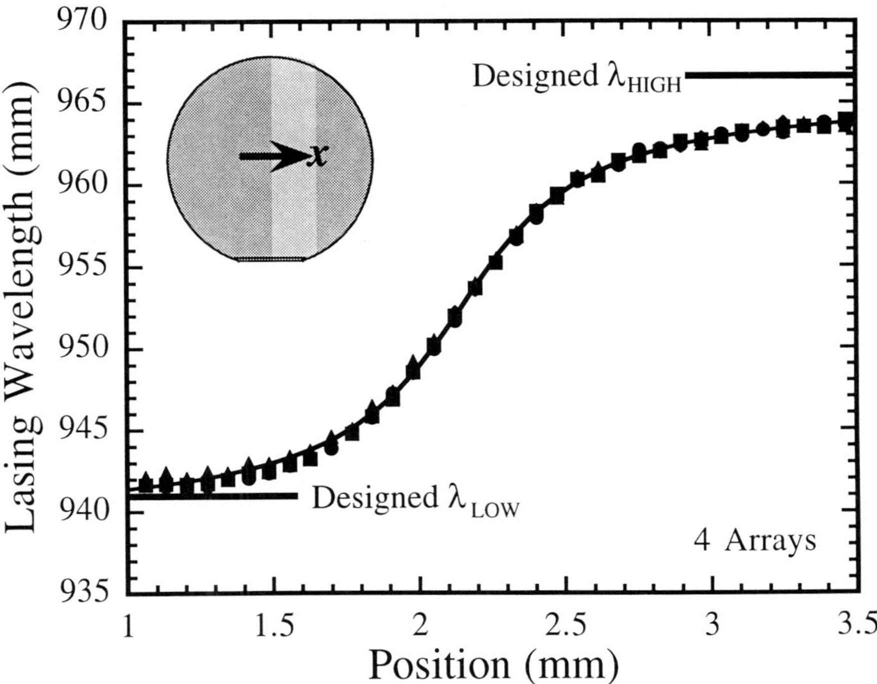

Fig. 12. Pulsed lasing wavelengths of four 35-element arrays. The designed high and low wavelengths are marked correspondingly.

The epitaxial structure of VCSELs grown by using this technique consists of 21.5 pairs of n-DBR, 25 pairs of p-DBR and an active region, where two identical InGaAs quantum wells are centred in a $Al_{0.5}Ga_{0.5}As$ cavity. The thickness of spacers is intentionally increased by 10% so that the cavity mode in the cold region is designed to be slightly longer than the QW wavelength of 950 nm. A $(5/4)\lambda$ GaAs wavelength-shifting layer is located at the first pair of the n-DBR and is grown at a higher temperature where Ga desorption occurs in the hot region but not in the cold region. 20 µm × 20 µm bottom-emitting VCSELs are fabricated and tested without heat sinks. A pulsed wavelength is measured to avoid any possible heating effects. The $L-I$ characteristics of both CW and pulsed operations are recorded.

Figure 12 shows the measured lasing wavelength under pulsed operation as a function of position in the direction perpendicular to the stripe of the substrate pattern with designed wavelengths marked. The pre-growth optical *in situ* measurements in Fig. 11 show that there is ~15% of Ga desorption rate in the hot region and no desorption in the cold region. The calculated lasing wavelengths in the hot region and cold region based on this information are approximately 941 nm and 966.7 nm, respectively, whereas the measured

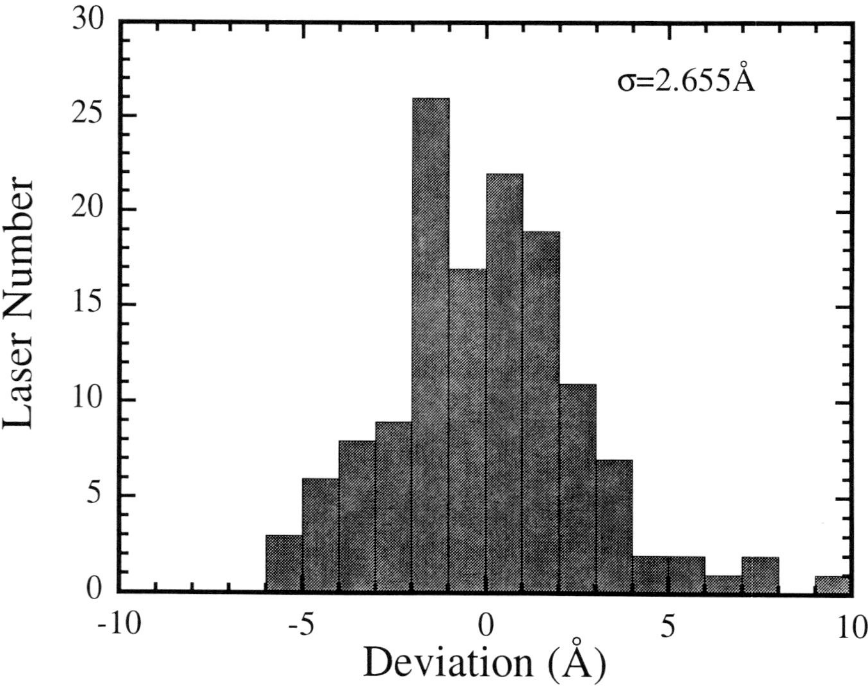

Fig. 13. Distribution of wavelength deviations from the ideal wavelength obtained by fitting all data points with a theoretical $\tan^{-1}$ function.

wavelengths are 941 nm and 964 nm, respectively. The lasing wavelength matches exactly in the hot region and presents only 0.28% of deviation in the cold region. The repeatability of the lasing wavelength is examined by fitting all of the wavelength data from four arrays with a single arctangent function, shown by the solid line in Fig. 12. By assuming the fitted curve is the ideal wavelength distribution, the deviation of the measured wavelengths is calculated with respect to the fitted curve, and the resulting deviation distribution is shown in Fig. 13. The σ (standard deviation) of the wavelength deviation is 2.66 Å. It is seen from Fig. 12 that the wavelength deviation is primarily from the short-wavelength end of an array, and in the transition region where applicable multiwavelength VCSEL arrays are made, the wavelength deviation is much smaller than the overall deviation. This shows the suitability of the patterned-substrate growth technique with the assistance of the location-resolvable optical *in situ* calibration method for making multiple-wavelength VCSEL arrays.

Figure 14 shows the CW lasing characteristics of one array. The peak power stays around 1 mW through the range of 944–958 nm, and the threshold current shows typical decreasing behaviour as the gain and FP peaks align better in wavelength. The laser performance can be further improved by using better current confinement structures [*30, 31*].

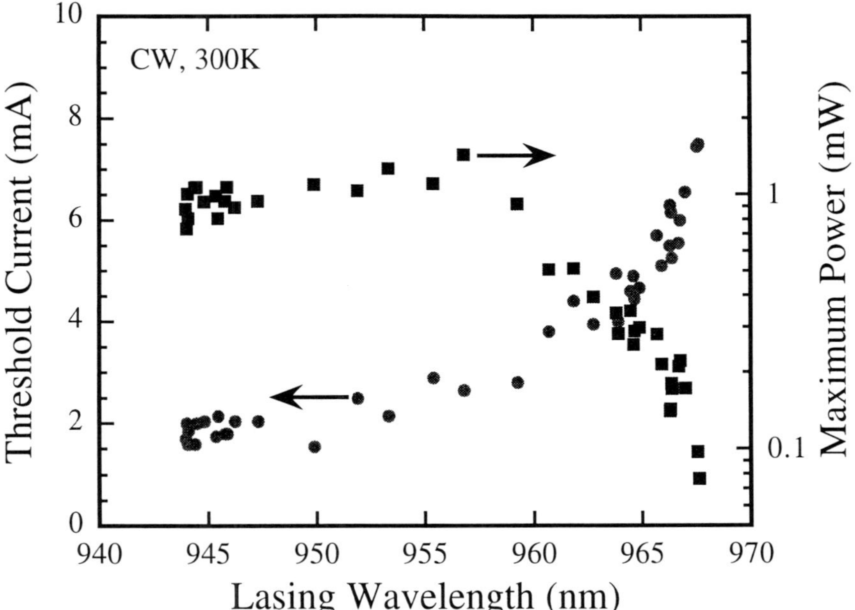

Fig. 14. Threshold current and maximum optical power vs. lasing wavelength under CW operation.

DISCUSSION

In this chapter, we have focused on two techniques for making multiwavelength VCSEL arrays. As mentioned, there are a number of other demonstrated techniques. In addition, it is not difficult to foresee more novel techniques to be reported in the future, e.g. selective area growth in MOCVD. There are obviously pros and cons of each technique, and specific applications for which each may be more suitable. For example, the two techniques discussed here are more suitable for obtaining a large number of lasers with small wavelength separations. This is because the gradient is typically small but continuous. On the other hand, using the methods in Ref. *17–19* or selective area MOCVD, only a small number of discrete wavelengths can be obtained, but the wavelength separation can be very large and the lasers can be positioned randomly (i.e. without sequential wavelength order). It is perhaps too early at this stage to predict which method will be more useful. However, it is vitally important for any given technique to result in precise and reproducible wavelengths with high yield. It is our view that being able to meet this criterion cost-effectively will determine the future of multiwavelength VCSEL and indeed all WDM applications.

CONCLUSIONS

In conclusion, progress in multiwavelength VCSEL arrays was reviewed. We described the basic design principle, and discussed in detail two fabrication techniques using MBE and their corresponding results. We further described a location-resolvable optical *in situ* calibration technique with which the growth rates on different regions of the same wafer can be accurately measured. By using both the patterned-substrate growth and the *in situ* calibration techniques, multiple-wavelength VCSEL arrays with accurate and repeatable lasing wavelengths were demonstrated. These devices will be useful for making multiple-wavelength VCSEL arrays used for cost-effective WDM applications.

ACKNOWLEDGMENT

We thank Dr L.E. Eng for his contribution to our patterned substrate MBE work. This work was supported by the ARPA ULTRA program, the Presidential Faculty Fellowship, the Joint Services Electronics Projects and the Packard Fellowship.

REFERENCES

1. K. Iga, F. Koyama, and S. Kinoshita, 'Surface emitting semiconductor kaser,' *IEEE J. of Quantum Electron.* **24**, pp. 1845–1855, 1988.

2. C.J. Chang-Hasnain, J.P. Harbison, C.-E. Zah and M.W. Maeda, L.T. Florez, N.G. Steffel and T.-P. Lee, 'Multiple wavelength tunable surface emitting laser arrays,' *IEEE J Quantum Electron* **27**, pp. 1368–1376, 1991.
3. J. Jewell, J.P. Harbison, A. Scherer, Y.H. Lee and L.T. Florez, 'Vertical cavity surface emitting lasers: design, growth, fabrication and characterization,' *IEEE J. Quantum. Electron.* **27**, pp. 1332–1346, 1991.
4. Randy Geels, Scott W. Corzine and Larry A. Coldren, 'InGaAs vertical cavity surface-emitting lasers,' *IEEE J. Quantum Electron.* **27**, pp. 1359–1367, 1991.
5. G. Hasnain, K. Tai, L. Yang, Y.H. Wang, R.J. Fischer, J.D. Wynn, B. Weir, N. K. Dutta, and A. Y. Cho, 'Performance of gain-guided surface emitting lasers with semiconductor distributed Bragg reflectors,' *IEEE J. of Quantum Electron.* **27**, pp. 1377–1385, 1991.
6. Y.J. Yang, T.G. Dziura, S.C. Wang, G. Du and S. Wang, 'Single-mode operation of mushroom structure surface emitting lasers,' *IEEE Photon. Tech. Lett.* **3**, pp. 9–11, 1991.
7. K. Tai, L. Yang, Y.H. Wang, J.D. Wynn and A.Y. Cho, 'Drastic reduction of series resistance in doped semiconductor distributed Bragg reflectors for surface emitting lasers,' *Appl. Phys. Lett.* **56**, pp. 2496–2498, 1990.
8. Y.A. Wu, C.J. Chang-Hasnain, and R. Nabiev, 'Single mode emission from a passive-antiguide-region vertical-cavity surface-emitting laser', *Electronics Lett.*, **29**, pp. 1861–1863, 1993.
9. K.D. Choquette, M. Hong, R.S. Freund, J.P. Mannaerts, R.C. Wetzel, and R.E. Leibenguth, 'Vertical-cavity surface-emitting laser diodes fabricated by in situ dry etching and molecular beam epitaxial regrowth,' *IEEE Photon. Technology Lett.* **5**, No. 3, pp. 284–287, 1993.
10. R.A. Morgan, G.D. Guth, M.W. Focht, M.T. Asom, K. Kojima, L.E. Rogers, and S.E. Callis, 'Transverse mode control of vertical-cavity top-surface emitting lasers', *IEEE Photon. Tech. Lett.* **4**, pp. 374–376, 1993.
11. D.L. Huffaker, D.G. Deppe, K.K. Kumar, and T.J. Rogers 'Native defined ring contact for flow threshold vertical cavity lasers,' *Appl. Phys. Lett.* **65**, pp. 97–99, 1994.
12 K.L. Lear, K.D. Choquette, R.P. Schneider, Jr., S.P. Kilcoyne and K.M. Geib, 'Selectively oxidised vertical cavity surface emitting lasers with 50% power conversion efficiency', *Electronics Letters* (2 Feb. 1995) **31**, no. 3, p. 208–209.
13. D.I. Babic, K. Streubel, R.P. Mirin, N.M. Margalit, J.E. Bowers, E.L. Hu, D.E. Mars, L. Yang and K. Carey, 'Room temperature continuous-wave operation of 1.54 μm vertical cavity lasers', *IEEE Photon. Tech. Lett.* **7**, pp. 1225–1227, 1995.
14. J.D. Walker, D.M. Kuchta and J.S. Smith, 'Vertical-cavity surface-emitting laser diodes fabricated by phase-locked epitaxy', *Appl. Phys. Lett.* (21 Oct. 1991) **59**, no. 17, 2079–2081.
15. G.M. Yang, M.L. MacDougal, P.D. Dapkus, 'Low threshold current VCSELs with enhanced resistance to heating', Technical Digest, Semiconductor Lasers, Advanced Devices and Applications, Keystone, Colorado, August 21–23, 1995.
16. A. Willner, C.J. Chang-Hasnain and J. Leight, '2-D WDM optical interconnections using multiple-wavelength VCSEL's for simultaneous and reconfigurable communication among many planes', *IEEE Photon. Tech. Lett.* **5**, 838–841, 1993.
17. F. Koyama, T. Mukaihara, Y. Hayashi, N. Ohnoki, N. Hatori, and K. Iga, 'Two-

dimensional multiwavelength surface emitting laser arrays fabricated by nonplanar MOCVD', *Electron. Lett.* **30**, 1947–1948, 1994.

18. T. Wipiejewski, M.G. Peters, and L.A. Coldren, 'Vertical cavity surface emitting laser diodes with post-growth wavelength adjustment', *IEEE Photon. Technol. Lett.* **7**, 727–729, 1995.

19. H. Saito, I. Ogura, Y. Sugimoto, and K. Kasahara, 'Monolithic integration of multiple wavelength vertical-cavity surface-emitting lasers by mask molecular beam epitaxy', *Appl. Phys. Lett.* **66**, 2466–2468, 1995.

20. L.E. Eng, K. Bacher, W. Yuen, J.S. Harris, and C.J. Chang-Hasnain, 'Multiple wavelength vertical cavity laser arrays on patterned substrates', *IEEE J. Select. Topics in Quantum Electron.* **1**, pp. 624–628, 1995.

21. W. Yuen, G.S. Li and C.J. Chang-Hasnain, 'Multiple-wavelength vertical-cavity surface-emitting laser arrays with a record wavelength span', *IEEE Photon. Tech. Lett.* **8**, pp. 4–6, 1996.

22. W. Yuen, G.S. Li, K. I. Ioakimidi, and C.J. Chang-Hasnain, 'Location-resolvable optical monitored growth of multiple-wavelength vertical-cavity laser arrays', submitted to *Electronics Letters*, September 1995.

23. R. Fischer, J. Klem, T.J. Drummond, R.E. Thorne, W. Kopp, H. Morkoc and A.Y. Cho, 'Incorporation rates of gallium and aluminium on GaAs during molecular beam epitaxy at high substrate temperatures', *J. Appl. Phys.* **54**, 2508–2510, 1983.

24. W.D. Goodhue, J.J. Zayhowski and K.B. Nichols, 'Planar quantum wells with spatially dependent thicknesses and Al content', *J. Vac. Sci. Technol.* **6**, 846–849, 1988.

25. D.E. Bossi, W.D. Goodhue, M.C. Finn, K. Rauschenbach, J.W. Bates and R.H. Reidecker, 'Reduced confinement antennas for GaAlAs integrated optical waveguides', *Appl. Phys. Lett.* **56**, pp. 420–422, 1990.

26. Y.M. Young, M.R.T. Tan, B.W. Liang, S.Y. Wang and D.E. Mars, 'In situ thickness monitoring and control for highly reproducible growth of distributed Bragg reflectors', *J. Vac. Sci. Technol. B* **12**, 1221–1224, 1994.

27. F.G. Bobel, H. Moller, A. Wowchak, B. Hertl, J. Van Hove, L.A. Chow and P.P. Chow, 'Pyrometric interferometry for real time molecular beam epitaxy process monitoring', *J. Vac. Sci. Technol. B* **12**, 1207–1210, 1994.

28. K. Bacher, B. Pezeshki, S. Lord and J.S. Harris, 'Molecular beam epitaxy growth of vertical cavity optical devices with in situ corrections', *Appl. Phys. Lett.* **61**, 1387–1389, 1992.

29. G.S. Li, W. Yuen, and C.J. Chang-Hasnain, 'Accurate molecular beam epitaxial growth of vertical cavity surface emitting laser using diode laser reflectometry', to be published in *IEEE Photon. Technol. Lett.*

30. J.W. Scott, B.J. Thibeault, D.B. Young, L.A. Coldren, and F.H. Peters, 'High efficiency submilliamp vertical cavity lasers with intracavity contacts', *IEEE Photon. Technol. Lett.* **6**, 678–680, 1994.

31. K.D. Choquette, R.P. Schneider, Jr., K.L. Lear, and K.M. Geib, 'Low threshold voltage vertical-cavity lasers fabricated by selective oxidation', *Electron. Lett.* **30**, pp. 2043–2044, 1994.

31 Compact blue-green laser sources

William J. Kozlovsky
IBM Research Division, Almaden Research Centre, San Jose, California 95120 USA

INTRODUCTION

Compact, all-solid-state blue-green laser sources are desired for numerous applications such as optical recording, printing, wafer inspection systems, projection displays, and chemical monitoring systems. Some of these applications require the use of blue light and so currently use argon-ion or helium–cadmium gas-discharge lasers. For other applications, although blue light is desired, the relatively high cost, large size, inefficient operation, and short lifetimes of the gas lasers are prohibitive, and so infrared diode lasers are used. Compact blue-green sources are now becoming possible using a variety of techniques that employ only solid-state components. This chapter will describe these various techniques, their advantages and disadvantages, and their relative level of technological maturity. Most of these techniques use infrared diode lasers as their primary light source.

Infrared GaAlAs diode lasers have evolved to offer high reliability and high output powers and are available with either single-frequency output from single-mode lasers and multimode output from broad-area or multistripe arrays. Methods of efficiently converting their power output to the blue-green regions depend on the output properties of the diode laser. The highest infrared powers are produced from wide-area emitters and multiple-stripe arrays. The output beam from these diodes arrays and wide-area emitters is typically not in a diffraction-limited Gaussian mode, which limits both the efficiency of direct non-linear frequency conversion as well as the usefulness of the converted beam, since many applications for a blue source require a diffraction limited beam. Diode laser arrays are therefore most effective when used as the energy source for exciting another solid-state laser material. Blue-green light can be produced directly by the solid-state laser material, if it uses multiple-step pump excitations of the diode light, or more typically the diodes can be used to pump longer-wavelength lasers that can be converted to shorter wavelengths with a variety of non-linear conversion techniques.

Single-spatial mode infrared diode laser output can be converted directly to shorter wavelengths using non-linear frequency conversion. To overcome the limitations of the lower powers available from single-mode diode lasers, either waveguide confinement or external resonator enhancement can be used to increase the efficiency of conversion. Even higher conversion efficiencies and powers are likely using these techniques with the recently developed master oscillator power amplifier (MOPA) diode lasers (O'Brien *et al.* 1993) that combine a low-power single-stripe laser with a tapered, broad-area amplifier to generate up to a few watts of single-mode output in the infrared.

Blue light can be directly produced using direct-injection wide-bandgap semiconductor lasers. Much progress in this area has been reported recently, although high CW powers and long lifetimes remain elusive. These sources should ultimately provide the simplest and lowest cost solution for providing blue light, which is a requirement for penetrating the highest volume consumer-based applications such as HDTV videodisk players. Until these blue diode lasers are widely available, sources based on nonlinear frequency conversion of infrared diodes will be able to meet the requirements of many other applications.

NON-LINEAR FREQUENCY CONVERSION

Non-linear frequency conversion techniques are nearly as old as the laser itself (Frankin *et al.* 1961). As laser light is passed through certain materials, the material responds by not only transmitting the original light, but also by generating light of a different wavelength. Common non-linear techniques for converting laser light of one wavelength to a shorter wavelength are second harmonic generation (SHG, also called frequency doubling) and sum-frequency mixing (SFM). In second harmonic generation, an input source of one wavelength is converted to an output beam of one-half the wavelength. Sum-frequency mixing is used to convert two inputs with wavelengths λ_1 and λ_2 to generate an output with wavelength $\lambda_3 = (\lambda_1^{-1} + \lambda_2^{-1})^{-1}$, or $\omega_3 = \omega_1 + \omega_2$ for ω being the frequency of the light. These second-order non-linear processes occur in certain non-centrosymmetric crystals that exhibit a non-linear response to the incident light waves. The efficiency of these conversion processes is proportional to the input intensities I_1 and I_2 of the two beams (where $I_1 = P_1/A_1$ and $I_2 = P_2/I_2$, where P is the power of the beams and A is the effective area of the beams) and the material's non-linear coefficient d_{eff}, and can be calculated from (Byer 1977):

$$I_3 = \frac{\omega_3^2 d_{\text{eff}}^2 l^2}{2\varepsilon_0 c^3 n_1 n_2 n_3} I_2 I_1$$

where ω is the frequency, n and l are the index of refraction and length of the non-linear material, ε_0 is the permittivity of free space, and c is the speed of light. This equation holds true for frequency doubling as well, since SHG is the degenerate case of SFM where $\lambda_1 = \lambda_2$, $n_1 = n_2$, and $I_1 = I_2$.

To generate usable amounts of the shorter-wavelength light, the interaction in the non-linear material must be cumulative: i.e., the wavefront of the original light and the newly generated light must propagate through the material at the same phase velocity. This phase-matching process is determined by $\Delta k = k_3 - k_2 - k_1 = 0$, where the magnitude of the wavevector $|k_i| = 2\pi n_i/\lambda_i$, for each beam $i = 1$, 2 and 3. To achieve this velocity matching, often a birefringence in the material (a different index of refraction in the material for the different polarizations) is used to offset the inevitable dispersion in the index of refraction of the material. Changing the indices of refraction of the non-linear crystal by adjusting either the propagation direction in the crystal or the temperature of the crystal can tune the wavelength where this offsetting of the dispersion with the birefringence occurs. Although birefringent phase-matching is often used in non-linear conversion, its use is limited to the range of wavelengths over which the material's birefringence is large enough to offset the dispersion.

A concept proposed well before the advent of birefringent phase-matching, but which has only recently been practically realized, is called quasi-phase matching (QPM). In second harmonic generation, for example, if the indices of refraction in the non-linear material are not adjusted to set Δk equal to zero, as the fundamental wave and the generated second harmonic wave slip out of phase by 180°, power flow between the two beams reverses direction, causing the second harmonic beam to be converted back to the fundamental. In quasi-phase-matching, the non-linear conversion is made cumulative by reversing the relative phases of the fundamental and second harmonic beams at every point where they become 180° out of phase. Therefore the conversion process continues to accumulate throughout the crystal, at a rate similar to that of true phase-matching. This process is graphically depicted in Fig. 1.

In practice, the phase reversal to achieve quasi-phase-matching is produced by reversing the ferroelectric domains of the non-linear material, which can be done near the surface by an ion exchange process, or through the bulk crystal by electric field poling or with heat gradients applied during the crystal growth. Quasi-phase-matching is particularly interesting since it allows a non-linear material to be used over a wider range of wavelengths than birefringent phase-matching. Since the quasi-phase-matched wavelength can be adjusted by

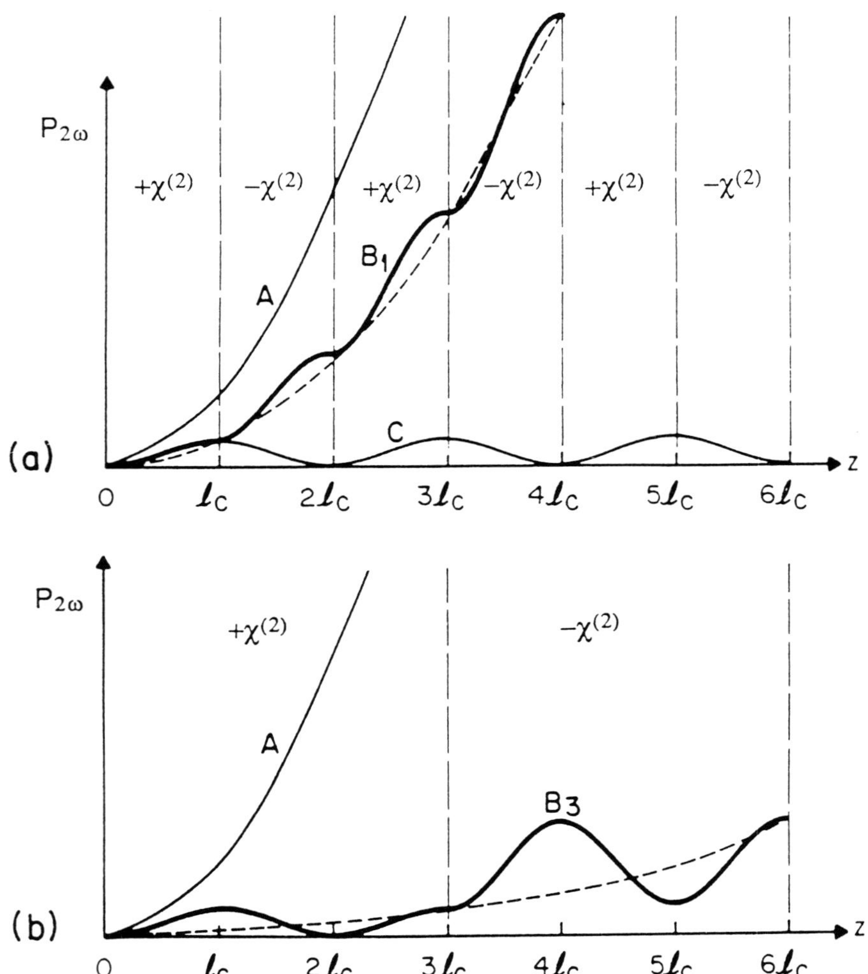

Fig. 1. Dependence of SHG power on length using quasi-phase-matching for (a) first-order QPM, where the domain is reversed every time the fundamental and second harmonic slip in phase by 18° (one coherence length, l_c), and for (b) third-order QPM, where the domain is reversed after three coherence lengths. Graph A shows the growth in power for a phase-matched interaction, B1 for first order QPM, and B3 for third-order QPM (figure from Lim 1992).

changing the period of the poling, the non-linear material can be theoretically used over its entire transparency range provided that domains of the proper length can be formed. Quasi-phase-matching also enables one to use crystal orientations with the highest non-linear coefficient, but which have insufficient birefringence to phase-match.

Even using highly non-linear coefficient materials, with either QPM or birefringent phase-matching, efficient conversion of continuous wave (CW) sources is particularly challenging. As an example, consider frequency doubling an 860-nm GaAlAs diode laser to the blue, which is phase-matched in potassium niobate ($KNbO_3$) near room temperature. Although $KNbO_3$ has a high non-linearity of 26 pm V^{-1}, a 100-mW 860-nm beam incident on a 7-mm-long $KNbO_3$ results in a blue 430-nm output of only 0.1 mW. Achieving even this low conversion efficiency requires proper focusing of the infrared beam into the non-linear crystal for producing the highest sustained power densities. A focusing looser than optimal decreases the power density in the centre of the crystal while focusing tighter than optimal produces an excess of beam diffraction which results in low power densities at the ends of the crystal.

As the numerical example above shows, frequency conversion of low-power, CW sources in a single pass through a non-linear crystal is too inefficient to be practical. To increase the conversion efficiency, there are several intensity enhancement methods that can be used; details of these schemes make up the bulk of the rest of this chapter. These methods are all based on either waveguide enhancement or resonator enhancement. Waveguide enhancement eliminates the diffraction-limiting effects of tight focusing by confining the light within a higher-density core (as happens in optical fibres) to maintain high power densities along the entire length of the non-linear crystal. Alternatively, the frequency doubling efficiency can be increased by placing the non-linear crystal inside an optical resonator, either a laser resonator or an external enhancement resonator, where the circulating power is greater than the power outside the resonator. Both intensity enhancement methods have been successfully used to produce overall conversion efficiencies of infrared to blue light in excess of 10%.

DIODE-PUMPED SOLID-STATE LASERS

The best-developed and highest-power blue-green all-solid-state lasers to date are based on diode-laser-pumping of other solid-state lasers. These high powers have been made possible by the availability of GaAlAs broad-area diode lasers and laser arrays that produce multiple watts of output. These high-power diode lasers are not easily directly converted to shorter wavelengths due to the poor wavefront and spectral quality of their multimode output (Risk and Lenth 1989) but they can be used as the excitation source for other solid-state laser materials such as Nd : YAG and Nd:vanadate due to the fortuitous overlap of the wavelengths produced by GaAlAs diode lasers (780–860 nm) and the strong absorption band of the neodymium ion (~810 nm).

Figure 2 shows two typical layouts and pumping geometries for diode-pumping of solid-state lasers. The Nd laser resonator produces a good spatial and spectral mode output that is ideally suited for non-linear frequency conversion, and so it thereby provides spatial and spectral mode conversion of the diode laser array output at the cost of having a slightly longer wavelength to convert (e.g., 1064 nm from a Nd:YAG laser vs. ~800 nm from the original diode laser output). The Nd ion also offers a long upper-state lifetime for storing energy, enabling Q-switched pulsed operation of the Nd laser for

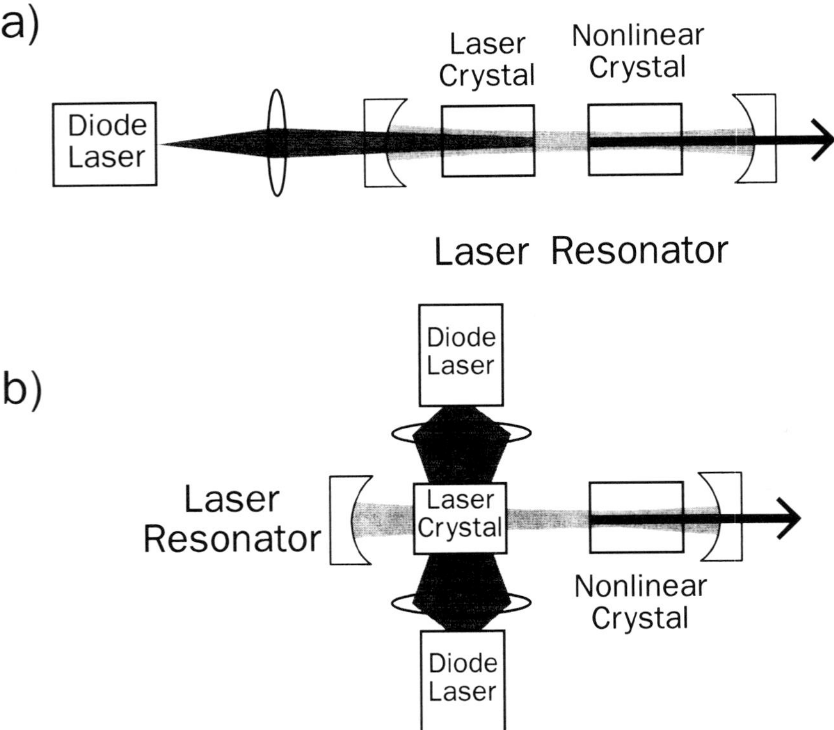

Fig. 2. Diode-laser-pumped solid-state laser configurations. (a) End pumping provides excellent overlap between the solid-state laser mode volume and the diode laser power, but usually limits the power output possible. (b) Side pumping of the solid-state laser rod or slab can be scaled to very high powers, but the lower overlap of the laser mode and diode light can make single mode and efficient operation challenging. The high mode quality and high circulating intensity of the solid-state laser are ideally suited for non-linear conversion to shorter wavelengths. Alternatively, fibre optic coupling of the diode laser output to the solid-state material can be used in either end-pumped and side-pumped configurations.

producing high peak power pulses of up to hundreds of kW in a few nanoseconds. Second harmonic generation in the non-linear material KTP can straightforwardly convert these high-power infrared pulses to the green at 532-nm in a single pass with efficiencies exceeding 30% (Baer *et al.* 1990).

Intracavity Frequency Doubling

Diode-pumped solid-state lasers operated continuously (CW) can also be efficiently converted to shorter wavelengths using a number of enhancement techniques which can be used for the pulsed laser sources as well. Most of these enhancement techniques utilize the fact that the 1064-nm power circulating inside the Nd laser resonator is much higher than the power that can be transmitted outside the Nd laser. Since the conversion efficiency increases with power of the fundamental, placing the non-linear crystal inside the laser resonator can therefore increase the conversion efficiency substantially. Intracavity frequency doubling with KTP and Nd:YAG has been used to generate 3.5 watts of 532-nm light (Liu *et al.* 1994).

Commercial versions of these laser sources are available from numerous vendors, at powers ranging from 10 mW to over a watt. These lasers have been designed to overcome the 'green noise problem', first noted by Baer (Baer 1986), that is caused by mode competition between axial modes of the neodymium laser and coupling effects on these modes from sum-frequency conversion. For example, the noise can be reduced using techniques that ensure single-axial mode operation. Although the watt-level green lasers can be nearly as large as a small air-cooled argon laser, their lifetimes are much longer and their efficiencies are much higher. Intracavity doubled lasers can also be designed to be very small; for instance, a compact frequency-doubled diode laser assembled in a 28 mm × 38 mm × 16 mm package has produced 10 mW of 532-nm output and was used for reading an optical disk at very high densities in a drive prototype (Kubota *et al.* 1993). Compact devices have also been fabricated by mounting together thin pieces of neodymium and non-linear optical materials inside the package that usually contains just the diode-laser.

Intracavity frequency doubling of other laser transitions has also been demonstrated. Nd:YAG lasers can be made to operate at 946 nm, which can be doubled to 473 nm using $KNbO_3$ (with either temperature tuning or angle tuning). This transition is not as efficient as the 1064-nm transition, due to both the lower gain cross-section and the fact that the transitions' lower level is close enough to the ground state to be thermally populated. In spite of these difficulties, this laser has demonstrated output powers of 1 W at 946 nm and 70 mW at 473 nm (Hanson 1995). Similarly, the 930-nm laser transition in $Nd:YAlO_3$ has been diode-pumped and intracavity frequency doubled to produce 15 mW of 465-nm light in $KNbO_3$ (Zarrabi *et al.* 1995). Shorter wavelengths can be generated by using Cr:LiSAF as the laser material, which has been pumped by 670-nm laser diodes and intracavity doubled with $KNbO_3$ to produce 13 mW tunable from 427 nm to 443 nm (Falcoz *et al.* 1995).

Intracavity Sum-Frequency Mixing

Shorter wavelengths than are possible with frequency doubling can be generated by sum-frequency mixing the high power circulating inside the laser resonator with the shorter wavelength diode laser light. For instance, 1064-nm Nd:YAG lasers have been mixed with diode laser output near 810 nm (Risk and Lenth 1989) in KTP to produce 459-nm light. Since the output from diode laser arrays is not useful for non-linear conversion, single-spatial mode diode lasers have been used as the source to mix with the intracavity Nd:YAG power, resulting in output powers for these devices in the blue of much less than a milliwatt. Although there has been much progress in achieving higher output powers of up to 400 mW from these single-stripe diode lasers (Jaeckel *et al.* 1992), and although MOPA diode sources can provide up to a few watts, the conversion efficiency possible using SFM in this configuration remains limited by the single-mode diode laser output power. Higher conversion efficiencies and blue output power up to 20 mW (Kea *et al.* 1993) have been realized by building up the diode laser power in the laser resonator, in a technique similar to that described below.

Up-conversion Lasers

Certain solid-state laser materials emit at a shorter wavelength than their pump source. These up-conversion laser materials exhibit an intermediate metastable state that allows them to absorb more than one pump photon to reach their excited state. Active ions demonstrating up-conversion include Pr, Er, and Tm, and these ions are typically doped into fluoride crystal hosts. High power densities from the pump laser ensures sufficient population of the excited state to reach threshold, and the most efficient operation occurs at low temperatures where the metastable state lifetime is longer. Room temperature operation with low threshold powers has been demonstrated in fibre up-conversion lasers, where the waveguide confinement increases the efficiency of populating the excited state. A diode-pumped Tm-doped ZBLAN fibre up-conversion laser recently demonstrated 106 mW of output power with optical-to-optical conversion efficiencies of up to 30% (Sanders *et al.* 1995). Although problems persist with the stability of the fibres used for up-conversion, further development of these laser sources is underway.

EXTERNAL-RESONATOR-ENHANCED FREQUENCY DOUBLING

The shortest wavelength possible using non-linear frequency conversion occurs by directly frequency doubling the diode laser source. GaAlAs single-mode diode lasers operating near 860 nm can be frequency doubled directly in potassium niobate ($KNbO_3$) near room temperature. As indicated previously, although $KNbO_3$ has a high non-linearity of 26 pm V^{-1} single-pass doubling experiments have produced only a few milliwatts of blue light. To increase the frequency doubling efficiency, the $KNbO_3$ can be placed in an external resonator to increase the infrared power in the non-linear crystal. Figure 3 shows the experimental schematic for such a system.

To maintain the high circulating fundamental power in the non-linear crystal, the frequency of the diode laser must be controlled to match a frequency of one of the longitudinal modes of the external resonator. As indicated in Fig. 3, this frequency control can be performed (using an FM-sideband technique to obtain the frequency error signal) by tuning the injection current in the diode laser, since the frequency of the diode laser tunes at nearly 3 GHz per milliampere change in current. By superimposing an rf current on the dc-injection current, the FM sidebands used for obtaining an error signal can be placed on the diode laser ('carrier') frequency. The level of the rf current sets the fraction of the IR power in the sidebands and therefore can be used to control the blue output and to rapidly switch between blue power levels. Such a system has been used to generate 54 mW of 429-nm light from a diode laser producing 160 mW of infrared power, with demonstrated switching times between two blue power levels of several nanoseconds (Kozlovsky and Lenth 1994). Although this frequency doubling approach is somewhat complex, commercial devices are available based on this approach that produce 10 mW of 430-nm light.

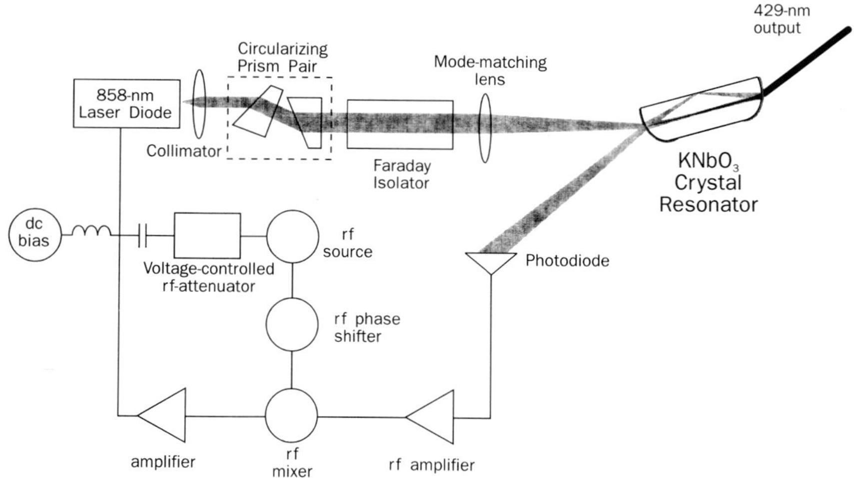

Fig. 3. Direct frequency doubling of a diode laser using an external enhancement resonator. The diode laser current is controlled to match the diode laser frequency to a mode of the enhancement resonator. A small rf current produces FM sidebands on the light to derive a frequency error signal for controlling the dc-injection current. This setup has produced 54 mW of 429-nm light from 160 mW of diode laser output power.

A somewhat simpler and less expensive frequency locking technique for external resonator-enhanced frequency doubling utilizes the diode lasers' sensitivity to optical feedback. A narrow range of weak feedback (e.g., 10^{-3} to 10^{-2}) with the proper feedback phase has been shown to stably lock the diode to an external resonator (Dixon 1994), and this technique has been used to generate 22 mW of blue from a 100 mW diode laser (Hemmerich *et al.* 1994). Alternatively, the diode laser can be anti-reflection coated and feedback from the cavity used to operate the diode laser stably in an external cavity mode. Such an extended cavity diode laser containing an enhancement resonator for frequency doubling has demonstrated 14 mW of blue output with 100 mA of injection current to the GaAlAs gain element (Kozlovsky and Risk 1994).

WAVEGUIDE FREQUENCY DOUBLING

The previous sections have described various techniques for efficiency enhancement using optical resonators, both laser resonators and external enhancement resonators. These techniques use bulk non-linear optical crystals and discrete optical elements that can be straightforwardly assembled into devices to create commercial products. Although these bulk devices are sufficient for some applications, many applications require laser sources that are both smaller and cheaper. Non-linear frequency conversion using waveguides offers the potential for significant reductions in cost, size and complexity, since the waveguides could be fabricated in batch processes much like semiconductors are manufactured today. Ideally the waveguide and diode laser would be designed in a compact package where that light from the diode laser could be coupled directly into the waveguide simply by mounting them together, as illustrated conceptually in Fig. 4.

Non-linear waveguides are formed when processes such as ion diffusion or proton diffusion are used to change the index of refraction of the non-linear material for defining a small core (several microns wide and deep in cross-section) which confines the infrared and second harmonic beams to a channel near the surface. The power densities of the fundamental and second harmonic beams in the waveguide can be quite high due to the small area of the confined beams. Since the beams are confined, the diffraction effects that limit the conversion efficiency of tightly focused beams in bulk interactions are eliminated. High power densities are therefore maintained throughout the length of the non-linear waveguide, enabling good non-linear conversion efficiencies for even low-power, CW laser sources, providing that there is good overlap between the modes of the fundamental and second harmonic beams in the waveguide.

The fundamental and second harmonic beams that propagate as waveguide modes in the non-linear material experience different indices of refraction than in the bulk, due to the fact that the core has a different index of refraction from

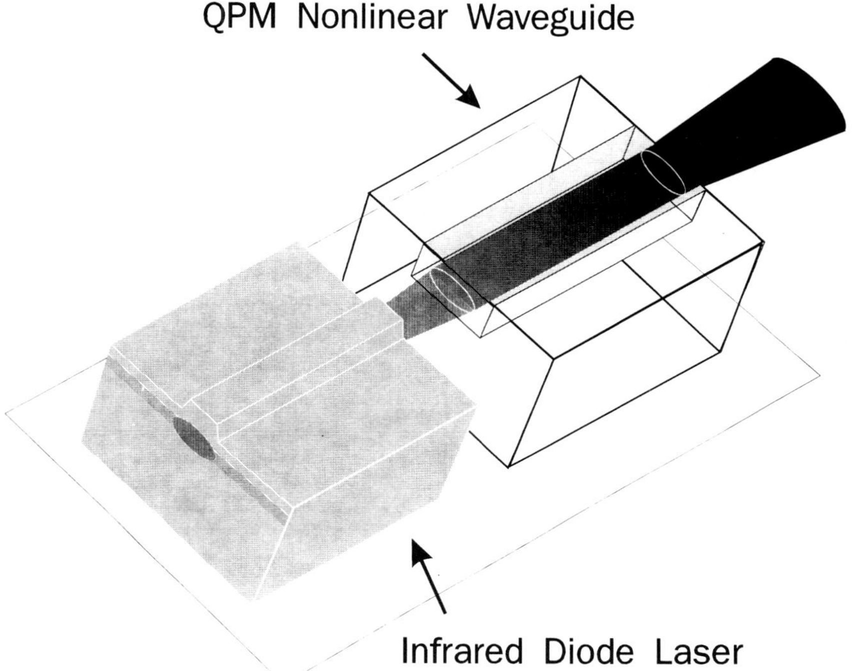

Fig. 4. Frequency doubling of a diode laser using a waveguide formed in a non-linear material. By coupling the diode laser output light directly into a QPM waveguide, a compact, reliable, and low-cost blue device should be possible (figure courtesy Bill Risk, IBM).

the bulk material that surrounds it. There are several different transverse modes that could be supported in a waveguide, and these modes have different effective indices of refraction since they 'sample' different fractions of the bulk and waveguide indices. Phase matching of the non-linear interaction in a waveguide involves matching the effective index of refraction for the second harmonic mode and the fundamental mode, and generation of higher-order waveguide modes in the second harmonic can be used to provide phase-matching. This modal phase-matching technique is not often used since it both reduces the conversion efficiency by reducing modal overlap and it produces a higher-order SHG mode that is not as useful as a TEM_{00} beam. These reasons are also applicable to Cherenkov frequency doubling, where a waveguide mode of the fundamental is used to generate a radiation mode of the second harmonic, producing a crescent-shaped SHG output beam at a shallow angle to the waveguide (Ito *et al.* 1993).

For producing the highest conversion efficiency in a waveguide to the second harmonic in a useful TEM_{00} output beam, quasi-phase-matching has proved particularly useful. By altering the length of the sections of periodic domain reversals, the wavelength that is quasi-phase-matched can be tuned, accounting for the different effective indices of refraction of the waveguide modes. Non-linear materials that have shown domain reversal under ion exchange, such as KTP, $LiTaO_3$ and $LiNbO_3$, also have proven to be readily processed to produce waveguides. The ion exchange processes usually produce shallow domain reversals of several microns near the surface of the material, which matches well to where the waveguide confines the light. Recently, electric field poling of these same materials has been demonstrated, resulting in straighter domain sidewalls, narrower feature sizes, and thicker regions of inversion (up to 1 mm in thickness) than is possible with the ion exchange periodic poling processes (e.g., Yamada *et al.* 1993).

Frequency doubling experiments using diode lasers and QPM waveguide frequency doublers have resulted in several mW of output in many non-linear materials. A coupling efficiency of 85% between the diode laser and the lithium tantalate waveguide was demonstrated in a compact device that produced 2.8 mW of blue (Kitaoka *et al.* 1995). KTP waveguides were used to frequency double a diode laser to produce 3.6 mW of 429-nm light (Eger *et al.* 1995). A MOPA diode laser frequency doubled in lithium niobate produced 5.1 mW at 488-nm from 163 mW of 976-nm power in the waveguide (Bortz *et al.* 1994).

The promise of cheaper, smaller blue-green sources based on waveguide frequency doubling in periodically poled non-linear materials has led to extensive efforts toward commercialization. Current development work is concentrating on repeatable and robust processes for the periodic poling and waveguide formation, as well as inexpensive, compact and manufacturable device designs that can provide efficient coupling of the diode laser output to the waveguide. Questions about the long-term stability of the non-linear waveguides at high power outputs and therefore high power densities in the waveguides are also being addressed.

BLUE DIODE LASERS

For many applications where blue light is desired, especially in consumer products such as CD-based optical storage where high volumes of low-cost blue sources are required, even waveguide frequency doubling of infrared diode lasers may be too complex or expensive. Ultimately, the most compact, rugged, and lowest-cost source of blue light should be a wide-bandgap semiconductor injection laser. Blue diode lasers have therefore been an active area of research for over twenty years. Recently, the development of improved materials, doping, and device technologies has accelerated progress, with the announcement in 1991 by researchers at 3M of a blue injection laser at 77 °C (Haase *et al.* 1991). Many other groups have announced results based on this Mg, Cd, ZnS, Se materials systems, and room temperature CW operation has been demonstrated for a few hours. Work is ongoing to solve the problems of defect origins and propagation in these soft materials to increase operating lifetimes. Research into other, harder material systems is also ongoing, and blue LEDs based on SiC and GaN are now commercially available. The rapid progress in GaN LEDs (Nakamura *et al.* 1994) has been encouraging, and has led to a recent announcement of pulsed GaN diode laser operation at room temperature.

SUMMARY

There are many possible approaches to an all-solid-state blue-green laser source. Most are based on converting the output of infrared diode lasers, which are well developed, high power, reliable and relatively inexpensive sources, to shorter wavelengths using non-linear conversion techniques. These technologies are in various stages of development: some have been commercially available for a few years, some are in the product development stage, and some are active areas of research.

The most mature sources are based on bulk non-linear conversion. Intracavity frequency doubling of neodymium-doped solid-state lasers makes possible compact devices that produce tens of milliwatts, or larger versions that produce up to several watts. To produce the shortest wavelength, the diode laser can be frequency doubled directly, and 10 mW of 430-nm light is available from such a device using an external enhancement resonator. These bulk non-linear conversion sources are well developed, straightforward and will likely continue to offer the highest power available in an all-solid-state source. However, these sources are somewhat complex and wider application would be made possible using devices that are cheaper and smaller.

A promising approach to smaller and cheaper sources is non-linear conversion of an infrared diode laser in a QPM waveguide. Active development of devices based on this technique makes it likely that a compact device producing several milliwatts in the blue should not be far off. Similarly, up-conversion fibre lasers might provide tens of milliwatts of blue-green power in an inexpensive, although somewhat larger device.

Research on wide-bandgap semiconductor materials and device fabrication has accelerated recently with the demonstration of direct injection lasers at room temperature. These advances make it possible that blue diode lasers might be commercialized in the not-too-distant future, which would provide the ultimate in compact and low-cost blue laser sources as required for high-volume, consumer-based applications.

BIBLIOGRAPHY

Baer, T. (1986), *J. Opt. Soc. Amer. B* **3**, pp. 1175–1180.

Baer, T.M., Head, D.F., Gooding, P. (1990) 'High peak power Q-switched Nd : YLF laser using a tightly folded resonator', in Conference on Lasers and Electro-optics, 1990 Technical Digest Series, p. 24. Optical Society of America, Washington.

Bortz, M.L., Field, S.J., Fejer, M.M., Nam, D.W., Waarts, R.G., Welch, D.F. (1994), *IEEE J. Quantum Electron.* **30**, 2953–2960.

Byer, R.L. (1977), 'Parametric oscillators and nonlinear materials' in *Nonlinear Optics* (ed. P.G. Harper and B.S. Wherette), pp. 47–160. Academic Press, New York.

Dixon, G.J., Tanner, C.E., Wieman, C.E. (1989), *Opt. Lett.* **14**, 731–733.

Eyer, D., Oron, M., Katz, M., Zussman, A. (1995), *J. Appl. Phys.* **77**, 2205–2207.

Falcoz, F., Balembois, F., Georges, P., Brun, A. (1995), *Opt. Lett.* **20**, 1274–1276.

Frankin, P.A., Hill, A.E., Peters, C.W., Weinreich, G. (1961), *Phys. Rev. Lett.* **7**, 118.

Haase, M.A., Qui, J., DePuydt, J.M., Cheng, H. (1991), *Appl. Phys. Lett.* **59**, 1272–1274.

Hanson, F. (1995), *Opt. Lett.* **20**, 148–150.

Hemmerich, A., Zimmermann, C., Hansch, T.W. (1994), *Appl. Opt.* **33**, 988–991.

Ito, H., Furiwara, T., Takyu, C. (1993), *Opt. Comm.* **99**, 237–240.

Jaeckel, H., Bona, G.-L., Buchmann, P., Meier, H.P., Vettiger, P., Kozlovsky, W.J., Lenth, W. (1992), *IEEE J. Quantum Electron.* **27**, 1560–1567.

Kea, P.N., Standley, R.W. Dixon, G.J. (1993), *Appl. Phys. Lett.* **63**, 302–304.

Kitaoka, Y., Mizuuchi, K., Yamamoto, K., Kato, M. (1995). *Rev. Laser Eng.* **23**, 788–794.

Kozlovsky, W.J., Lenth, W. (1994), *Opt. Lett.* **19**, 195–197.

Kozlovsky, W.J., Risk, W.P. (1994), *Appl. Phys. Lett.* **65**, 525–527.

Kubota, S., Oka, M., Masuda, H. (1993) in Advanced Solid-State Lasers and Compact Blue-Green Lasers Technical Digest, p. 290. Optical Society of America, Washington.

Lim, Eric J. (1992) Quasi-Phasematching for Guided-Wave Nonlinear Optics in Lithium Niobate, Stanford University, Stanford.

Liu, L.Y., Oka, M., Wiechmann, W., Kubota, S. (1994), *Opt. Lett.* **19**, 189–191.

O'Brien, S., Welch, D.F., Parke, R.A., Mehuys, D., Dzurko, K., Lang, R.J., Waarts, R., Scifres, D. (1993), *IEEE J. Quantum Electron.* **29**, 2052–2057.

Nakamura, S., Mukia, T., Senoh, M. (1994), *Appl. Phys. Lett.* **64**, 1687–1689.

Risk, W.P., Lenth, W. (1989), *Appl. Phys. Lett.* **54**, 789–791.

Sanders, S., Waarts, R.G., Mehuys, D.G., Welch, D.F. (1995), *Appl. Phys. Lett.* **67**, 1815–1817.

Yamada, M., Nada, N., Saitoh, M., Watanabe, K. (1993), *Appl. Phys. Lett.* **62**, 435–436.

Zarrabi, J.H, Garilovic, P., Singh, S. (1995), *Appl. Phys. Lett.* **67**, 2439–2441.

Index

Note - Page numbers in *italic* refer to figures and tables; **bold** page numbers refer to major discussions.